# Carbohydrates
Structure and Biology

Thieme
Organic Chemistry
Monograph Series

# Carbohydrates

## Structure and Biology

Jochen Lehmann
translated by Alan H. Haines

305 Figures
12 Tables

1998
Thieme
Stuttgart · New York

Prof. Dr. Jochen Lehmann
Institute of Organic Chemistry
University of Freiburg
Albertstr. 21
D-79104 Freiburg

Translated by
Dr. Alan H. Haines
School of Chemical Sciences
University of East Anglia
Norwich NR4 7TJ
England

*Cover*
The chemical structure on the cover is used with the consent of Dr. M. Petitou, who was prominantly involved in the synthesis of this molecule, the biologically active part of Heparin. This illustrates impressively the close relationship between biology and chemisty in the carbohydrate field.

This book is an authorised translation of the German edition published and copyrighted 1976, 1996 by Georg Thieme Verlag, Stuttgart, Germany. Title of German edition: Kohlenhydrate Chemie und Biologie

*Die Deutsche Bibliothek –*
*CIP-Einheitsaufnahme*

*Lehmann, Jochen:*
Carbohydrates : structure and biology ; 12 tables / Jochen Lehmann. Transl. by Alan H. Haines. – Stuttgart ; New York : Thieme, 1998
(Thieme organic chemistry monograph series)
Einheitssacht.: Chemie der Kohlenhydrate <dt.>



Typesetting: ScreenArt GmbH & Co. KG
D-72827 Wannweil (System 3b2)
Printed in Germany by Gutmann, Talheim

Georg Thieme Verlag, Stuttgart
ISBN 3-13-110771-5 (Flexicover)
ISBN 3-13-110781-2 (Hardcover)

Thieme, New York
ISBN 0-86577-795-0 (Flexicover)
ISBN 0-86577-790-X (Hardcover)

1 2 3 4 5 6

*Dedicated to A. B. Foster*

# Foreword to the English translation of the 2nd edition

20 Years have passed since the Thieme textbook *Chemistry of the Carbohydrates Monosaccharides and Derivatives* appeared. This timespan has been distinguished by intense scientific activity in an area of research that according to preference is denoted either as glycobiology[1] or glycobiochemistry. The great scientific growth in this area combined with the knowledge of the biological significance of carbohydrates of all types for all living things[2] led also to a stimulation of research into the organic chemistry of carbohydrates and to the specific application of the most modern physical methods for structural investigation, especially NMR spectroscopy and mass spectrometry. From a biological standpoint the especially interesting carbohydrates for the most part are not monosaccharides but rather their condensation products, oligosaccharides, polysaccharides, and conjugates with peptides, proteins or lipids, the so-called glycoconjugates. Accordingly, the organic chemistry of the carbohydrates as regards analysis and also synthetic methods also orientated itself predominantly to the higher molecular weight carbohydrate structures. A very large part of all recent innovations in organic methodology, the extraction and isolation, and the physical analysis of oligosacharides and their derivatives can also be attributed quite reasonably to this change in emphasis.

The main focus of molecular biological research gradually shifted during the past decade from investigations of covalent molecular transformations to those concerned with non-covalent molecular interactions and which are described with the catchphrase *recognition phenomena*. The remaining, still unanswered questions of molecular biology belong to this category. Organic chemistry, essentially dealing with the linking and breaking of covalent bonds, is developing in the wider areas as a scientific aid to biology. This is true to a large extent for chemical research on natural products and biopolymers and with it, naturally, also on carbohydrates. I have attempted to take this development into account.

Since in the first edition oligosaccharides, polysaccharides and glycoconjugates were disregarded in favour of the *Monosaccharides and Derivatives*, this deficiency has now been compensated. However, in view of the new emphasis of the book on biological aspects, the necessity to compromise on a convenient volume size and overall production cost, and the general availability of texts or reviews covering many of the fundamental aspects of monosaccharide chemistry, it was decided somewhat reluctantly to omit the original chapter 2 on *Chemical Aspects* which appeared in the German 2nd edition.

Modern carbohydrate chemistry, and with it glycobiology, is inconceivable without highly developed physical analysis, especially NMR spectroscopy with all its variants as well as mass spectrometry. These extensive special areas, which would be beyond the scope of a text book on carbohydrates, are covered in detail in monographs and review articles. For this reason, a detailed consideration in this new edition is unnecessary and only references to the application are given (section 1.3). Furthermore, most chemistry students receive a grounding in physical methods for structural analysis at an early stage in their study through textbooks of organic chemistry.

Chapter 2 *Biological Aspects* did not appear in the first edition of the book but it forms the most important part of this English translation of the second edition. I believe that the integrated treatment of the theme *Carbohydrates* (see ref 3) follows a trend in the teaching of modern biosciences and that only such a treatment can give an *understanding* of a class of substances and their function which extends across the disciplines. It seems appropriate to give the Biological Aspects a special emphasis in this Forward to the new edition.

As one of the four main classes of natural products, carbohydrates, together with nucleic acids, proteins, and lipids, are an essential component of living organisms. Even most viruses, which for a long time were thought to contain only proteins and nucleic acids, contain carbohydrates. Certainly for reasons of their general biological distribution, it is both necessary and important in a textbook on carbohydrates not to neglect their biological aspects.

In many areas of biochemistry, for example metabolism, carbohydrates have been the object of classical investigations. The unique ability of enzymes to act as structure specific biocatalysts for glycoside hydrolysis was already being researched in the previous century and at the start of the present century the process of glycolysis was already explained and recognised as one of the most important metabolic pathways of all cells. However, that alone did not suffice for carbohydrates to be regarded as especially interesting biologically and to gain them a place with proteins and nucleic acids, the protagonists of modern molecular biology. Since the organic chemistry of carbohydrates was also, essentially, the biological chemistry of carbohydrates, it was for a long time a special art in the hands of relatively few outsiders. This situation has changed fundamentally over the last 20 years. Many areas of biological carbohydrate chemistry have developed because of intensive research so that, as Jean Montreuil remarked in a 1980 review article (see ref 4), one can talk, for example, of the molecular biology and molecular pathology of glycoconjugates.

The growing abundance of important primary literature and comprehensive secondary literature makes it necessary to examine the expanding research areas, which are by no means clarified in all spheres, in the framework of a textbook for the rising generation of chemically and biologically oriented scientists. I have endeavoured, therefore, beside the defined valid area of classical carbohydrate metabolism, to deal with current views on the biogenesis and biological function of carbohydrates in all their forms. Thereby, especially instructive examples are drawn from the different areas of microorganisms, animals, and plants. Chapter 2 is arranged into four sections. Section 2.1 provides the necessary basic knowledge for an understanding of the *Biological Aspects* and corresponds in this sense to the role of Chapter 1 for the whole book. It covers above all a description of structure and the associated special properties. Sections 2.2 and 2.3 which build on this are concerned with the more dynamic aspects of carbohydrate biochemistry, the function of carbohydrates in biological processes and the metabolism of carbohydrates. In section 2.3, the biosynthesis of polysaccharide and glycoconjugates is given special consideration.

I believe that enzymes, as bioreagents, have achieved an established role in preparative organic chemistry today. Preparative carbohydrate chemistry is no exception. For this reason, a section 2.4 is dedicated to biochemical methods for the preparation of carbohydrates.

Since a textbook must impart to the reader coherent information about a no doubt limited but still extensive subject area, an in-depth treatment has not been attempted. Therefore, it has been our aim to refer to the already cited secondary literature and, in especially important cases, also to the primary literature. Because of the nature of the subject, if literature references are especially abundant, it should be possible to tell where the emphasis of present research lies, or where the reader can obtain experimental details.

The appearance of this new edition is a result of the initiative of Thieme Verlag. Fundamental doubts on my part, above all what the treatment of modern carbohydrate chemistry involves, whose practical side I have outgrown more and more in the past years, was balanced by the encouragement of some of my professional colleagues as well as the incentive through the general relevance of carbohydrate research, which has caused the circle of those interested to increase greatly in recent years.

The old pocket book was regarded as a valuable aid in university teaching and for general further education. It seemed that a new edition would be welcome.

It is reassuring if as an author one can ensure critical judgement. I am pleased that Prof. K. Jann and Dr. H.G. Padeken were willing to undertake the thankless task of careful inspection of the manuscript. In doing this, they contributed greatly to the improvement of its quality. I thank them for this favour to a friend. Dr. Padeken helped additionally advising and arranging at the thorough revision of the final version.

My thanks go to Herrn Dr. L. Ziser from my research group for important scientific discussions and especially to Frau Helga Lay for the composition of good, clear illustrations from my rudimentary sketches. With great patience she fulfilled my constant wish to make alterations. Frau Dr. S. Huschens produced the final form of the illustrations with much technical sensitivity. I am very satisfied and I thank her. Finally, as a competent and unerring critic on the question of style, my wife has also brought an influence to bear on the form of the manuscript, for which I thank her. In the end phase of the book project, Herr V. Eberl was an understanding and patient contact, who very kindly encouraged me.

The production of the book was helpfully attended by the staff of Georg Thieme Verlags, Dr R. Springer, and Dr. J. Richmond. For bringing the project, despite all difficulties, to a satisfactory conclusion, and for her helpfulness and her understanding, I thank Frau Dr. E. Hillen.

Freiburg, Autumn 1996 Jochen Lehmann

## References


1 Dwek, R. A. *Chem. Rev.* **1996**, *96*, 683.
2 Varki, A. *Glycobiology* **1993**, *3*, 97.
Glycoproteins: New Comprehensive Biochemistry, Vol. 29 a, Montreuil, J.; Vliegenthart, J. F. G.; Schachter, H.; Elsevier: Amsterdam, 1995.
3 Sharon, N. *Sci. Am.* **1980**, *243 (5)*, 80.
4 Montreuil, J. *Adv. Carbohydr. Chem. Biochem.* **1980**, *37*, 157.


*Note from the translator*:

This book is dedicated by the author (J. L.) to A. B. Foster and it is fitting that the translation was undertaken, at the request of the author, by A. H. H. since both author and translator learned many of their skills and had their enthusiasm raised in carbohydrate chemistry while members of Allan Foster's research group in the Department of Chemistry at the University of Birmingham.

Norwich, England Summer 1997 Alan H. Haines

# Foreword to the 1st edition

Carbohydrate chemistry, once a much worked on specialist area, especially in Germany, seemed likely to be removed from the curriculum of students of chemistry and natural sciences with the development of more restricted courses of study. The emergence of new directions of research places necessarily new emphasis in teaching and it is not easy for the classical area of organic chemistry and with it carbohydrate chemistry to deal with this step in development.

Burdened with an almost impossible abundance of compounds with very special properties for organic chemical substances, carbohydrate chemistry is difficult to classify in a modern teaching of organic chemistry oriented on theoretical foundations. The chemistry of carbohydrates remains therefore, as it ever was, a special area. Nevertheless, its right to exist in the curriculum is particularly justified where the organic chemistry forms the foundations of a biological oriented study. It is a fact that carbohydrates are the raw materials produced in very large amounts for the construction of the living world and the energy source for driving it.

This monograph is a simple and short source of information about the chemistry of this important class of natural products, in which the monosaccharides have priority as key compounds of carbohydrate chemistry. It was my intention, by extensive limitation in the description of the classes of substances to make room for the more important presentation of principles. In this sense the book should be seen as a key to an understanding of the secondary literature.

For the checking of the manuscript, for discussions and suggestions, I thank Herren Dr. K. Himmelspach (Max Planck Institut for Immunobiology, Freiburg), Dr. D. Hofmann (F. Hofmann La Roche & Co. AG., Basel) as well as Herrn. Dipl. Chem. E. Schröter (Chemistry Laboratory of the University of Freiburg). I also thank Georg Thieme Verlag for the energetic help with the finishing of the manuscript ready for printing and for the friendly readiness to accommodate my wishes.

Freiburg, July 1976 Jochen Lehmann

# Contents

# Abbreviations

| | |
|---|---|
| A | Adenine |
| Ac | Acetyl |
| ACh | Acetyl choline |
| ADH | Alcohol dehydrogenase |
| ADP | Adenosine 5'-diphosphate |
| AIDS | acquired immunodeficiency syndrome |
| All | Allyl |
| AMP | Adenosine 5'-monophosphate |
| ap | anti-periplanar |
| ATP | Adenosine triphosphate |
| Bn | Benzyl |
| BOC | *tert*-Butoxycarbonyl |
| BPG | D-2,3-Diphosphoglycerate |
| Bu | Butyl |
| Bz | Benzoyl |
| C | Cytosine |
| cAMP | Adenosine 3',5'-cyclic monophosphate |
| CAN | Cerium ammonium nitrate |
| Cbz oder Z | Benzyloxycarbonyl |
| CDP | Cytidine 5'-diphosphate |
| CGTase | Cyclodextrin glucanotransferase |
| CM | Carboxymethyl |
| CMP | Cytidine 5'-monophosphate |
| CoASH | Coenzyme A |
| Cp | Cyclopentadienyl |
| CTP | Cytidine 5'-triphosphate |
| d | deoxy |
| Da | Dalton |
| DAST | Diethylaminosulphur trifluoride (Diethylaminotrifluorosulphuran) |
| DCC | 1,3-Dicyclohexylcarbodimide |
| DBU | 1,8-Diazabicyclo[5.4.0]undec-7-ene |
| DDQ | 2,3-Dichlor-5,6-dicyano-1,4-benzoquinone |
| DEAD | Diethyl azodicarboxylate |
| DEAE | Diethylaminoethyl |
| DG | 1,2-Diacylglycerol |
| DHAP | 1,3-Dihydroxyacetone phosphate |
| DIBAH | Diisobutylaluminiumhydride [Bis(2-methylpropyl)alane] |
| DMAP | 4-Dimethylaminopyridine |
| DMF | *N,N*-Dimethylformamide |
| DMIPS | Dimethylisopropylsilyl |
| DMS | Dimethyl sulphide |
| DMSO | Dimethyl sulphoxide |
| DMTST | Dimethyl(methylthio)sulphonium triflate |
| DMTr | Bis[4-methoxyphenyl]phenylmethyl |
| DNA | Deoxyribonucleic acid |
| DNP | 2,4-Dinitrophenyl |
| Dol | Dolichol or Dolichyl |
| DOPA | L-3,4-Dihydroxyphenylalanine |
| DPMS | Diphenylmethylsilyl |
| DTBMS | Di-*tert*-butylmethylsilyl |
| *E* | Extinction |
| $\varepsilon$ | Extinction coefficient |
| EE | 1-Ethoxyethyl |
| EF | Elongation factor |
| EDTA | Ethylenediaminetetraacetic acid |
| EGF | Epidermal growth factor |
| ESR | Electron spin resonance |
| ER | Endoplasmatic reticulum |
| Et | Ethyl |
| f. a. b. | Fast atom bombardment |
| FAD | Flavine adenine dinucleotide (oxidised form) |
| $FADH_2$ | Flavine adenine dinucleotide (reduced form) |
| FBP | Fructose 1,6-bisphosphate |
| FMN | Flavine mononucleotide |
| Fmoc | 9-Fluorenylmethoxycarbonyl |
| F1P | Fructose 1-phosphate |
| F6P | Fructose 6-phosphate |
| G | Guanine |
| GAP | Glyceraldehyde 3-phosphate |
| GAPDH | Glyceraldehyde 3-phosphate dehydrogenase |
| GDP | Guanosine 5'-diphosphate |
| GLC | Gas liquid chromatography |
| GMP | Guanosine 5'-monophosphate |
| G1P | Glucose 1-phosphate |
| G6P | Glucose 6-phosphate |
| GTP | Guanosine 5'-triphosphate |
| Hb | Haemoglobin |
| HDL | High-density lipoprotein |
| HIV | Human immunodeficiency virus |
| HMDS | Hexamethyldisilazane |
| HMPTA | Hexamethylphosphoric triamide |
| HPLC | High pressure liquid chromatography |
| Hyl | 5-Hydroxylysine |
| Hyp | 4-Hydroxyproline |

IgG Immunglobulin G
InsP1 or IP1 Inositol 1-phosphate
InsP3 or IP3 Inositol 1,4,5-triphosphate
IP Isopropyl
IPDMS Isopropyldimethylsilyl
IPTG Isopropyl β-D-thiogalactoside
IR Infrared

kDa Kilodalton
$K_M$ Michaelis constant
$K_I$ Inhibiton constant

LAH Lithium aluminium hydride
LDH Lactate dehydrogenase
LDL Low-density lipoprotein

MCPB 3-Chloroperoxybenzoic acid
MHC Major histocompatibility complex
Ms Methanesulphonyl
MS Mass spectrometry

NA Neuraminic acid
$NAD^+$ Nicotinamide adenine dinucleotide (oxidised form)
NADH Nicotinamide adenine dinucleotide (reduced form)
$NADP^+$ Nicotinamide adenine dinucleotide phosphate (oxidised form)
NADPH Nicotinamide adenine dinucleotide phosphate (reduced form)
NBS *N*-Bromosuccinimide
NIS *N*-Iodosuccinimide
NMR Nuclear magnetic resonance
NPS 2-Nitrophenylsulphenyl

PAGE Polyacrylamide gel electrophoresis
PAPS 3'-Phosphoadenosine 5'-phospho-sulphate
PCC Pyridinium chlorochromate
PdIns Phosphatidylinositol
PdInsP2 Phosphatidylinositol-4,5-bisphosphate
PEP Phosphoenolpyruvate
3PG 3-Phosphoglycerate
PGI Phosphoglucose isomerase
Pi inorganic phosphate
PK Pyruvate kinase
PLP Pyridoxal 5-phosphate
PMB 4-Methoxybenzyl
PP Pyrophosphate
ppm parts per million
PRPP 5-phosphoribosyl-α-pyrophosphate
Pv Pivaloyl
Pyr Pyridine

RER Rough endoplasmic reticulum
RNA Ribonucleic acid
RP Reverse phase
Ru5P Ribulose 5-phosphate
Ru1,5P Ribulose 1,5-bisphosphate
Rul Ribulose

S Svedberg unit
SAM *S*-Adenosylmethionine
sc syn-clinal
SDS Sodium dodecylsulphate

T Thymine
TBDMS *tert*-Butyldimethylsilyl
TEA Triethylamine
*t*-Bu *tert*-Butyl
TFA Trifluoracetyl or Trifluoroacetic acid
THF Tetrahydrofuranyl or Tetrahydrofuran
THP Tetrahydropyranyl or Tetrahydropyran
TLC Thin layer chromatography
TMTr Tris(4-methoxyphenyl)methyl
TMS Trimethylsilyl or Tetramethylsilane
TMV Tobacco mosaic virus
TPP Triphenylphosphine or Thiamine pyrophosphate
Tr Triphenylmethyl (Trityl)
Tris Tris(hydroxymethyl)aminomethane (2-amino-2-hydroxymethyl-1,3-propanediol)
Ts *p*-toluenesulphonyl
TTP Thymidine 5'-triphosphate

U Uracil
UDP Uridine 5'-diphosphate
UDPG Uridine 5'-(α-D-glucopyranosyl diphosphate)
UMP Uridine 5'-monophosphate
Un Undecaprenyl
UTP Uridine 5'-triphosphate
UV Ultraviolet
$V_{max}$ Maximum velocity

Xul Xylulose
Xu5P Xylulose 5-Phosphate

* Abbreviations for monosaccharides (excepting some irregular notations) and amino acids are not considered (see sections 1.4 and 2.1.1). Other textbooks should be consulted regarding the amino acids.

Chapter 1

# General

## 1.1 Introduction

As with other expressions in organic chemistry, the term “carbohydrate” as a collective term should be viewed historically. Although in general use, the definition of a carbohydrate as a hydrate of carbon, with the overall formula $C_n(H_2O)_n$ in the case of monosaccharides or $C_n(H_2O)_m$ for oligosaccharides or polysaccharides, is inaccurate for many carbohydrates, whilst other compounds which correspond to this definition (for example lactic acid or acrylic acid) are not carbohydrates. Of all the four large classes of natural products, which essentially all organisms synthesise, the carbohydrates predominate in quantitative terms, compared with proteins, lipids and the nucleic acids (see section 2.1). The class of substances termed carbohydrates embraces a series of sub-groups, which can be derived from the basic formula of carbohydrates. To these belong the polyhydroxy alkanes and polyhydroxy carboxylic acids, in so far as they are compounds with a straight chain C-skeleton as in monosaccharides, and glycosides, oligosaccharides, and polysaccharides since they are condensation products of monosaccharides with compounds of other classes or with themselves.

All carbohydrates have in common an accumulation of functional groups in relation to their molecular size which is shown by no other class of substances and, of these groups, hydroxy groups predominate. Interaction between the different groups as well as competition between them are responsible for the often complex properties of carbohydrates in their chemical reactions. A result of this polyfunctionality is the large number of possible stereo- and constitutional-isomers found in relatively small molecules of this class. The majority of carbohydrates occurring in nature possess more than 4 asymmetric carbon atoms . The number of stereoisomers in the case of *n* asymmetric C-atoms is $2^n$ which, for a condensation product **A** formed from 2 monosaccharides, each with 5 asymmetric C-atoms, leads theoretically to $2^{10} = 1024$ definable species.

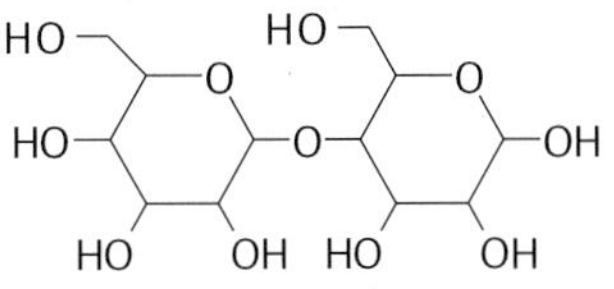

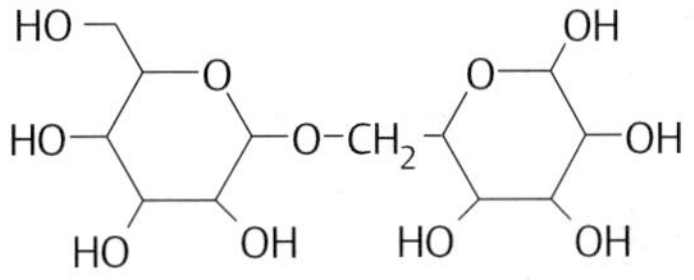

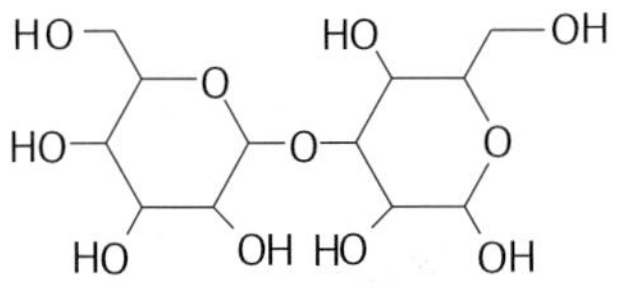

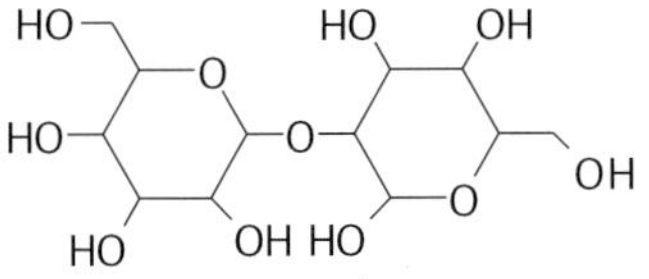

Since, in addition, the constitution can also be changed, for instance through interchanging the positions of bonding (variants **B - E**), the number of isomers is increased fivefold as a result of the 5 possible bonding positions to a total of 5120. However, nature makes only very limited use of these varied possibilities (see section 2.1).

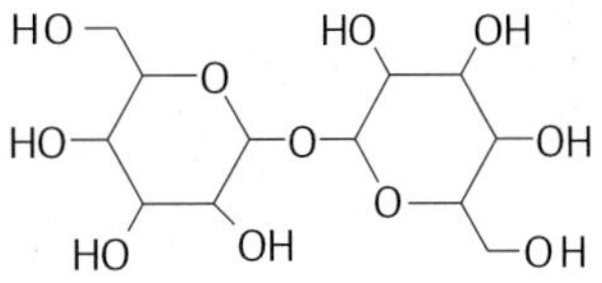

The monosaccharides are regarded as typical carbohydrates. Of these, the most widely distributed and most intensively investigated are those with an aldehyde function (aldoses) and 6 C-atoms (aldohexoses). D-Glucose, mostly in the form of its oligomeric and polymeric condensation products and probably the most abundant organic chemical compound occurring in the biosphere, is an aldohexose. Since all remaining carbohydrates may be regarded as derivatives of monosaccharides, especially D-glucose, it is sensible to treat the principles of carbohydrate chemistry predominantly from the point of view of this group of compounds.

## 1.2 Structure of mono- and oligosaccharides

In this book, the term structure is taken to mean the entire description of the molecular shape. It is to be borne in mind that the structure of a molecule can change, for example, through rotation about a single bond, which can occur even at room temperature. In order to take this into account, we separate the formal constitution and configuration of a molecule from its conformation. The two former are stable under normal conditions but, under similar conditions, the latter is subject to continual change from a thermodynamically relatively stable, and therefore statistically relevant, resting position.

The *constitution* of a molecule, when the molecular formula is known, describes the linkage of the atoms to one another, that is the type and sequence of the covalent bonds between the individual atoms. The *configuration* of a molecule with a definite constitution gives the spatial disposition of the atoms, or groups of atoms as they are fixed by covalent chemical bonds.

Molecules with different constitutions, which correspond only with respect to their overall formula, are called constitutional isomers and give rise to constitutional isomerism. Molecules with the same constitution which differ only in their configuration are called stereoisomers and lead to stereoisomerism. Stereoisomers are either enantiomers, when they are related as object and mirror image, or else they are diastereomers.

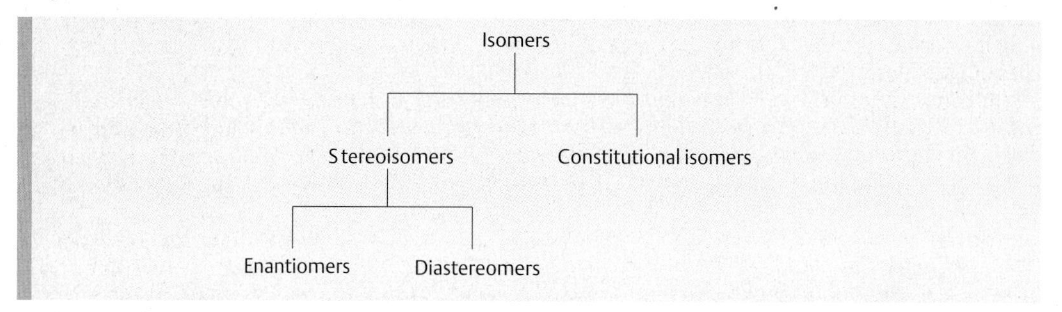


Strictly speaking, molecules are only identical if their structures are superimposable, that is if their conformation is also the same. Normally this rule is not applied, since one considers not only just a molecule, but rather an average structure from a large number, spontaneous conformational changes compensating each other.

### 1.2.1 Constitution and configuration

Monosaccharides of the general formula $C_n(H_2O)_n$ are polyhydroxy-aldehydes (*aldoses*) or -ketones (*ketoses*) with predominantly unbranched C-skeletons. Technically, the length of the C-chain is not limited. However, naturally occurring monosaccharides with more than six C-atoms are comparatively rare. The parent name of a monosaccharide is derived from the number of C-atoms in the basic skeleton. An aldose with 3, 4, 5, or 6 C-atoms etc., is designated as a *triose, tetrose, pentose, hexose*, etc. Ketoses are named correspondingly *triulose, tetrulose, pentulose, hexulose*, etc. In the case of ketoses, the carbonyl group can be situated at different positions in the C-chain.

| **Aldoses** | **Ketoses** |
|---|---|
| | $CH_2-OH$ |
| $CHO$ | $C=O$ |
| $(CH-OH)_n$ | $(CH-OH)_{n-1}$ |
| $CH_2-OH$ | $CH_2-OH$ |
| ↑ | ↑ |
| $CHO$ | $CH_2-OH$ |
| $CH-OH$ | $C=O$ |
| $CH_2-OH$ | $CH_2-OH$ |
| Glyceraldehyde | 1,3-Dihydroxyacetone |

Monosaccharides possess at least one asymmetric C-atom and, on this basis, 1,3-dihydroxyacetone cannot be designated as a monosaccharide, although it clearly is the parent 3-carbon compound for the ketose series and therefore is included in the foregoing diagram representing the family of structures in the carbohydrates. With growing chain length the number of stereoisomers increases since each additional C-atom is asymmetric. Aldoses and ketoses with the same number of carbon atoms are, by definition, constitutional isomers and each form independent families with distinctly different chemical properties.

All monosaccharides, with the exception of glyceraldehyde, 1,3-dihydroxyacetone, and tetrulose can, in principal, form stable 5- and 6-membered rings through formation of an intramolecular hemiacetal. The tendency for ring formation is so great that, in so far as they crystallise, all monosaccharides exist as cyclic hemiacetals. In solution, the acyclic carbonyl-form of all free monosaccharides do indeed exist but in the case of many, only in vanishingly small concentrations.

Two model compounds show how large the energy gain is through cyclisation. 4-Hydroxybutanal and 5-hydroxypentanal are in equilibrium with the corresponding hemiacetal forms in aqueous 1,4-dioxan[1]. As measured by UV spectroscopy, the content of open chain aldehyde amounts in the first case to 11.4, and in the latter case to 6.1%. This corresponds to a difference in free energy of 5.02 kJ/mol (1.2 kcal/mol) and 6.69 kJ/mol (1.6 kcal/mol), respectively, between cyclic and acyclic forms (see p. 4).

The energy gain through cyclisation, especially for 6-membered rings, can be raised considerably if substituents are present. The reversible formation of cyclic hemiacetals from the corresponding polyhydroxy-aldehydes or -ketones is, according to definition, a change in constitution, which is often denoted as *ring-chain tautomerism*. The instability of the acyclic form and the position of equilibrium, readily attained in solution, makes it impossible to isolate *D-glucose* as 2,3,4,5,6-pentahydroxyhexanal. However, it is possible to detect the aldehydo-form in solution both chemically through specific reactions and also with the help of physical measurements[2].

If one wishes to denote that a monosaccharide exists as a 5- or 6-membered hemiacetal ring, then the parent name relating to this compound is expanded. A hexose then becomes a *hexofuranose* or a *hexopyranose*, by comparison with the heterocycles *furan* and *pyran*. Cyclisation leads to the formation of a new asymmetric C-atom and a doubling of the number of stereoisomers. The hemiacetal hydroxy group is named a *lactol group* to distinguish it from the remaining hydroxy groups and the C-atom to which it is attached is named the *anomeric* C-atom or the anomeric centre. Despite its formal similarity to the other hydroxy groups, the lactol group differs from them considerably in chemical terms, as does the anomeric C-atom from the other C-atoms.

As already stated at the outset, monosaccharides can deviate to a greater or lesser degree from the general formula $C_n(H_2O)_n$. Widespread in nature, for example, are *glycuronic acids* which carry a carboxy group, instead of a primary hydroxy group, or *glyconic acids* in which the formyl group is oxidised.

4-hydroxybutanal (11.4%) ⇌ hemiacetal (88.6%)

5-hydroxypentanal (6.1%) ⇌ hemiacetal (93.9%)

Equally, there are *deoxymonosaccharides*, in which one or more hydroxy groups are missing or *amino-deoxymonosaccharides* in which one or more hydroxy groups are replaced by amino groups. Sulphur containing monosaccharides are also found, as are monosaccharides with branched C-skeletons. Even though nature is comparatively economical with the given possibilities of structural variation (Section 2.1.1), there are hardly any limits today for the organic chemist with regard to variation of constitution and configuration in monosaccharides. Occasionally, the only justification to name a certain compound a monosaccharide is the structural element of the cyclic hemiacetal. In this connection, it is important to mention that the vast majority of all monosaccharides occurring in nature exist not as *free* hemiacetals but as acetals, that is in the form of *glycosides, oligosaccharides* or *polysaccharides*.

### 1.2.2 Glyceraldehyde as a standard

An essential characteristic of all carbohydrates is their *optical activity*, which is a consequence of their molecular chirality. *Glyceraldehyde*, which can be regarded as the simplest monosaccharide, exists in two enantiomeric forms. Although already described by Emil Fischer, together with 1,3-dihydroxy-acetone, as an easily fermentable oxidation product of glycerol, and named glycerose[3], $C_3H_6O_3$, D- or L-glyceraldehyde (first prepared by A. Wohl and F. Momber[4] in 1914) was first defined many years later as the configurational reference compound for all monosaccharides[5] and the configurative relationship (*genetic relationship*) established between D-glyceraldehyde (then still called d-glyceraldehyde) and the naturally occurring positive-rotating grape sugar (*d*-glucose). The configuration of grape sugar had been determined by E. Fischer in 1891 and correspondingly the configuration of each monosacharide that could be correlated to it at that time[6], though configurative inconsistencies appeared. Because both enantiomeric trioses were still unknown in pure form, M. Rosanoff in 1906 related the configurations of the tetroses *erythrose* and *threose* and replaced the Fischer *d,l*-notation with a $\delta,\lambda$-nomenclature[7]. Membership to one or another series of enantiomers was decided through the asymmetric C-atom at the furthest distance from the carbonyl group. If in the projection formula the hydroxy group stood on the right, then it was a member of the $\delta$- or D-family, and if to the left of the $\lambda$- or L-compounds. Although originally thus defined, the configuration of a chiral compound, D or L, may not be equated with the sense of its optical activity, (+) or (–). Both appear independently as prefixes before the name of the substance, for example *(+)-D-glyceraldehyde*. With his suggestion, Rosanoff separated conclusively the sign of the rotation (+) or (–) of a compound from its membership of one or the other class of configuration.

The agreement between arbitrarily determined and true (absolute) configuration of the naturally occurring (+)-D-glyceraldehyde was confirmed in 1951 by Bijvoet, Peerdeman, and van Bommel, as it was possible to ascertain the absolute configuration of sodium rubidium (+)-tartrate[8]. Since the configurations of *(+)-glyceraldehyde* and the naturally occurring (+)-tartaric acid can be correlated by chemical transformation, the absolute configuration of (+)-glyceraldehyde, its mirror image form, and all of the monosaccharides derived from them is secured. The chirality of glyceraldehyde depends on the asymmetry of the central C-atom, as is made clear with the tetrahedral formulae:

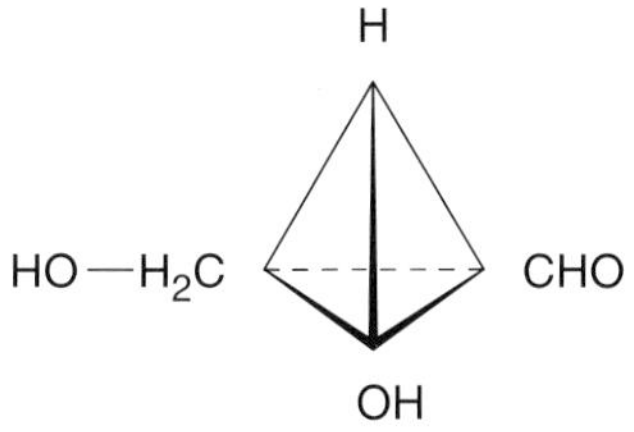

*(+)*-D-Glyceraldehyde

The tetrahedron shown has no symmetry plane. It can be rotated into different positions $\mathbf{A_1}$ to $\mathbf{A_6}$. The configuration remains always the same:

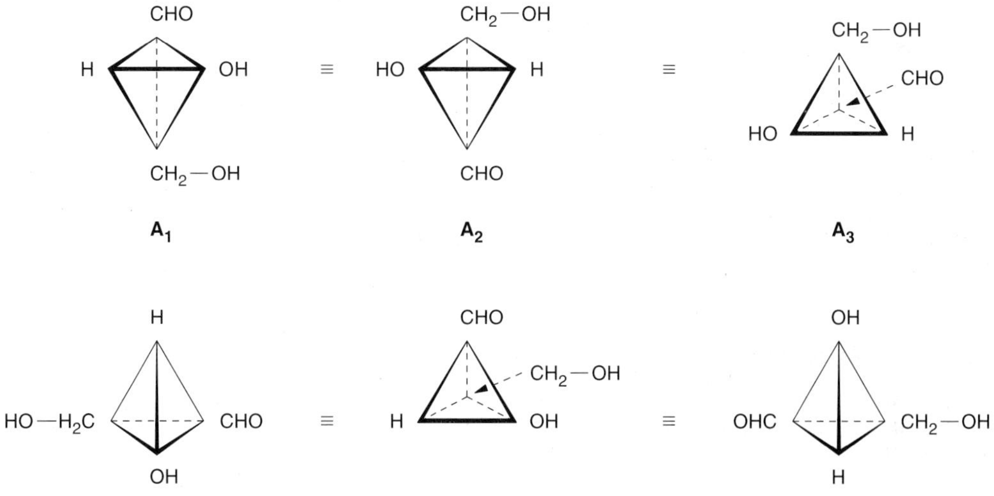

$\mathbf{A_1}$ ≡ $\mathbf{A_2}$ ≡ $\mathbf{A_3}$

$\mathbf{A_4}$ ≡ $\mathbf{A_5}$ ≡ $\mathbf{A_6}$

### 1.2.2.1 Fischer convention

The necessity of an unequivocal, graphic representation of the monosaccharides arose with clarification of their configuration. Graphically and typographically the tetrahedral formulae **A** and also the spatial formula **A'** derived therefrom are too complicated, especially with monosaccharides with several asymmetric C-atoms:

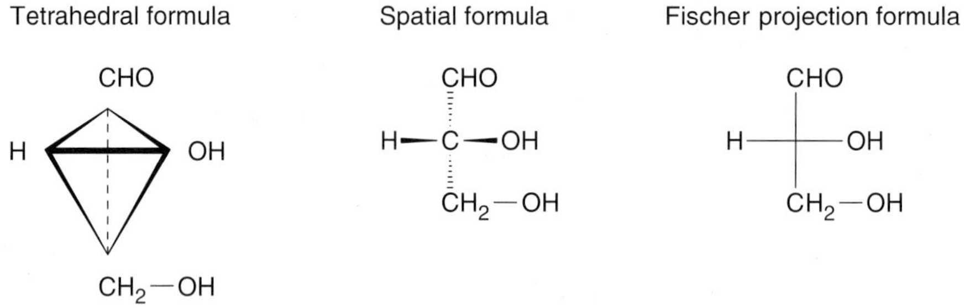

| Tetrahedral formula | Spatial formula | Fischer projection formula |
|---|---|---|
| A | A' | A'' |

Both the tetrahedral formula and also the spatial formula can be represented in different orientations without changing the spatial relationship of the bonds and the attached atoms or groups of atoms. This is not the case with the simple projection formula **A''** introduced by E. Fischer. The projection formula depends on the so-called Fischer convention. The C-chain is arranged vertically so that the end with the highest oxidation number is at the top. In general this is a carbonyl group in the case of monosaccharides or a carboxy group in the case of amino acids or other substituted carboxylic acids. Each single C-atom is then viewed separately and is imagined to lie in the plane of a table. The two bonds to the neighbouring C-atoms are made to point under the plane of the table, the other ones to an hydroxy group or another functional group and to the H-atom above the plane of the table, as is shown (see above) in this example of glyceraldehyde with the spatial formula **A'**. In the projection formula **A''**, all bonds are then *projected* onto the plane of the table. The intersection symbolises the *imagined* C-atom. According to Fischer, the orientation of the functional group to the right or left decides whether a certain substance is a member of the D- or L-series. A projection formula based on this convention may not be rotated and turned arbitrarily without falsifying the configuration. Only rotation by 180° in the plane of the table is allowed.

The projection formula is especially advantageous for the graphical representation of higher monosaccharides. The relative relationship of the asymmetric C-atoms one to another is very easily represented. This is one of the most important reasons why, for monosaccharides and amino acids, the Fischer style of writing formulae has remained up to the present day.

### 1.2.2.2 The Cahn, Ingold and Prelog sequence rules

The configuration of many compounds wih asymmetric C-atoms cannot be assigned with the help of the Fischer convention, as for example when it cannot be clearly decided how to write the projection formula:

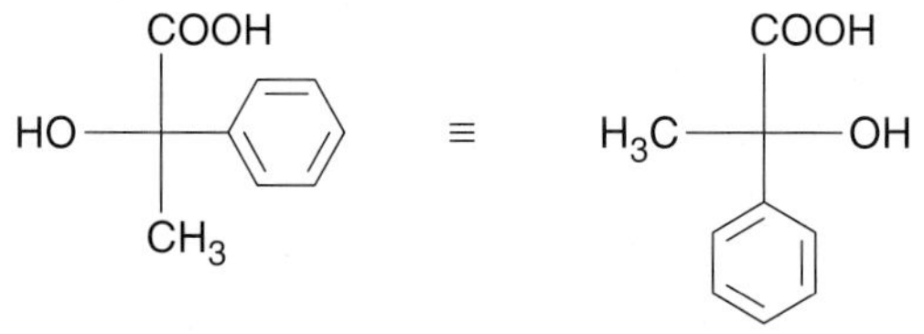

This difficulty can be avoided through use of the Cahn, Ingold, and Prelog (CIP) convention, which depends upon the determination of a priority sequence for the ligands (*sequence rule*)[9] on each single asymmetric centre. The priority of a ligand arises above all from the atomic number (ordinal number in the periodic table of the elements) of the atom bound directly to the asymmetric C-atom in question (first degree ligand). When two or more such ligands are the same, the priority is decided by the next furthest removed atom (second degree ligand ). If these atoms afford no choice, one goes on to the next furthest removed atom, and so on, always following along branches with atoms of the highest atomic number:

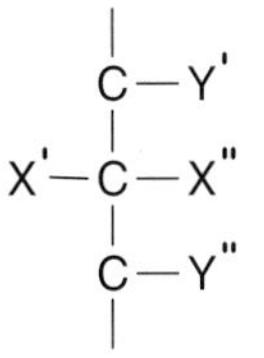

X = first degree ligand
Y = second degree ligand

Increasing atomic number means increasing priority. Double or triple bonded atoms are counted twice or three times, respectively.

increasing priority →

| | | | | | | |
|---|---|---|---|---|---|---|
| H | < | $CH_2$—OH | < | CHO | < | OH |
| $CH_3$ | < | $C_6H_5$ | < | COOH | < | OH |

The sequence of decreasing priority is then observed by viewing along the bond axis from the asymmetric C-atom to the ligand of least priority.

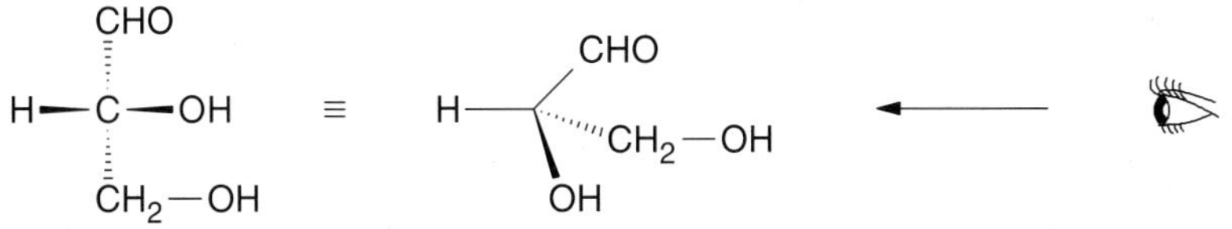

(*R*)-configuration

If the priorities of the remaining three ligands decrease in a clockwise manner, then the asymmetric C-atom is designated *R* (*rectus*), and if the sequence is anticlockwise then the designation is *S* (*sinister*). Accordingly, *D-(+)-glyceraldehyde* possesses an *R*-configuration, as does consequently also the configuration-determining asymmetric C-atom (that with the highest positional number) of each D-monosaccharide. Since introduction of the CIP convention into carbohydrate chemistry brings no improvement over the Fischer convention, it has not been generally accepted in this area. In this book also, as in general is customary, monosaccharides will be denoted according to the Fischer-Rosanoff convention.

### 1.2.3 The family of monosaccharides

The basic forms of the monosaccharides derive from glyceraldehyde or from tetrulose (disregarding 1,3-dihydroxyacetone since tetrulose is the first ketose with an asymmetric C-atom) by insertion of a CH-OH unit. With each unit the number of asymmetric carbon atoms grows by one and the number of stereoisomers doubles, resulting in families of diastereoisomers:

- 2 diasteromeric tetroses
- 4 pentoses
- 8 hexoses
- 16 heptose, etc.

Since ketoses have one asymmetric carbon atom fewer than aldoses with a similar number of C-atoms, this family of diastereoisomers begins first with the two pentuloses *D-ribulose* and *D-xylulose*. D- or L-Glyceraldehyde (or D- or L-tetrulose) are the basic forms of the monosaccharide D- or L-series, respectively. Since each D-monosaccharide has an enantiomeric L-partner, and they differ from each only as object and mirror image, it is sufficient to consider just the monosaccharide family of the D-series, since all considerations can be applied analogously to the L-series.

## D-Aldoses

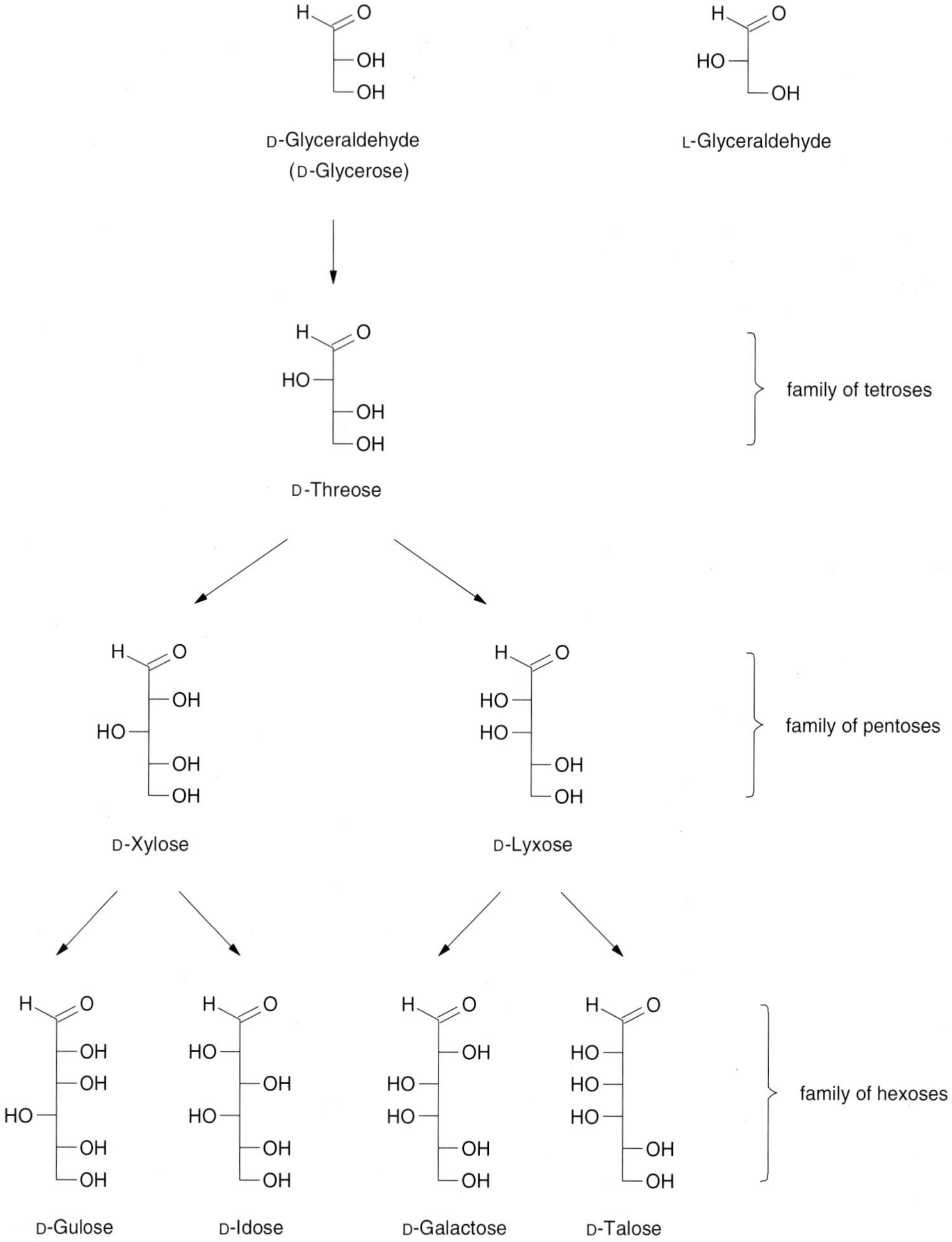

D-Glyceraldehyde
(D-Glycerose)

L-Glyceraldehyde

D-Threose

family of tetroses

D-Xylose

D-Lyxose

family of pentoses

D-Gulose

D-Idose

D-Galactose

D-Talose

family of hexoses

## D-Aldoses

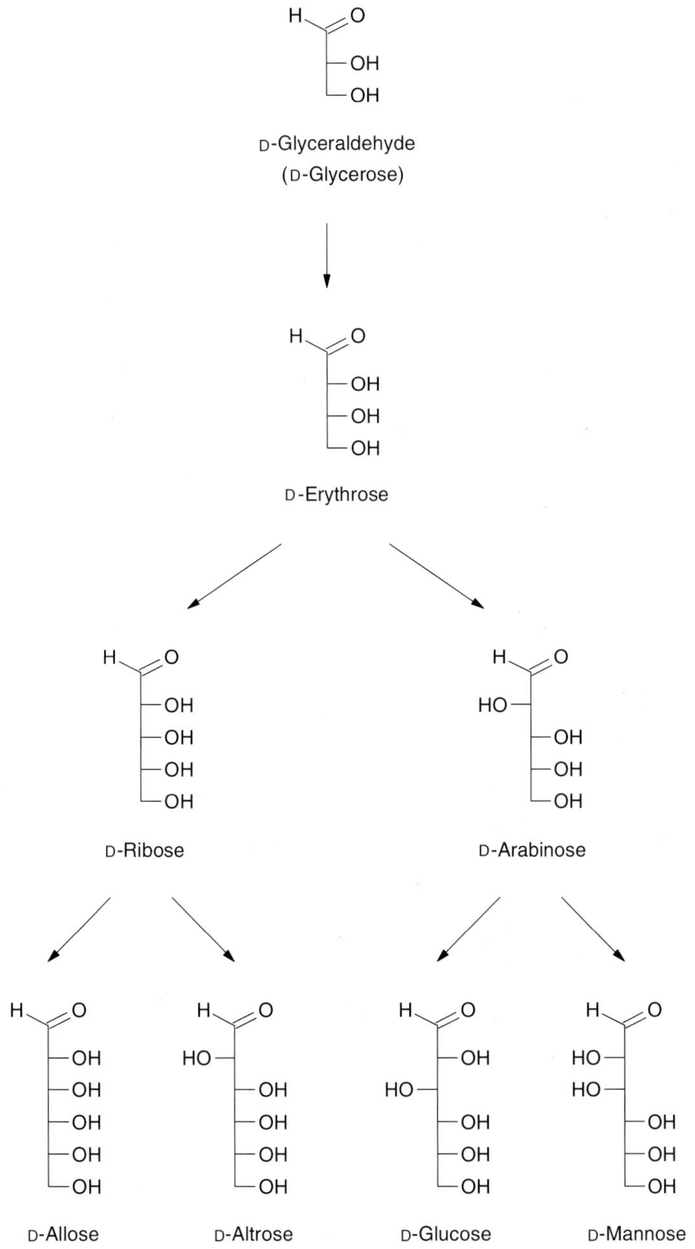

D-Glyceraldehyde
(D-Glycerose)

D-Erythrose

D-Ribose

D-Arabinose

D-Allose

D-Altrose

D-Glucose

D-Mannose

## D-Ketoses

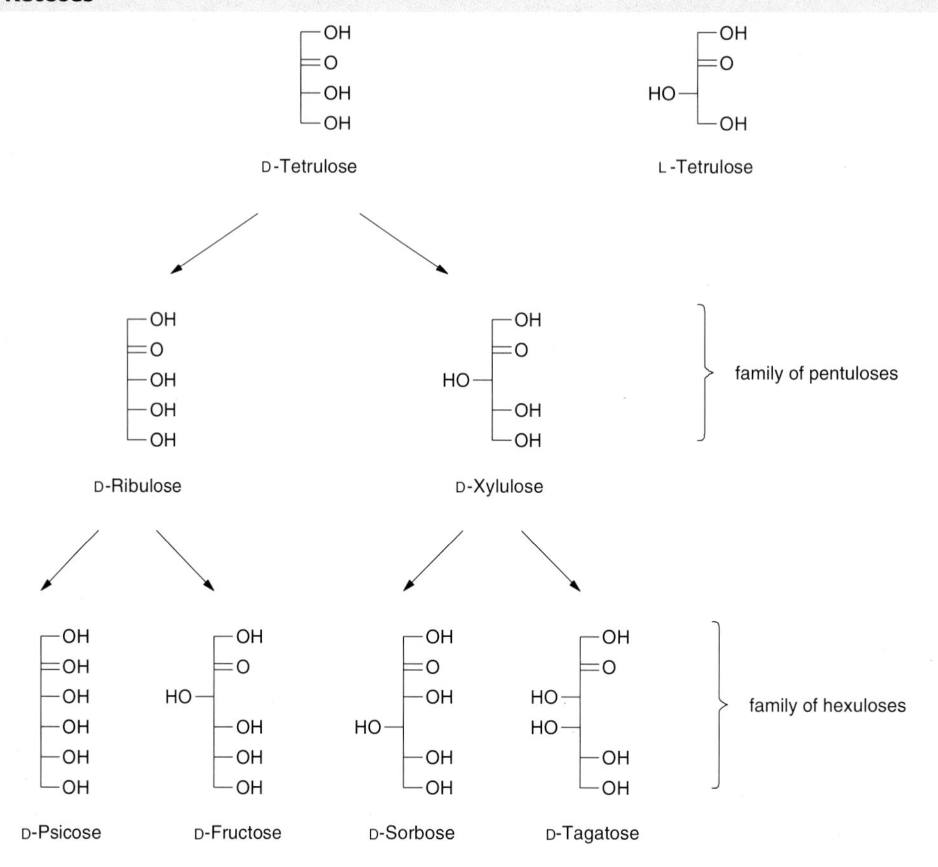


The C-chain of the monosaccharide is numbered, beginning from the top C-atom in the Fischer projection formula, consecutively downwards:

D-Glucose D-Fructose

**constitutional isomers**

The exchange of two ligands on an asymmetric C-atom of a monosaccharide chain is called epimerisation. Epimerisation results in a change in configuration. According to which C-atom is involved in the configurational change, one talks of 2-, 3-, 4,- or 5-epimerisation (see p. 11).

*D-Glucose* and *D-galactose* are epimers at the 4-position. Epimeric compounds or epimers within the family of the tetroses, pentoses, or hexoses are not related as object and mirror image, and they are therefore diastereomers. Monosaccharides up to 6 C-atoms (hexoses and hexuloses) carry trivial names

D-Glucose —— 2-epimerisation ⇌ —— D-Mannose

D-Glucose —— 4-epimerisation ⇌ —— D-Galactose

D-Glucose —— 5-epimerisation ⇌ —— L-Idose

which are still employed and which also serve in the form of a prefix as a symbol of configuration. Systematic nomenclature is dealt with in section 1.4 (p. 54).

### 1.2.4 Ring formula and configuration at the anomeric centre

As already explained in section 1.2.1, the acyclic form of a monosaccharide is of higher energy than the *cyclic* furanose or pyranose form arising from the intramolecular addition of a hydroxy group onto the carbonyl group. Originally, starting from the Fischer projection formula, the cyclic form was represented as the *hemiacetal* with a long, looped bond (p. 12).

This style of representation suggested by B. Tollens[10] indeed still possesses the clarity of the Fischer projection formula but, regarding the definition of the configuration of the hemiacetal-C-atom and the reproduction of the actual bonding situation, is a rather unfortunate choice and therefore is no longer found in the literature. The Fischer projection formula of the monosaccharide must be viewed in three dimensions. The vertically arranged C-chain forms a curve, whose plane is perpendicular to the paper and whose convex side is turned towards the observer (**A**).

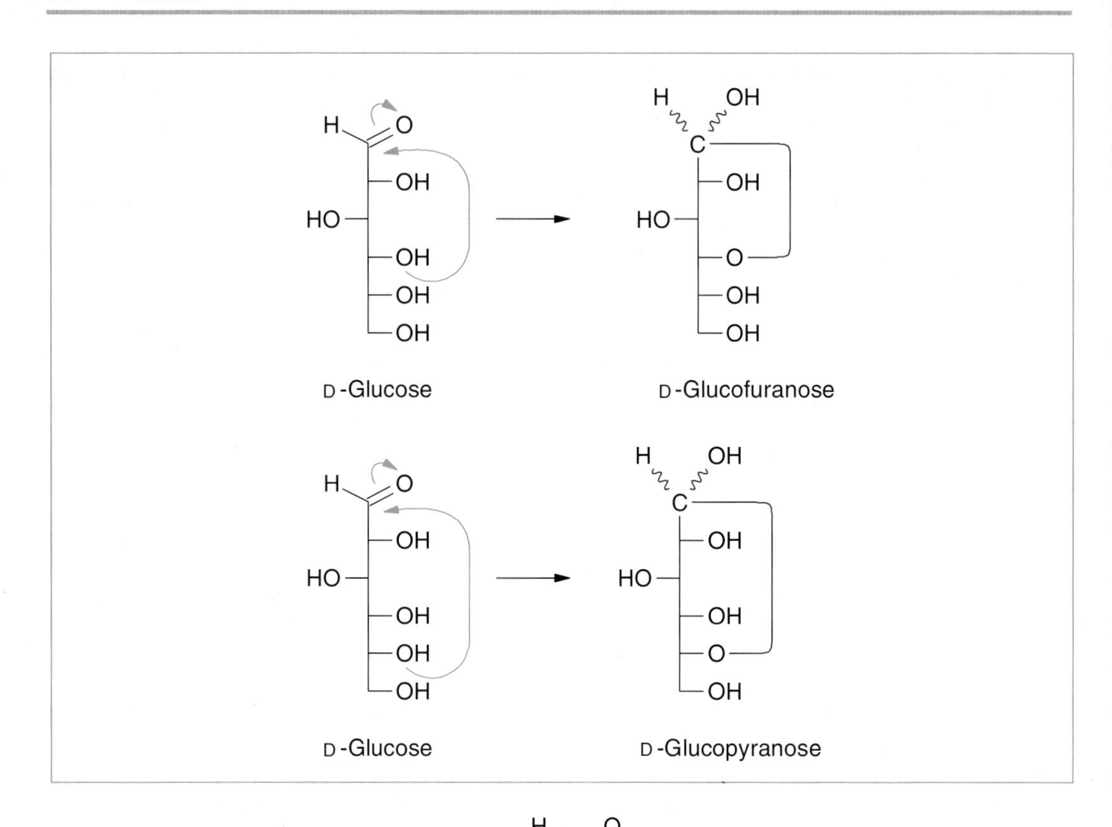


H O
1
2 OH
HO 3
4 OH
5 OH
6 OH

**A**

≡ ≡

H OH
H 2 1 O
HO 3 H
HO
4 OH
H 6 OH
5
H

≡

H OH
H H
OH
OH H O
HO
H OH

**A'** **A''**

D-Glucose

As with any spatial formula, that of D-glucose **A** may be rotated and turned randomly. If the *curved carbon chain* is first laid in the plane of the paper as in **A'** and then tipped over into the horizontal (**A"**), a pictorial representation results which allows a clear recognition of the potential ring system of a monosaccharide with the *cis*- and *trans*-relationships of the substituents. For ring closure to yield a *furanose* or *pyranose*, the hydroxy group in the 4- or 5-position, respectively, must only be brought near to the carbonyl group for hemiacetal formation. This is achieved by a twofold substituent exchange (p. 14) at C–4 or C–5, respectively, which leads to no change in configuration at these centres. Alternatively, correct positioning of the 4-OH or 5-OH may be viewed in terms of rotation about the C–3 to C–4, or C–4 to C–5 bond, respectively.

The formula thus drawn of a *furanose* or *pyranose* shows the ring horizontal, planar and perpendicular to the plane of the table. The style of representation was introduced by W. N. Haworth and reproduces constitution and configuration in a relatively realistic manner[11] (p. 15).

Another form of representation of the *Haworth-projection formula*, originated with J. A. Mills. In this case, the ring plane lies in the plane of the table. The position of the substituents in relation to the ring plane is indicated by dotted (under the plane) or bold-face bonds (above the plane)[12].

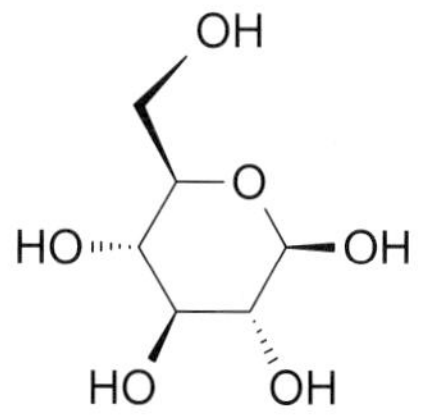

β-D-Glucopyranose

Today, both the *Mills formulae* and also the Haworth projection formulae are often replaced, especially in the case of 6-membered rings, by the clear conformational formulae (Section 1.2.5.2, p. 21).

The cyclisation of a monosaccharide through hemiacetal formation converts the carbonyl C-atom into an additional asymmetric C-atom, the so-called anomeric carbon atom. Anomer means, roughly, *overriding epimer.* Thus, the term anomeric centre indicates the special position of the corresponding C-atom. The diastereomers newly arising from hemiacetal formation are anomers, in the case of aldoses 1-epimers, with ketoses 2-epimers. The configurational change at the anomeric C-atom is an anomerisation.

The anomers are designated α or β according to the position of the hydroxy group on the anomeric C-atom (the anomeric hydroxy group is called the lactol group). In the Haworth projection formulae, the ring plane lies perpendicular to the plane of the table. In the case of furanoses, one places the ring O-atom behind, in the case of pyranoses behind and to the right, of the table plane. The lactol group is always on the right. In the D-series, with the β-configuration the hydroxy group projects above the ring plane and with the α-configuration below the ring plane (p. 16).

The α-L-form of a monosaccharide is always the mirror image of the α-D-form. Accordingly, for example, α-L-glucopyranose and α-D-glucopyranose are enantiomers.

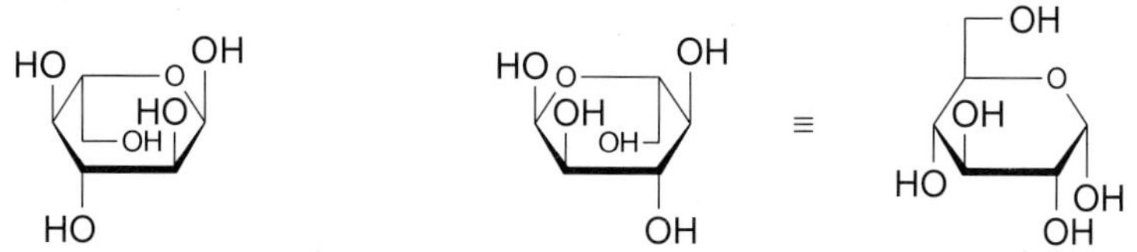

α-L-Glucopyranose

α-D-Glucopyranose

The anomeric configuration can be very simply ascertained today with physical methods (polarimetry and NMR spectroscopy). Anomeric configuration was determined for the first time by J. Böeseken in 1913 on α- or *β-glucopyranose*[13]. The α-form, which crystallises from aqueous solution, increased the electrical conductivity of boric acid solution. The β-form, in contrast, showed no effect shortly after

**Furanose**

1. Substituent exchange on C-4

D-Glucose

2. Substituent exchange on C-4

Cyclisation

D-Glucofuranose

**Pyranose**

1. Substituent exchange on C-5

D-Glucose

2. Substituent exchange on C-5

Cyclisation

D-Glucopyranose

## D-Aldofuranoses

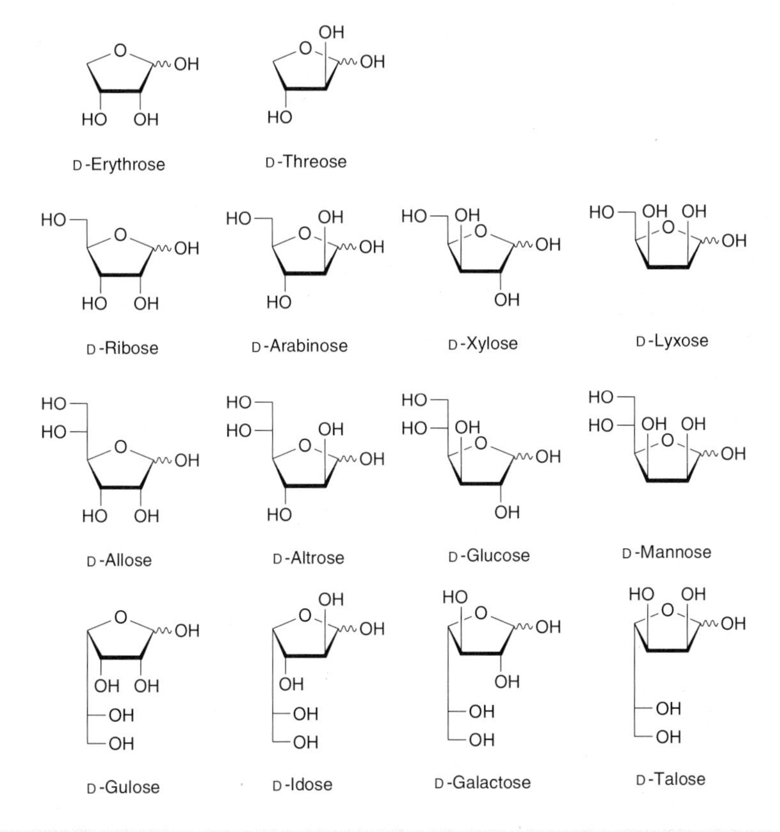


## D-Aldopyranoses

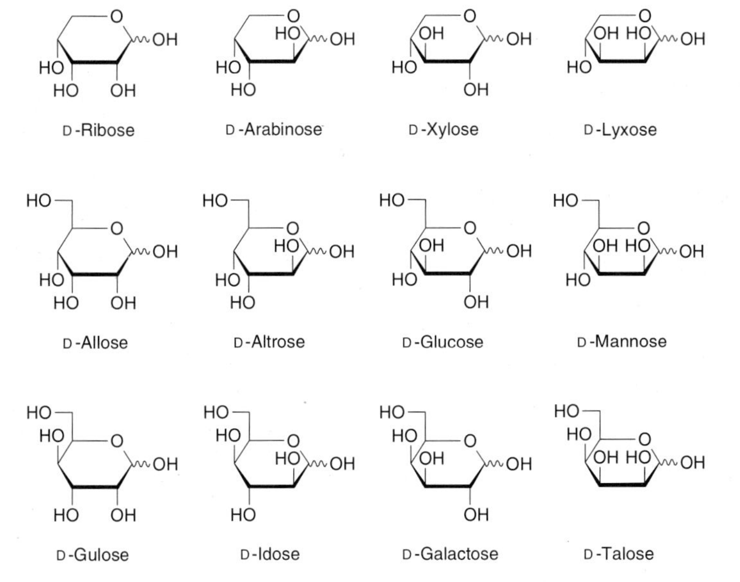

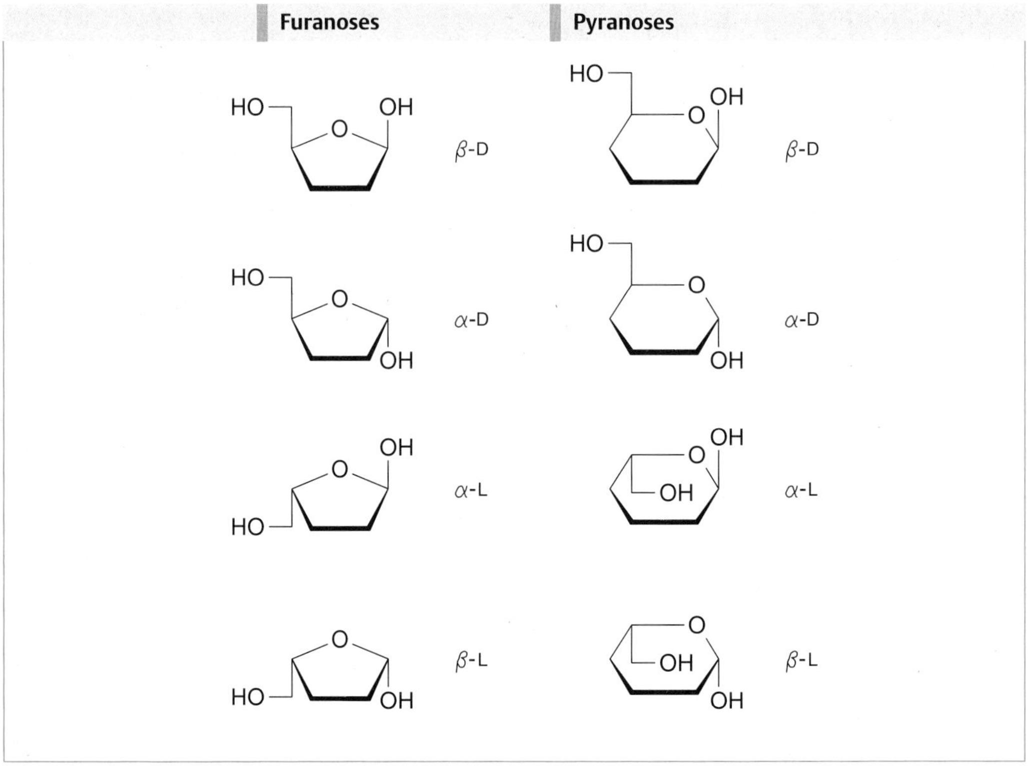

| Furanoses | | Pyranoses | |
|---|---|---|---|
| | β-D | | β-D |
| | α-D | | α-D |
| | α-L | | α-L |
| | β-L | | β-L |

dissolution. This proved the *cis*-orientation of the hydroxy groups in the 1- and 2-positions in the case of α-D-glucopyranose. Complexation is only possible with 1,2-*cis*-diols and not with the 1,2-*trans*-diols.

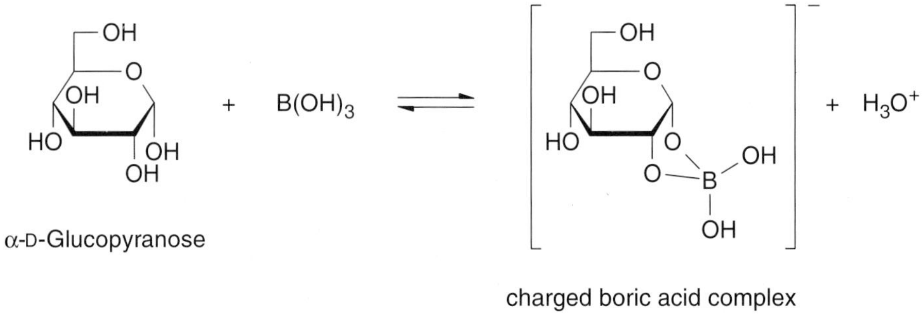

α-D-Glucopyranose + $B(OH)_3$ ⇌ [charged boric acid complex]$^-$ + $H_3O^+$

α-D-Glucopyranose

charged boric acid complex

## 1.2.5 Conformation[14,15]

Constitution and configuration of a monosaccharide are fixed. In order to change them a chemical reaction is required, that is a relative high activation energy or a catalyst. On the other hand, conformational interchange in solution or in the liquid state already proceeds at room temperature. The *activation energy* lies, in general, below 63 kJ/mol. The fact that one can talk of definite conformations in the case of many monosaccharides, especially of the cyclic ones, and that one can also assign these unequivocally with physical methods, especially NMR spectroscopy, lies in a sufficiently high energy barrier for conformational interconversion and, above all, in the energetic preference of some forms (but mostly only one form) among an innumerable number. Energy differences of a few kJ/mol suffice to allow a stable conformation to dominate to almost 100% over less stable ones in the equilibrium in solution. This is especially true for pyranoses.

In the crystalline state, monosaccharides and their derivatives are conformationally homogeneous. No example is known to the present day of conformational polymorphism. The stable conformations of a monosaccharide and the relative proportions of them to one another in solution are, apart from influences of the environment, dependent on the underlying constitution and configuration. They are, so to speak, the fine structure of a compound and the representation that comes nearest to the true shape of the molecule.

Thus, the physical properties and also the behaviour of the monosaccharides in their chemical reactions can be traced first of all to the predominant conformation or conformations. Essentially, there are two factors, in part interacting, which relatively stabilize a certain conformation of a monosaccharide in comparison with other possible conformations:

1. steric (*van der Waals*) and
2. polar interactions (*electrostatic interactions*).

They may be usefully summarised by the term "non-bonded interactions". Under certain circumstances with favourable conditions, intramolecular hydrogen bonds can also play a role.

The idea that rotation about a single bond is not entirely free was already appreciated at a relatively early stage[16]. It found its first first quantitative expression in the experimental determination of the so-called ethane barrier as about 12.6 kJ/mol by J. D. Kemp and K. S. Pitzer in the 1930's[17]. The small energy barrier increases with increasing spatial requirements of the substituents. The destabilising interaction designated as *torsional-* or *Pitzer-strain* really relates only to the C-H bonds opposing one another in space. The Pitzer-strain is superimposed on repulsive van der Waals interactions, and repulsive or also attractive polar interactions in the case of large polar substituents. Today, it is usual to regard the different interactions through space collectively.

Rotation about a single bond causes the distance to change between substituents which are not on the same C-atom or other multivalent atom. On rotation through 360°, two substituents X on a C,C-bond experience, once, a moment of closest approach **B**, and also a moment of greatest separation **A**. Conformation **B** possesses at the same time high potential energy and low stability:

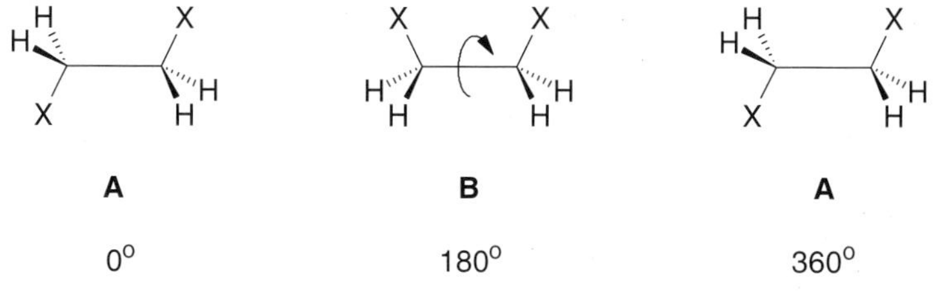

A B A

0° 180° 360°

If only the substituents X exert an interaction on each other, then it can be reasonably assumed that conformation **A** alone represents the molecule in question. With more space filling substituents, collective interactions need to be considered, according to their importance, and also those interactions between H-atoms and larger substituents.

It is often difficult to represent clearly in the plane of the paper the spatial relationships in a molecule. Thus far, spatial formulae have served for illustration. To some extent, however, exact distances and angle relationships cannot be demonstrated in this manner. For the pictorial representation of conformations, Newman projection formulae are well suited in combination with various alternatives of the notation. If the grouping X-C-C-X of a molecule is rotated about the C-C bond axis, six significant arrangements are obtained which may be illustrated with the help of *Newman projection formulae* and spatial formulae and to which clear descriptions have been assigned[18] (p. 18).

The angle in the projection formula between a bond in front and a bond behind the plane of the paper is the *dihedral angle*:

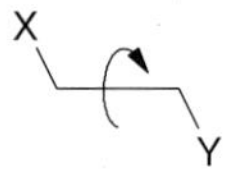

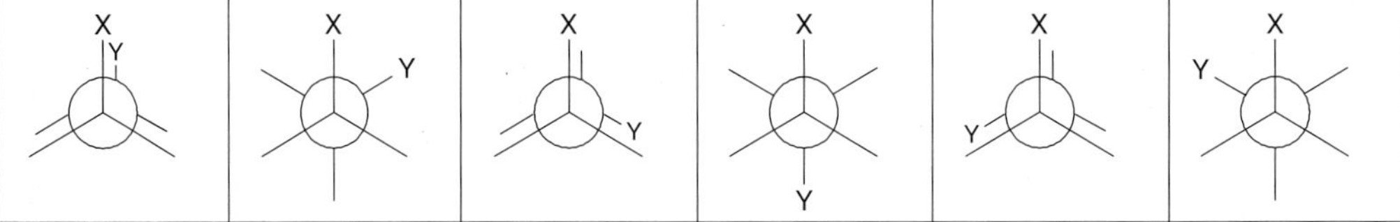

| | | | | | |
|---|---|---|---|---|---|
| **Description 1** | | | | | |
| ± *syn-periplanar* (± sp) | + *syn-clinal* (+ sc) | + *anti-clinal* (+ ac) | ± *anti-periplanar* (± ap) | – *anti-clinal* (– ac) | – *syn-clinal* (– sc) |
| **Description 2** | | | | | |
| *syn* | *gauche* | — | *anti* | — | *gauche* |

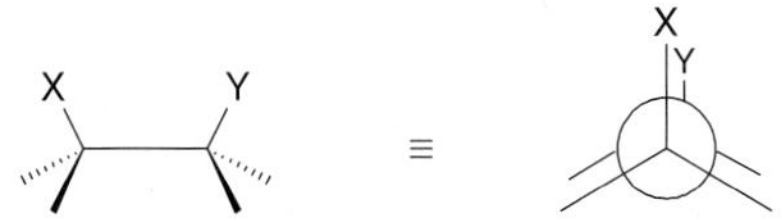

spatial formula ≡ Newman projection formula

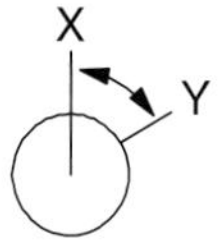

dihedral angle

Although every dihedral angle between 0 and 360° is theoretically possible, only three are significant:

| Dihedral angle (°) | Description | |
|---|---|---|
| 0 | eclipsed | *syn* |
| 60 | staggered | *gauche* |
| 180 | staggered | *anti* |

In general, steric interaction is the decisive factor in the relative destabilisation or stabilisation of certain conformations of a molecule. The interaction is stronger, the larger the sum of the radii of the clashing atoms or groups of atoms. It is possible that only in saturated aliphatic hydrocarbons is space-filling responsible for the *non-bonded interactions*. Polar substituents exert an electrostatic interaction on each other which overlies the pure van der Waals interaction and which is proportional to the polarity of the C-substituent bond in question. In the case of 1,2-dichloroethane, dipolar repulsion in the *gauche* conformation is regarded as the dominant factor compared with a relatively weak steric interaction and is responsible for the preference for the *anti* conformation[18].

The following series gives a qualitative impression of the steric *bulk* of different atoms and groups of atoms:

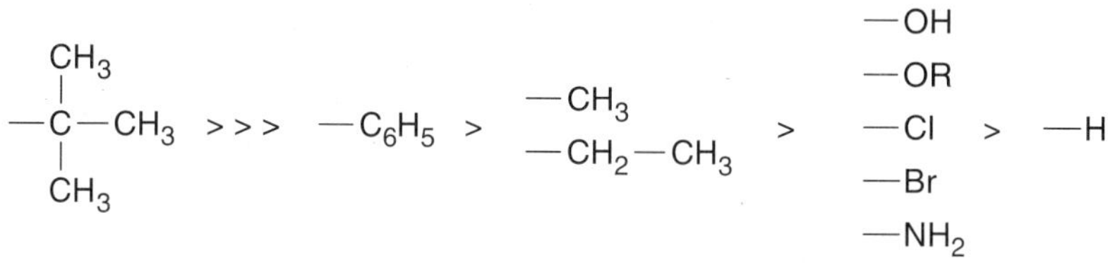

$$-C(CH_3)_3 \;>>>\; -C_6H_5 \;>\; \begin{matrix} -CH_3 \\ -CH_2-CH_3 \end{matrix} \;>\; \begin{matrix} -OH \\ -OR \\ -Cl \\ -Br \\ -NH_2 \end{matrix} \;>\; -H$$

Non-bonded interactions with a hydrogen atom need only be considered in the case of a *syn*-relationship. From the staggering of the ligands as shown, it may be determined that an unbranched C-chain prefers to adopt a planar zig-zag arrangement, which is also true when it carries heteroatoms. Only in this manner can an *anti*-arrangement be achieved of the conformationally influential C-residues.

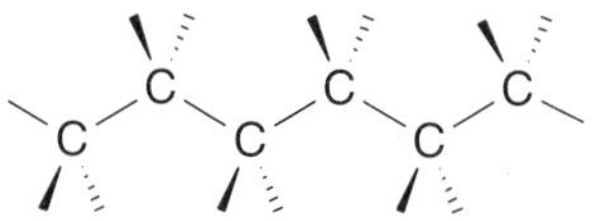

A comparison of the diastereoisomeric 2,3-butanediols in each stable zig-zag arrangement makes it clear that the *threo*-form is the thermodynamically less stable stereoisomer, because of an additional *gauche* interaction of the hydroxy groups:

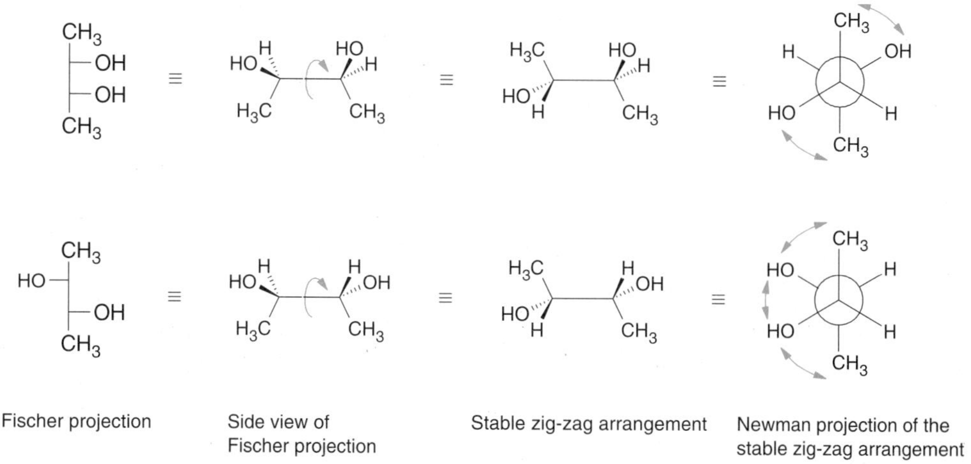

Fischer projection | Side view of Fischer projection | Stable zig-zag arrangement | Newman projection of the stable zig-zag arrangement

The spatial formulae of both stereoisomeric 2,3-butanediols make it clear, moreover, that the *Fischer form*, that is the conformation which forms the basis of the the Fischer projection formula, is the most unstable of all of the possible conformations.

#### 1.2.5.1 Open-chain monosaccharide derivatives

It is often found that translation of the Fischer projection formula of an acyclic polyol into a planar zig-zag representation proves difficult. A simple method has been suggested by S. Signorella and L. F. Sala[19]:

1. The *glycol groups* in the Fischer projection formula are separated into *threo*- and *erythro*-arrangements and the formula is viewed from one side (left).
2. The zig-zag chain is drawn with the $R_1$ group placed below and on the left.

3. If the first group (a) in the Fischer projection formula is on the far side, away from the viewer, then in the planar zig-zag representation it is placed below the plane of the paper.
4. *threo*-Arrangements result in both groups lying on the same side of the planar zig-zag representation and *erythro*-arrangements on opposite sides.

As an example, the D-*galacto*-arrangement is illustrated both as the Fischer projection formula and also in the planar zig-zag representation.

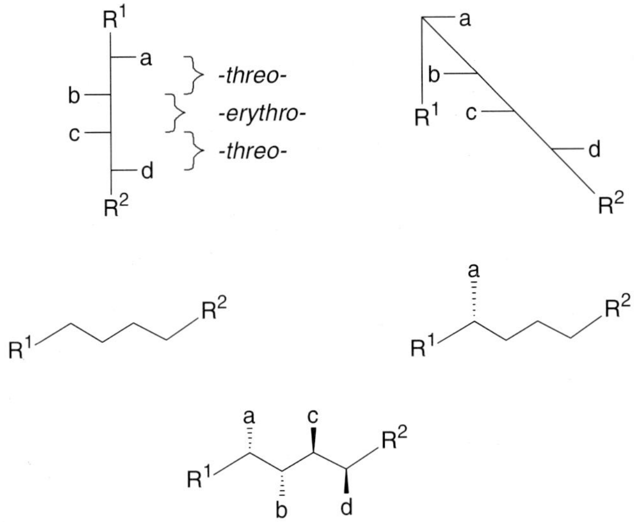

Certain configurations of longer chain acyclic monosaccharide derivatives can, however, lead to deviation from the planar zig-zag conformation as can be shown with the help of NMR spectroscopy[15]. If, for example, two space-demanding groups occupy a 1,3-*syn*-orientation, that is they are parallel to one another in the zig-zag conformation, then the carbon chain can deviate by rotation about a C,C-bond:

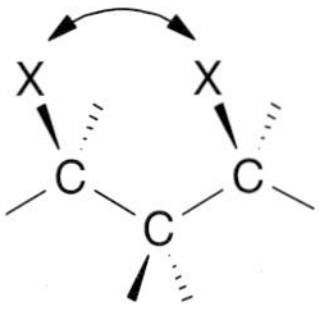

This rotation leads, as shown by the example of *D-glucitol*, to a transformation to the *sickle* conformation:

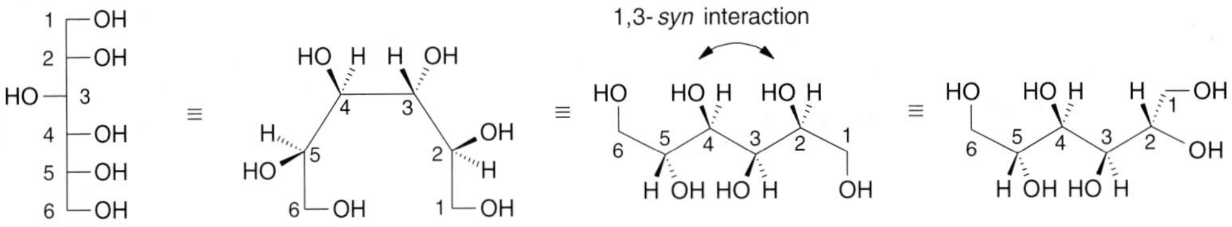

D-Glucitol

Side view of the Fischer projection

Zig-zag conformation

Sickle conformation

The sickle conformation has been demonstrated for acyclic glucose derivatives and also in the configuratively related xylose derivatives. Acyclic mannose and galactose derivatives show no 1,3-*syn* interaction in the stable zig-zag conformation. Both NMR investigations and X-ray structural analysis showed that acyclic derivatives with *galacto* and *manno* configurations, both in solution and also in the crystalline state, prefer the planar zig-zag conformation of the C-chain. All acyclic monosaccharide derivatives with stable zig-zag conformations have in common in the Fischer projection formula a pairwise alternating series of asymmetric C-atoms:

........... -left - right - right - left - left - right - right- .........

Each configuration which matches such a series is equivalent to a stable zig-zag conformation. A knowledge of the most stable conformation of an acyclic monosaccharide derivative is helpful in the planning and prediction of certain chemical reactions of these compounds, especially when two hydroxy groups react with formation of a cyclic product (cyclic acetal, cyclic ester) or cyclic intermediates (periodic esters). In such cases, reaction proceeds very readily if there is a favourable orientation of the hydroxy goups in the stable zig-zag conformation of the C-chain. That is mostly the situation in the case of a normally destabilising *1,2-gauche-* (preferred periodate cleavage of the 3,4-bond in mannitol; *threo*-position) or 1,3-*syn*-relation (formation of *2,4-O-benzylidene-glucitol*[20]):

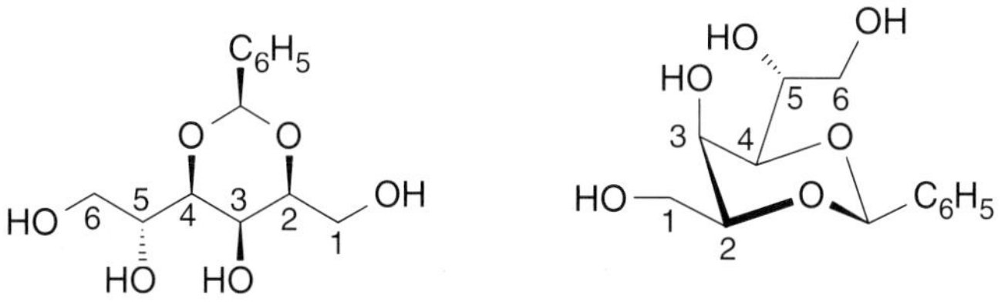

2,4-O-Benzylidene-D-glucitol

Other open-chain monosaccharide derivatives such as mercaptals (*S,S*-acetals) and dialkyl acetals are comparable with the monosaccharide alcohols.

### 1.2.5.2 The pyranoid ring system

Free pentoses and hexoses as well as hexuloses and also longer chain monosaccharides are especially stable as 6-membered cyclic hemiacetals. In extreme cases, for example in the case of glucose with a specially unstable acyclic form and specially stable pyranose forms ($\alpha$ and $\beta$), the difference in free energy amounts to about 29.4 kJ/mol. As already shown (Section 1.2.1, p. 2), the energy gain in the transition of the acyclic to the cyclic form of 5-hydroxypentanal is sufficient to place the equilibrium greatly in favour of the hydropyran ring, and apparently the relative stability of the hydropyran ring increases with the number, size and type of the substituents. Especially, an electronegative substituent, for instance an hydroxy group in the 2-position, amongst other factors, favours the formation of the hemiacetal. In water, *xylopyranoses* and *glucopyranoses* predominate to the extent of 94% and 99%, respectively, over the corresponding isomeric furanoses and acyclic forms. In addition, the positions of substituents on the pyranose ring, which in its shape is comparable to the cyclohexane ring, are important factors influencing the relative stability of a pyranose. Differences arise through the ring oxygen atom. The C,O-bond (142 pm) is around 10% shorter than a C,C-bond. The bond angle at the O-atom (112 – 114°) is larger than the tetrahedral angle (more *s*-character). In principle, however, both ring systems correspond in their conformations. Cyclohexane as a ring model for the pyranose ring can assume a multitude of different conformations of which the following form a representatve selection:

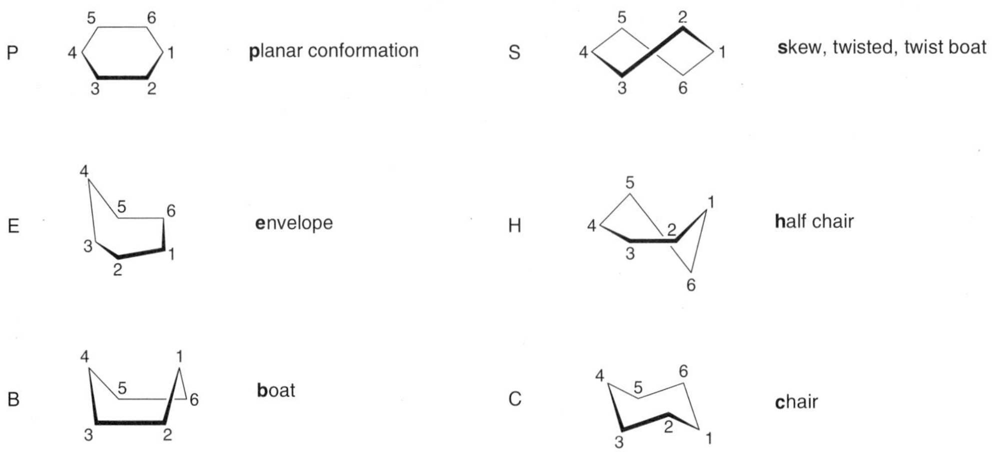

P planar conformation

S skew, twisted, twist boat

E envelope

H half chair

B boat

C chair

Numerous investigations have shown that the *chair* conformation C is by far the most stable of the six conformations shown. Indeed, the B- and S-conformations are also free from ring strain on the basis of deformation of the tetrahedral angle, yet the C-conformation alone has no ligands on the ring which occupy a sterically unfavourable *syn-periplanar* position to one another. This means that cyclohexane at room temperature is exclusively in the chair form. This goes also for the vast majority of pyranoses.

The C-conformation of cyclohexane has two types of C,H-bonds:

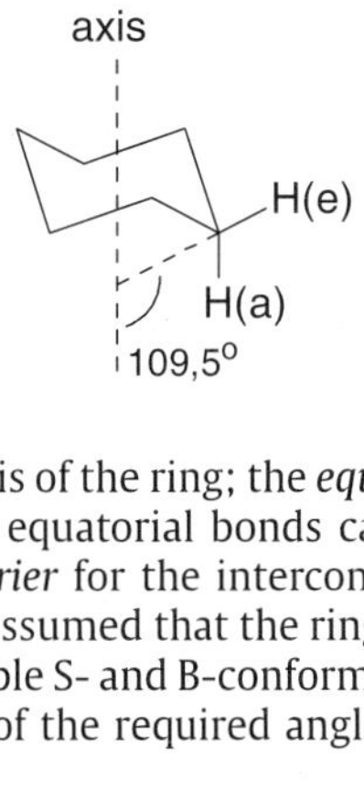

axis H(e) H(a) 109,5°

The *axial (a)* bonds run parallel to the axis of the ring; the *equatorial (e)* bonds are inclined to the axis at about the tetrahedral angle. Axial and equatorial bonds can be interchanged if the ring skeleton changes its conformation. The *energy barrier* for the interconversion of one chair conformation into another amounts to about 42 kJ/mol. It is assumed that the ring inversion takes place through the high energy H-conformation as well as the flexible S- and B-conformations. The high energy barrier between C- and S-conformation is mostly a result of the required angle deformation energy for the transition (Fig. 1.1).

C-Conformations are correspondingly rigid. In the case of the S → B conformational interconversion, only torsional (Pitzer) strain must be overcome. The rings are correspondingly flexible. The ready change of the flexible conformations one into another is termed *pseudorotation*, a term met with especially in the case of five-membered rings. All bonds which are axial in one C-conformation are equatorial in the other C-conformation, and vice versa. This can be seen in the two Newman projections of cyclohexane, in which the C-atoms are numbered.

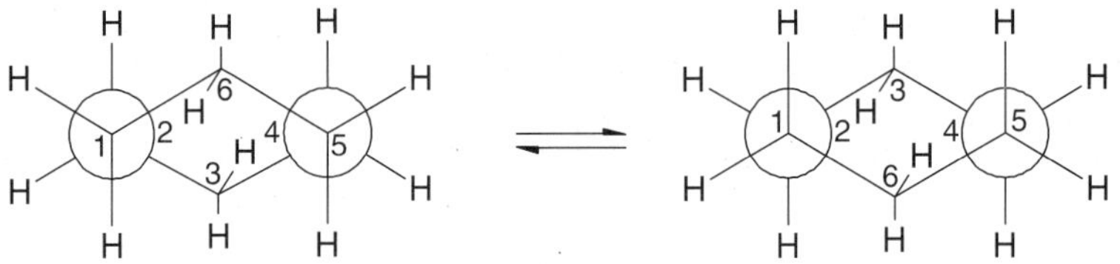

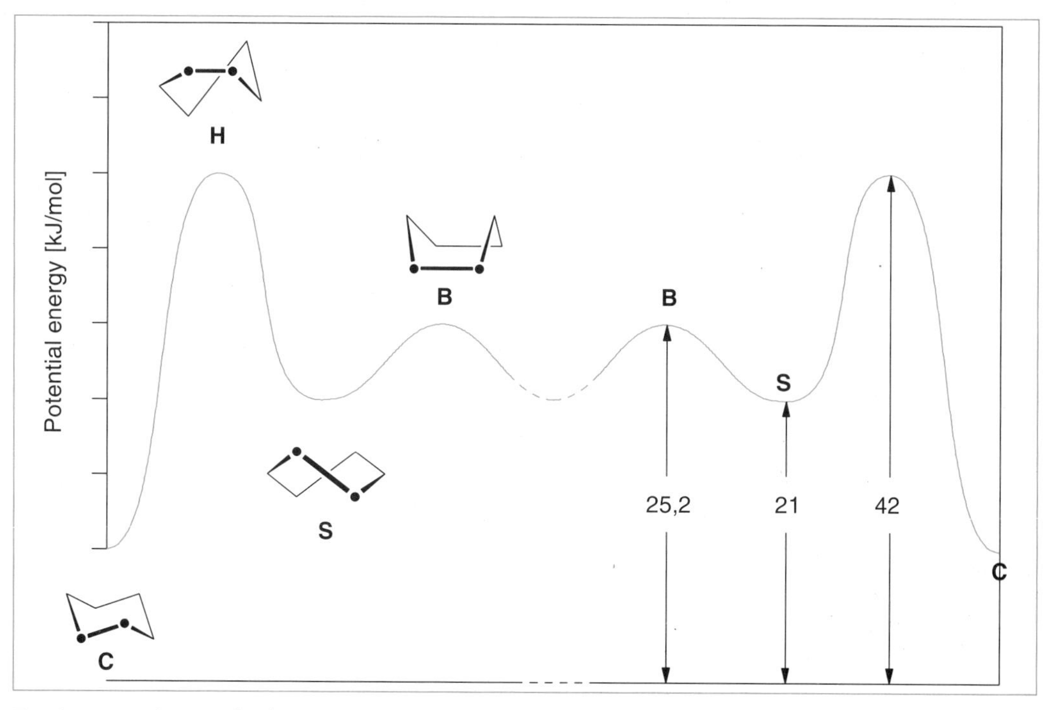


**Fig. 1.1** Energy barriers for the conversion of one C-conformation into another C-conformation in a 6-membered ring system.

Both chair conformations are equivalent as long as the substituents, which in the shown example are H-atoms, are the same. Replacement of a hydrogen atom by another group leads to two non-superimposable, that is different C-conformations. 2-Hydroxytetrahydropyran (the numbering used in the diagram is not correct but it corresponds to that in a pyranose), which serves as a model for a pyranose, occurs in two C-conformations:

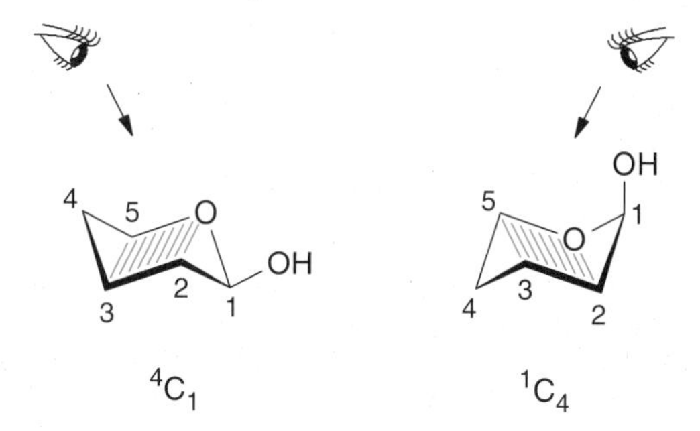


To distinguish the two conformations, the pyranose rings are numbered as usual. A reference plane is drawn through four atoms of the ring so that the C-atom with the lowest positional number lies out of the plane. A ring atom which to the observer lies above this reference plane is placed in the form of its ordinal number before the abbreviation of the ring conformation as a superscript, whilst the one below the plane is written as a subscript following this letter. In the formulae shown, each of the C-atoms 2, 3 and 5 as well as the O-atom lie in the reference plane. In one case, C-atom 1 lies under and C-atom 4 above the plane, leading to the symbol $^4C_1$. In the other case, the position is reversed. In order that an unequivocal *above* and *below* the reference plane may be defined, the pyranose ring must be observed

so that the series of ordinal numbers appears *clockwise*, the observer then being *above* the plane. The remaining ring conformations H, B, and S can also be designated unambiguously in this manner.

The *Haworth projection formula* is easily translated into a ${}^4C_1$ conformation. The chair ring is first drawn, then the bonds are attached and afterwards the substituents are added accordingly, upwards or downwards, as indicated by the Haworth projection formula:

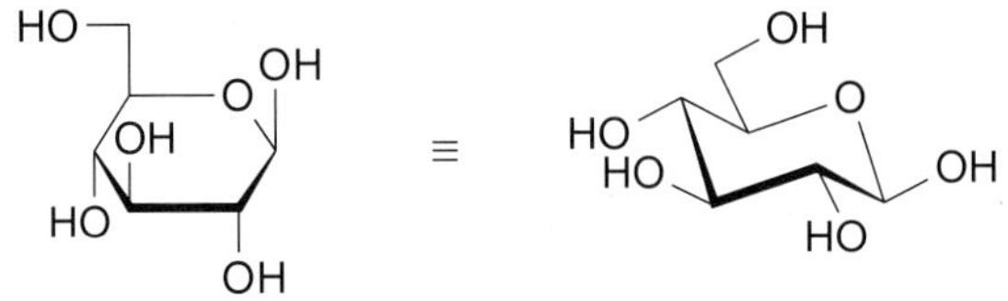

β-D-Glucopyranose

*Cis,trans*-relationships in the Haworth projection formula correspond to the following *axial-equatorial* relationships in the C-conformation:

| | | | |
|---|---|---|---|
| 1,2-*cis* | e-a or a-e | 1,3-*trans* | e-a or a-e |
| 1,2-*trans* | e-e or a-a | 1,4-*cis* | e-a or a-e |
| 1,3-*cis* | e-e or a-a | 1,4-*trans* | e-e or a-a |

The interactions already described for acyclic monosaccharide derivatives (Section 1.2.5.1, p. 19) apply equally for cyclic derivatives. Starting from the justifiable assumption that all pyranoses exist in a chair form, in so far as they are not fixed through additional ring formation in another conformation, then the relative partial interactions summarised in Table 1.1 result beween the different equatorial and axial groups on the pyranose ring. The energy values are taken relative to an assumed interaction of zero between two correspondingly oriented H-atoms.

**Table 1.1** Partial interactions between axial and equatorial groups on the pyranose ring

| Interaction | Description | | relative interaction energy (kJ/mol)[a] |
|---|---|---|---|
| 1,3-*syn* | carbon-oxygen | $C_a : O_a$ | 10.5 |
| | oxygen-oxygen | $O_a : O_a$ | 6.3 |
| | carbon-hydrogen | $C_a : H_a$ | 3.78 |
| | oxygen-hydrogen | $O_a : H_a$ | 1.89 |
| 1,2-*gauche* | carbon-oxygen | $C_1 : O_2$ | 1.89 |
| | oxygen-oxygen | $O_1 : O_2$ | 1.89 |

[a] The interactions are of the destabilising type. The energy values are relative to an assumed zero interaction between correspondingly oriented H-atoms. They are valid for aqueous solutions at room temperature.

The values were ascertained especially by Angyal et al. with the help of equilibrium measurements on a large number of model compounds[16]. The semi-quantitative calculations of the partial interaction energies extend only to 1,3-diaxial interactions and 1,2-*gauche* interactions, and interactions of a secondary size, for example between H-atoms or 1,2-*gauche* interactions between H-atoms and substituents of another type, are neglected. A simplification is also in the assumption that a pyranose ring possesses the geometry of a cyclohexane ring. The interaction of a specific substituent is accordingly independent of its position on the ring. An exception is made in the case of the substituent at the anomeric centre. Although they should be employed with certain reservations, the *Angyal parameters* indicated in Table 1.1 may be utilised to advantage for the calculation of relative free energies. They allow a quantitative statement about the relative stabilities of monosaccharides. The strongest effects, as with the acyclic derivatives, are a result of 1,3-*syn*-interactions.

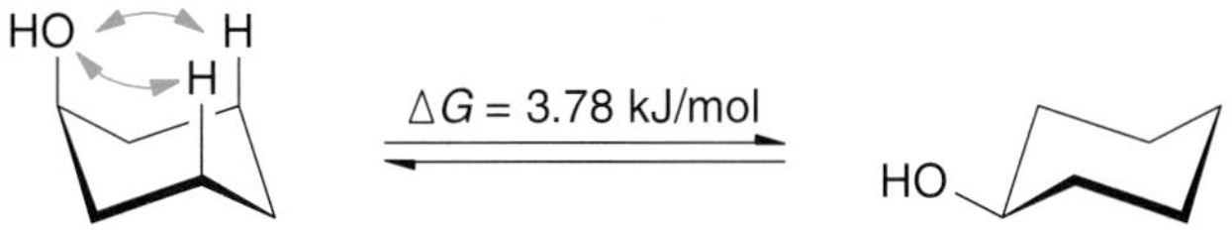

$\Delta G = 3.78$ kJ/mol

2 ($O_a : H_a$)

Each *axial* substituent on the pyranose ring gives rise, therefore, to a destabilisation since it must interact at least with one *syn-axial* H-atom. The effect increases greatly with the steric demands of the *syn-axial* partner, which becomes evident if one compares the *syn-axial* interaction between O- and H-atoms ($O_a : H_a = 1.89$ kJ/mol) on the one hand and O- and C-atoms ($O_a : C_a = 10.5$ kJ/mol) on the other. A hexopyranose will also endeavour to adopt a conformation which allows an *equatorial* rather than an axial position for the hydroxymethyl group. Thus, in the D-series, a $^4C_1$ conformation will be preferred.

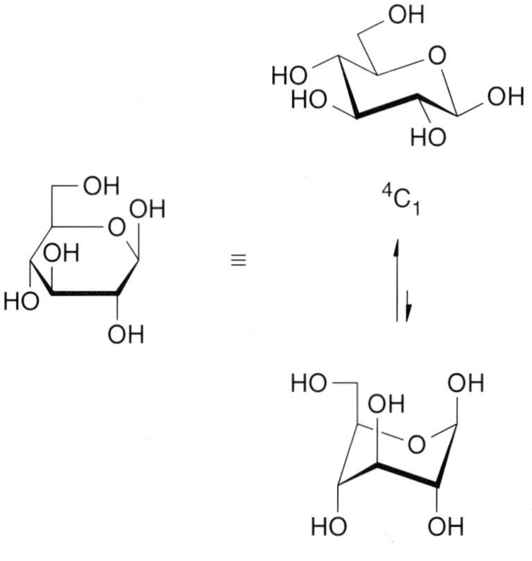

$^4C_1$

$^1C_4$

This rule is valid with the exception of the *idopyranose*, in which case an *equatorial* position for the hydroxymethyl group ($^4C_1$) must be bought at the expense of three *axial* hydroxy groups, and for this reason the alternative $^1C_4$ conformation is of comparable stability.

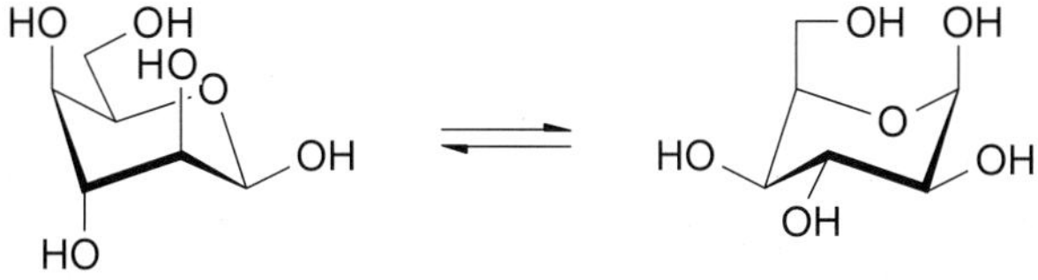

$^4C_1$ (18.2 kJ/mol) $\rightleftharpoons$ $^1C_4$ (16.1 kJ/mol)

### 1.2.5.3 The anomeric effect[21–23]

Measurements of optical rotation, NMR spectra, and also gas chromatographic analysis of suitable derivatives indicate that D-pyranoses in solution equilibrium are often more stable in their α-configuration than would be expected from a summation of steric interaction energies. Data for three different anomeric pairs are shown in Table 1.2.

**Table 1.2** Interaction energies $\Delta/\alpha \rightleftharpoons \beta$ in kJ/mol

| | Calc. from partial steric interaction energies (Tab. 1.1, p. 24) | Calc. from the equilibrium of the anomers in aqueous solution | Anomeric effect (kJ/mol) |
|---|---|---|---|
| 36% α- ⇌ 64% β- D-Glucose | $2\,(O_a:H_a) = 3.8$ | 1.47 | 2.30 |
| 69% α- ⇌ 31% β- D-Mannose | $2\,(O_a:H_a) - (O_1:O_2) = 2.3$ | − 1.88 | 4.18 |
| 47,5% α- ⇌ 52,5% β- 2-Deoxy-D-*arabino*-hexopyranose | $2\,(O_a:H_a) = 3.8$ | 2.09 | 3.56 |

The preference for the sterically *unfavourable* axial position, especially by electronegative substituents on the anomeric C-atom of pyranoses, was discussed as an unusual feature for the first time by J. T. Edward[24] and given the term *anomeric effect* many years later by R. U. Lemieux[25]. The apparently reversed influence of the lactol group in comparison to the other substituents on the ring, the same anomeric effect, is not always of equal magnitude, as can be seen from the values in Table 1.2, but depends especially on the substituents on the neighbouring C-atom 2. With an axial substituent, the anomeric effect is strengthened, with an equatorial one it is weakened. If there is no substituent, or if the influence of an axial hydroxy group in the 2-position is compensated by an axial hydroxy group in the 3-position, the effect is midway between these two extremes. It is now recognised that hydroxy groups in positions other than the 2- or 3-positions can also influence the anomeric effect; their influence is, however, far less and in most cases can be neglected. The anomeric effect depends on the type of substituent. Distinct differences in the anomeric effect of pyranoses arise by alkylation or acylation of the non-anomeric hydroxy groups. Thus, the concentration of the α-anomer increases in the equilibrium mixture with each methyl group from that of *D-mannose* (69%) to that of *2,3,4,6-tetra-O-methyl-D-mannose* (86%).

Still more impressive than with free hexopyranoses in aqueous solution, appears to be the influence of the anomeric effect on anomeric ratios in the equilibrium mixture of *methyl hexopyranosides, penta-O-acetyl-hexopyranoses and tetra-O-acetyl-hexopyranosyl halides* in the presence of an equilibrium-inducing catalyst. Very clearly, the α-form with the axial anomeric substituent is preferred here over the *β*-form.

Although regarded as a speciality of pyranoid systems, and accordingly as a phenomenon of carbohydrate chemistry, the anomeric effect is recognised today as a stereoelectronic effect of general significance, which always occurs if at least two hetero atoms are bound to a tetrahedral centre. The general grouping regarded as essential for an anomeric effect is:

C-**X**-C-**Y** X = N, O, S; **Y** = Br, Cl, F, N, O, S

Of course, this generalised anomeric effect[26] can also be applied, though less strikingly, to the orientation of a substituent on the lactol group of a monosaccharide and also in the case of glycosides. This important factor, especially for the conformation of oligosaccharides, is called the *exo-anomeric effect*[27]. An additonal complication, under certain structural conditions, is that the anomeric effect is reversed, that is the anomeric substituent prefers not the axial but the *equatorial* position on the pyranosyl system. This stereoelectronic effect is called the *reverse anomeric effect*[28]. It is observed especially clearly with a quaternary N-atom on the anomeric C-atom, as in the case of of the peracetylated α-D-xylopyranosylpyridinium bromide. Here, as in the case of the corresponding peracetylated *β*-D-xylopyranosyl bromide[15], the otherwise unfavourable $^1C_4$ conformation is enforced in solution by the reverse and normal anomeric effect, respectively:

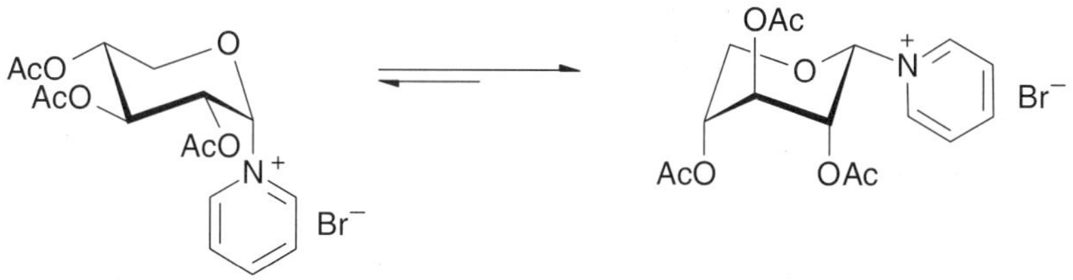

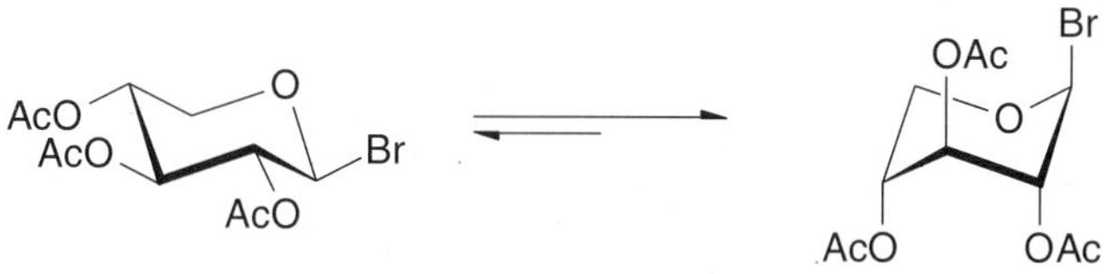

Systematic investigations on the glycoside model 2-methoxyoxan (2-methoxytetrahydropyran) illustrate the dependency of the anomeric effect on the solvent[26]. With increasing polarity of the solvent, the influence of the anomeric effect on the equilibrium position between the alternative conformations decreases. Similar observations have been made with free *D-glucopyranoses* whose *α*/*β*-anomeric ratio is 36/64 in water and 45/55 in pyridine.

There are also peculiarities at the anomeric centre of pyranoses and pyranosides in the length and angle of the bonds involved. First indications were obtained from investigations of the crystal structure of various pyranoses, which showed a clear shortening of the anomeric $C_1$-$O_1$ bond compared to *normal* C,O-bonds[29]. Fundamentally, the acetal single bond appears to be shortened in pyranoses and pyranosides compared to the average normal bond length of 143 pm and the angle at the anomeric C-atom $O_5$-$C_1$-$O_1$, according to configuration, is with *axial* residues larger, and with *equatorial* residues smaller than the tetrahedral angle:

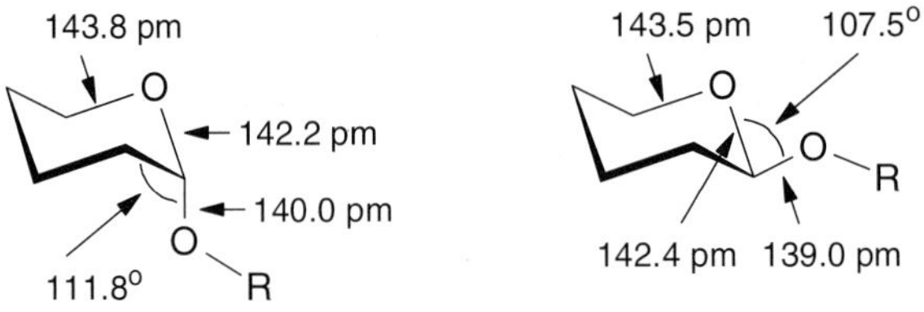

Crystal structure investigations on the acylated *glycosyl chlorides* and *fluorides* reveal still stronger differences in the geometric parameters about the anomeric C-atom[30]. Here, differences of bond lengths in the case of *axial* compared to *equatorial* orientations of the halogen atoms are especially striking. A lengthening of the *axial* $C_1$,F-bond with a shortening of the $C_1$,$O_5$-single bond stands in contrast to a shortening of the equatorial $C_1$,F-bond with a lengthening of the associated $C_1$,$O_5$-bond. At the same time, there is a clear widening of the $C_1$-$O_5$-$C_5$ angle in the case of an axial F-atom:

After 40 years discussion about the anomeric effect together with related effects, and about their causes, the opponents from two camps have finally decided on a type of compromise, which attempts to consider all, often contradictory, experimental data. One side traces the anomeric effect to an electrostatic interaction, which in the end is a destabilising interaction between two dipoles:

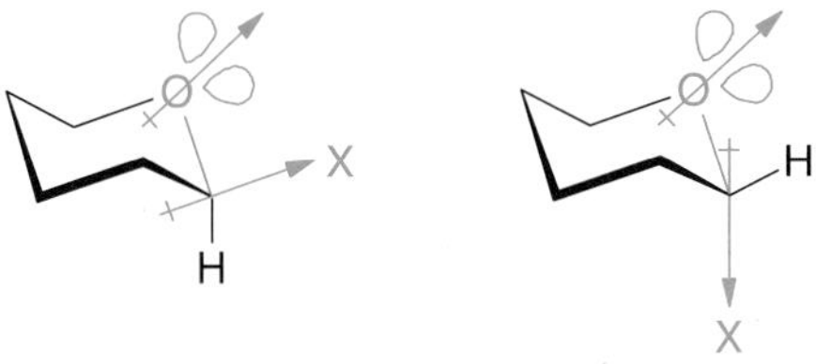

The other side proposes a stabilising influence which depends on the antiperiplanar arrangement of a free electron pair (donor) to a $\sigma$-bond, whereby the $\sigma^*$-orbital (acceptor) receives electrons:

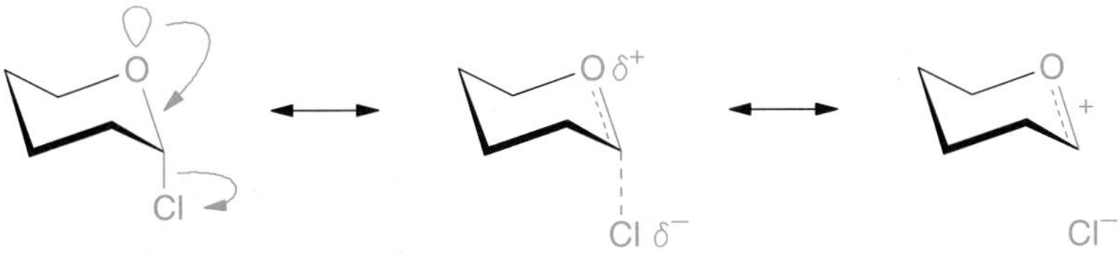

Delocalisation of the free electron pair on the O-atom into the antibonding $\sigma^*$-orbital of the neighbouring bond leads to stabilisation of the shown anti-periplanar arrangement in a glycopyranosyl chloride.

The former explanation, which is also described as the *Edward-Lemieux effect* or pictorially as the *rabbit ear effect* fits with the influence of the solvent on the anomeric effect and makes the reverse anomeric effect understandable. The latter, which is also described as *no bond-double bond resonance*, agrees with the change of bond lengths according to donor (shortening) or acceptor (lengthening) properties of the participating heteroatoms.

The electrostatic interaction exists between the bond of the anomeric C-atom-substituent X and the bonds joining C-atoms 5 and 1 to the ring O-atom. In case **A** the polar bond $C_{anomer}$,X in the Newman projection is between the orbitals of the free electron pairs on the ring oxygen atom (staggered) but in case **B** the bond is oriented *anti* to one orbital and *gauche* to the other. In case **A**, the resulting dipoles lead to destabilisation and in case **B** to stabilisation of the conformations represented. It is understandable that the substituent on the C-atom 2 according to location and polarity, has a greater or lesser influence on the anomeric effect. Thus, a *compressed* electrostatic interaction is to be expected if the anomeric substituent X is *equatorial* and the substituent on C–2 (OH) is *axial*:

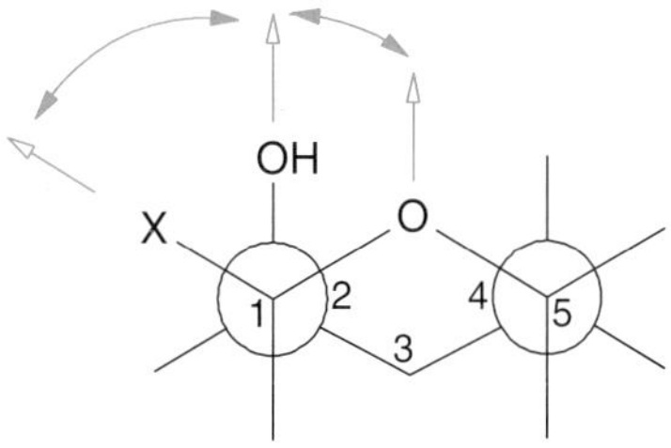

The anomeric effect is proportional to the electronegativity of the atom directly bonded to the anomeric C-atom and inversely proportional to the polarity of the solvent. Pyranose derivatives in which the dipole on the anomeric C-atom is reversed by introduction of an electropositive group show an *inverse anomeric effect*. As an example, the peracetylated α-D-xylopyranosylpyridinium bromide is shown, which prefers the $^1C_4$ conformation despite the axial acetoxy groups:

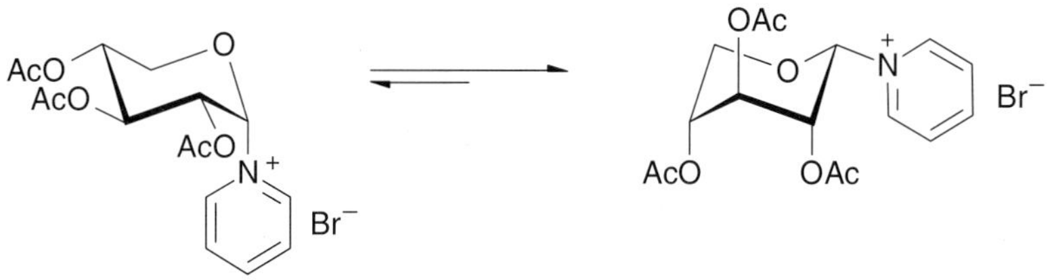

The second basis for the anomeric effect is of a totally different type. Here an $n \rightarrow \sigma^*$ hyperconjugation is invoked, which leads to a stabilising no bond-double bond resonance (see above). Such stabilisation is only possible with an *anti-periplanar* orientation of an electron pair on the ring O-atom and the anomeric $\sigma$-bond. Clearly, resonance can also occur between the $C_1,O_5$-$\sigma$ bond and an electron pair on the anomeric O-atom which is anti-periplanar to it. This resonance is the cause of the *exo*-anomeric effect. In the case of an acetal group in normal glycosides, the resonance effect leads to the shortening of both C,O-bonds. If the hetero atoms have differing electronegativities, then the less electronegative element becomes the *n*-donor, and vica versa, with the corresponding consequences for bond lengths and relative stabilites of the conformers.

It should be noted that whilst a rationalization of the anomeric effect based on electrostatic (dipole-dipole) interactions accounts for the decrease of the effect with increase in solvent polarity, it is inadequate to account for the bond length changes discussed above.

### 1.2.5.4 Intramolecular hydrogen bonds and binding electrostatic effects

Equilibrium measurements with the help of IR spectroscopy on *cis*- and *trans*–1,3-O-benzylideneglycerol in dilute carbon tetrachloride solution show intramolecular hydrogen bonds between the hydroxy groups and ring O-atoms. The stabilisation by the intramolecular hydrogen bonds is so great that the *cis*-compound exists only in the conformation shown, whereas the *trans*-compound lies to a considerable extent in the axial conformation[31–33]:

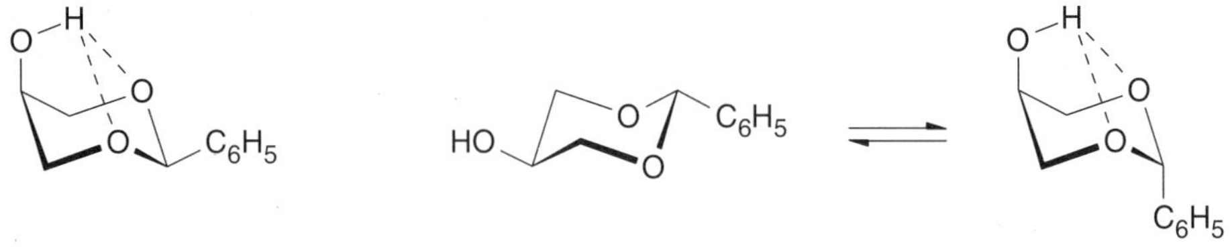

*cis* *trans*

For *methyl 3-deoxy-β-L-erythropentopyranoside* , in solvents which are not capable of forming hydrogen bonds with the hydroxy groups of the glycoside, the $^4C_1$ conformation is preferred, even though this possesses two destabilising 1,3-*syn*-hydroxy groups:

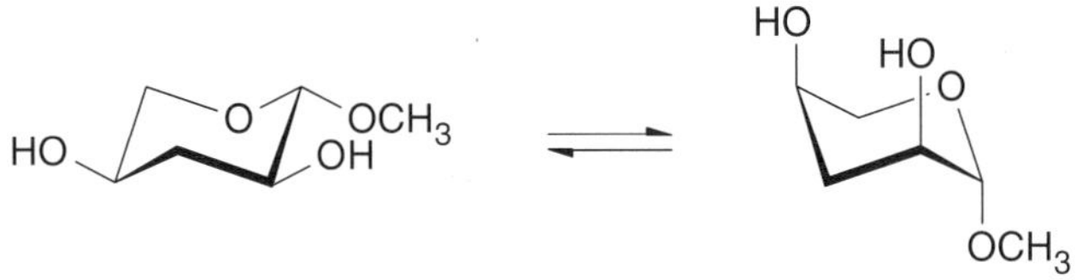

$^1C_4$ $^4C_1$

Doubtless, stabilisation is brought about here not only through the anomeric effect but also, above all, through intramolecular hydrogen bonding, which can occur both between the hydroxy groups and also between hydroxy groups and the ring O-atom. However, as soon as the compound is dissolved in solvents which are either polar or protic, as for example pyridine, dimethyl sulphoxide or water, then the hydrogen bonds are no longer intramolecular but between glycoside and solvent. Stabilisation through intramolecular H-bonds is removed and the equilibrium displaced in favour of the *all equatorial* $^1C_4$ conformation. This displacement is assisted, moreover, through the expected strong solvation which results in an enlargement of the effective steric radius of the hydroxy groups. *Methyl 2-deoxy-α-D-ribopyranoside*, which adopts a $^4C_1$ conformation in water and a $^1C_4$ conformation in chloroform, behaves in a very similar manner[16]:

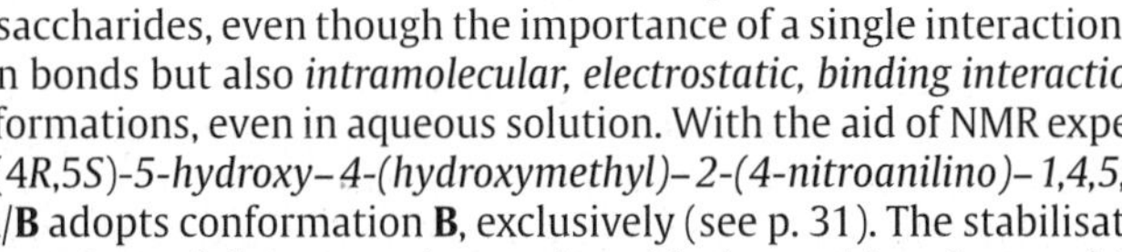

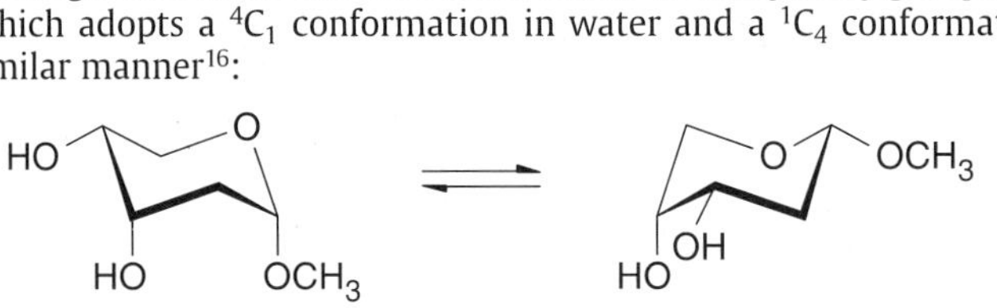

$^1C_4$ $^4C_1$

The stabilisation of certain conformations, even in an aqueous environment, is a significant factor, especially with polysaccharides, even though the importance of a single interaction is difficult to assess.

Not only hydrogen bonds but also *intramolecular, electrostatic, binding interactions* are able to stabilise unexpected conformations, even in aqueous solution. With the aid of NMR experiments, it has been demonstrated that (4*R*,5*S*)-*5-hydroxy-4-(hydroxymethyl)-2-(4-nitroanilino)-1,4,5,6-tetrahydropyrimidine hydrochloride* **A**/**B** adopts conformation **B**, exclusively (see p. 31). The stabilisation results probably through binding interactions of the oxygen lone pairs with the positive charge of the guanidinium ion. The 5-epimer (4*R*,5*R*)-*5-hydroxy-4-(hydroxymethyl)-2-(4-nitroanilino)-1,4,5,6-tetrahydropyrimidine hydrochloride* **A'**/**B'** exists, probably for the same reasons, only in conformation **A'**[34].

Although many different factors may influence the conformer equilibrium in pyranoid systems, above all it is the relative stability of an equatorial substituent which produces the preference of such

conformations which exhibit the largest number of correspondingly oriented substituents. Table 1.3, in which the anomeric pairs of D-hexoses and D-pentoses are listed collectively with their preferred conformations in aqueous solution, illustrate this point in a convincing manner[35]. On the other hand, it is still informative to consider the distribution of the anomeric pyranoses and furanoses in aqueous equilibrium (Table 1.4). From this it is very clear that the pyranoid system possesses a thermodynamic superiority and that, despite the anomeric effect and except for a few exceptions, an *equatorial* group is also preferred at the anomeric centre in an aqueous environment[14, 35].

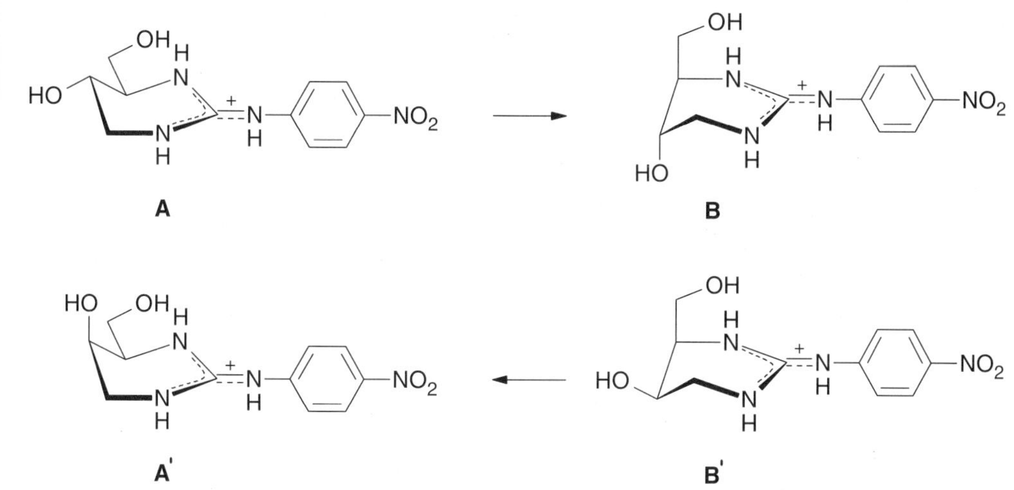

### 1.2.5.5 Non-pyranoid ring systems[36]

Basically, the single significant alternative to pyranoid ring systems for monosaccharides is the 5-membered hydrofuran ring. Important natural products, for example sucrose or nucleosides, contain monosaccharides fixed in the 5-membered ring form. 7-Membered rings form spontaneously only if the possibility for cyclisation to hydropyran or hydrofuran rings is blocked, as for instance in the case of *2,3,4,5-tetra-O-methyl-D-glucose*[36]. The 4-membered oxetan ring is never observed in free monosaccharides.

A signifcant difference between 6- and 5-membered ring systems in the case of alicyclic compounds and also with monosaccharides is the lack of a rigid conformation for the 5-membered ring. The chair conformation of the pyranose ring denotes minimal interaction energy and, moreover, these energy minima are separated from one another by a high energy barrier to interconversion. A pyranose derivative in a chair conformation is therefore clearly defined. Cyclopentane, just as *tetrahydrofuran*, is conformationally flexible. Two forms are distinguished in general: the *envelope* form E and the *twist* form T, which both possess about the same energy. According to orientation, the substituents are labelled *quasi-axial a'* or *quasi-equatorial e'* (p. 33).

The barriers to interconversion are so small that one cannot really speak of two chemically and physically defined species. A planar conformation, which is still depicted today for simplicity, certainly does not exist. In the case of furanoid monosaccharide derivatives, the energy minimum of a conformation is determined, therefore, almost exclusively by the steric demands of the substituents, not by the stability of a certain ring form. Viewing such a situation in an approximate manner, as for example in the case of *D-gulofuranose*, the crowding of *cis*-arrangements can be regarded as so sterically unfavourable that *D*-gulopyranose, which carries at least two destabilising axial substituents in both chair conformations, still dominates in solution equilibrium, being present to the extent of over 99%.

**Table 1.3** Interaction energies

| Aldose | Conformation | | Interaction energy (calc.) (kJ/mol) | |
|---|---|---|---|---|
| | found (NMR) | calc. | $^4C_1$ | $^1C_4$ |
| α-D-Allose | $^4C_1$ | $^4C_1$ | 16.38 | 25.47 |
| β-D-Allose | $^4C_1$ | $^4C_1$ | 12.39 | 36.41 |
| α-D-Altrose | $^4C_1$, $^1C_4$ | $^4C_1$, $^1C_4$ | 15.33 | 16.17 |
| β-D-Altrose | $^4C_1$ | $^4C_1$ | 13.77 | 25.47 |
| α-D-Galactose | $^4C_1$ | $^4C_1$ | 11.97 | 36.46 |
| β-D-Galactose | $^4C_1$ | $^4C_1$ | 10.5 | 32.55 |
| α-D-Glucose | $^4C_1$ | $^4C_1$ | 10.08 | 27.51 |
| β-D-Glucose | $^4C_1$ | $^4C_1$ | 8.61 | 33.60 |
| α-D-Gulose | | $^4C_1$ | 16.8 | 19.95 |
| β-D-Gulose | $^4C_1$ | $^4C_1$ | 12.21 | 22.84 |
| α-D-Idose | $^4C_1$, $^1C_4$ | $^4C_1$, $^1C_4$ | 18.27 | 16.17 |
| β-D-Idose | | $^4C_1$, | 17.01 | 25.47 |
| α-D-Mannose | $^4C_1$ | $^4C_1$ | 10.5 | 23.31 |
| β-D-Mannose | $^4C_1$ | $^4C_1$ | 12.39 | 32.13 |
| α-D-Talose | $^4C_1$ | $^4C_1$ | 14.91 | 24.78 |
| β-D-Talose | | $^4C_1$ | 16.8 | 33.60 |
| α-D-Arabinose | $^1C_4$ | $^1C_4$ | 13.44 | 8.61 |
| β-D-Arabinose | | $^4C_1$, $^1C_4$ | 12.18 | 10.08 |
| α-D-Lyxose | $^4C_1$, $^1C_4$ | $^4C_1$, $^1C_4$ | 8.61 | 10.92 |
| β-D-Lyxose | $^4C_1$ | $^4C_1$ | 10.5 | 14.91 |
| α-D-Ribose | $^4C_1$, $^1C_4$ | $^4C_1$, $^1C_4$ | 14.49 | 14.91 |
| β-D-Ribose | $^4C_1$, $^1C_4$ | $^4C_1$, $^1C_4$ | 10.5 | 13.02 |
| α-D-Xylose | $^4C_1$ | $^4C_1$ | 8.19 | 15.12 |
| β-D-Xylose | $^4C_1$ | $^4C_1$ | 6.72 | 16.38 |

**Table 1.4** Thermodynamic stability of pyranoses and furanoses in water

| monosaccharide | α-pyranose | β-pyranose | furanoses | | |
|---|---|---|---|---|---|
| | | | α | together | β |
| Glucose | 36 | 64 | | <1 | |
| Mannose | 67 | 33 | | <1 | |
| Galactose | 27 | 73 | | <1 | |
| Allose | 18 | 70 | 5 | | 7 |
| Altrose | 27 | 40 | 20 | | 13 |
| Talose | 40 | 29 | 20 | | 11 |
| Gulose | <22 | <78 | | <1 | |
| Idose | 31 | 37 | 16 | | 16 |
| Arabinose | 61 | 35 | 2 | | 2 |
| Lyxose | 71 | 29 | | <1 | |
| Ribose | 21.5 | 58.5 | 6.5 | | 13.5 |
| Xylose | 35 | 65 | | <1 | |

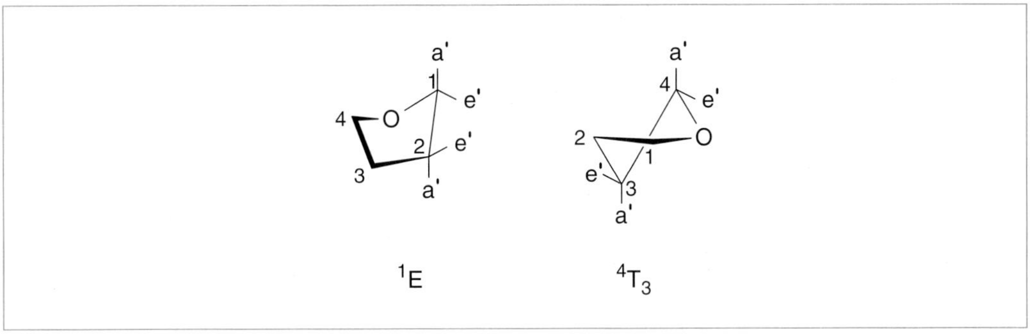

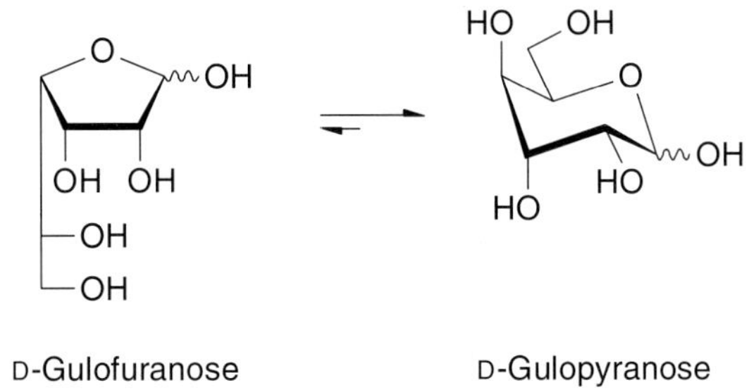

$^{1}E$ $^{4}T_{3}$

D-Gulofuranose D-Gulopyranose

Without the predominance of an individual form, 10 E and 10 T conformations alone can be formulated for a furanoid ring, whose description begins, similar to the pyranoid system, with the number of one atom lying out of the reference plane formed by 4 (E) or 3 (T) atoms. For the twist conformation, according to the position of the remaining atom above or below the reference plane, the conformational symbol T is given a further, corresponding number index. In solution, because of the fast conformational interchange (*pseudorotation*) in furanose derivatives, no unambiguous structure can be designated.

There is an an anomeric effect in furanose derivatives just as in the pyranoid systems. Thus, a hydroxy group on the anomeric C-atom will prefer a quasi-axial position and, if possible, a *trans*-arrangement to the substituent on the C-atom 2. The example of *5-O-methyl-β-D-glucofuranose* - here 6-ring formation is not possible - shows a 1,2-*trans*-arrangement:

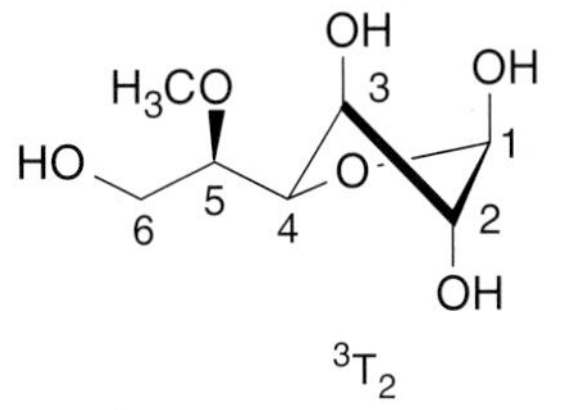

$^{3}T_{2}$

Unlike substituents on a pyranosyl ring, for which an ideal staggered orientation is possible, substituents on a furanoid ring system can never achieve a complete staggering. Even by pseudorotation, rapid changing *syn*-overlapping cannot be avoided. Principally, the consequences of this are the great difference of thermodynamic stability between pyranoses and furanoses (Table 1.4, p. 32).

### 1.2.5.6 Oligosaccharides

Linkage of two or more pyranoid or furanoid monosaccharides together with formation of an acetal bond leads to di- or oligosaccharides and, in the case of a sufficiently but not clearly defined molecular size, also to polysaccharides. The oligosaccharides have, in most cases, a free lactol group and do not differ in this respect from a monosaccharide. These oligosaccharides are also named reducing oligosaccharides because of their easily oxidisable hemiacetal group. In many oligosaccharides an acetal bond is formed by condensation between two lactol groups. In such instances there is, necessarily, no longer a free lactol group and these compounds are non-reducing oligosaccharides. To this class belong such widespread natural products as the disaccharide *sucrose or trehalose.*

If one assumes that almost all oligo- and polysaccharides consist of pyranosyl units linked together, which were treated as relatively rigid conformations on p. 21–25, then one can imagine an oligosaccharide molecule as a more or less twistable ribbon of separated pyranosyl-platelets. In fact, similar to peptides, oligosaccharides possess a secondary structure organised from stiff structural elements, which has precedence over the primary, covalent structure. The latter, which is here assumed to be known, results from the sequence and branching analysis of the oligosaccharide, by methods which are considered in Section 1.3 (p. 37).

Fundamentally, there are two very different steric possibilities for the linkage between two hexopyranoses. Either a primary hydroxy group is involved in the acetal oligosaccharide bond **A**, or the linkage takes place with a secondary hydroxy group (**B** and **C**). The linkage formed by formal condensation between two lactol groups **D** must be viewed as a special case of the type of linkage **B** and **C**.

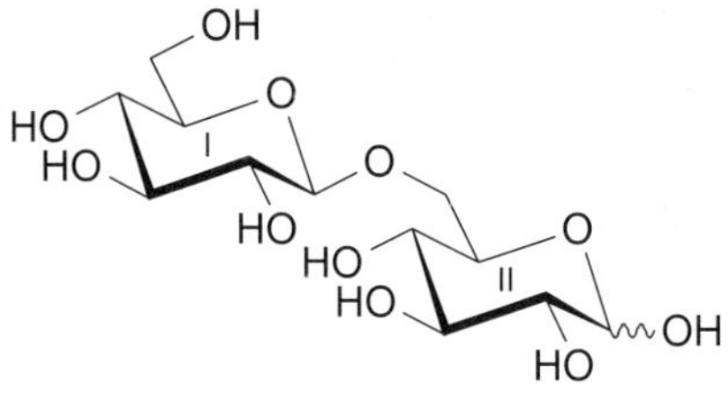

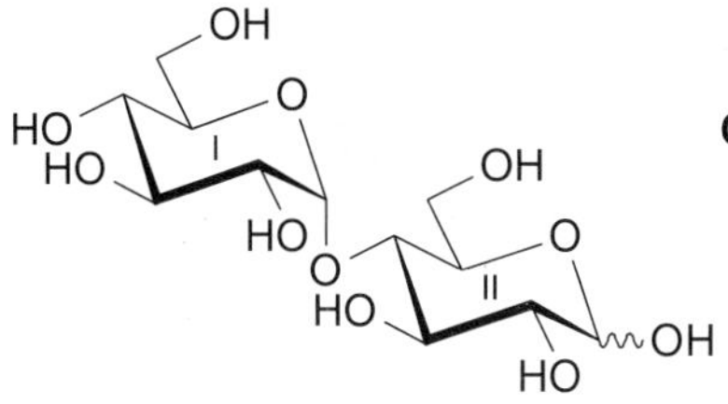

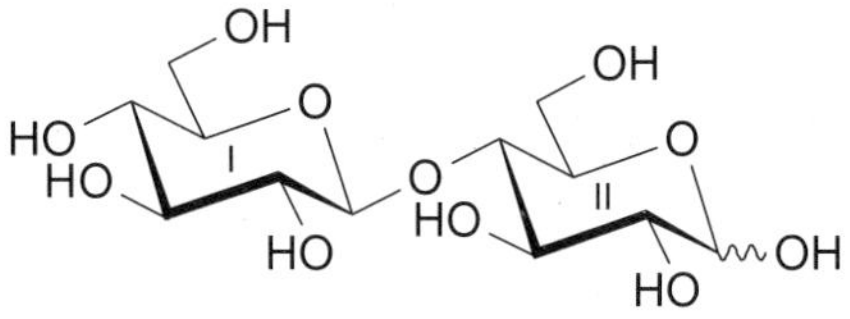

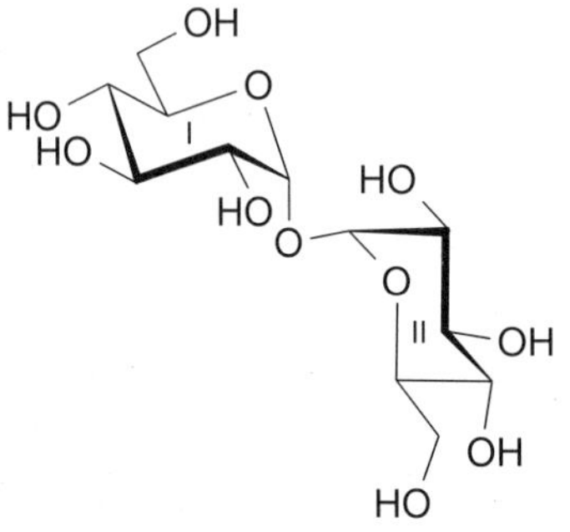

Formally, a further chain lengthening can take place randomly and leads to tri-, tetra, and pentasaccharides, etc. Chain lengthening of structure **D** always leads to the formation of a non-reducing oligosaccharide. The enlargement of structures **A** to **D** through further glycosylation leads to arbitrary structural variants. If in a monosaccharide unit more than one of the non-acetal hydroxy groups is glycosylated, then chain branching is said to occur. If the monosaccharide unit is a hexose, then there is a maximum of four possibilities for chain lengthening or branching. In natural oligosaccharide structures, no use is made of this maximal option for substitution. It should, on steric grounds, also be very difficult to prepare synthetically such maximally branched oligosaccharides. Ether linkages between monosaccharide units are unknown in nature.

In contrast to pyranoid monosaccharides, with an oligosaccharide, even if it consists only of pyranoid monomers as shown in the general formulae **A** to **D**, one cannot speak of a preferred conformation. The conformational possibilities for **A** are especially varied. The movement of both rings with respect to one another has three degrees of freedom of rotation about non-restricted single bonds: rotation about

**1.** $C_5$-$C_6$ in monosaccharide II
**2.** $C_1$-$O_1$ in monosaccharide I
**3.** $O_6$-$C_6$ in monosaccharide II ($O_6$ = in monosaccharide II = $O_1$ in monosaccharide I).

**Example:** Maltose (B)

Direction of view: along $C_{1'} \rightarrow O_{1'}$ bond

Direction of view: along $C_4 \rightarrow O_{1'}$ bond

Possibilities for movement are very much more limited in molecules **B** to **D**, where rotation about $C_5$-$C_6$ of monosaccharide II is not a factor influencing the relative disposition of the two pyranoid rings.

If the numerous non-bonded interactions and hydrogen bonds are disregarded, which are often very difficult to assign and which depend on the anomeric configuration, the positions of linkage, and also on the structure of the monosaccharide units concerned, then there remains, as an important conformation-determining factor, the *exo*-anomeric effect[37] which has already been considered in the case of monosaccharides. Essentially, it determines the conformation about the $C_1$-$O_1$ bond in all four disaccharide segments **A** to **D**. In structure **D**, two *exo*-anomeric effects need to be considered.

The relative position of the pyranosyl planes is determined by the dihedral angles $\phi$ and $\varphi$. In **D**, $\phi$ and $\varphi$ are interchangeable. **B** represents a $\alpha(1\rightarrow4)$-linked disaccharide, specifically in this case maltose. The definition of the dihedral angle is represented pictorially on p. 18. In both the Newman projection along $C_{1'}$-$O_{1'}$ ($\phi$) and also along $C_4$-$O_{1'}$ ($\varphi$), the sign of the defining *syn*-clinal arrangement ($H_{1'}/C_4$) or $H_4/C_{1'}$, respectively) is negative.

With crystalline maltose, the angle $\phi$, largely determined by the *exo*-anomeric effect, is $-32°$. This still allows an $n\rightarrow\sigma^*$ overlap between a free orbital on $O_{1'}$ and the bond which is anti-periplanar to it, $C_{1'}$-$O_{5'}$. In maltose, $\varphi$ is $-13°$, which must be regarded as a steric compromise after consideration of all relevant factors. The established definitions for **B** and the points taken into consideration can be transferred also to other linkage positions and to other configurations of the bridgehead C-atoms. Thus, for cellobiose, **C**, the energy minimum lies at $\phi = +31°$ and $\varphi = -26°$. It is remarkable that calculated values

**Example:** Cellobiose (C)

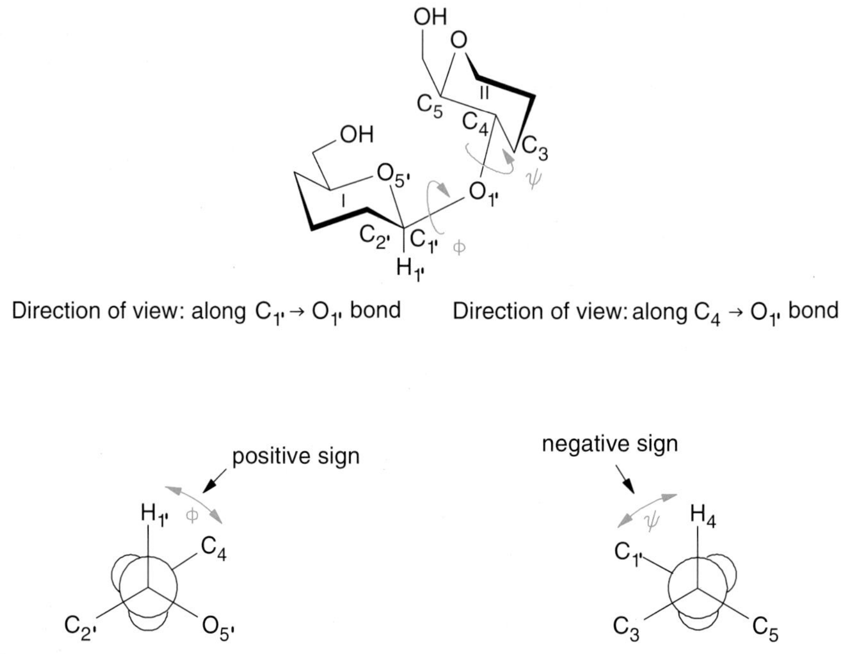

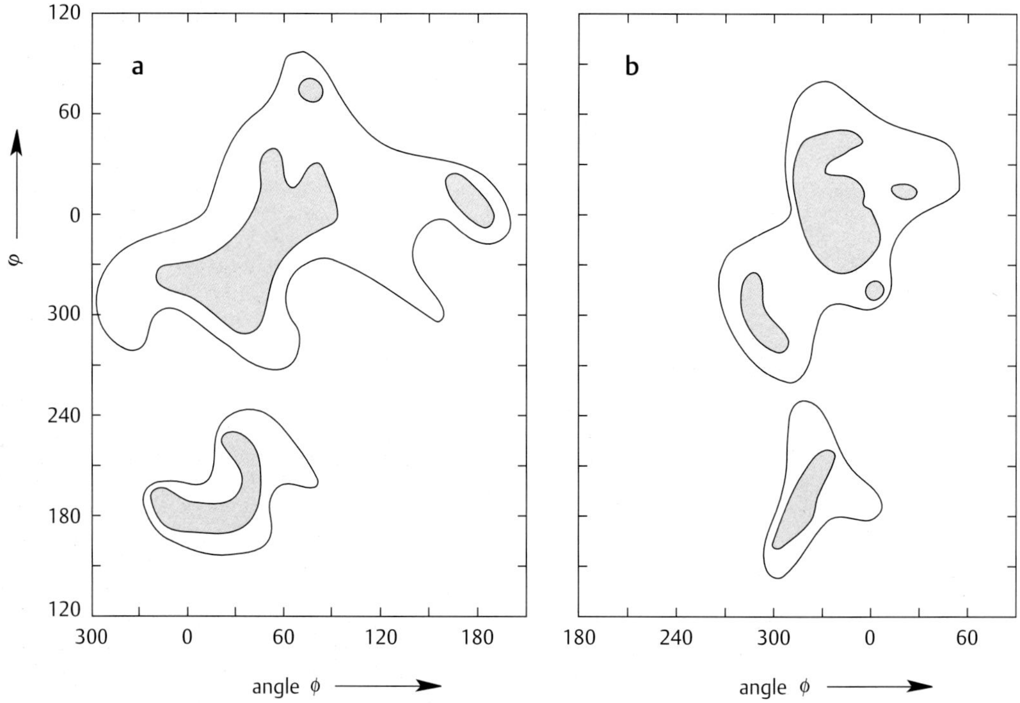

Direction of view: along $C_{1'} \rightarrow O_{1'}$ bond    Direction of view: along $C_4 \rightarrow O_{1'}$ bond

Fig. 1.2 Graph of the conformational energy for **a** cellobiose and **b** maltose. □ $E \leq 21$kJ/mol above $E_{pot}$-minimum. □ $E \leq 42$kJ/mol above $E_{pot}$-minimum.

and those ascertained from crystal structure analysis differ very little from one another[38] and that very narrow boundaries exist for allowed conformations (Fig. 1.2 a, b).

With structures of type **A**, conditions become less strict and also more confused since here rotation about the $C_5$-$C_6$ bond must be considered. The angle $\omega$ determines the conformational orientation about this bond, which is found only in hexopyranoses.

Preferred conformations of oligosaccharides, which also have validity for the corresponding polymers, may differ more or less from one another according to whether determined in solution by NMR spectroscopy or in the crystalline state by X-ray analysis. For biological assessment of oligosaccharide conformations in the future, high resolution NMR spectroscopy will be the method of choice[39].

An astonishing result of all conformational analyses on oligosaccharides carried out so far is that compounds of the type **B**, **C**, or **D** (p. 34) appear to be truly rigid molecules, even in solution. The *exo*-anomeric effect together with van der Waals interactions and H-bonds between the functional groups of neighbouring pyranose residues allow the chain of monosaccharide units relatively little freedom of movement. With increasing molecular weight the description of definite molecular conformations, certainly in the dissolved state, becomes more difficult. Whether for large oligosaccharides and soluble polysaccharides one can speak of definite tertiary structure, in the sense of the structure of globular proteins, remains open to question. It would be difficult for the exclusively hydrophilic carbohydrates, in an aqueous environment, to stabilise higher-ordered structures through intramolecular non-covalent bonding forces, which surely would be of the hydrogen bonding type.

## 1.3 Methods for structural investigation

Structural investigations on carbohydrates have a long history. They began, if one does not include investigation of the constitution of monosaccharides, with the determination of the relative configuration of glucose by E. Fischer and, through correlation, of the numerous other monosaccharides. The methodology rested upon relatively simple chemical reactions, polarimetric measurements and logical conclusions[40]. Essentially, these methods changed little at first. Only in the second half of this century were the modern *chromatographic* separation methods (GLC[41-46], TLC[47], HPLC[48] and GPC[49]) developed to unimagined perfection, after the breakthrough in the 1940's with paper chromatography. The most advantageous additions, especially for carbohydrate chemistry, proved to be high resolution *$^1$H- and $^{13}$C-NMR spectroscopy*[50-54], the modern applications of *mass spectrometry*[55,56] and, as the newest research area, the computer-based structure representations (computer modeling) of carbohydrates[57]. The technique of *X-ray diffraction* on single crystals should not be forgotten as the most direct and unambigous method of structural determination on solid compounds. The modern analytical techniques allowed chemical methods to gain a place just as preparatory aids for special derivatisation or specific fragmentation of bio-oligomers and biopolymers. Because of the diversity above all in the physical methods and the abundance of methodical variants for the separation of substances, only some indications of the possibilities for application will be given in this section.

### 1.3.1 Separation of carbohydrates

A prerequisite for structural investigation is the availability of a chemically homogeneous preparation. All physical methods for structural elucidation require less than 10 mg of a pure substance and some can manage with $^1/_{1000}$ thereof or even less. Since most of the monosaccharides and their simple derivatives have been known for many years, and since really new discoveries (apart from unusual structural variants) are hardly to be expected, chromatographic comparison is usually sufficient for the identification of a compound. Identity of an unknown compound and a known comparison compound is proved with some certainty if, with different separation methods, no difference can be detected, and if this is also the case with the same type of derivative of the two compounds. Detection limits can be lowered significantly by introduction of a chromophoric group or a radioactive label. In the nanomolar region, mono and also higher molecular weight oligosaccharides may be identified as dansylhydrazones by *TLC*[58] or *HPLC*[59] *methods*. Almost just as sensitive for detection are *N*-(4-nitrophenyl)glycosylamines. They have the advantage of chemical stability over the light- and oxidation-sensitive dansylhydrazones. The derivatives are suitable for the preparative isolation of very small amounts of monosaccharides, up to larger oligosaccharides. Homogeneous compounds can be isolated

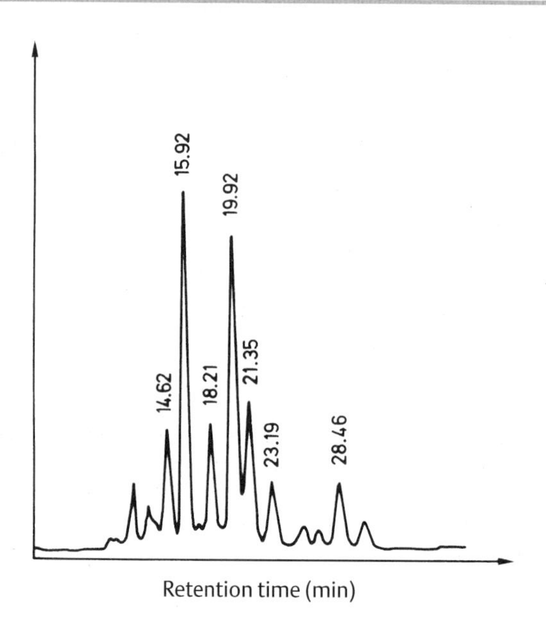


**Fig. 1.3** Separation of a derivatised *oligosaccharide mixture* from human $\alpha_1$-acidic glycoprotein on an RP-column (at $\lambda = 375$ nm).

both by TLC[60] and also by HPLC[61] methods and, if necessary, supplied for further structural investigations. Fig. 1.3 shows the separation of an oligosaccharide mixture from human $\alpha_1$-acidic glycoprotein on a RP-(reversed phase) column (detection at $\lambda = 375$ nm). All oligosaccharides are present as *N*-(4-nitrophenyl)glycosylamines. The compound with the retention time of 19.92 min is a decasaccharide derivative (Fig. 1.3).

▪ **Gas-liquid chromatography**. In contrast to separation of labelled carbohydrate derivatives by TLC and HPLC (radioactive labelled derivatives are also suitable in such cases), gas-liquid chromatography (GLC), which is unequalled in its resolving power, has the disadvantage of being unsuited for preparative isolation of compounds in pure form. Coupled with a mass spectrometer, when the technique is known as *GC-MS*, GLC is used today for the identification of complex monosaccharide mixtures. The carbohydrate derivatives used for the separation by GLC must be volatile. Apolar silicon rubber coatings on mineral carriers are often used as stationary phases and the mobile phase is mostly nitrogen. Eluted substances are detected and registered by means of a flame ionisation detector (FID) with an attached pen-recorder, but the compounds are destroyed.

*Trimethylsilyl- (TMS-) ethers* of carbohydrates are especially suited as volatile derivatives, but methyl ethers, isopropylidene derivatives and, with smaller molecules, also acetates can be separated by gas chromatography. The advantage of TMS derivatives is their high temperature stability (up to 350 °C) and their simple and above all rapid method of preparation[62]. The reaction with the silylation mixture is complete in a few minutes and the reaction mixture can be applied directly to the column without work-up. Even tetrasaccharides are amenable to gas chromatography as their per-trimethylsilyl derivatives.

Protection of the anomeric hydroxy group of free mono- or oligosaccharides by silylation allows the detection of anomers and in general these may be easily separated. Sweeley et al. demonstrated the practicality of gas chromatography for the component analysis of glycosphingosides and gangliosides in a methanolysis mixture from about 1 mg of a sphingolipid which was separated into its constituents after trimethylsilylation (Fig. 1.4)[42].

The separation of anomeric glycosides can be clearly seen (for example **7** and **8**). A correlation can be performed by co-chromatography with authentic, pure derivatives. This type of comparative chromatography, not only in the case of GLC, is the most used method today for the rapid identification of monosaccharides. Gas chromatograms can also be evaluated quantitatively by measurement of the area under the peaks. The application of gas chromatography in combination with mass spectrometry for the structure determination of polysaccharides will be covered in section 1.3.3 (p. 45).

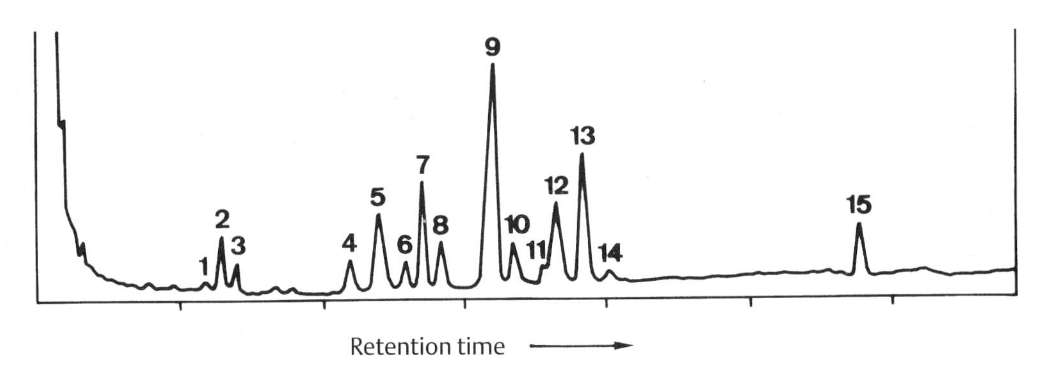


**Fig. 1.4** Gas chromatogram of a mixture obtained by methanolysis of a *glycosphingolipid* after O-trimethylsilylation: **1,2,3** Fucose (6-deoxygalactose); **4,5,6** galactose; **7,8** glucose; **10, 12** *N*-acetylgalactosamine; **11, 13, 14** *N*-acetylglucosamine; **15** *N*-acetylneuraminic acid; **9** mannitol (as internal standard).

■ **Gel permeation chromatography** GPC is a method applicable to polymers, which has not achieved the same significance for carbohydrates as for proteins. Cross-linked polyacrylamides of different grades of cross-linking serve as gels, which are known in the trade as biogels. Particularly, so-called homologous series of oligomers or polymers, for example mixtures of maltodextrins, can be readily separated on biogels. The case illustrated (Fig. 1.5) involves the separation (Biogel P2) of a mixture of oligomers arising by enzymatic transglycosylation on 4-nitrophenyl $\beta$-D-glucopyranoside[63].

Separation takes place with polydisperse mixtures essentially, and in the case of polymolecular mixtures, exclusively by molecular size. With higher molecular weight heterogeneous homoglycan mixtures, gel chromatography leads to the recovery of components of fairly uniform molecular mass. Gel permeation chromatography is ideally suited for the de-salting of oligo- or polysaccharide preparations.

### 1.3.2 NMR spectroscopy

Today, high molecular weight, complex oligosaccharides rather than monosaccharides are the focus of structural investigations with high resolution NMR spectroscopy, and the potential of the method in this field is not yet fully exploited. Some indications of this fact and to sources in the literature were given earlier. For an understanding of the general application to carbohydrates it is appropriate to examine first the simpler spectra of the monosaccharides.

■ **$^1$H-NMR spectroscopy** affords three types of measurements: the chemical shift, the spin-spin coupling constants and, through integration of the signals, the relative number of the registered nuclei in a sample. These three experimentally obtained data first of all serve to determine the conformation and configuration of a molecule. Secondly, they confirm the constitution of a molecule often already found by chemical means.

Although measurements on *$^{13}$C-* and to a lesser extent in the case of F-derivatives *$^{19}$F-nuclei* are required ever more frequently for structural determination, proton magnetic resonance spectroscopy is now, as before, the method of choice for structural investigation of carbohydrates and their derivatives in solution. It is, except for limited enzymatic techniques, the only method for determination of anomeric configuration in oligosaccharides.

The chemical shift of a proton signal, that is the position of the signal on the relative $\delta$ scale, whose origin $\delta = 0$ is the signal of the 12 H-nuclei in *tetramethylsilane (TMS)* in organic solvents or of the 9 H nuclei in sodium 3-trimethylsilyl-1-propanesulphonate (sodium 2,2-dimethyl-2-silapentane-5-sulphonate, DSS) for aqueous solutions, is dependent on the chemical environment of the protons concerned. In a molecule such as $\beta$-D-glucopyranose there are above all two types of proton which differ in their immediate chemical environment. The one type are bound to C-atoms and the other to O-atoms. The hydroxy protons are generally unsuited for structural investigations since they afford broad signals which are variable in their chemical shift. They can be removed in order to simplify the spectrum. Either

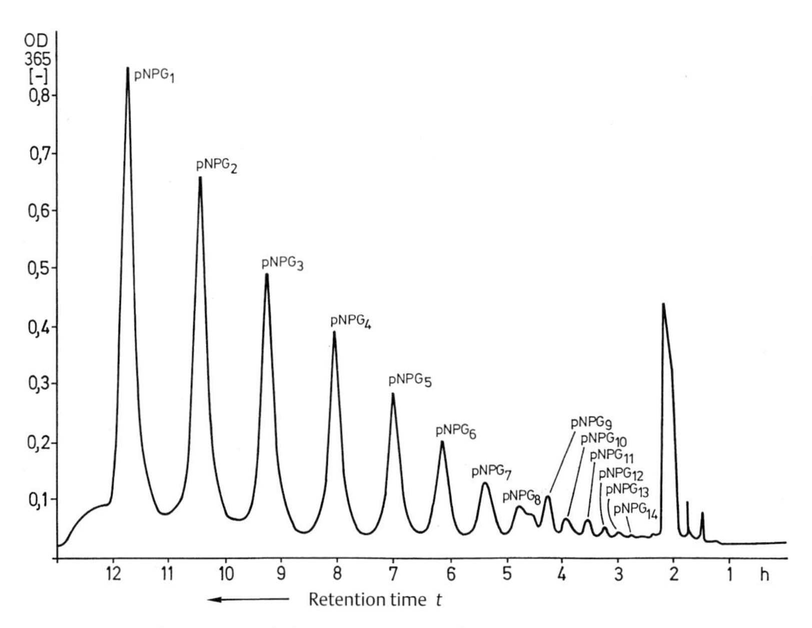


**Fig. 1.5** GPC separation of a mixture of oligomers on Biogel P2 arising from enzymatic transglucosylation on 4-nitrophenyl $\beta$-D-glucopyranoside (pNPG$_1$).

a derivative of the compound is prepared, for example the pentaacetate of $\alpha$- or $\beta$-D-glucopyranose, whose spectra are shown in Fig. 1.6 a,b and the measurement performed in an organic solvent, for example deuteriochloroform ($CDCl_3$), or the protons of the hydroxy groups are exchanged with deuterons and measurements are made in $D_2O$ or another suitable solvent. In both cases, only protons bound to C-atoms are detected in this manner and the spectrum is thereby simplified.

The chemical shift depends on the atoms bound to the corresponding C-atom. Differences are to be expected between the protons on $C_2$ to $C_6$ on the one hand as well as the proton on $C_1$ on the other and, naturally, the acyl protons. The protons on $C_2$ to $C_6$ must be more strongly shielded than the proton on the anomeric C-atom, $H_1$. The latter appears, correspondingly, at lower field. The situation is not so clear with the remaining approximately chemically equivalent protons (2 x $H_6$, $H_5$, $H_4$, $H_3$, $H_2$). A correlation on the basis of their chemical shift alone cannot be made directly. The signals partly overlap each other to such an extent, as for example the signals $H_5$ and $H_6$' of *penta-O-acetyl-α-D-glucopyranose* (Fig. 1.6 a), that they are hard to resolve. The very similar positions of the signals for the five acetyl groups is understandable. Their integration divided by 15 is always a very good and exact scale for the quantitative relation of all protons among one another, and the integration of all signals can be an indication as to the purity of the sample and a support for the suggested constitution of a compound. Differences in the chemical shift of chemically similar protons, for example the respective $H_1$ in *penta-O-acetyl-α-* (Fig. 1.6 a) and *β-D-glucopyranose* (Fig. 1.6 b) can also be produced through steric differences (geometric shift), for example different alignment on the pyranose ring. Thus, the general rule holds that signals for *equatorial* protons appear at lower field than those for equivalent *axial* protons. This may be clearly seen by comparison of both spectra of Fig. 1.6, a and b, regarding shifts for the anomeric protons, which indicate $H_1$-$\alpha$ = 6.33 ppm (Fig. 1.6 a) and $H_1$-$\beta$ = 5.72 ppm (Fig. 1.6 b).

Investigation of interaction between protons on neighbouring C-atoms affords useful auxiliary information in $^1$H NMR spectroscopy. With vicinal protons, the size of the reciprocal splitting of proton signals brought about by mutual magnetic influence of the nuclei is a function of the dihedral angle $\phi$ and is designated as the coupling constant $J$. The relationship is contained in the *Karplus equation*[64], whose simplified form is

$$J = J_o \cos^2\phi - 0.28$$

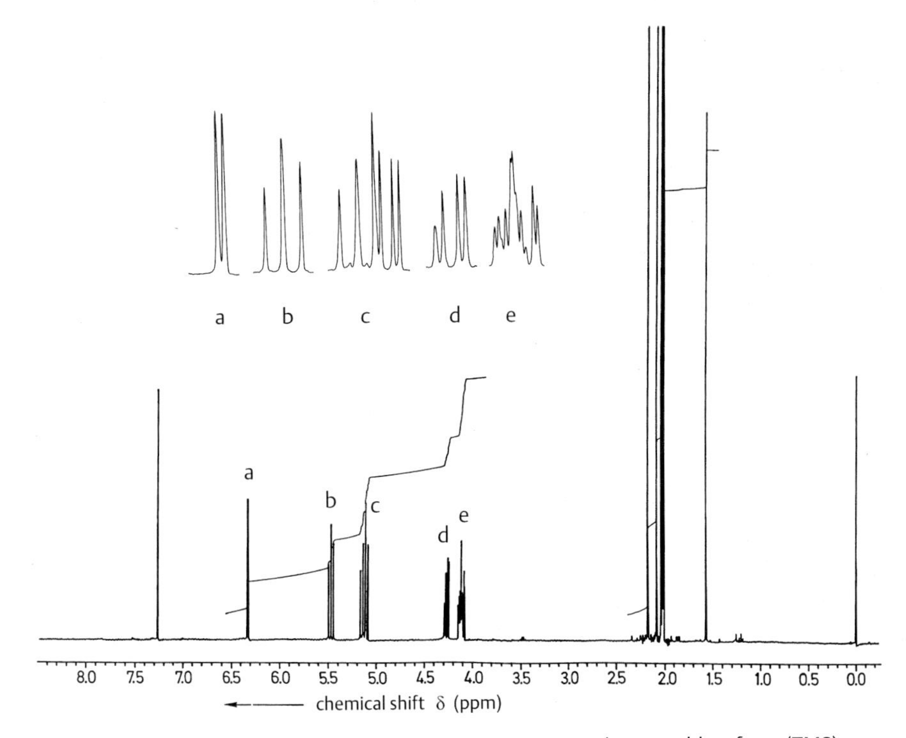


**Fig. 1.6 a** $^1$H-NMR spectrum of *penta-O-acetyl-α-D-glucopyranose* in deuteriochloroform (TMS).

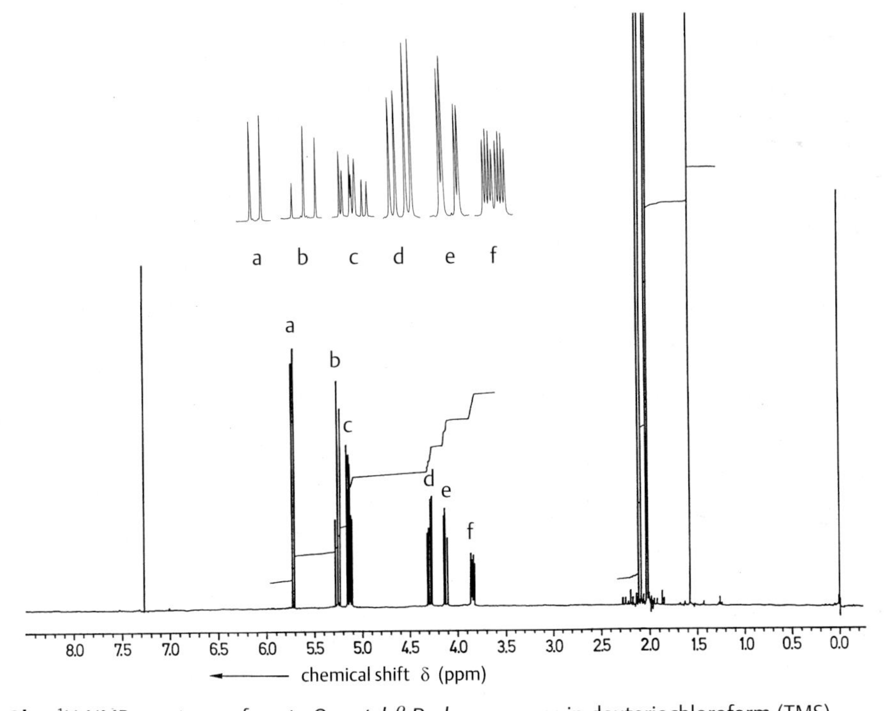


**Fig. 1.6 b** $^1$H-NMR spectrum of *penta-O-acetyl-β-D-glucopyranose* in deuteriochloroform (TMS).

and which can be represented graphically. The application to the configurational and conformational analysis of carbohydrates was soon recognised and exploited by R. U. Lemieux and was then utilised generally[65]. The size of $J_o$ when $\phi$ is between 0 and 90° is 8.5 Hz and with $\phi$ between 90 and 180° is 9.5 Hz. In other words, in a pyranoid system, a coupling constant of 1 to 4 Hz is equivalent to $\phi = 60°$, and a value of 6 to 10 Hz to $\phi = 180°$. The former corresponds to an *axial-equatorial* or *equatorial-equatorial* orientation and the latter to an *axial-axial* orientation.

The analysis of a spectrum usually begins with a search for a signal that can be unambiguously assigned to a proton and then for identical coupling constants (Fig. 1.6 a, b). If we begin with the anomeric proton $H_1$ of the α-compound (Fig. 1.6 a), it is identified through its chemical shift, in this case 6.33 ppm (see above), and through the fact that it has only $H_2$ as a neighbour, which leads because of the *gauche* relationship to the small splitting of the signal to a doublet with $J_{1,2} = 3.75$ Hz. $H_2$ couples additionally with the next neighbour, $H_3$, wherefore both $J_{2,1} = J_{1,2} = 3.75$ as well as $J_{2,3}$ (10.35 Hz) are to be sought. The group of four signals with $\delta = 5.10$ ppm as their mid-point corresponds to $H_2$ because $J_{2,3}$ (10.35 Hz) is found in the $H_3$ signal ($\delta = 5.48$ ppm). In the same manner, the signals for $H_4$ and $H_5$ can be found. $H_5$ shows a higher splitting pattern since it couples both with $H_4$ and both of the protons $H_6$ and $H_{6'}$, which for their part display a geminal coupling of $J_{6,6'} = 12.45$ Hz.

The $\beta$-compound (Fig. 1.6 b) can be recognised immediately by the large coupling constant $J_{1,2}$ of the $H_1$-signal, which is typical for a 1,2-*diaxial* disposition of the H-atoms. The other signals are assigned in an analogous manner to that used for the α-anomer. In this spectrum the clear splitting of the $H_5$-signal is noteworthy. It can readily be understood from a consideration of the spectrum of a monosaccharide, which is relatively simple to decipher, that the spectrum of an oligosaccharide is much more complex. The assignment of the signals through decoupling experiments and by analysis of two-dimensional spectra requires considerable effort and in complex cases can only be successfully performed by specialists in such techniques. The spectrum of a naturally occurring disaccharide glycoside[53], in which only the anomeric protons of both monosaccharide residues can be located straight away, affords an impression of the signal complexity in such compounds (Fig. 1.7).

■ **$^{13}$C-NMR spectroscopy** has great similarity with corresponding detection methods for $^1$H nuclei, above all if the spectra are measured with the aid of Fourier-transform instruments. However, the methods are sufficiently different in their application that they ideally complement each other. The information gathered from $^1$H spectra is especially limited in the case of investigations on higher molecular weight oligo- or even polysaccharides, while the well-defined proton decoupled $^{13}$C-spectra can furnish a wealth of information on the chemical and physico-chemical properties of high molecular weight carbohydrates. Thanks to their well separated chemical shifts, the signals are well resolved and are suited especially for very sensitive structure proof, without a complete assignment being necessary. A $^{13}$C spectrum is like a fingerprint and as such unchangeable. For mono- and simpler oligosaccharides there are already tables which are similar to the classical melting point tables[52, 66] (Fig. 1.8 a–d).

In most cases, the use of heavy water ($D_2O$) as solvent is advantageous. As examples of polysaccharide spectra, those of two glucans are shown in Fig. 1.9 a and 1.9 b. Distinguishing C-atoms which carry a substituent, for example C-4 in an amylose chain, from those which do not, for example C-4 in the glucose residues of amylopectin that form chain ends, can be specially informative. In contrast, the difference in the chemical shift between substituted and unsubstituted $C_6$ are very small.

The chemical shift $\delta$ (in ppm), often given the subscript c and therefore written $\delta_c$, is for the most part larger than that of TMS (tetramethysilane) which, in the case of aqueous measurements, is used as an external standard. As internal standards, simple water soluble compounds such as 1,4-dioxan or acetone can be used, which for their part are in turn standardised against TMS. The region of measurement stretches over about 300 ppm. Aliphatic hydrocarbons lie at high field, unsaturated compounds, arenes, amines alcohols are distributed over a middle range, and carbonyl compounds are found at low field. The $^{13}$C chemical shifts for some methyl hexopyranosides are collected in Table 1.5 (p. 45). As in the case of proton resonance, the chemical shift of a C-atom is influenced by the electron density about the nucleus concerned. It is therefore reasonable that the signal of the anomeric C-atom, with a relative low electron density, lies at a lower field than the signals of the remaining C-atoms of the molecule. The $^{13}$C signal is sensitive to changes in substitution at the carbon atom concerned. Thus, the replacement of a hydroxy group by a H-atom has a considerable effect: the $^{13}$C signal for $C_2$ in *methyl α-D-glucopyranoside* lies at 72.2 ppm, whereas the corresponding signal in *methyl 2-deoxy-α-D-glucopyranoside (methyl 2-deoxy-α-D-arabino-hexopyranoside)* is at 39.1 ppm.

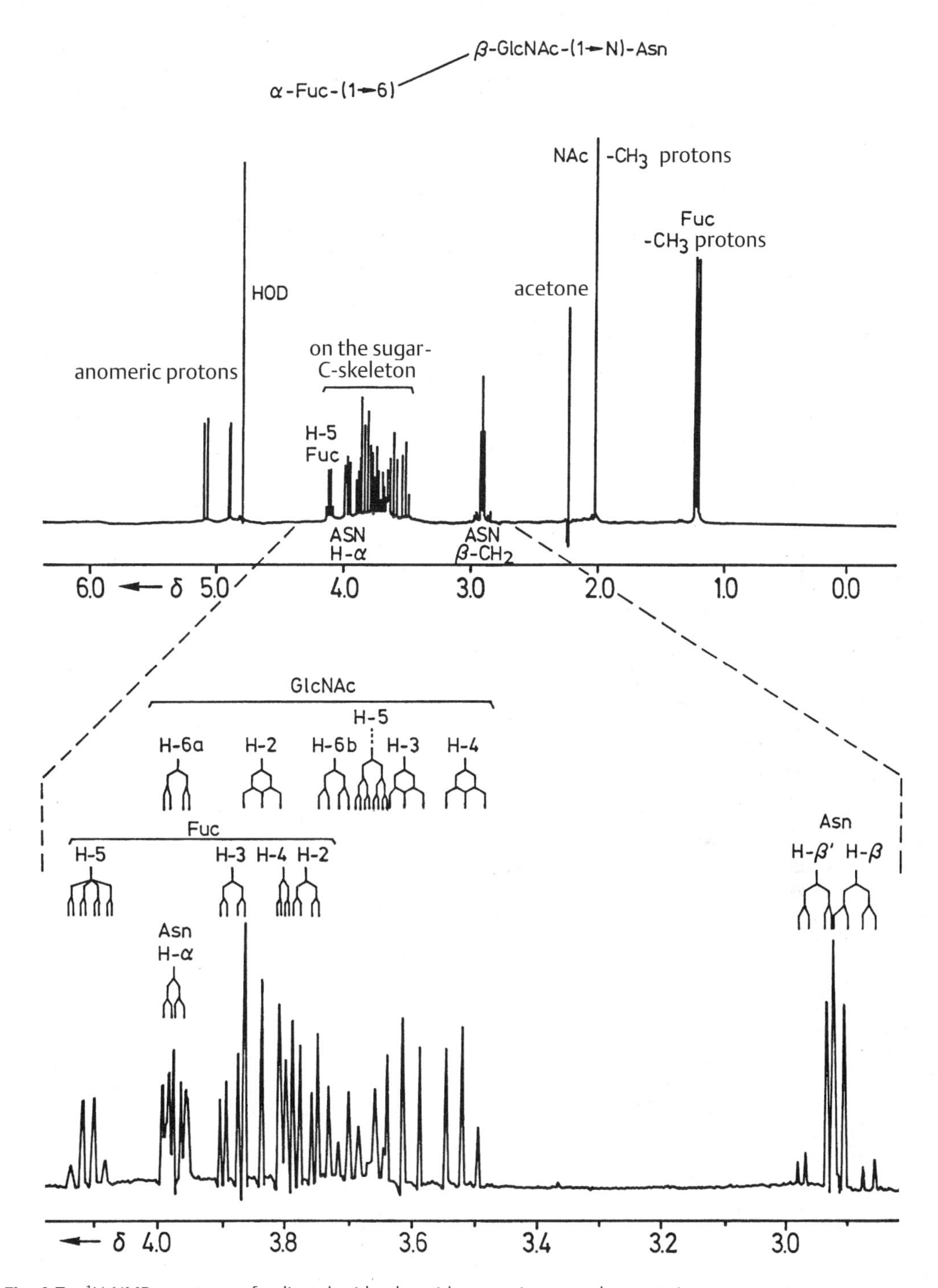


**Fig. 1.7** ¹H-NMR spectrum of a disaccharide glycoside occurring as a glycoprotein component.

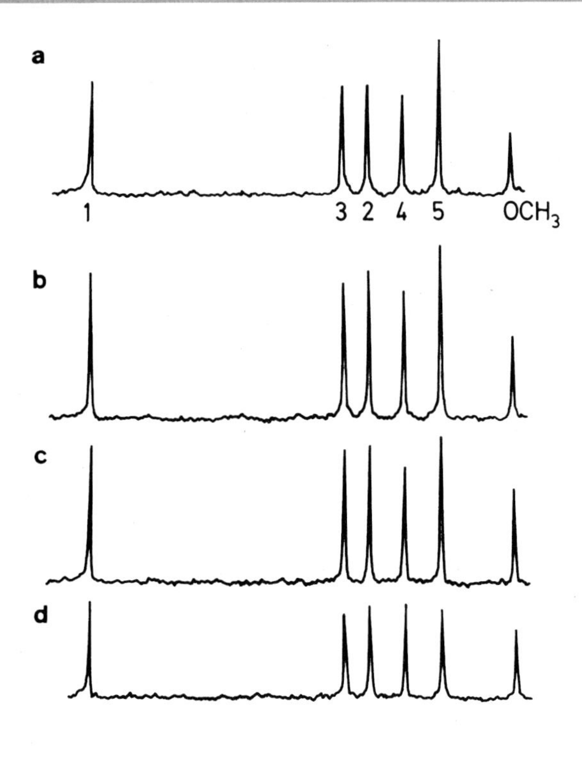


**Fig. 1.8** $^{13}$C-NMR Spectrum of methyl *β*-D-xylopyranoside in deuterium oxide ($D_2O$) at 22.63 MHz with different pulse widths.
Pulse width: **a** 90° **b** 63°
**c** 45° **d** 27°
**1–5** $C_1$–$C_5$

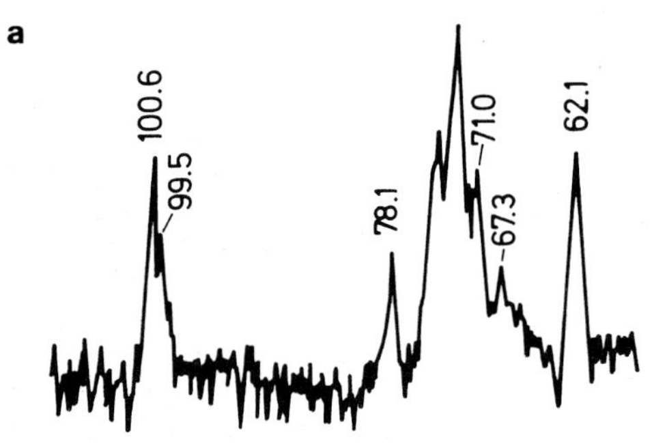


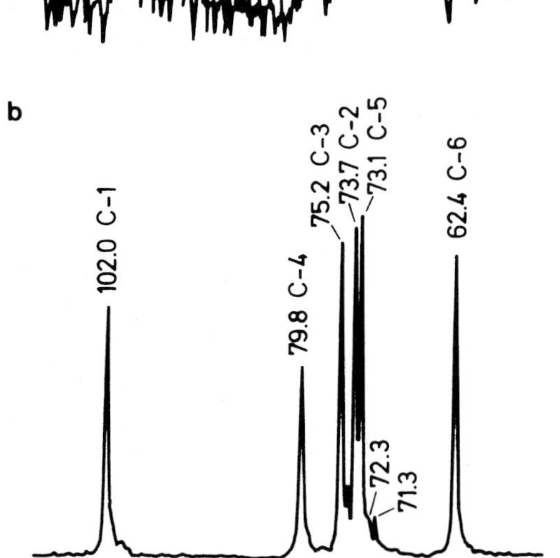


**Fig. 1.9** $^{13}$C-NMR Spectrum of
**a** *β*-limit dextrin of rabbit liver glycogen (aqueous solution, ambient temperature, shifts from TMS).
**b** amylopectin from wax-barley (in deuterium oxide at 70 °C; shifts based on external TMS).

**Table 1.5** $^{13}$C-NMR Signals of some methyl glycosides[52] (ppm)

| D-Pyranoside | $OCH_3$ | C-1 | C-2 | C-3 | C-4 | C-5 | C-6 |
|---|---|---|---|---|---|---|---|
| α-Glc | 55.9 | 100.0 | 72.2 | 74.1 | 70.6 | 72.5 | 61.6 |
| β-Glc | 58.1 | 104.0 | 74.1 | 76.8 | 70.6 | 76.8 | 61.8 |
| α-Gal | 56.0 | 100.1 | 69.2 | 70.5 | 70.2 | 71.6 | 62.2 |
| β-Gal | 58.1 | 104.5 | 71.7 | 73.8 | 69.7 | 76.0 | 62.0 |
| α-Man | 55.9 | 101.9 | 71.2 | 71.8 | 68.0 | 73.7 | 62.1 |
| β-Man | 56.9 | 101.3 | 70.6 | 73.3 | 67.1 | 76.6 | 61.4 |
| α-Xyl | 56.0 | 100.6 | 72.3 | 74.3 | 70.4 | 62.0 | |
| α-2d-Glc | 56.9 | 100.8 | 39.1 | 70.8 | 73.6 | 74.6 | 63.3 |

Equally, the chemical shift is strongly influenced by different spatial orientations of the directly bound substituent and also by the substituent on neighbouring C-atoms. Thus, a shift of the signal to higher field is observed if an *equatorial* substituent is changed into an *axial* one. The change is particularly clear with the anomeric C-atom, as shown by a comparison of the $C_1$ signal of α- and β-hexopyranosides (Table 1.5).

In addition, a shift mostly takes place to higher field if a change occurs in the orientation of the substituent on the neighbouring C-atom from equatorial to axial. Likewise, a shift to higher field occurs if an H-atom bound to the C-atom concerned occupies a 1,3-*diaxial* relationship to a hydroxy group. From these empirical rules an interesting dependence can be derived of the chemical shift on the conformational stability. The greater the conformational stability of a pyranose or a pyranoside, the smaller is the sum of the $^{13}$C-chemical shifts.

### 1.3.3 Mass spectrometry

Mass spectrometry serves as an aid in the determination of end groups, branching, and linkage points in oligosaccharides. Just as with NMR spectroscopy, mass spectrometry, the second most important method for the determination of carbohydrate structures, is an independent, highly developed research area. Within the context of this book, only indications can be given to some of the important applications and, as for the rest, to the secondary literature. Progress reports appearing at regular intervals give accounts of constantly developing experimental methods , starting with the classical ionisation of volatile molecules through electron impact (EI) or chemical ionisation (CI)[55, 67], about the field desorption technique (FD)[68], up to the modern methods of fast atom bombardment (FAB)[56] and electrospray ionisation (ESI)[69].

The determination of the primary or covalent structure of an oligo- or polysaccharide[70] begins in a conventional way with complete methylation of the substrate. For this reason the numerous variants of structure determination on permethylated carbohydrates carry the description methylation analysis. The principle may be represented in the example of a galactomannan from plant seeds. The analysis of the polysaccharide requires different component steps:

1. *Total hydrolysis* of the polysaccharide and quantitative determination of the monosaccharides after their chromatographic separation,
   *Result:* Gal and Man = 1:5.
2. Permethylation.
3. Total hydrolysis, chromatographic separation and $^1$H-NMR spectroscopy.
   *Result:* 2 tetramethyl hexoses (Man*p* and Gal*f*), 2 trimethyl hexoses (Man*p*), 2 dimethyl hexoses (Man*p*).
   *Conclusion:* 2 non-reducing end groups (Man*p* and Gal*f*), 2 normal members of a chain, and 2 branching members (each Man*p*).
4. Reduction of the monosaccharide derivatives with $NaBD_4$ followed by peracetylation (separation by GLC).
   *Result: (GLC)* 4 compounds with different retention times, decreasing in the order 2 diacetates A and B, 1 triacetate C, 1 tetraacetate D, in the molar ratios A:B:C:D = 1:1:2:2,
   *Result (MS)* **A** 1,5-di-, **B** 1,4-di-, **C** 1,5,6-tri-, **D** 1,2,5,6-tetraacetate.

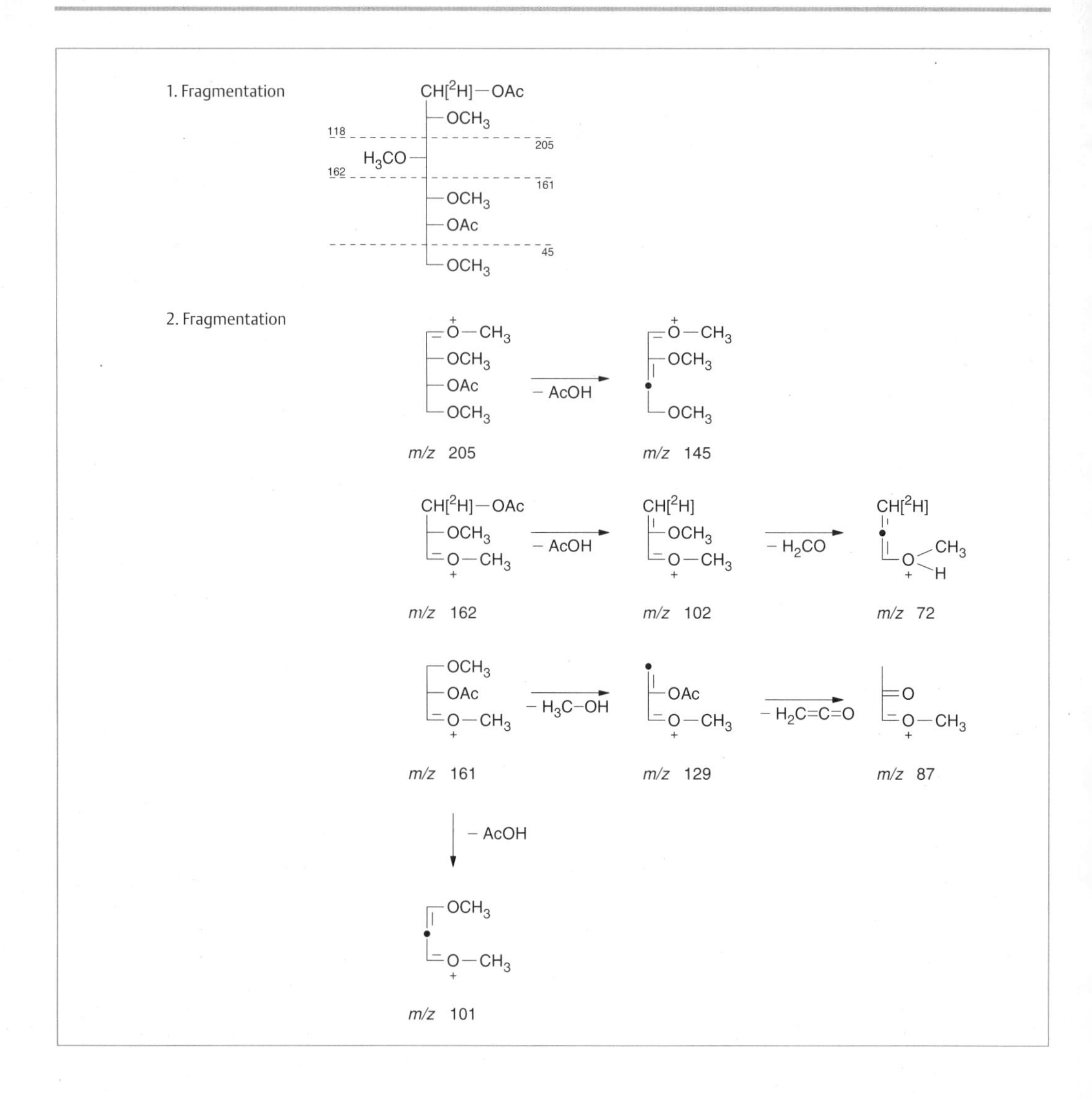


The mass spectrum of *partially methylated alditol acetates*, additionally deuterated at $C_1$, with the retention times and molar composition originally determined, provides an idea of the polymer structure. In our case the main chain consists of (1→6)-linked mannose residues (**C** and **D**), of which the half **D** carries in the 2-position either a mannopyranosyl- **A** or galactofuranosyl- residue **B**. The substitution pattern results from typical mass spectra, which point to certain component fragments, according to simple rules:

**a** formation of the smallest ion possible
**b** preferred cleavage between two methoxylated C-atoms
**c** little tendency for cleavage at an acetoxylated C-atom.

Component **A** (*1,5-di-O-acetyl-2,3,4,6-tetra-O-methyl-1-²H-mannitol*) with a total mass of 323 Da, is chosen as an example. After ionisation by electron impact, the molecule decomposes into five primary fragments, which yields signals with the higher intensities (205, 162, 161, 118, and 45).

Three of these (205, 162 and the isotope isomer 161) lose mainly acetic acid and methanol through $\beta$-elimination and produce the just as typical secondary fragments (145, 129, 102 or 101) in the spectrum.

**I**

$$\left[ \text{Man}_p\text{—}(1 \rightarrow 6)\text{—}\underset{\substack{2 \\ \uparrow \\ 1 \\ \text{Gal}_f}}{\text{Man}_p}\text{—}(1 \rightarrow 6)\text{—}\underset{\substack{2 \\ \uparrow \\ 1 \\ \text{Man}_p}}{\text{Man}_p}\text{—}(1 \rightarrow 6)\text{—}\text{Man}_p\text{—}(1 \rightarrow 6) \right]_n$$

**II**

$$\left[ \text{Man}_p\text{—}(1 \rightarrow 6)\text{—}\underset{\substack{2 \\ \uparrow \\ 1 \\ \text{Man}_p}}{\text{Man}_p}\text{—}(1 \rightarrow 6)\text{—}\underset{\substack{2 \\ \uparrow \\ 1 \\ \text{Gal}_f}}{\text{Man}_p}\text{—}(1 \rightarrow 6)\text{—}\text{Man}_p\text{—}(1 \rightarrow 6) \right]_n$$

**III**

$$\left[ \text{Man}_p\text{—}(1 \rightarrow 6)\text{—}\underset{\substack{2 \\ \uparrow \\ 1 \\ \text{Gal}_f}}{\text{Man}_p}\text{—}(1 \rightarrow 6)\text{—}\text{Man}_p\text{—}(1 \rightarrow 6)\text{—}\underset{\substack{2 \\ \uparrow \\ 1 \\ \text{Man}_p}}{\text{Man}_p}\text{—}(1 \rightarrow 6) \right]_n$$

The information obtained would fit three different polymers **I** to **III**, which differ in their repeat units (see above).

Chemical degradation provides help here. Since furanosides are relatively acid-labile, the galactofuranosyl residues may be largely removed selectively from the permethylated polysaccharide and the positions so liberated then substituted by ethylation. Partial hydrolysis of the fully derivatised polymer affords, among other things, the trisaccharide **A'** and the disaccharide **B'**. Both compounds are acetylated after reduction with $NaBH_4$. The compounds are subjected to mass spectrometric analysis which proves that both the disaccharide **B'** and also the trisaccharide **A'** carried a 2,3,4-tri-*O*-methyl-mannose residue at each of their reducing ends. In the case of derivative **B'**, a cation ($m/z$ 261) is apparent, corresponding to the 6-*O*-acetyl-2-*O*-ethyl-3,4-di-*O*-methyl-mannosyl residue and, for derivative **A'**, the 2,3,4,6-tetra-*O*-methyl-mannosyl residue is found and, as a cation at $m/z$ 451, the 6-*O*-acetyl-3,4-*O*-di-*O*-methyl-mannosyl residue still substituted with this same tetra-*O*-methyl-mannosyl moiety. These fragments are only compatible with structures **A'** and **B'** and therewith only in accord with the primary structure III.

The underlying fragmentation schemes for glycosides or oligosaccharides proceed from the ionisation of the ring O-atom through removal of an electron. This cation radical undergoes different decompositions, A, B, C, D and E, of which path A, leading to formation of the cyclic glycosyl cation, is probably the most significant (see p. 49).

Further fragmentation of the primary fragments gives information about the substitution pattern[55]. Despite highly developed methods, branched heteroglycans and higher molecular weight complex oligosaccharides still present the analytical chemist with problems which are difficult to solve. Only by collating information from various sources can reliable deductions be made regarding the covalent structure of a carbohydrate. It is probable that structural investigations of the type used in the long established and automated sequence analysis of proteins or nucleic acids will never be possible in the field of carbohydrates.

**Trisaccharide derivative A'**

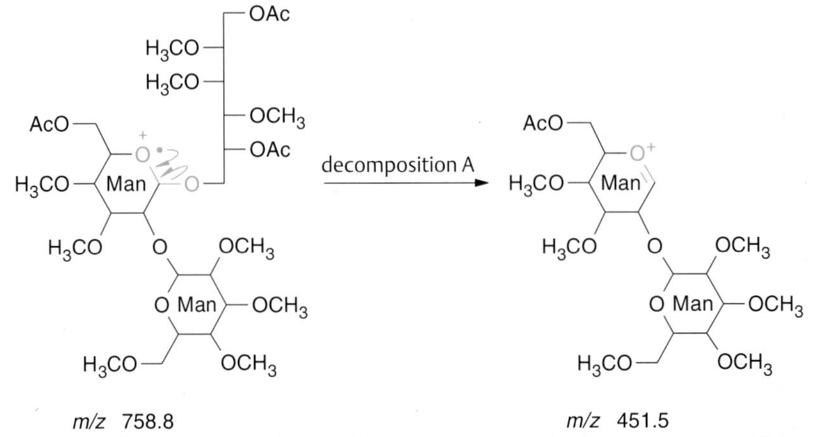

decomposition A

*m/z* 758.8 → *m/z* 451.5

**Trisaccharide derivative B'**

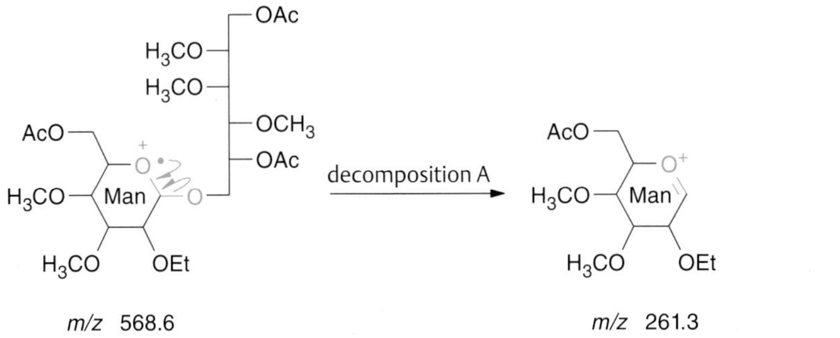

decomposition A

*m/z* 568.6 → *m/z* 261.3

The analysis of multi-antennary oligosaccharide structures, whose oligomeric attenae differ structurally among one another, becomes extremely difficult. The fictitious model of a tetra-antennary molecule with differently structured oligomeric branches should give an impression of the possibilities and of the difficulties in elucidating such structures (see p. 50).

Fortunately, nature does not utilise all of the possibilities given to her with regard to linkage types, type of chain members and number and type of branches, but provides definite often recurring structures, thanks to the limited number of available glycosyl transferases.

### 1.3.4 Polarimetry

The dependence of optical rotation on the structure of an optically active substance has been known for decades and found its first semi-quantitative expression in the rules of isorotation which C. S. Hudson derived from the results of a large number of measurements on pyranosides as well as on glyconic acid lactones at the beginning (1909) of this century[71]. The real reason for the often very high optical rotation of carbohydrates was not then recognised because of ignorance of the true, three-dimensional structure of carbohydrates. A series of large discrepancies between measured rotations and those determined according to Hudson's rules led to the relationships between structure and optical rotatory power being thought over anew on the basis of modern conformational considerations. Modern ideas about connections between optical rotation and structure can be traced back especially to the investigations of Whiffen[72], Brewster[73] and Lemieux et al.[27,74,75]. In order to be able to compare the rotation values of different compounds, it is appropriate to relate these to the corresponding *molecular mass M*. The *molecular rotation [M]* results from the *specific rotation [α]* as follows:

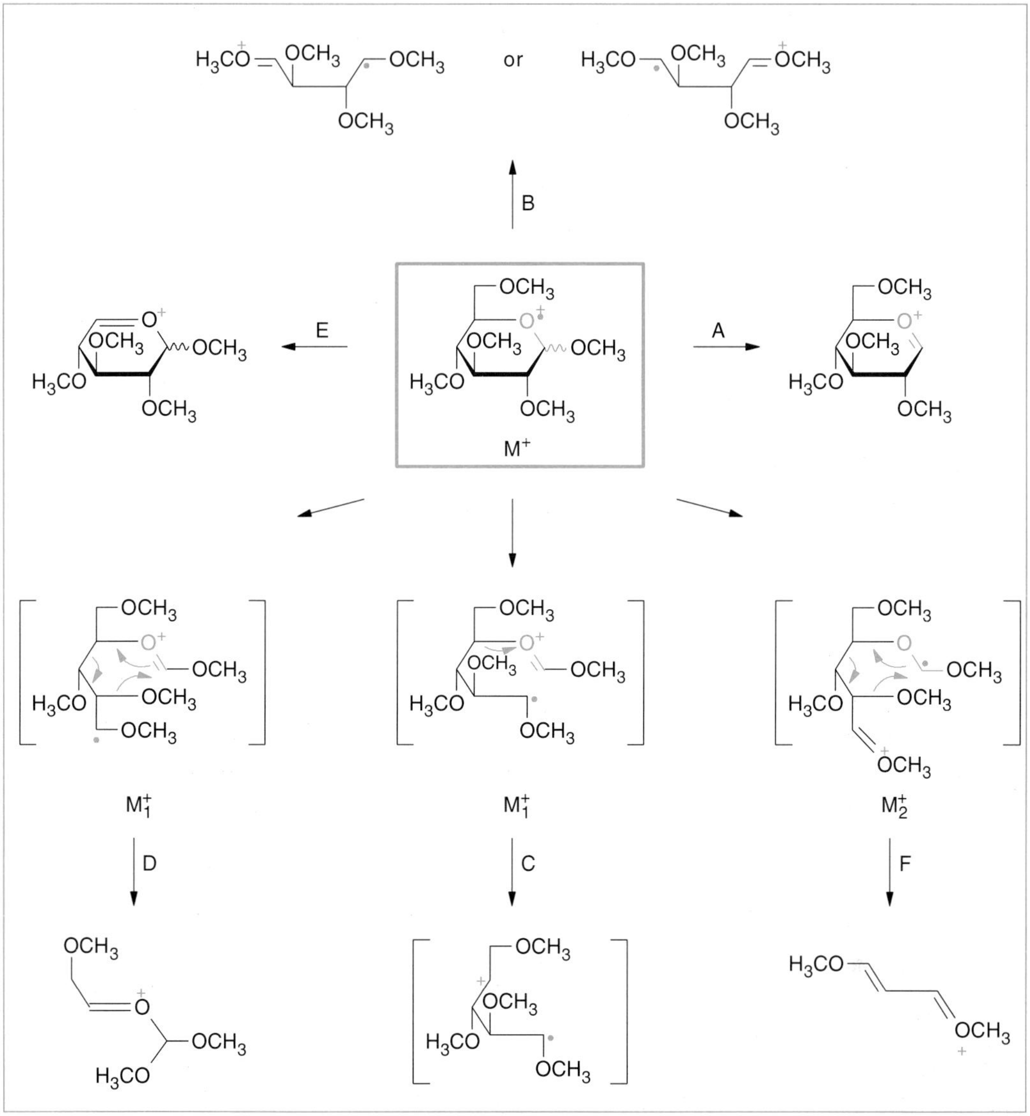

$$[M] = M\,[\alpha] / 100$$

The *specific rotation* is determined from the measured rotational angle α at a determined wavelength of monochromatic light divided by the concentration *c* in g/100 mL and the length of the sample cell *l* in dm. Since the value of rotation is strongly dependent on the wavelength λ of the light used and on the temperature *T*, both must be stated. It is equally important to give the concentration and the solvent used. The *Hudson rules of isorotation* apply the *principle of optical superposition*, formulated by van't Hoff, to the anomeric C-atom. A monosaccharide is divided into two parts, each of which contributes to the total rotation, a part **A** (anomeric C-atom with the ligand X) and **B** (the rest of the molecule):

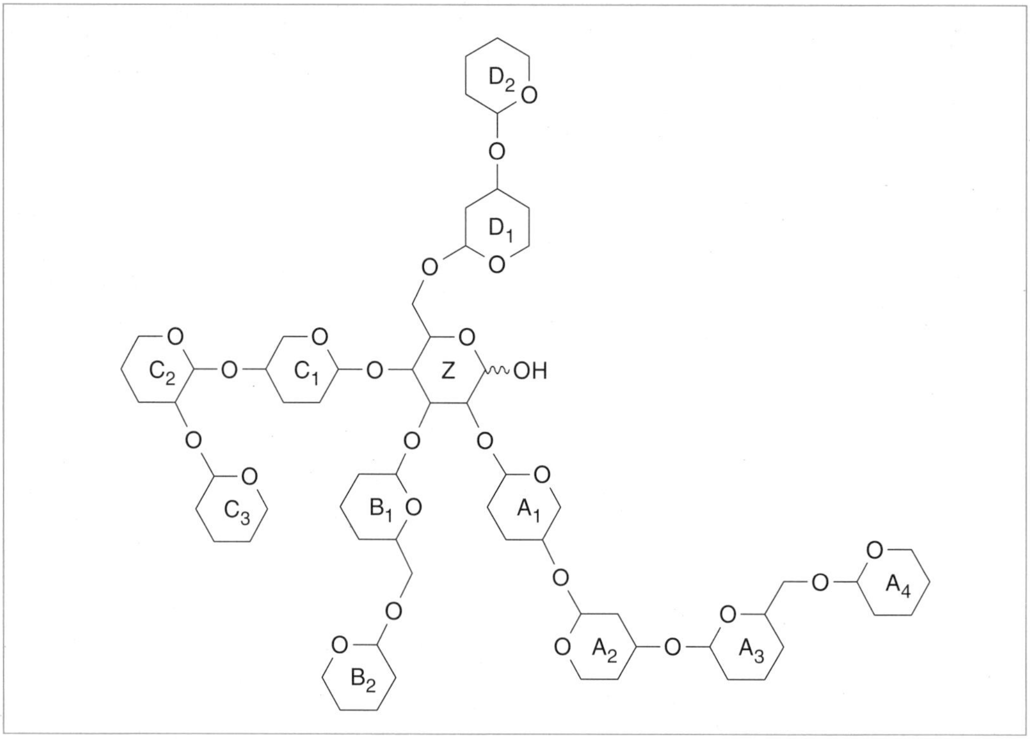

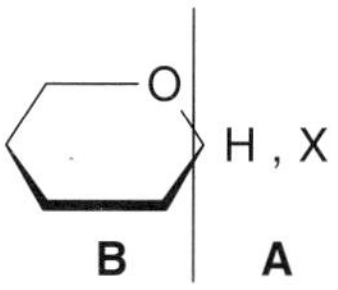

For the α-D-compound, the total rotation is composed of **B** + **A**, and for the β-anomer, **B** - **A**. Correspondingly, but with reversed signs, this also applies to the L-compound. From a pair of anomers, for example *methyl α-D-glucopyranoside [M]* = 318 ° and *methyl β-D-glucopyranoside [M]* = –68 °, **A** and **B** may be calculated:

| | | |
|---|---|---|
| **B** + **A** = 318°, | → | **B** = 125°, |
| **B** – **A** = – 68°, | | **A** = 193° |

**A** can now be used to calculate the rotation of other methyl pyranosides, and **B** to calculate the rotation of other D-glucosides. For a good many glycosides the rule is applicable but for many, however, it is useless, especially if the anomeric C-atom carries a strongly polar substituent, if the monosaccharide derivative contains chromophoric groups, or if the C-atom 2 has an axial substituent.

The Hudson rule is, in the end, only directed at the anomeric C-atom and does not consider in detail the contribution of the remaining asymmetric C-atoms, which are collected in **B**. If one wants to resolve further the contribution **B** of the pyranosyl ring to the optical rotation, it is appropriate first to define chiral structural elements as part of the entire molecule. Analysis of countless optical rotation measurements on pyranoses and pyranose derivatives has shown that a large optical rotation is always to be expected if the molecule incorporates to a maximum extent, as structural elements, an uninterrupted spiral arrangement of easily polarisable bonds, or elements with easily polarisable free electron pairs:

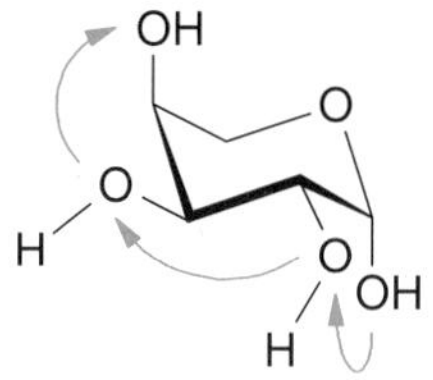

β-L-arabinopyranose
$[\alpha]_D = +202°$

An example of a compound with an extremely high rotation is *β-L-arabinopyranose*. The direction of rotation of the spiral decides the sign of the rotation. With this assumption useful, empirical rules may be formulated. Fixed C,C-single bonds with polar substituents are viewed as chiral structural elements. The *glycol group* in a pyranose or cyclohexane ring is such a grouping:

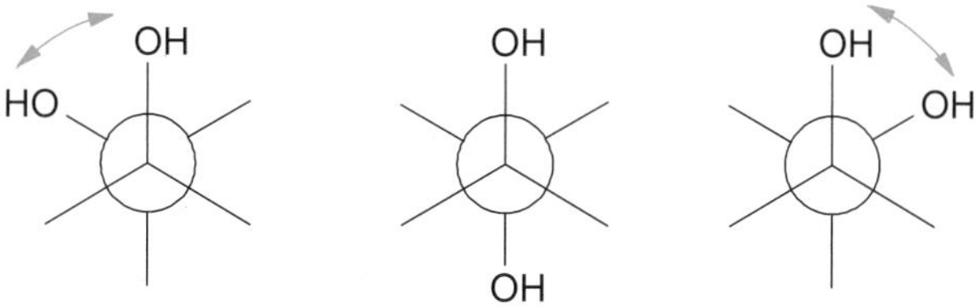

| – *syn-clinal* | ± *anti-periplanar* | + *syn-clinal* |
|---|---|---|
| (– *sc*) | (± *ap*) | (+ *sc*) |
| optically active | optically inactive | optically active |

Only +sc and -sc are chiral. Both produce, according to the direction of rotation, a molar contribution of +45 or –45°. The optical rotation of inositols may be determined by simple summation of these diol increments. However, in contrast to the cyclohexane ring, the pyranose ring is itself asymmetric. This fact is allowed for by additional structural elements, in which the ring-O-atom is involved.

projection along the C,C-single bond — + 10° — – 10°

projection along the C,O-single bond — + 60° — – 60°

projection along the C,O-single bond — + 115° — – 115°

If one then further includes the asymmetry of the $C_5$-$C_6$ bond in hexopyranose derivatives as +30°, then the molar rotation of asymmetric hydropyran derivatives may be readily calculated with relatively few parameters. An assumption is a regular conformation of the ring (dihedral angles of 60° and 180°).

As an example, the ${}^4C_1$ conformation of $\beta$-L-arabinopyranose may be split up into its chiral structural elements, the increments added to give the calculated molar rotation, and this then compared with the experimentally found molar rotation (see p. 53).

Even though in a few cases the deviation between the calculated and measured rotation *[M]* is too large by application of this very simple method, yet at least a decision can be reached regarding the anomeric configuration in the case of all pyranoses and simple pyranosides. With simple cyclohexanes or cyclohexane derivatives, the assignment of absolute configuration is also possible by the Whiffen-Brewster-Lemieux rules. Racemates may be resolved by formation and separation of diastereoisomeric mixtures. The diastereoisomeric per-*O*-acetylated thioglucosides **A** and **B** can be separated chromatographically. They differ solely in the chiral, acetylated *glycol grouping*[76]. The smaller rotation value corresponds to the -sc-arrangement of the acetoxy groups in **A**, the higher to the +sc-arrangement in **B**.

**A**

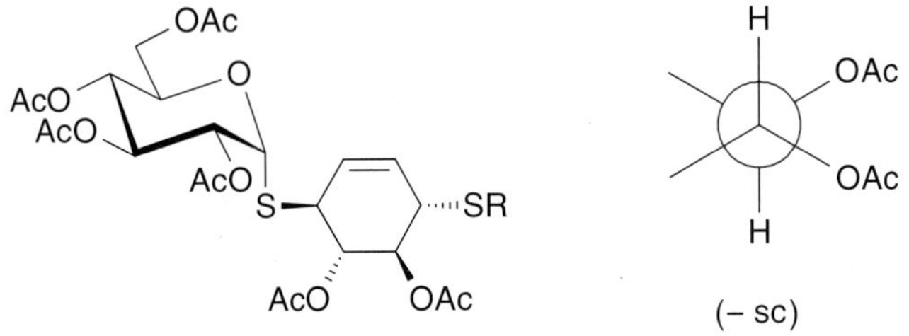

$$[\alpha]_D^{20} = +114.1^\circ \ (c = 1;\ CHCl_3)$$

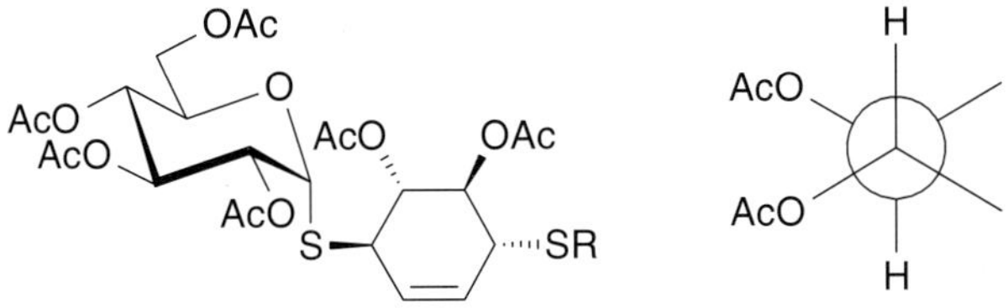

(– sc)

**B**

$$[\alpha]_D^{20} = +157.7^\circ \ (c = 1;\ CHCl_3)$$

(+ sc)

By the enzymatic galactosylation of the racemate of a substituted cyclohexene, only one enantiomer is converted, whereby the galactoside **A** is formed and then isolated. The absolute configuration of the aglycone results from the positive rotation of the *(3,5/4,6)–3,6-diazido–4,5-dihydroxycyclohexene* isolated after enzymatic hydrolysis[77] (see p. 54):

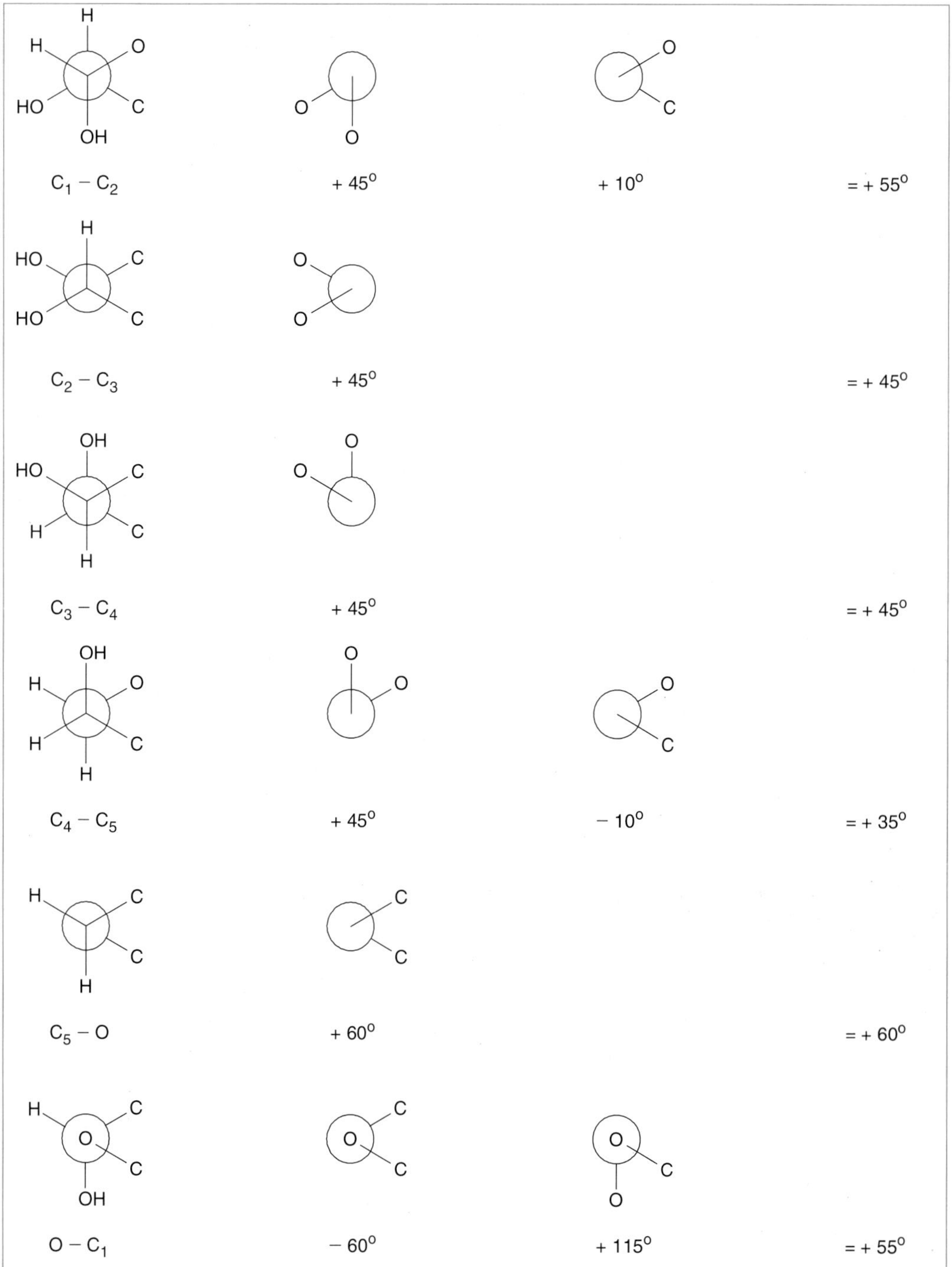

| Bond | Increment | Increment | Sum |
|---|---|---|---|
| $C_1 - C_2$ | + 45° | + 10° | = + 55° |
| $C_2 - C_3$ | + 45° | | = + 45° |
| $C_3 - C_4$ | + 45° | | = + 45° |
| $C_4 - C_5$ | + 45° | − 10° | = + 35° |
| $C_5 - O$ | + 60° | | = + 60° |
| $O - C_1$ | − 60° | + 115° | = + 55° |

| | | |
|---|---|---|
| $[M]_D$ | calculated by summation of the chiral increments | + 295° |
| $[M]_D$ | experimental | + 303° |

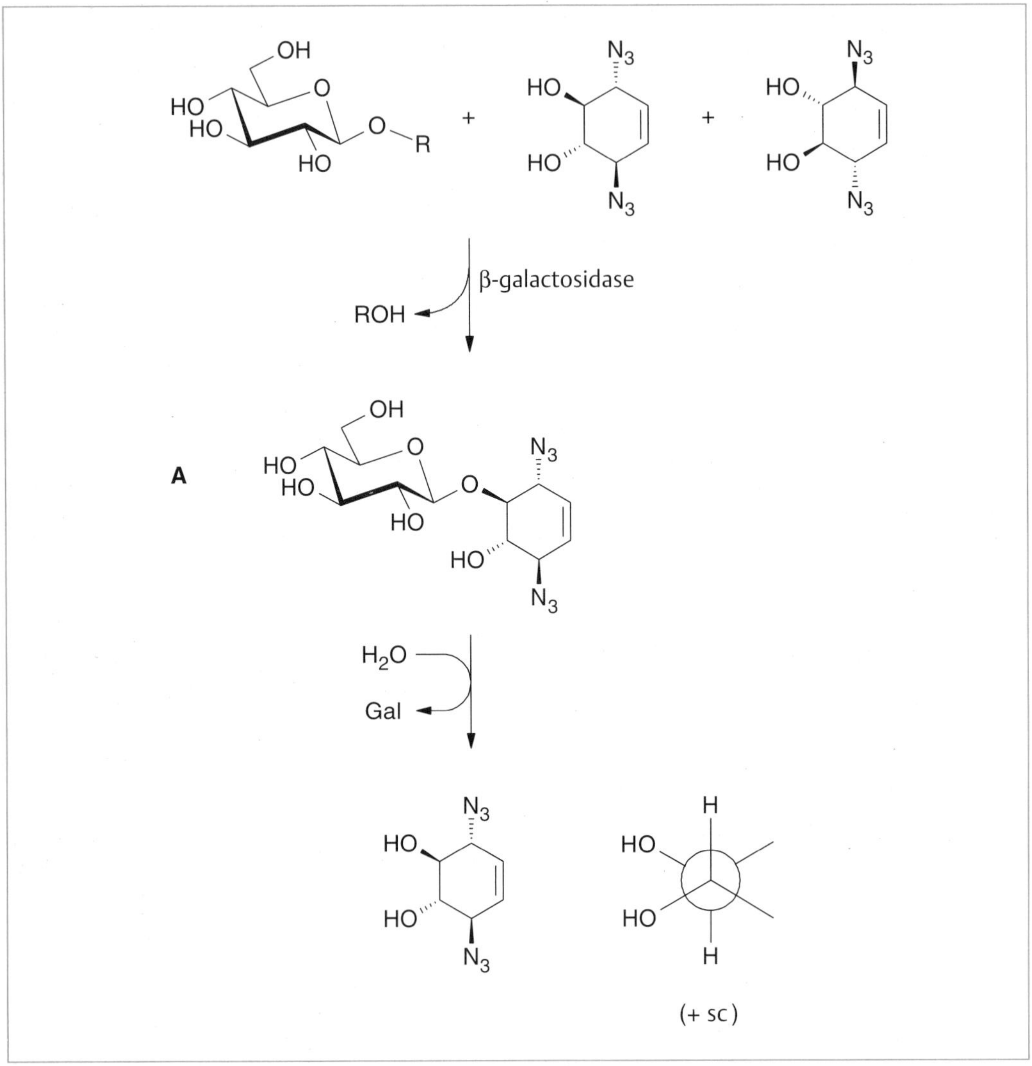

## 1.4 Systematic nomenclature

Decades of commission work were necessary to produce a universally agreed document for the naming of carbohydrates of all possible manifestations. The *Nomenclature of Carbohydrates published by the IUPAC-IUBMB Joint Commission on Biochemical Nomenclature JCBN* is a collection of about 40 main rules and almost 100 printed pages and is understood to be in a constant state of renewal. This openness with regard to change and amendment is necessary with such a fast a developing area. The basis for the rules were already laid in 1969. The so-called *Tentative Rules* are published in several journals[78] and a revision of the 1971 document was published in 1997[79]. The nomenclature of the cyclitols[80] is not included in this book, but that of the glycoconjugates remains reserved for chapter 2.

Carbohydrate chemistry is especially rich in trivial expressions. Although the pressure for the systematic nomenclature becomes greater with the increase of new syntheses and new discoveries of carbohydrates and carbohydrate derivatives, traditional and often self-invented trivial names remain, particularly for use in the laboratory. This is understandable if one tries to construct, for instance, the systematic name of a complex compound composed of two monosaccharides and a cyclitol derivative, as in the antibiotic *kanamycin A*:

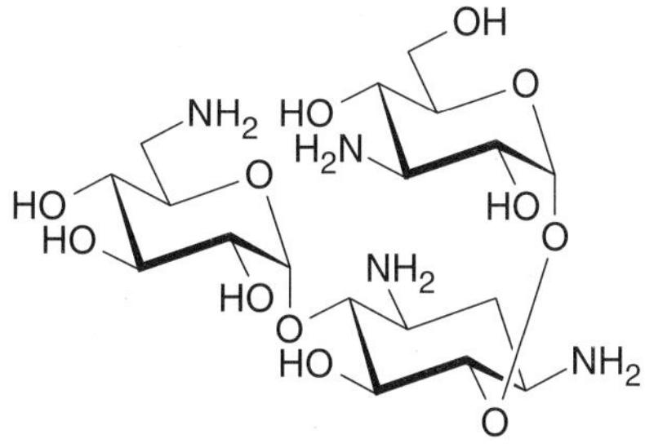

*Kanamycin A*

Thus, in the field of mono- and oligosaccharides, old familiar names have lasted and, furthermore, are also used. In an appendix to *Nomenclature of Carbohydrates*[79], the trivial names of carbohydrates and derivatives are assembled together with their systematic equivalents and short symbols in, as is expressly emphasised, a non-limiting list. The basis of the systematic nomenclature is the trivial names of the monosaccharides up to 6 C-atoms. From these parent monosaccharides are derived the corresponding configurational prefixes (see p. 56, I).

The chain of asymmetric C-atoms can be interrupted by non-asymmetric groups (e. g. methylene or keto groups), without the configurational prefix being changed. As described earlier, there are formally no limits to the C-chain. In the case of more than 4 asymmetric C-atoms, that is a chain length of more than 6 C-atoms, composite configurational prefixes are used, and the number of the C-atoms must also appear in the name. One focusses therein exclusively on the asymmetric C-atoms in the series. Prefixes are assigned in order to the chiral centres in groups of four, beginning with the group proximal to C-1. The prefix relating to the group of carbon atom(s) furthest from C-1, which may contain less than four atoms, is cited first (p. 56, II).

In general, monosaccharides carry only one carbonyl group and correspondingly they are called aldoses or ketoses. If one keto group should divide the chain into two equal halves with the same number of asymmetric C-atoms, then alphabetical order must be followed with reference to the first configurational prefix.

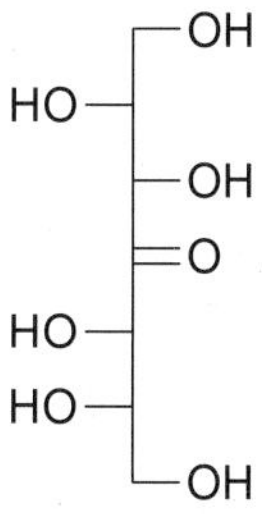

L-*gluco*-hept-4-ulose

*not*

D-*gulo*-hept-4-ulose

Two aldehydic carbonyl groups make the monosaccharide into a *dialdose*, two ketonic carbonyl groups into *diketose*. The combination of an aldehydic and a ketonic carbonyl group gives a ketoaldose (not aldoketose) or also *aldosulose* (see p. 57, I).

Replacement of one or more hydroxy groups in a monosaccharide by an H-atom leads to a *deoxy monosaccharide*. Replacement of a hydroxy group by an amino group affords an *amino-deoxy-monosaccharide*, in which the position of substitution must always be specified (p. 57, II).

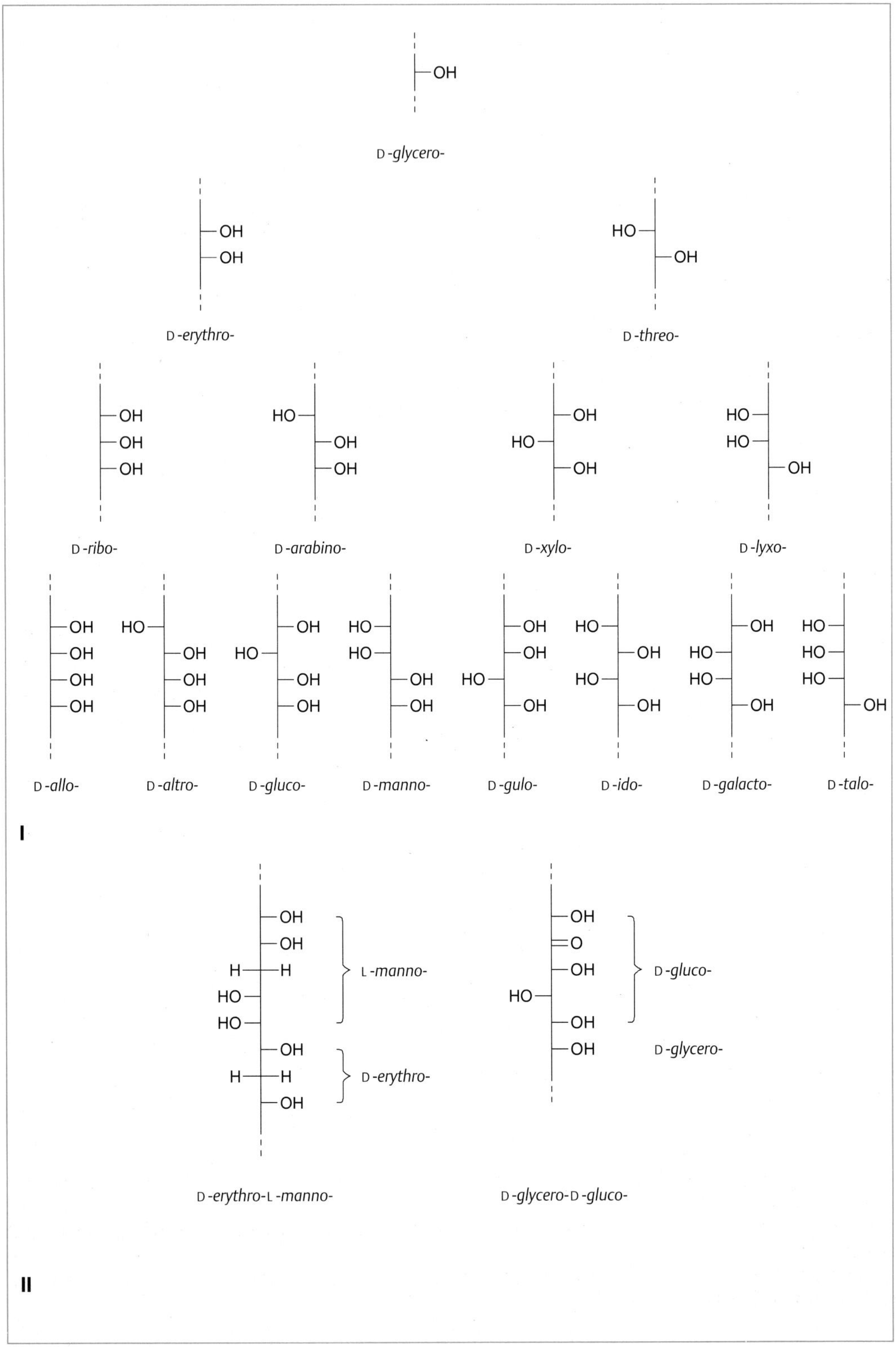

D-*glycero-*

D-*erythro-* D-*threo-*

D-*ribo-* D-*arabino-* D-*xylo-* D-*lyxo-*

D-*allo-* D-*altro-* D-*gluco-* D-*manno-* D-*gulo-* D-*ido-* D-*galacto-* D-*talo-*

**I**

L-*manno-*

D-*erythro-*

D-*erythro*-L-*manno-*

D-*gluco-*

D-*glycero-*

D-*glycero*-D-*gluco-*

**II**

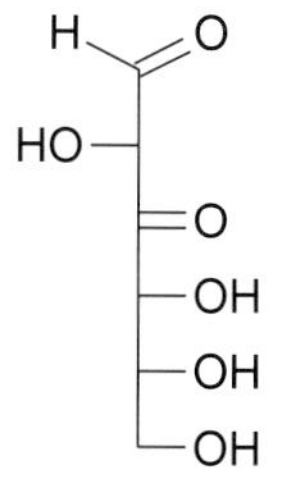

D-*arabino*-hexos-3-ulose

**I**

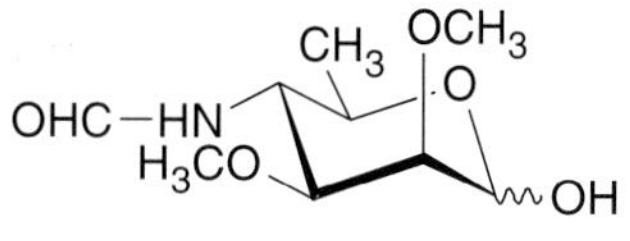

4,6-dideoxy-4-formamido-2,3-di-*O*-methyl-D-mannopyranose

**II**

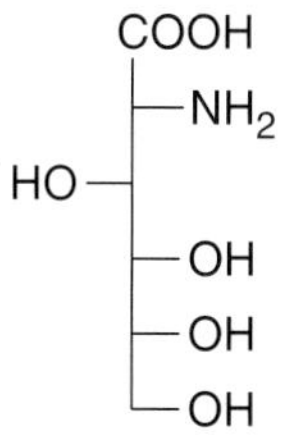

**III** 2-amino-2-deoxy-D-gluconic acid

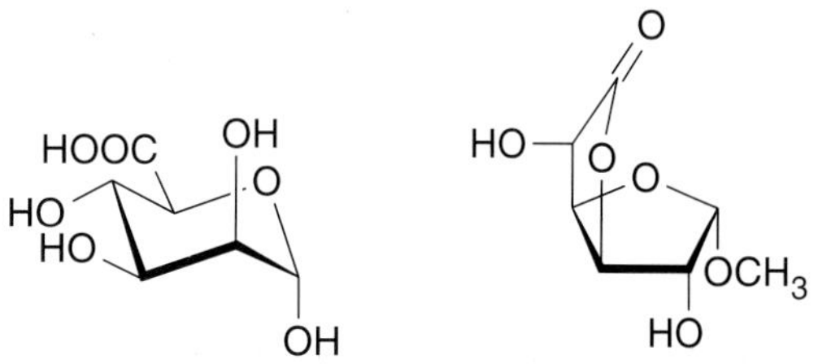

α-D-mannopyranuronic acid

methyl α-D-glucofuranosidurono-6,3-lactone

**IV**

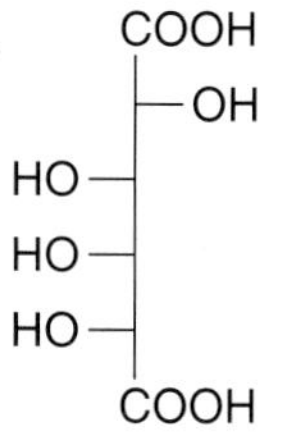

L-altraric acid

*not*

L-talaric acid

**V**

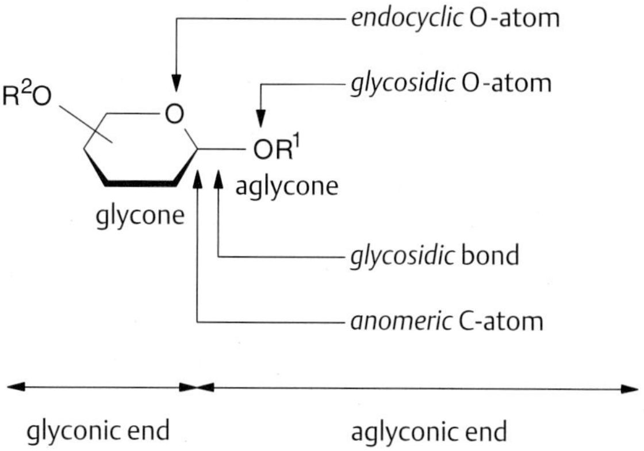

*endocyclic* O-atom

*glycosidic* O-atom

aglycone

glycone

*glycosidic* bond

*anomeric* C-atom

glyconic end

aglyconic end

**VI**

Polyhydroxyalkanes are called *alditols*. In *aldonic acids*, the aldehydic group of an aldose is oxidised to a carboxy group (above, III).

A *uronic acid* is an aldose in which the primary hydroxy group has been converted into a carboxy group (above, IV).

*Aldaric acids* are dicarboxylic acids derived from aldoses by replacement of both terminal groups by carboxy groups (above, V).

*Glycosides* are mixed acetals which are formally derived by dehydration between a lactol group of a monosaccharide or oligosaccharide and the hydroxy group of another component. The bond between the two components is called a glycosidic bond (p. 57, VI).

*Oligosaccharides* are compounds in which monosaccharides are linked to one another by glycosidic bonds. The borderline between oligo- and polysaccharides is not strictly defined. If an exact structure can be described, then even with high molecular weight compounds, the term oligosaccharide may be used.

Polysaccharides are called *glycans*. If they consist of only one type of monosaccharide then they are described as homoglycans, but if the polysaccharide contains different types of monosaccharide they are called heteroglycans. Unlike the genetically determined globular proteins, polysaccharides are not mono-disperse. In general they are poly-disperse or at all events, in the case of regular construction from so-called repeat units, they are polymolecular. For frequently occurring groups in monosaccharides abbreviations are permissible and desirable:

| | | | |
|---|---|---|---|
| Acetyl | Ac | Trimethylsilyl | $Me_3Si$ |
| Benzyl | Bn | *tert*-Butyltrimethylsilyl | $Bu^tMe_2Si$ |
| Benzoyl | Bz | Phenyl | Ph |
| Ethyl | Et | Triflyl (Trifluormethyl) | Tf |
| Methyl | Me | Tosyl | Ts |
| | | Trityl (Triphenylmethyl) | Tr |

The numbering of a group or residue follows the general rules of organic chemistry but where the numbering of the parent monosaccharide is not already defined, as for instance in a polyol, the correct name is that which gives the lowest set of locants:

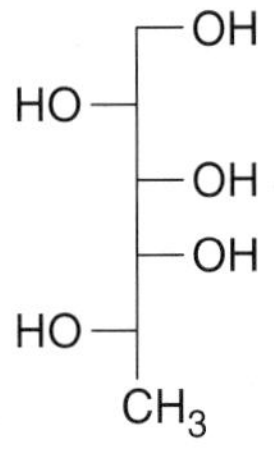

1-deoxy-D-galactitol

*not*

6-deoxy-L-galactitol

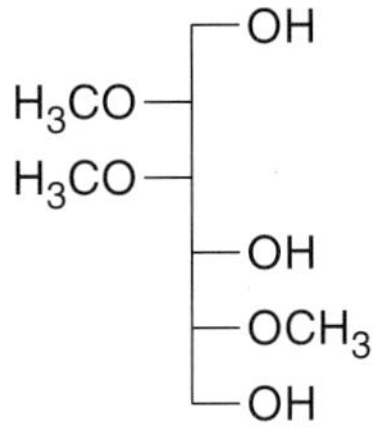

2,3,5-Tri-*O*-methyl-D-mannitol

*not*

2,4,5-Tri-*O*-methyl-D-mannitol

In general, when only one configurational prefix is used, the anomeric reference atom is the C-atom which determines the absolute configuration, that is the highest numbered asymmetric C-atom of a monosaccharide. If multiple configurational prefixes are used, then the reference atom is the highest-numbered atom of the group of chiral centres next to the anomeric centre that is involved in the heterocyclic ring and specified by a single configurational prefix. The anomeric configuration symbol is designated $\alpha$ if the lactol O-atom in the Fischer projection is *cis* to the oxygen attached to the anomeric reference atom. In the $\beta$-form these oxygen atoms are formally *trans*. A series of examples illustrates the relationships (see p. 59, I).

The Greek letters $\alpha$ and $\beta$ can only be used if the anomeric C-atom has a lower positional number than the anomeric reference atom. In the case of dicarbonyl compounds, ring closure is also possible involving a carbonyl group with a higher number than that of the anomeric reference atom. In such cases the configuration at the anomeric carbon is indicated by the Cahn-Ingold-Prelog convention with the symbol *R* or *S*, as appropriate (see p. 59, II).

It should be remembered that numbering is governed by the alphabetical order of the configurational prefix, that is *D-gluco* before *L-gluco*. Although often omitted, systematic nomenclature can also be used for the usual aldohexoses, which must contain the anomeric symbol, the configurational prefix,

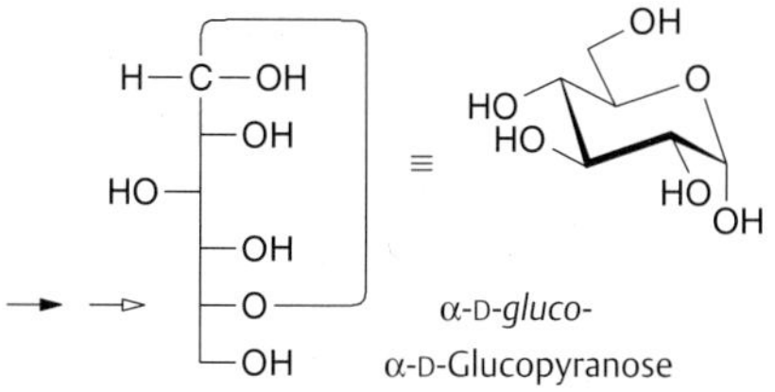

α-D-*gluco*-
α-D-Glucopyranose

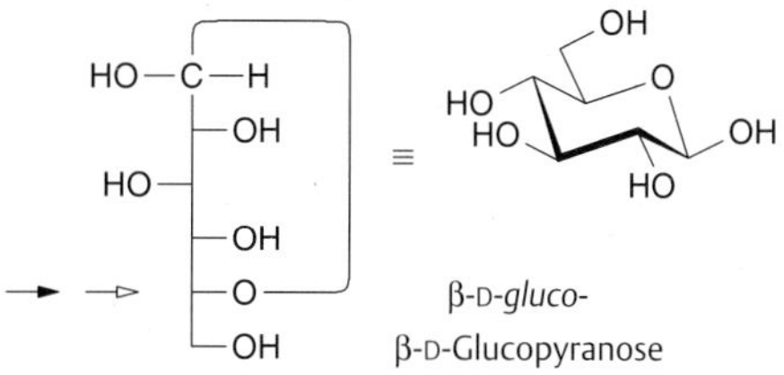

β-D-*gluco*-
β-D-Glucopyranose

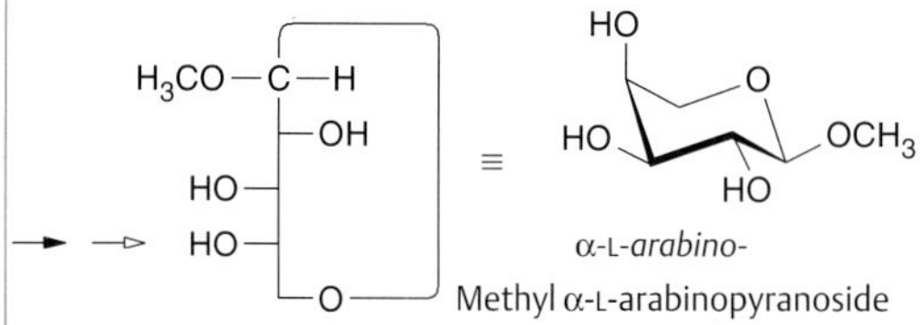

α-L-*arabino*-
Methyl α-L-arabinopyranoside

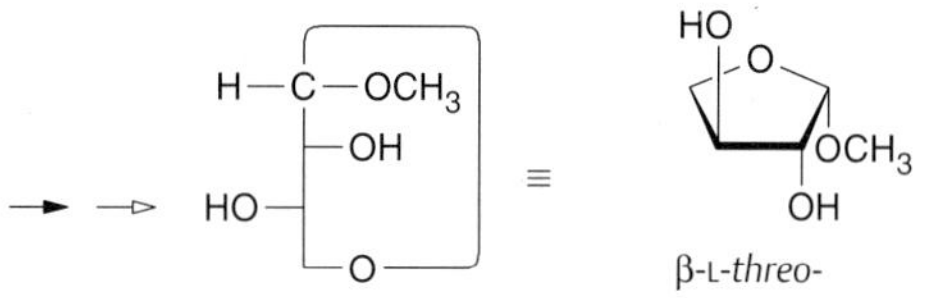

β-L-*threo*-
Methyl β-L-threofuranoside

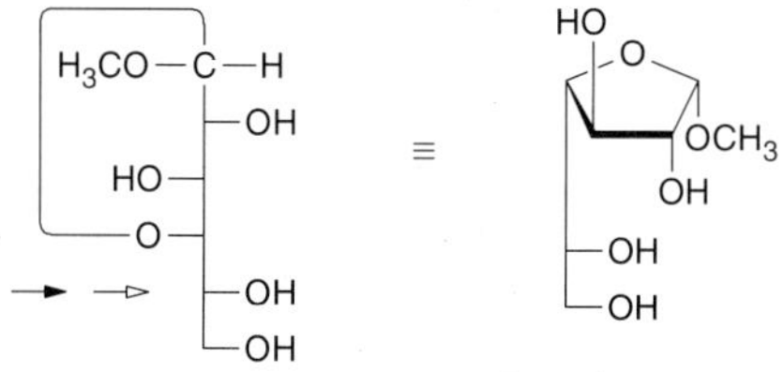

β-D-*galacto*-
Methyl β-D-galactofuranoside

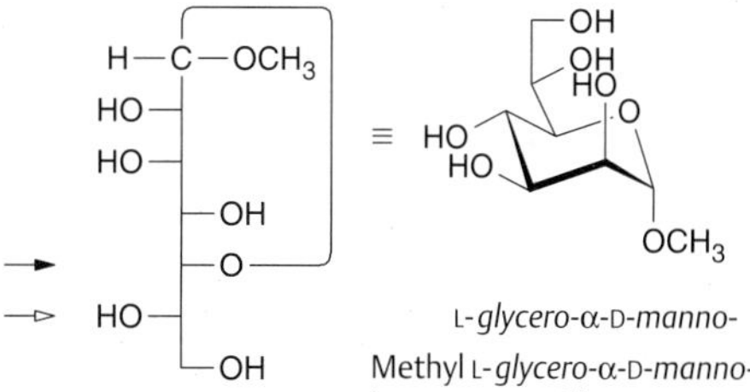

L-*glycero*-α-D-*manno*-
Methyl L-*glycero*-α-D-*manno*-heptopyranoside

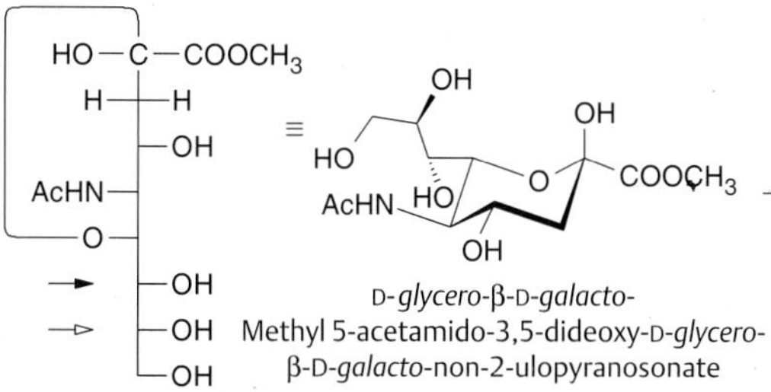

D-*glycero*-β-D-*galacto*-
Methyl 5-acetamido-3,5-dideoxy-D-*glycero*-β-D-*galacto*-non-2-ulopyranosonate

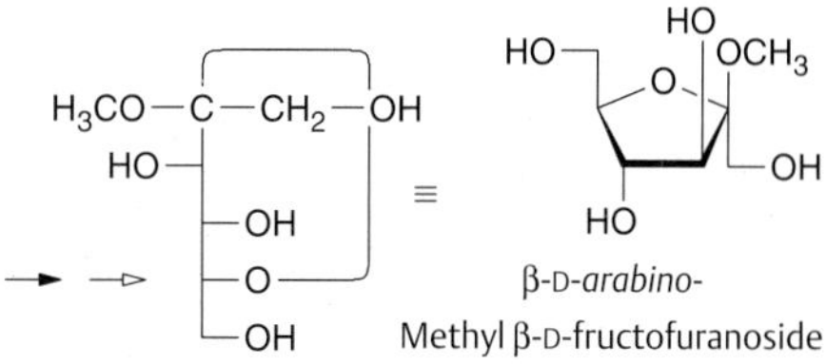

β-D-*arabino*-
Methyl β-D-fructofuranoside

I

→ denotes the anomeric reference atom
⇾ denotes the configurational atom

II

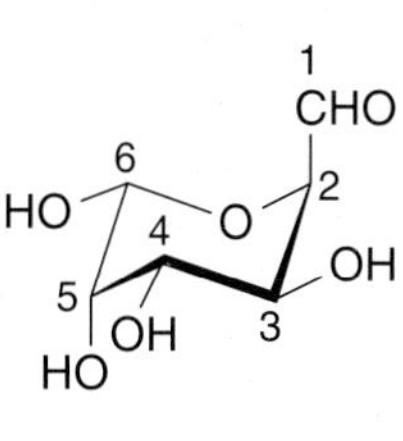

(6*R*)-D-*gluco*-hexodialdo-6,2-pyranose

the number of C-atoms, and the symbol for the absolute configuration. Thus, *α-D-galactopyranose* is correctly named *α-D-galacto-hexopyranose*. The systematic nomenclature is indispensable in complicated cases.

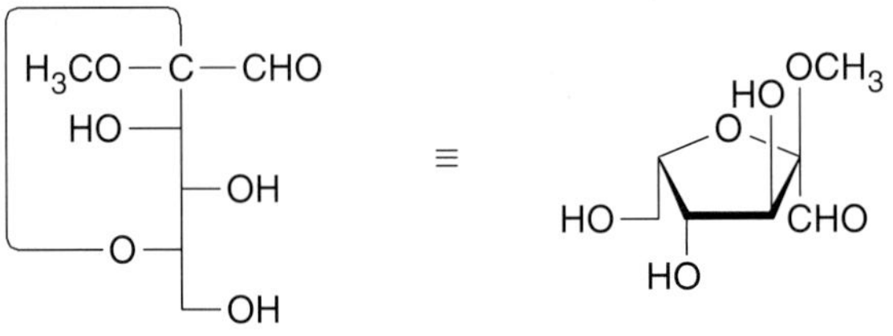

Methyl α-L-*xylo*-hexos-2-ulo-2,5-furanoside

The designation of alkyl or aryl groups as substituents on the anomeric carbon O- (or also S-) atom is placed before the anomeric symbol $\alpha$ or $\beta$. Where there is an anomeric S-atom, the descriptor *thio* is included with the position of substitution. Compounds with such substiuents are glycosides. The following four compounds are selected as representative of monocarbonyl compounds:

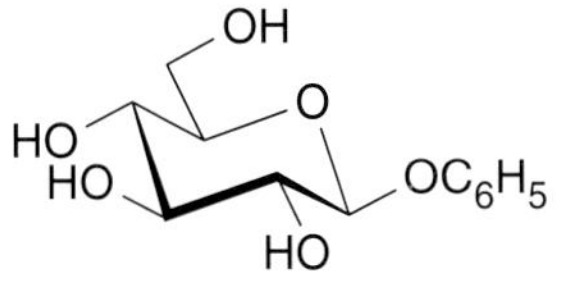

Phenyl β-D-*gluco*-hexopyranoside

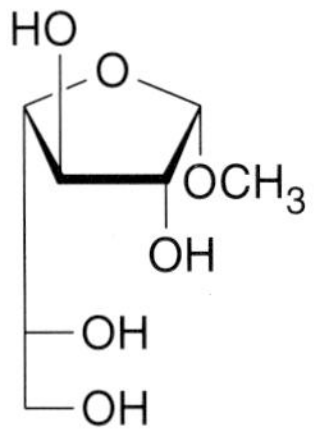

Methyl α-D-*galacto*-hexofuranoside

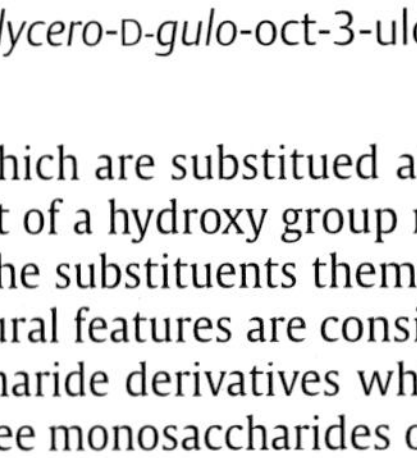

Methyl α-D-*glycero*-D-*gulo*-oct-3-uloseptanoside

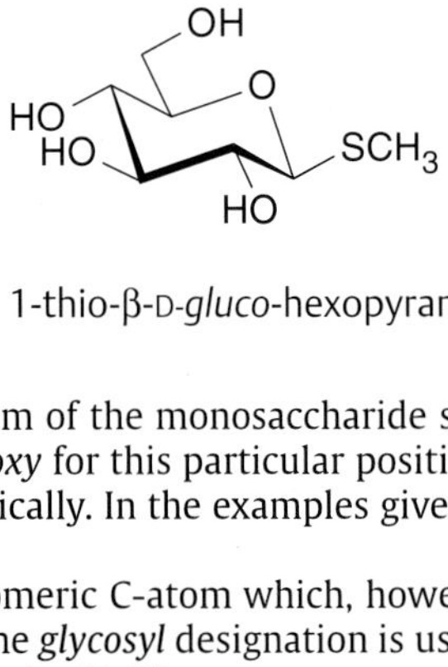

Methyl 1-thio-β-D-*gluco*-hexopyranoside

Compounds which are substitued at a non-anomeric carbon atom of the monosaccharide skeleton with replacement of a hydroxy group must carry the descriptor *deoxy* for this particular position. This descriptor with the substituents themselves are arranged alphabetically. In the examples given (p. 61, I), various structural features are considered.

For monosaccharide derivatives which carry a group on the anomeric C-atom which, however, are not glycosides, free monosaccharides or 1-thio-monosaccharides, the *glycosyl* designation is used. This is valid for all halogens, amines, phosphates, sulphates, and nitrates (p. 61, II).

Acyl groups are exceptions. In this case the parent name is retained (p. 61, III).

Monosaccharides which carry a double bond as part of their C,C-backbone are named by inserting into the name of the fully saturated derivative *en* (or *eno* for euphony), preceded by the corresponding locant of the lower-numbered C-atom involved in the double bond, immediately after the stem name that designates the chain length of the sugar.

Frequently, such unsaturated monosaccharide derivatives are cyclic enol ethers which can be thought of as being derived from intramolecular ethers. In carbohydrates, intramolecular cyclic ethers

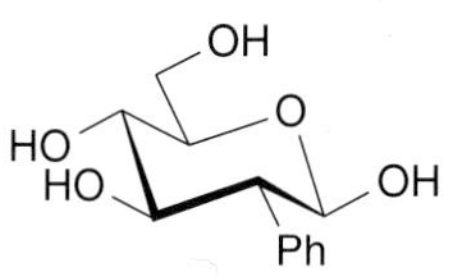

2-Deoxy-2-phenyl-β-D-glucopyranose
or
2-Deoxy-2-*C*-phenyl-β-D-glucopyranose
or
*(2R)*-2-Deoxy-2-phenyl-β-D-*arabino*-hexopyranose

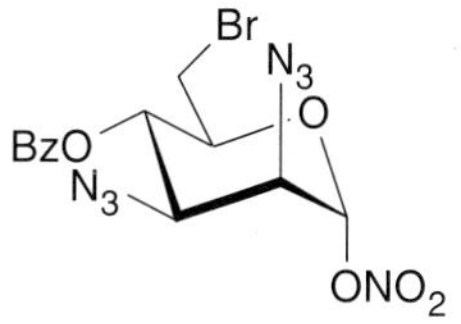

2,3-Diazido-4-*O*-benzoyl-6-bromo-2,3,6-trideoxy-α-D-mannopyranosyl nitrate

I II

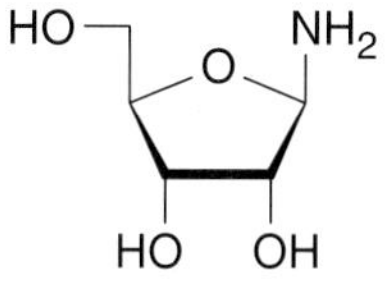

β-D-ribofuranosylamine

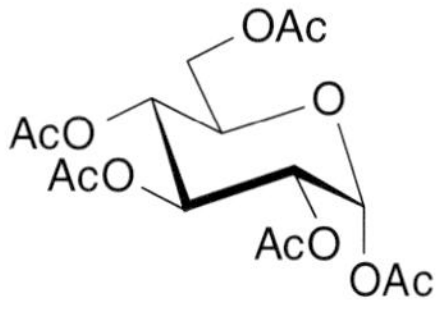

1,2,3,4,6-Penta-*O*-acetyl-α-D-glucopyranose

II III

are named as anhydro compounds. There are also anhydro compounds which include the anomeric hydroxy group. The preceding locants specify the ring size. The prefix *anhydro* is employed to describe intramolecular ethers (p. 62).

Similar to anhydro derivatives, the intramolecular esters of glyconic acids (lactones) have locants to describe the ring size:

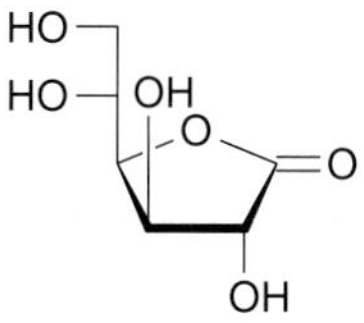

D-Glucono-1,4-lactone
(D-Gluconic acid γ-lactone)

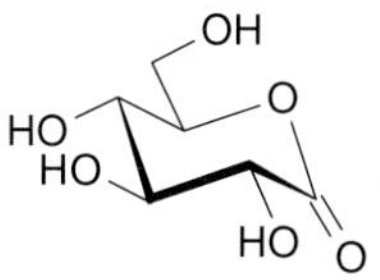

D-Glucono-1,5-lactone
(D-Gluconic acid δ-lactone)

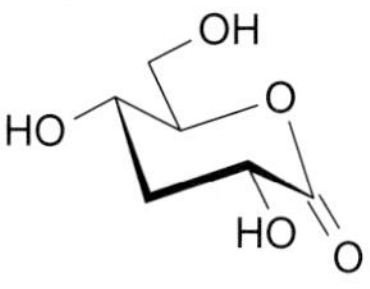

3-Deoxy-D-*ribo*-hexono-1,5-lactone

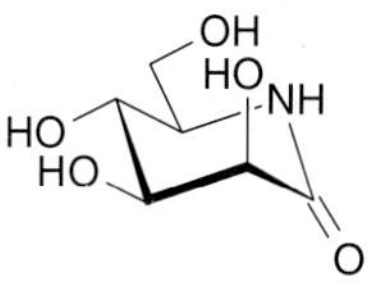

5-Amino-5-deoxy-D-mannono-1,5-lactam

2,6-Anhydro-1-deoxy-D-*altro*-hept-1-enitol
(alphabetic preference over:
2,6-Anhydro-7-deoxy-D-*talo*-hept-6-enitol)

1,5-Anhydro-2-deoxy-
D-*arabino*-hex-1-enitol

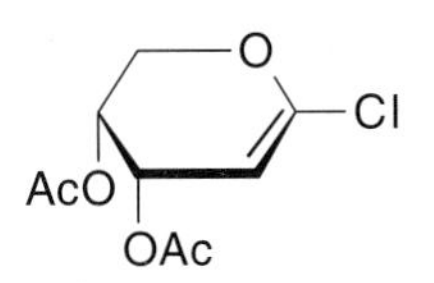


3,4-Di-*O*-acetyl-2-deoxy-D-*erythro*-
pent-1-enopyranosyl chloride

2,3-Dideoxy-α-D-erythro-
hex-2-enopyranose

Methyl 3,4-dideoxy-β-D-glycero-
hex-3-en-2-ulopyranoside

As monosaccharide derivatives, the names of *glycuronic acids* are formed by replacing the ending *-ose* with *-uronic acid*, and glycosides of the uronic acids are named by replacing the *-oside* of the glycoside name by *-osiduronic acid*. Salts or esters are named using the ending *-uronate*:

Methyl (phenyl-4-deoxy-β-L-*threo*-
hex-4-enopyranosid)uronate

Cyclic acetals, which are formed by the reaction of monosaccharides or monosaccharide derivatives with aldehydes or ketones, are named in accordance with O-substituents. If in the course of cyclic acetal formation an asymmetric C-atom is formed and the configuration is known, then the stereochemistry is indicated using the appropriate *R* or *S* symbol (see p. 63, I).

Monosaccharides can form acyclic acetals and the corresponding thioacetals. The names of these compounds are made by combining the name of the monosaccharide or monosaccharide derivative and the ending *acetal*, *monothioacetal*, or *dithioacetal* (see p. 63, II).

Previously, the organic group of the alcohol component was named but more and more frequently in carbohydrate chemistry one is confronted with the necessity of distinguishing a monosaccharide not as the parent structure but as a residue in a molecule of different origin. Depending on from which position of the monosaccharide the linkage arises, and which element forms the bridge, the locant, the bridging element and then the ending *yl* is to be used. The fragment *yl* always signifies the replacement of a hydrogen atom in a monosaccharide (p. 64, I). Linkage through the anomeric C-atom without loss of

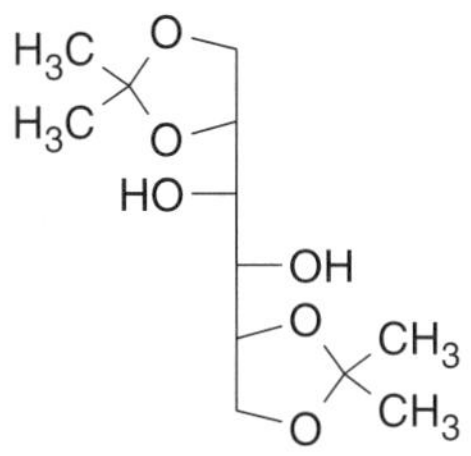

1,2:5,6-Di-*O*-isopropylidene-D-mannitol

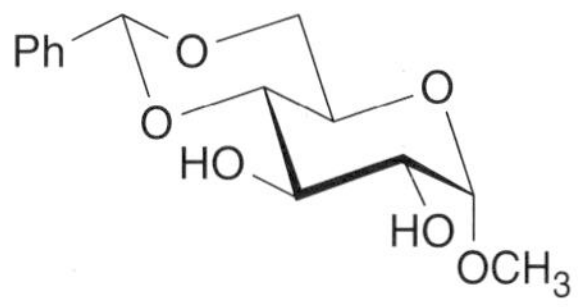

Methyl *(R)*-4,6-*O*-benzylidene-α-D-glucopyranoside

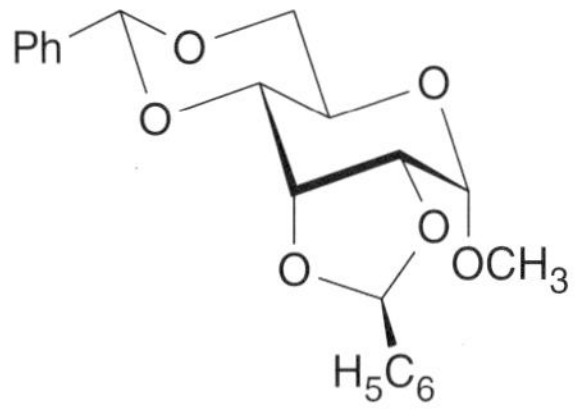

Methyl *(S)*-2,3:*(R)*-4,6-di-*O*-benzylidene-α-D-allopyranoside

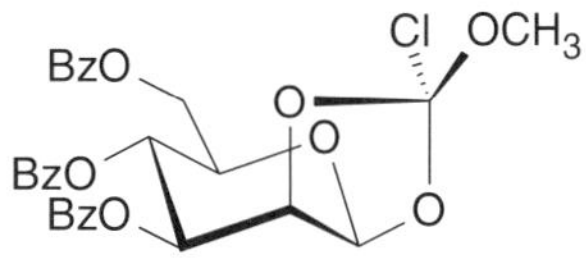

3,4,6-Tri-*O*-benzoyl-[*(S)*-1,2-*O*-chloro(methoxy)methylene]-β-D-mannopyranose

I

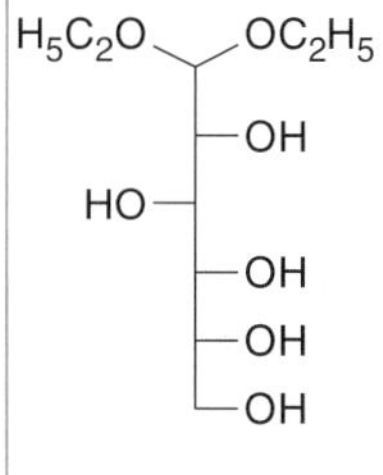

D-Glucose diethyl acetal

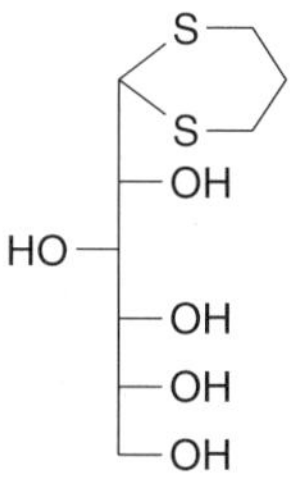

D-Glucose propane-1,3-diyl dithioacetal

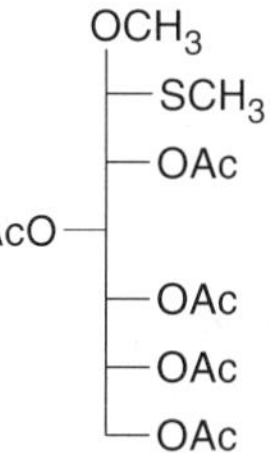

*(1R)*-2,3,4,5,6-Penta-*O*-acetyl-D-glucose dimethyl monothioacetal

II

the anomeric hydroxy group is somewhat unusual, as the last formula shows. In the past years monosaccharide derivatives in which the ring-O has been replaced by a N- or C-atom have acquired importance (p. 64, II). The term *aza*, as in the penultimate formula, can only be used for replacement of a C-atom. The prefix *carba* as generally used denotes the replacement of a hetero- by a C-atom. In our case, the heteroatom is oxygen. Such derivatives are called *pseudo*-monosaccharides. The inserted carbon atom receives the index 5a if one wishes to retain the monosaccharide numbering.

For many years disagreement has existed about the systematic nomenclature of oligosaccharides. It is generally agreed today that the *non-reducing end-groups* will be written as a residue with the ending *yl*, with labelling of the linkage from the anomeric C-atom to the bridgehead atom on the *reducing* side. Clearly, as in the case of cane sugar (sucrose or saccharose), it can also be that there are no *reducing* ends. In this case the alphabetic arrangement is applied (p. 65).

The systematic representations of higher oligosaccharides are, as the example of the tetrasaccharide shows, very unclear and more and more often are replaced by short symbols. Proposals concerning this type of nomenclature are also to be found in chapter 2, and the joint IUPAC and IUBMB document

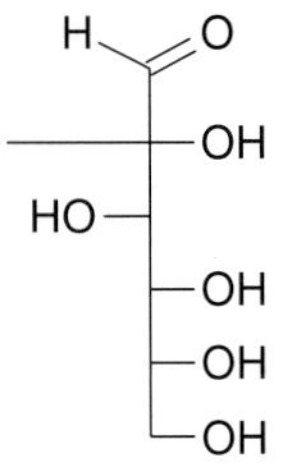

D-Glucos-2-C-yl-

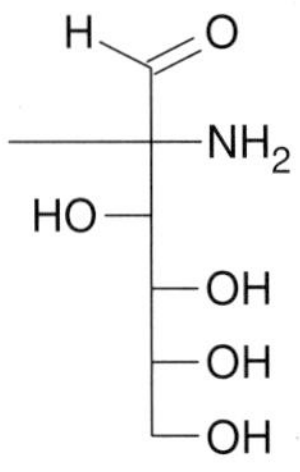

2-Amino-2-deoxy-
D-glucos-2-C-yl-

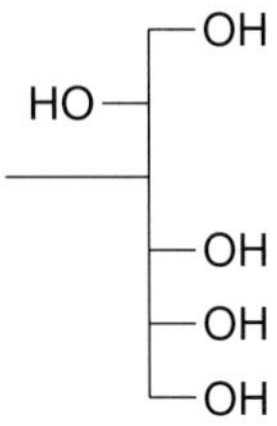

3-Deoxy-D-mannitol-3-yl-

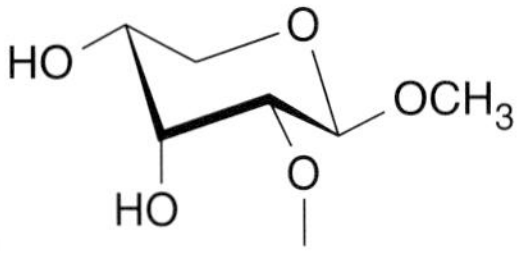

(Methyl β-D-ribopyranosid-2-O-yl)-

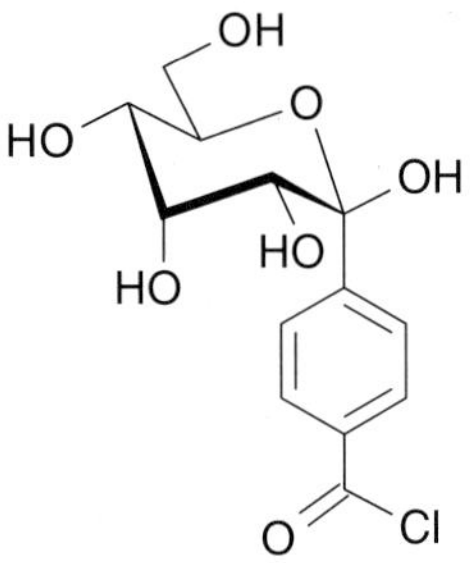

4-(1-hydroxy-α-D-allopyranosyl)benzoyl chloride

I

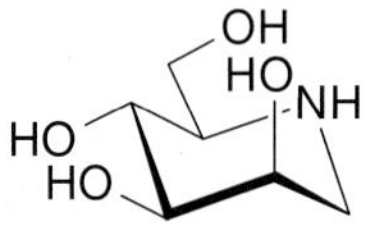

1-Amino-1,5-anhydro-1-deoxy-D-mannitol
or
1,5-Dideoxy-1,5-imino-D-mannitol
(1-Deoxymannonojirimycin)

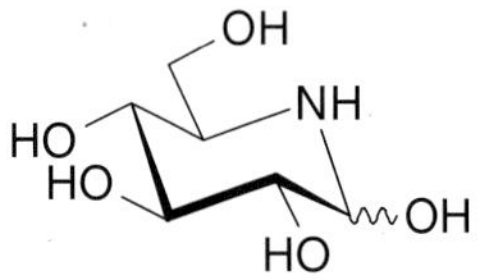

5-Amino-5-deoxy-D-glucopyranose
(Nojirimycin)

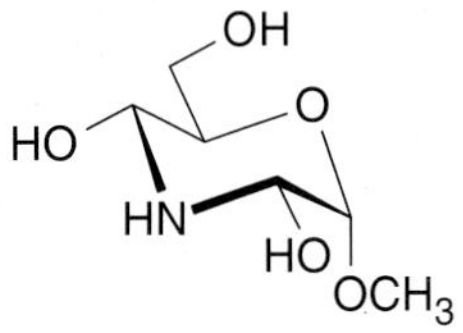

Methyl 3-deoxy-3-aza-α-D-*ribo*-hexopyranoside

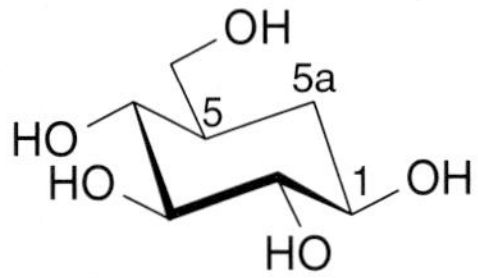

II

5a-Carba-β-D-glucopyranose

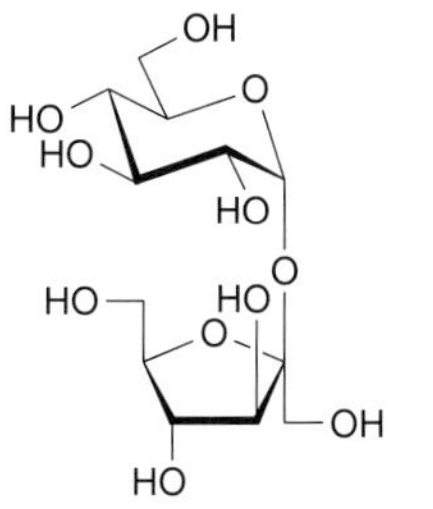

β-D-Fructofuranosyl α-D-glucopyranoside
[β—D—Fru*f*—(2 ↔ 1)—α—D—Glc*p*]
(sucrose, saccharose)
*not*
α-D-glucopyranosyl β-D-fructofuranoside

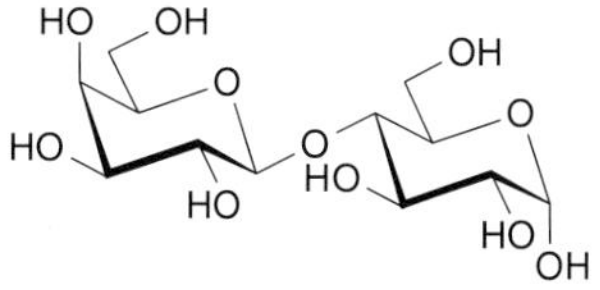

β-D-Galactopyranosyl-(1 → 4)-α-D-glucopyranose
*or*
4-*O*-β-D-galactopyranosyl-α-D-glucopyranose
[β—D—Gal*p*—(1 → 4)—α—D—Glc*p*]
(α-lactose not α-D-lactose)

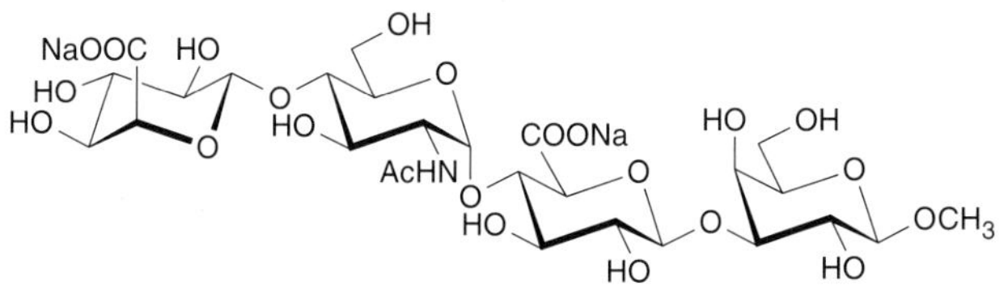

Methyl (sodium α-L-idopyranosyluronate)-(1 → 4)-(2-acetamido-2-deoxy-α-D-glucopyranosyl)-(1 → 4)-(sodium β-D-glucopyranosyluronate)-(1 → 3)-β-D-galactopyranoside
{$Na_2$[α—L—Ido*p*A—(1 → 4)—α—D—Glc*p*NAc—(1 → 4)—β—D—Glc*p*A—(1 → 3)—β—D—Gal*p*OMe]}

"Nomenclature of Carbohydrates (Recommendations 1996)"[79] should be consulted for the most recent versions of naming higher oligosaccharides, polysaccharides, glycoproteins, glycopeptides and peptidoglycans, including the use of symbols.

## References


1 Hurd, C. D.; Saunders, W. H. *J. Am. Chem. Soc.* **1952**, *74*, 5342.
2 Los, J. M.; Simpson, L. B.; Wiesner, K. *J. Am. Chem. Soc.* **1956**, *78*, 1564.
3 Fischer, E. *Ber. Dtsch. Chem. Ges.* **1890**, *23*, 2114.
4 Wohl, A.; Momber, F. *Ber. Dtsch. Chem. Ges.* **1914**, *47*, 3346.
5 Wohl, A.; Freudenberg, K. *Ber. Dtsch. Chem. Ges.* **1923**, *56*, 309.
6 Fischer, E.; *Ber. Dtsch. Chem. Ges.* **1891**, *24*, 2683.
7 Rosanoff, M. A. *J. Am. Chem. Soc.* **1906**, *28*, 114.
8 Bijvoet, J. M.; Peerdeman, A. F.; van Bommel, A. J. *Nature*, **1951**, *168*, 271.
9 Cahn, R. S.; Ingold, C.; Prelog, V. *Angew. Chem.* **1966**, *78*, 413; *Int. Ed.*, **1966**, *5*, 385.
10 Tollens, B.; *Ber. Dtsch. Chem. Ges.* **1883**, *16*, 922.
11 Haworth, W. N. *The Constitution of the Sugars*, Arnold: London, **1929**.
12 Mills, J. A., *Advan. Carbohydr. Chem.* **1956**, *10*, 1.
13 Böeseken, J., *Ber. Dtsch. Chem. Ges.* **1913**, *46*, 2612.
14 Stoddart, J. F., *Stereochemistry of Carbohydrates*, Wiley-Interscience. New York, 1971.
15 Durette, P. L., Horton, D. *Advan. Carbohydr. Chem.* **1971**, 26, 49.
16 Eliel, E. L.; Allinger, N. L.; Angyal, S. J.; Morrison, G. A. *Conformational Analysis*, Wiley & Sons: New York, **1965**.
17 Kemp, J. D.; Pitzer, K. S. *J. Am. Chem. Soc.* **1937**, *59*, 276.
18 Klyne, W.; Prelog, V. *Experientia*, **1960**, *16*, 521.
19 Signorella, S.; Sala, L. F. *J. Chem. Ed.* **1991**, *68*, 105.
20 Barker, S. A.; Bourne, E. J. *Advan. Carbohydr. Chem.* **1952**, *7*, 137.
21 Tvaroska, I.; Bleha, T. *Advan. Carbohydr. Chem. Biochem.* **1989**, *47*, 45.

22 *The Anomeric Effect and Associated Stereoelectronic Effects*; Thatcher, G. R.J., Ed.; ACS Symposium Series 539; American Chemical Society; Washington, D.C., **1993**.
23 Kirby, A. J.; *The Anomeric Effect and Related Stereoelectronic Effects at Oxygen*, Springer: Berlin, 1983.
24 Edward, *J. T. Chem. Ind. (London)*, **1955**, *1102*.
25 Lemieux, R. U. In *Molecular Rearrangements*, Vol. 2; Mayo P. de, Ed., Interscience: New York, 1964, p. 709.
26 de Hoog, A. J.; Buys, H. R.; Altona, C.;. Havinga, E *Tetrahedron*, **1969**, *25*, 3365.
27 Lemieux, R. U.; Pavia, A. A.; Martin, J. C.; Watanabe, K. A. *Can. J. Chem.*, **1969**, *47*, 4427.
28 Lemieux, R. U; Morgan, A. R. *Can. J. Chem.*, **1965**, *43*, 2205.
29 Berman, H. M.; Chu, S. S. C.; Jeffry, G. A. *Science*, **1967**, *157*, 1576.
30 Kothe, G.; Luger, P.; Paulsen, H. *Acta Crystallogr. Sect.B*, **1979**, *35*, 2079.
31 Barker, S. A.; Brimacombe, J. S.; Foster, A. B.; Whiffen, D. H.; Zweifel, G. *Tetrahedron*, **1959**, *7*, 10.
32 Dobinson, B.; Foster, A. B. *J. Chem. Soc.* **1961**, *2338*.
33 Baggett, N. Bukhari, M. A.; Foster, A. B.; Lehmann, J.; Webber, J. M. *J. Chem. Soc.* **1963**, 4157.
34 Lehmann, J.; Rob, B. *Liebigs Ann. Chem.* **1994**, 805.
35 Angyal, S. J. *Angew. Chem., Int. Ed. Engl.* **1969**, *8*, 157.
36 Angyal, S. J. *Adv. Carbohydr. Chem. Biochem.* **1984**, *42*, 15.
Angyal, S. J. *Adv. Carbohydr. Chem. Biochem.* **1991**, *49*, 19.
37 Pérez, S.; Marchessault, R. H. *Carbohydr. Res.* **1978**, *65*, 114.
38 Rees, D. A.; Morris, E. R.; Thom, D.; Madden, J. K. In *The Polysaccharides*, Vol. 1, Aspinall; G. O.; Academic Press: New York, **1982**, Chapter 5.
39 Bock, K. *Pure & Appl. Chem.* **1983**, *55*, 605.
40 Fieser, L. F.; Fieser, M. *Organic Chemistry*; 3rd edition; Reinhold: New York, **1956**, Chapter 14.
41 Brobst, K. M. *Methods Carbohydr. Chem.* **1972**, *6*, 3.
42 Sweeley, C. C.; Tao, R. V. P. *Carbohydr. Chem.*, **1972**, *6*, 8.
43 Sloneker, J. H. *Carbohydr. Chem.*, **1972**, *6*, 20.
44 Jones, H. G. *Carbohydr. Chem.* **1972**, *6*, 25.
45 Seymor, F. R. *Carbohydr. Chem.* **1993**, *9*, 59.
46 Leontein, K.; Lönngren, J. *Carbohydr. Chem.* **1993**, *9*, 87.
47 Wing, R. E.; BeMiller, J. N. *Methods Carbohydr. Chem.* **1972**, *6*, 42 and 54.
48 McGinnis, G. D.; Fang, P. *Methods Carbohydr. Chem.* **1980**, *8*, 33.
49 Whistler, R. L.;. Anisuzzman, A. K. M *Methods Carbohydr. Chem.* **1980**, *8*, 45.
50 Hall, L. D. *Adv. Carbohydr. Chem. Biochem.* **1974**, *29*, 11.
51 Gorin, P. A. J. *Adv. Carbohydr. Chem. Biochem.* **1981**, *38*, 13.
52 Bock, K.; Pedersen, C. *Adv. Carbohydr. Chem. Biochem.*, **1983**, *41*, 27.
53 Vliegenthart, J. F. G.; Dorland, L.; Van Halbeek, H. *Adv. Carbohydr. Chem. Biochem.* **1983**, *41*, 209.
54 Friebolin, H. *Basic One- and Two-Dimensional NMR-Spectroscopy*, VCH: Weinheim, **1993**.
55 Lönngren, J.; Svensson, S. *Adv. Carbohydr. Chem. Biochem.* **1974**, *29*, 41.
56 Dell, A. *Adv. Carbohydr. Chem. Biochem.* **1987**, *45*, 19.
57 *Computer Modeling of Carbohydrate Molecules*; French, A. D.; Brady, J. W. *ACS Symposium Series 430*, Washington D.C., **1990**.
58 Avigad, G. *J. Chromatography*, **1977**, *139*, 343.
59 Alpenfels, W. F. *Anal. Biochem.*, **1981**, *114*, 153.
60 Läufer, K.; Lehmann, J.; Petry, S.; Scheuring, M.; Schmidt-Schuchardt, M. J. *Chromatogr. A*, **1994**, *684*, 370.
61 Kurth, H.; Lehmann, J. *Biomed. Chromatogr.*, **1986**, *1*, 58.
62 Sweeley, C. C.; Bentley, R.; Makita, M.; Wells, W. W. *J. Am. Chem. Soc.* **1963**, *85*, 2495.
63 Wallenfels, K.; Földi, P.; Niermann, H.; Bender, H.; Linder, D. *Carbohydr. Res.* **1978**, *61*, 359.
64 Karplus, M. *J. Chem. Phys.* **1959**, *30*, 11.
65 Hall, L. D. *Adv. Carbohydr. Chem.* **1964**, *19*, 51.
66 Bock, K.; Pedersen, C.; Pedersen, H. *Adv. Carbohydr. Chem. Biochem.* **1984**, *42*, 193.
67 Kochetkov, N. K.; Chizhov, O. S. *Adv. Carbohydr. Chem.* **1966**, *21*, 39.
68 Schulten, H.-R. *Int. J. Mass Spectrom. Ion Phys.*, **1979**, *32*, 97.
69 Whitehouse, C. M. Dryer, R. N., Yamashita, M., Fenn, J. B. *Anal. Chem.* **1985**, *57*, 675.
70 Aspinall, G. O. In *The Polysaccharides: The Polysaccharides*, Vol. 1, Aspinall. G. O., Ed.; Academic Press: New York, 1982, Kap. 3.
71 Hudson, C. S. *J. Am. Chem. Soc.* **1909**, *31*, 66.
72 Whiffen, D. H *Chem. Ind. (London)*, **1956**, *964*.
73 Brewster, J. H. *J. Am. Chem. Soc.* **1959**, *81*, 5483.
74 Lemieux, R. U.; Pavia, A. A. *Can. J. Chem.* **1968**, *46*, 1453.
75 Lemieux, R. U.; Martin, J. C. *Carbohyd. Res.* **1970**, *13*, 139.
76 Bar-Guilloux, E.; Defaye, J.; Lehmann, J.; Nardin, R.; Robic, D.; Urbahns, K. *Carbohydr. Res.* **1993**, *250*, 1.
77 Lehmann, J.; Rob, B. *Carbohydr. Res.* **1995**, *276*, 199.
78 *Biochem. J.* **1971**,*125*, 673; *Biochemistry*, **1971**, *10*, 3983; *Biochim. Biophys. Acta*, **1971**, *244*, 223; *Eur. J. Biochem.*, **1971**, *21*, 455; *Eur. J. Biochem.* **1972**, *25*, 4; *J. Biol. Chem.* **1972**, *247*, 613.
79 IUPAC-IUB Joint Commission on Biochemical Nomenclature (JCBN), Nomenclature of carbohydrates (Recommendations 1996), *Pure Appl. Chem.* **1996**, *68*, 1919; *Carbohydr. Res.* **1997**, *297*, 1.
80 *Biochem. J.* **1976**, *153*, 23; *Eur. J. Biochem.* **1975**, *57*, 1; *Biochem. J.* **1989**, *258*, 1; *Eur. J. Biochem.* **1989**, *180*, 485.

Chapter 2

# Biological Aspects

## 2.1 Occurrence and general biological importance

Carbohydrates occur in all living things but not all organisms make this class of compounds in equal amounts. It may be assumed that the major amount of carbohydrates is to be found in the plant kingdom and it is there also that they are in evidence most conspicuously. The *de novo* biosynthesis of organic material in green plants found on land produces directly about $2 \times 10^{11}$ t of carbohydrate per year and a considerably greater quantity than this already exists in living matter in the biosphere.[1] The largest part of this, about 90%, is found in trees. The extent of biosynthesis by algae[2] in the oceans is difficult to assess and therefore is not further considered here. The bulk of carbohydrates has a comparatively simple biological importance:

1. Carbohydrates serve as mechanically stable supports and frameworks for both single cells and higher organisms and so are responsible for an optimal structural integrity of living creatures. In the case of vertebrates, they provide the means of frictionless movement and shock absorption.
2. Carbohydrates are an easily mobilised C-reserve for the higher organisms and, thereby, are a source of energy and of building blocks for the production of other classes of natural products. As has recently become apparent, a large class of oligomeric carbohydrates, conjugated with proteins or lipids, so-called glycoconjugates, possess a differentiation role in general biological processes.
3. Glycoconjugates serve very frequently to change the chemical and, still more, the physiochemical properties of cell constituents or the so-called extracellular matrix. They can contribute to raising hydrophilicity or can bring about changes in charge density. They can also be molecular markers by which cells recognise each other, or which allow cells to orient themselves in their surroundings. Examples from all spheres of this general biology of carbohydrates are dealt with in this chapter.

### 2.1.1 Naturally occurring monosaccharides

By far the greatest part of all naturally occuring carbohydrates consists not of free monosaccharides but of their condensation products: oligosaccharides, polysaccharides, or glycosides, the latter with a non-carbohydrate component as the aglycone residue.

A remarkable feature of monosaccharides as polyfunctional compounds is an abundance of stereoisomers. Their number doubles with each additional C-atom. Structural variants, in which the usual hydroxy group is replaced by a hydrogen atom, an alkyl residue, or by an amino group, or by the introduction of a carboxy group, increases the variety still more. Since lengthening of the carbon chain of monosaccharides is, in principle, unlimited, the number of natural monosaccharides can be immeasurably large. It now seems certain that in this regard nature acts moderately, especially in relation to the monosaccharide building units in higher organisms. Bacteria and lower organisms, such as fungi and algae, are much richer in variants[3], a fact which is also biologically based (compare sections 2.1.2.1, p. 73 and 2.1.2.2, p. 80). In total about 250 different, naturally occuring monosaccharides have been identified.

Common to all living things is the almost exclusive use of hexoses and pentoses. Only rarely are monosaccharides found with a chain of more than nine C-atoms. It is also interesting that of the theoretically possible number of diastereoisomers, only very few compounds are utilised biologically. Presumably it is no accident that conformational (thermodynamic) relative stability seems of overriding importance. Thus, of the eight diastereoisomeric aldohexoses, only three, namely D-glucose, D-mannose, and galactose, the latter in the D- or L-configuration, appear in nature.

In Fig. 2.1 are represented the 22 (24 with the 2 enantiomers) monosaccharides which are utilised for the biosynthesis of most of the oligomeric and polymeric carbohydrates. It is noteworthy that all of the compounds that are able to form a pyranose ring contain, at the most, one axial, that is destabilising hydroxy group.

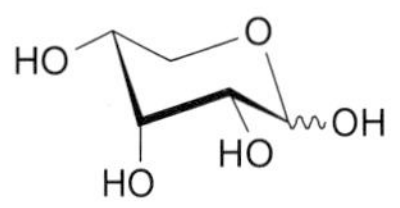

D-Ribose (Rib)

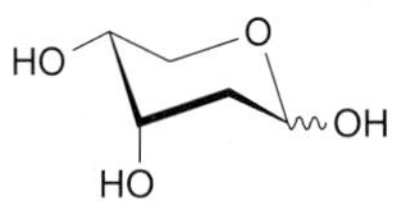

2-Deoxy-D-ribose (dRib)

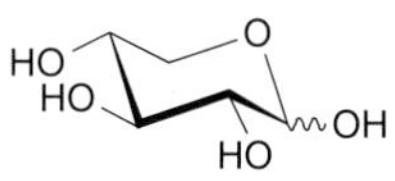

D-Xylose (Xyl)

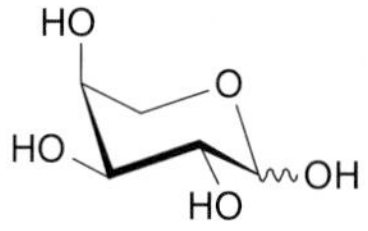

D- and L-Arabinose (Ara)

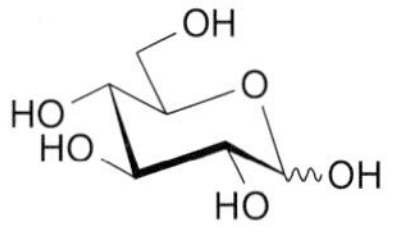

D-Glucose (Glc)

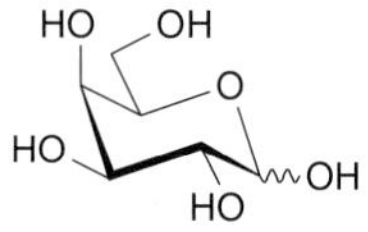

D- and L-Galactose (Gal)

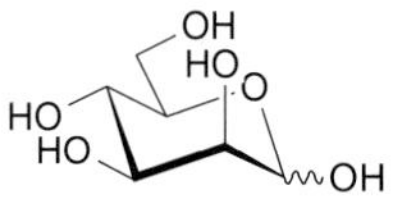

D-Mannose (Man)

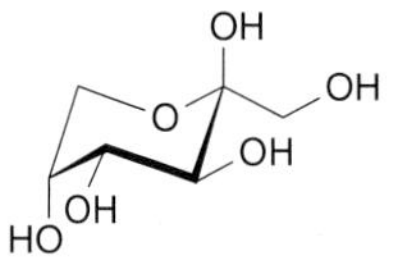

D-Fructose (Fru)

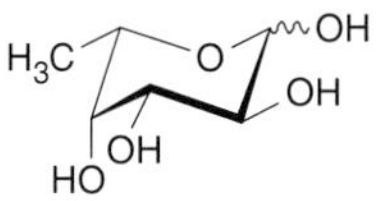

L-Fucose (Fuc)

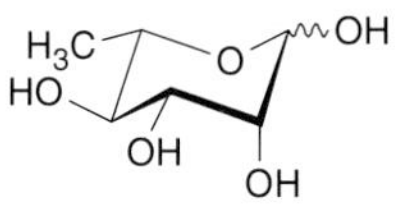

L-Rhamnose (Rha)

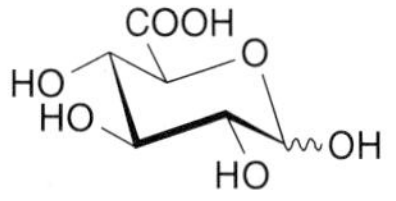

D-Glucuronic acid (GlcA)

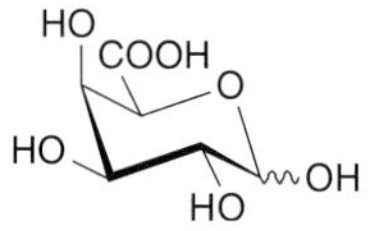

D-Galacturonic acid (GalA)

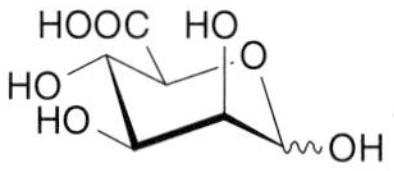

D-Mannuronic acid (ManA)

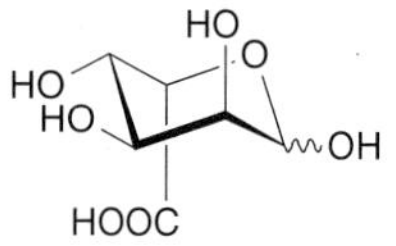

L-Guluronic acid (GulA)

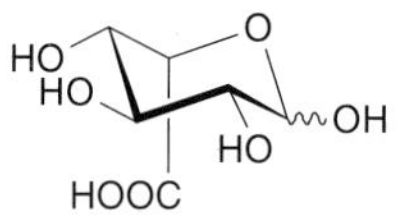

L-Iduronic acid (IdoA)

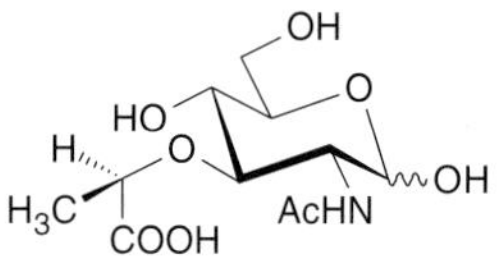

*N*-Acetylmuramic acid (MurAc)

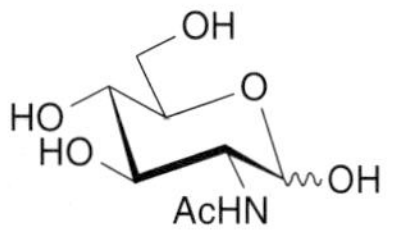

*N*-Acetyl-D-glucosamine (GlcNAc)

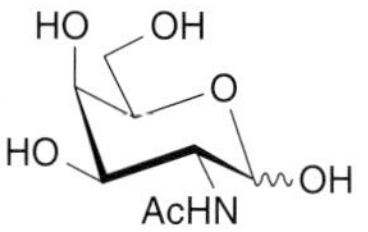

*N*-Acetyl-D-galactosamine (GalNAc)

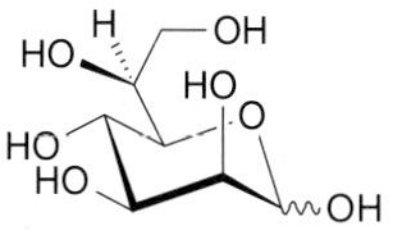

L-*Glycero*-D-*manno*-heptose (Hep)

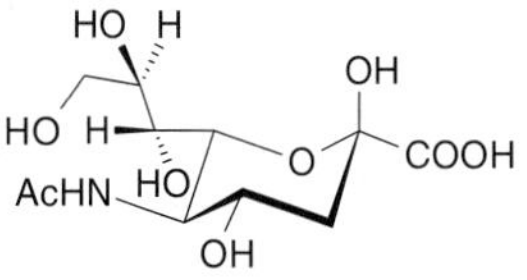

5-Acetamido-3,5-dideoxy-D-*glycero*-D-*galacto*-nonulosonic acid, *N*-acetylneuraminic acid (NeuAc)

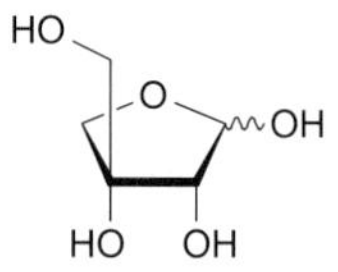

D-Apiose (Api)

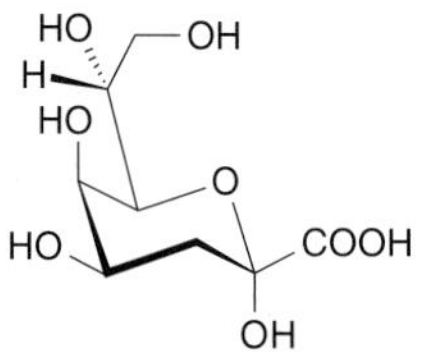

3-Deoxy-D-*manno*-octulosonic acid (Kdo)

Some naturally occuring monosaccharides arise through post-biosynthetic modification as in the the 5-epimerisation of D-glucuronic acid into L-iduronic acid or D-mannuronic acid into L-guluronic acid, or in the 3-*O*-alkylation of D-glucosamine with the 1-carboxyethyl residue to give muramic acid.

Other post-biosynthetic modifications such as alkylation, acylation (sulphation and phosphorylation included) may have biological importance, but shall, for the moment be excluded. A valid rule seems to be that an increasing development of the living being is accompanied by an increasing restriction to fewer monosaccharides as building units and further, that the relative quantiy of carbohydrate in the organism decreases. The human body consists of about 1% carbohydrate as compared to 15% protein and lipid. The rest is inorganic material (5%) and principally water.

It is tempting to speculate on the selection of certain monosaccharides for the formation of oligomeric and polymeric carbohydrates and to make a comparison with the 20 proteinogenic amino acids. This comparison is surely not permissible, since the functions of the biologically active proteins of strictly determined structure are not to be compared with those of the carbohydrates. While the amino acids are more or less evenly distributed in all living things and all proteins, many monosaccharides are found – here ribose and deoxyribose are particularly noteworthy – only in certain biopolymers and many are found exclusively in bacteria, in plants, or in animals.

Certainly, it can be assumed that the functional groups of the monosaccharides, especially the carboxy groups of glycuronic and glyculosonic acids, have important functions to fufil, such as the binding of metal ions, and hydroxy groups and acetamido groups are especially suited to form hydrogen bonds, and thereby to provide high hydrophilicity. Also, the conformational rigidity of the pyranosyl ring and the possibility of variation in the anomeric configuration, by which the conformation of the condensation products is strongly influenced, is surely of importance . This is well illustrated by the physicochemical and biological differences between cellulose and amylose.

The systematic *nomenclature* of *monosaccharides* is very complex and somewhat convoluted[4] and, transferred to the larger oligosaccharides or even heteropolysaccharides, is not really practicable. For this reason, use is generally made today of a shorthand nomenclature for the monosaccharides which occur regularly in natural carbohydrates.[5] The *shorthand* nomenclature contains the following information:

1. The type of monosaccharide, described by the first three letters. For example, Ara = arabinose; Gal = galactose, etc. This rule is broken only for glucose (= Glc), because the abbreviation Glu is reserved for the amino acid glutamic acid. Sugar alcohols are characterised by *ol*: ribitol = Rib-*ol*. Glycerol, for the aforementioned reason, cannot be denoted by "Gly" (glycine) but is designated by *Gro*.
2. The ring size *p* for pyranose and *f* for furanose, as the first suffix printed in *italics*.
3. The configuration at the anomeric centre $\alpha$ and $\beta$ as well as the absolute configuration D and L as a prefix.
4. For a 2-amino-2-deoxy group and its acetylated form, a second suffix N or NAc, respectively, after the ring symbol.
5. For a carboxy group, A (acid) as the last suffix.
6. The linkage between two monosaccharides in the case oligosaccharides or polysaccharides, according to the general nomenclature, is given with the position numbers and a direction arrow in brackets. If monosacharides are linked through their anomeric centres, then a double-headed arrow is used (sucrose, for example, $\alpha$-D-Glc*p*-(1$\leftrightarrow$2)-$\beta$-D-Fru*f*).
7. Branching of the carbon chain in monosaccharides and alkylation or acylation. Examples are the ethyl ester of D-glucuronic acid in the pyranose form D-Glc*p*A6Et, the 4-*O*-sulphate of $\beta$-D-galactopyranose $\beta$-D-Gal*p*4$SO_3^-$, and the branched-chain sugar 2-C-methyl-D-xylose D-Xyl2CMe.

Special cases of monosaccharides are *Hep* (L-*glycero*-D-*manno*-heptose), the ulosonic acid *Kdo* (3-deoxy-D-*manno*-oct-2-ulosonic acid) and *NeuAc* (5-acetamido–3,5-dideoxy-D-*glycero*-D-*galacto*-non-2-ulosonic acid) as well as muramic acid *MurAc* (2-acetamido-2-deoxy-3-*O*-[(*R*)-1-carboxyethyl]-D-glucose), resulting from the alkylation of GlcNAc.

⇐ **Fig. 2.1** The 24 most important monosaccharides occurring in oligosaccharides, polysaccharides, and other biopolymers with the generally used abbreviations in brackets (some unusual monosaccharides in biopolymers of microorganisms are not considered). Only L-Ara and D-Gal are shown. The structure shown for NeuAc is the usual one. Acetylation or general acylation on other positions of *Neu* are also found. In such cases the position number is given. NeuAc is correctly designated Neu-5-Ac. Substituted neuraminic acids are designated also as sialic acids (Sia).

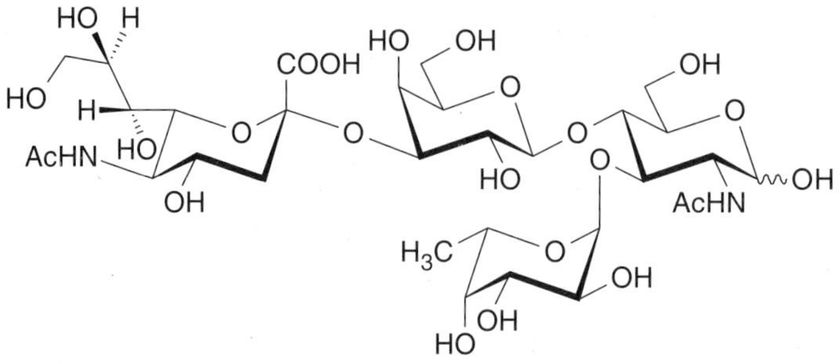

**Fig. 2.2** The structural formula of the carbohydrate fragment comprising four monosaccharides which is responsible for the attachment of leucocytes onto endothelial cells.

Suggestions have been made[6] for the designation of conjugates between carbohydrates on the one hand and proteins and peptides on the other, as well as between carbohydrates and lipids[7].

The tetrasaccharide (sialyl Lewis x), shown as a structural formula in Fig. 2.2, is a ligand for the receptor mediated binding of leucocytes onto endothelial cells and is represented in shorthand notation in the following manner:

α-D-Neu*p*Ac(2→3)*β*-D-Gal*p*(1→4)[α-L-Fuc*p*(1→3)]-D-Glc*p*NAc

or somewhat more clearly but in more space demanding manner with the branching on a separate line:

```
α-D-NeupAc(2→3)β-D-Galp(1→4)-D-GlcpNAc
                               3
                               ↑
                               1
                            α-L-Fucp
```

It is usual, as with monosaccharides, to place the reducing end (synonymous with the aglyconic end) on the right hand side. Often the shorthand method is perceived as still too involved and awkward so that further reductions, for example leaving out the hyphens and the ring symbols, replacement of the bond arrows with a comma or nothing at all, and leaving out the anomeric position have become usual. In this form the above shown tetrasaccharide appears as follows:

```
NeuAcα3Galβ4GlcNAc
           |
         Fucα3
```

Formulae are simplified enormously through the use of geometrical signs. Since there is no unanimity in the international literature over such symbols[8,9], we wish to propose, for the most important monosaccharides with pyranose rings which occur in higher organisms, the geometrical representation shown in Fig. 2.3.

With these nine monosaccharides nearly all oligosaccharides and polysaccharides occuring in mammalian organisms can be described. In the symbol is placed the anomeric configuration $\alpha$ or $\beta$ and the glycosylated position in the next monosaccharide forming the aglycone. The branched tetrasaccharide named above would, accordingly, be adequately represented by the formula shown in Fig. 2.4.

Much confusion arising in the past is associated with the nomenclature of carbohydrates linked to proteins, peptides, or lipids. Following a suggestion of the *IUPAC-IUB Joint Commission on Biochemical Nomenclature (JCBN)*, such compounds are designated as glycoconjugates (Fig. 2.5).

Glycoproteins, glycopeptides, peptidoglycans, glycolipids and liposaccharides are subsiduary groups of the glycoconjugates.

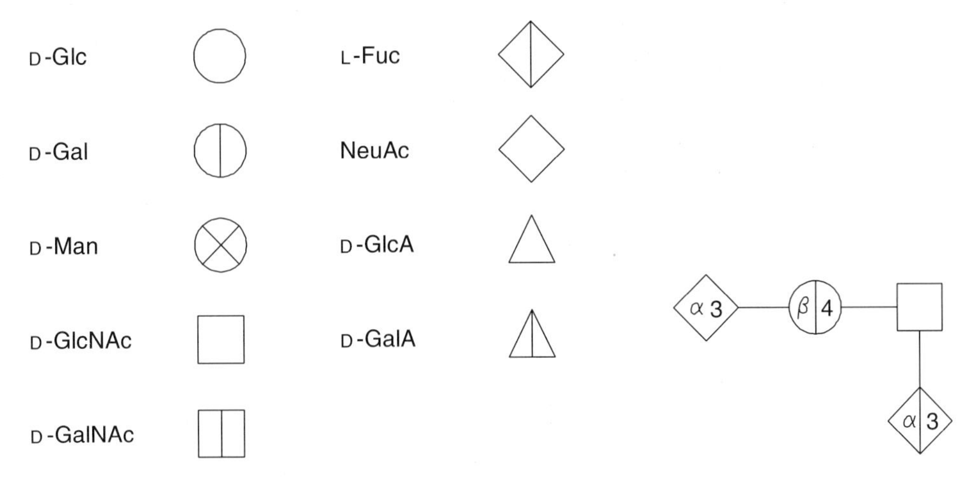


**Fig. 2.3** Geometric symbols for nine monosaccharides distributed in higher organisms. It is to be assumed that all compounds are pyranosides.

**Fig. 2.4** The tetrasaccharide sialyl Lewis X described with the geometric symbols.

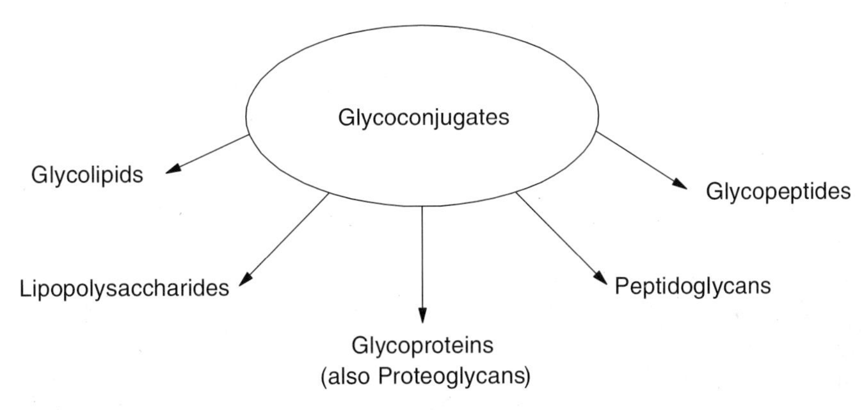


**Fig. 2.5** All compounds in which carbohydrate components are covalently linked with other types of natural products are collectively named under the general term glycoconjugates.

■ **Glycoproteins** contain carbohydrates covalently and glycosidically bound to protein. The carbohydrates can be monosaccharides, oligosaccharides, or polysaccharides and also their derivatives (eg acetate, phosphate, sulphate). The carbohydrate part can consist of one, several, or many residues. Artificial glycosylated proteins, as model compounds for experimental purposes, are called *neoglycoproteins*[10].

*Proteoglycans* are a sub-class of the glycoproteins, in which the carbohydrate residues are polysaccharides containing amino sugars. These polysaccharides are named glycosaminoglycans.

In many glycoproteins (from serum or membranes) the carbohydrates are found as oligosaccharides, either linear or branched. The highly branched oligosaccharides can be composed of up to 20 monosaccharide residues.

Glycoproteins are also known which carry only monosaccharides or disaccharides. To such a class of compounds belong, for example, collagen or the "anti-freeze"-glycoproteins of arctic fish. Many glycoproteins carry oligosaccharides which are built from repeating units. Thus, for example, chains frequently comprise poly-*N*-acetyllactosamine. A proteoglycan can carry up to 100 glycosaminoglycan chains. The glycosaminoglycans are always linear. They contain up to 200 disaccharide repeat units.

**Glycopeptides** contain carbohydrates, mostly oligosaccharides, which are bound to oligopeptides. The linkage is of the glycosidic type through O, S or N. The amino acids of the peptide have the D or L configuration. They are not natural components of living organisms, but are either degradation products or chemically synthesised compounds.

**Peptidoglycans** are glycosaminoglycans which are cross-linked with peptides.

**Glycolipids** are glycosylated lipids in which the lipid moiety is diacylglycerol or ceramide. The carbohydrate part can be a monosaccharide, disaccharide, or a complex branched oligosaccharide. The nomenclature is here, in part, very confusing and certainly needs simplification and standardisation. The abbreviations used, even for simple glycolipids, are hardly intelligible for non-specialists:

| Structure | Trivial designation | Abbreviation |
|---|---|---|
| Galα4Galβ4GlcβCer | Globotriaose | $GbOse_3$ |
| Galβ3GlcNAcβ3Galβ4GlcβCer | Lactotetraose | $LcOse_4$ |

The question of nomenclature is the subject of a proposal to the Commission for Biochemistry (CBN). It can be imagined that the descriptions of higher molecular weight gangliosides, whose carbohydrate portion by definition always contains neuraminic acid derivatives, are in general still harder to interpret. Two short-nomenclature methods for gangliosides by H. Wiegandt and L. Svennerholm are used by specialists in the field:

| Structure | Nomenclature after | |
|---|---|---|
| | L. Svennerholm | H. Wiegandt |
| **NeuAc**α3Galβ4GlcβCer | $G_{M3}$ | $G_{Lac1}$ |
| **NeuAc**α3Galβ3GalNAcβ4Galβ4GlcβCer<br>\|<br>**NeuAc**α8**NeuAc**α3 | $G_{T1}$ | $G_{NT3}$ |

The abbreviation suggested by Svennerholm seems to be the most popular, despite its peculiar logic. The index M, D, or T gives the number of the neuraminic acid residues. The number index indicates 5 minus the number of neutral sugars. Thus, the first example has one neuraminic acid residue (M) and two neutral sugars (5 – 2 = 3). This extraordinary simplification is only possible because the nuclear structure of the gangliosides is constant with respect to the bond positions and anomeric configurations. According to Wiegandt the index Lac indicates the lactose unit, NT the lacto*N*-tetraose, to which one (Index 1) or three (Index 3) neuraminic acid molecules, respectively, are bound.

The short symbolism is intended only for relatively simple gangliosides and is not generally applicable. There are glycosphingolipids with very many complex carbohydrate structures which are not accommodated by this simple scheme.

**Lipopolysaccharides** are the last large group of glycoconjugates. Systemisation has not yet progressed far with this class of compounds. It is generally assumed that the term lipopolysaccharide includes all polymeric carbohydrate derivatives which are anchored in a cell membrane through a lipid, for example *lipid A* or *phosphatidylinositol*.

## 2.1.2 Polysaccharides, protection, and mechanical stability

The cells of microorganisms and higher organisms are surrounded by polymeric compounds which confer mechanical stability and, with that also, protection. There are fundamental differences in these extra-cellular materials between bacteria, plants, and animals (Fig. 2.6).

While the bacteria cell wall – outside the cytoplasmic membrane – which maintains contact with the most hostile outer world is comparatively thick, (up to 30% of the dry weight of the organism) and in this way offers protection, one can not speak of an individual cell wall in the case of an animal cell.

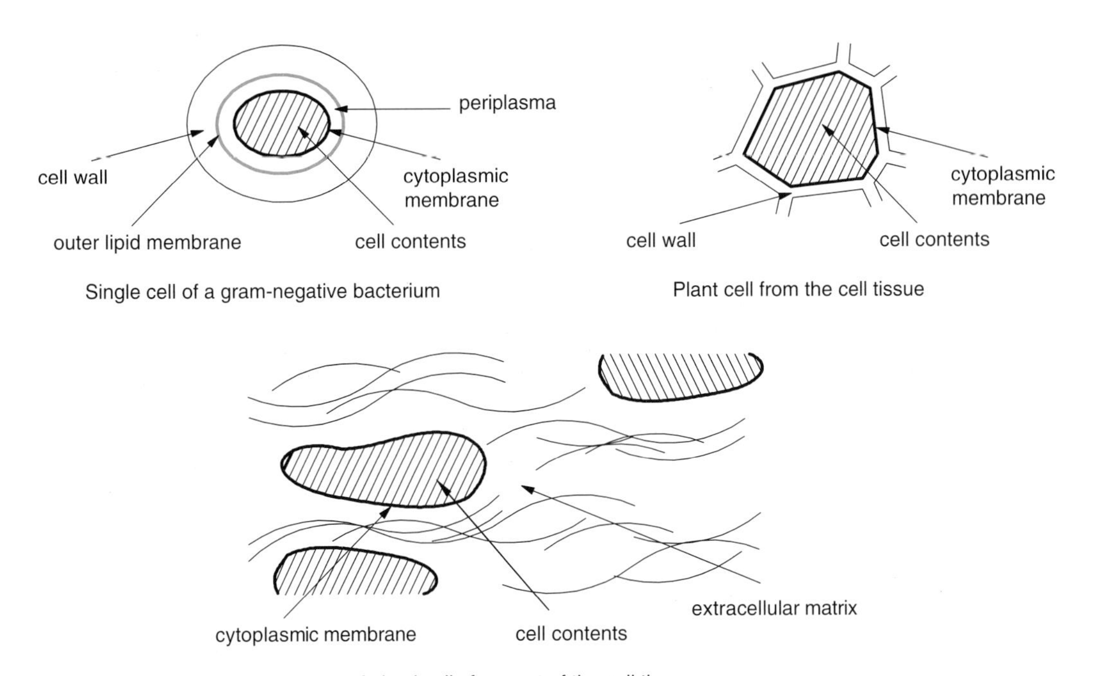


**Fig. 2.6** The construction principles of the cell exteriors are different according to the biological requirements.

Here, the cells are embedded as a unit in the extra-cellular matrix which can be very different chemically and functionally, according to the cell type. The cell wall of a mature plant cell is especially stable, mechanically. It gives not only protection to the single cell but also carries the whole organism. Polysaccharides in particular play a dominant role in the structure of the plant cell wall. Also, bacterial cell walls consist for the greater part of carbohydrates. On the other hand, the extra cellular matrix of animal cells consists, for the largest part, of protein which, depending on function, is supported by associated polysaccharide or covalently attached carbohydrate units. Fundamentally, the principle of *ferroconcrete* or *fibreglass* applies for the support and protection of all cells: linearly ordered thread-like polymers are embedded in a more or less amorphous matrix of other, mostly short-chain polymers. In this way, both mechanical tensile stability as well as stability to compression is guaranteed.

### 2.1.2.1 The bacterial cell wall

Bacteria as well as other single cell organisms such as algae, yeasts and fungi contain in the cytoplasm dissolved ions and low molecular weight compounds in a concentration which greatly exceeds that of the outside medium. The resultant osmotic pressure of several atmospheres would burst the cell asunder were it not for containment by the especially stable bacterial cell wall. There are two different types of cell wall (Fig. 2.7) by which bacteria have been divided since 1884 into two classes, the gram-positive and the gram-negative bacteria.

The classification is based on a colour test devised by C. Gram[11]. If bacteria of both types are treated with certain reagents and dyes, then they are all coloured an intense violet after washing out the excess dyes. A further treatment with ethanol washes out the colour from one type (gram-negative) but the other type (gram-positive) remains coloured. Even today it is not exactly settled as to what the different colour-fastness depends on. With very few exceptions[12], bacteria with an outer lipid-containing membrane are, in general, gram-negative and those without are gram-positive.

■ **Peptidoglycan murein**. As shown in Fig. 2.7, both bacterial types share a similar peptidoglycan which, in reference to its solidity, is named murein (latin murus = wall). In the case of gram-positive

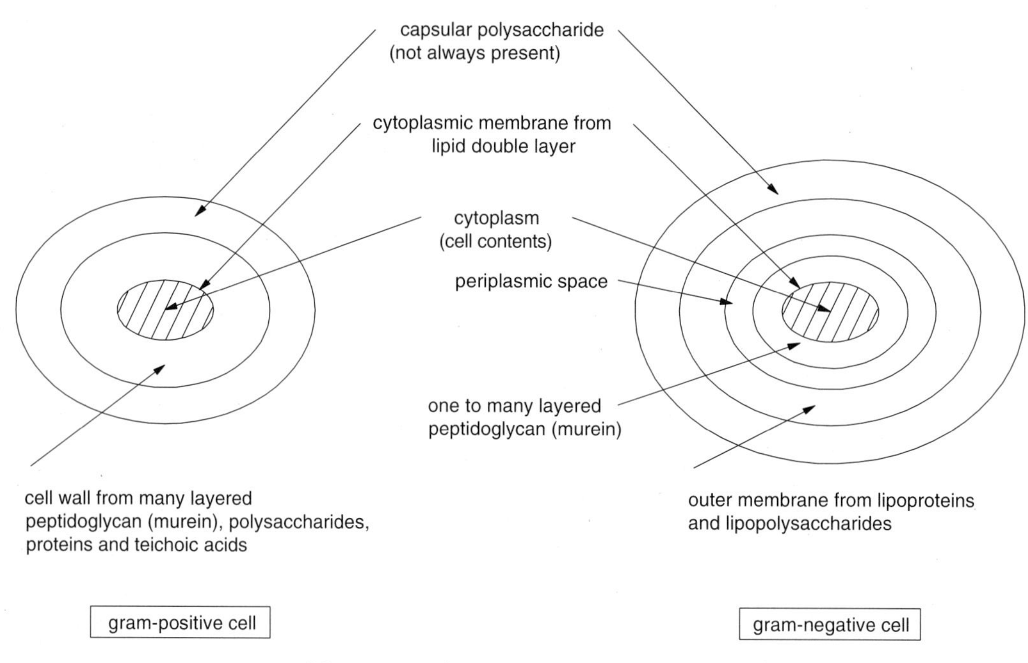


**Fig. 2.7** Bacteria possess two different outer layers, which allow them to be divided into gram-negative and gram-positive cells.

bacteria the murein constitutes 50% of the cell wall mass, but with the gram-negative bacteria it is only about 10%. The glycan chains are similar for all mureins with some rare exceptions, but there are more variations in the case of the cross-linked peptides in which *unnatural* D-amino acids are also found. The latter are valuable as a defence mechanism of the microrganism against breakdown by peptidases. The α,ω-diamino acid, which mostly occupies the third position (Fig. 2.8) and which is necessary for cross-linking, varies frequently. Instead of L-lysine one of the alternative amino acids L-ornithine, L-hydroxylysine, or *meso*-diaminopimelic acid (2,6-diaminoheptanedioic acid) is found here.

The general building principle is very useful and is suited to the task of the bacterial cell wall. Murein builds a stable network (sacculus) in which the cell finds support. The cross-linked network can be effected not only in one plane, that is two-dimensionally, and so form only one layer, as in the case of most of the gram-negative bacteria, but also between different layers, giving rise to the thick peptidoglycan outer coat of the gram-positive bacteria.

The glycan murein consists of repeat units of a disaccharide containing GlcNAc and MurAc (Fig. 2.9). All of the monosaccharides are $\beta$-(1→4)-linked. Except for the lactyl residue in MurAc, the glycan, possesses, with few exceptions, the mechanically very stable linear structure of chitin (see section 2.1.2.2, p. 80) and thereby also its capability to form stable hydrogen bonds between different chains.

The building principle of the murein-glycans has obviously stood the test of time during the evolution of the bacteria and has been conserved until the present day. Insignificant and very rare divergences are found only in the MurAc-units. The amino group can be free or acylated with glycolic acid, or can form an internal amide with the neighbouring lactyl residue. Occasionally the *muramic acid* can possess the D-*manno* instead of the D-*gluco* configuration. The lactic acid residue in MurAc, attached by a stable ether linkage, forms, so to speak, the bridgehead for the covalent connection of two glycan chains via an oligopeptide. The structure of the muropeptides from *Staphylococcus aureus* is reproduced here as an example (Fig. 2.10).

Numerous antibiotics are able to hinder specifically the biosynthesis of the mureins (see section 2.3.8.1, p. 207) and in this way to prevent bacterial growth. The best known are the penicillins, which prevent the last step of cross-linking in formation of the network, the joining of two peptide ends through transpeptidisation. Since peptidoglycans only occur in bacteria, inhibitors of their biosynthesis

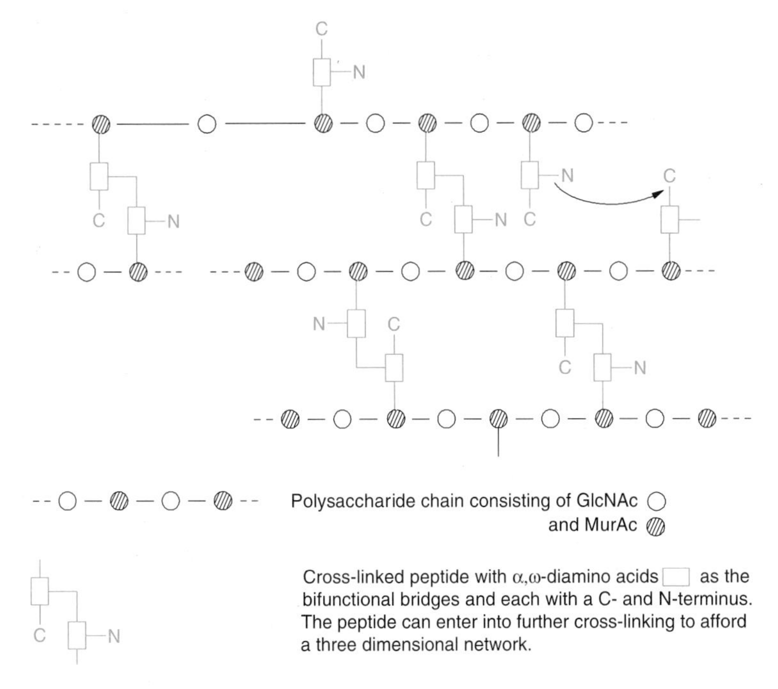

Polysaccharide chain consisting of GlcNAc
and MurAc
Cross-linked peptide with α,ω-diamino acids as the bifunctional bridges and each with a C- and N-terminus. The peptide can enter into further cross-linking to afford a three dimensional network.


**Fig. 2.8** Schematic representation of the synthesis of a single layered *peptidoglycan*.

**Fig. 2.9** The repeat unit of the glycan strand in *murein*.

are not toxic to other organisms. It is this fact that leads to the enormous success of the $\beta$-lactam antibiotics[13]. The history of investigation of the peptidoglycans is not just connected with that of the penicillins but also with that of the enzyme lysozyme, one of the chief weapons of animal organisms against bacterial infection (see section 2.3.4, p. 186).

■ **Teichoic acids**[14] (Greek teichos = wall) are typical for gram-positive bacteria and they are not found in gram-negative bacteria. Phosphoric acid is the characteristic structural element of these higher polymers. Low molecular weight carbohydrate fractions are connected to each other by esterification with phosphoric acid. In this sense teichoic acids are not polysaccharides but polyesters of phosphoric acid with carbohydrates as the diol components and thereby in their basic structure are to be compared to the nucleic acids[15]. It is astonishing that the teichoic acids, which occur in relatively large amounts, were only first discovered in 1950, through investigations by J. Baddiley and his research group[16]. Most frequently diesters of the alcohols glycerol and ribitol (Fig. 2.11) are to be found, but monosaccharides or oligosaccharides can be incorporated also at regular intervals, according to the principal of the re-

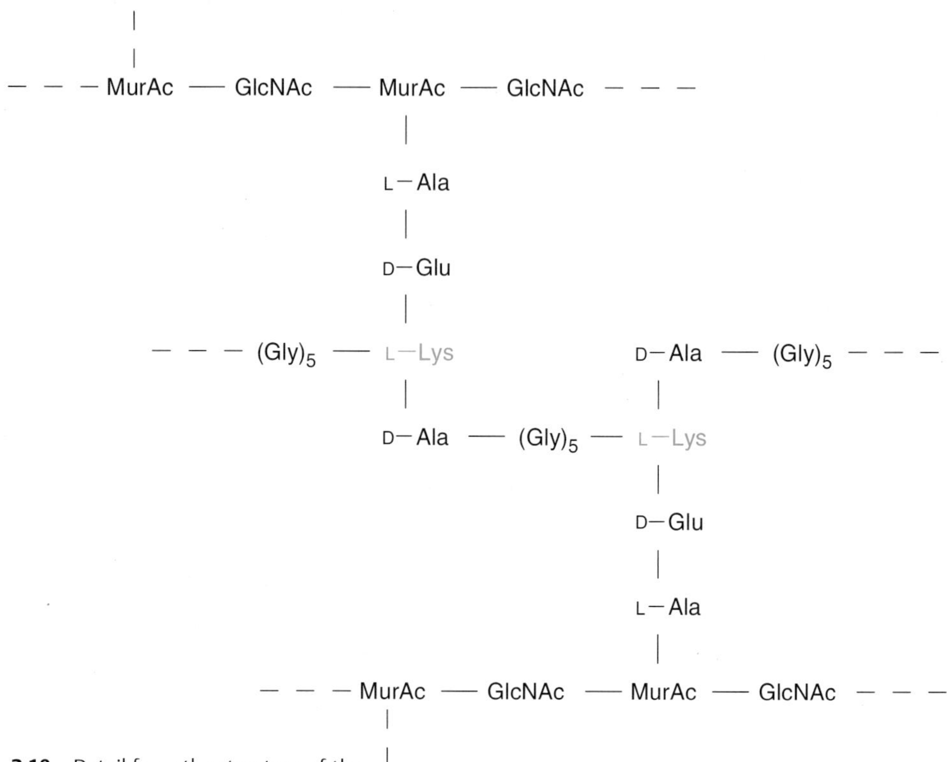


**Fig. 2.10** Detail from the structure of the peptidoglycan from *Staphylococcus aureus*.

peating unit (Fig. 2.11). As structural variation, hydroxy groups of the carbohydrate component in the chain are sometimes glycosylated or acylated with amino acids.

If gram-positive bacteria are allowed to grow in a phosphate deficient medium, then the available phosphate is first used for the biosynthesis of the nucleic acids. If it is not sufficient for the biosynthesis of teichoic acids then *teichuronic acids* are formed instead. In the teichuronic acids, the negative charge, which is evidently essential for the organism, is held by a glycuronic acid. However, there are also gram-positive bacteria, for example *Micrococcus lysodeicticus*, which, in principle, only synthesise teichuronic acids. Apparently, one of the functions of the teichoic acids and teichuronic acids is to concentrate the divalent cations, above all $Mg^{2+}$ directly outside the plasma membrane. The teichoic acids of the cell wall of gram-positive bacteria should not be confused with the *lipoteichoic acids* (Fig. 2.12), which are embedded in the plasma membrane by their lipophilic residue. Teichoic acids and lipoteichoic acids participate in the regulation of cell growth (degradation and synthesis of the peptidoglycans).

Teichoic acids and teichuronic acids in the cell wall are linked covalently with the peptidoglycan. The position of linkage always appears to be in the 6-position of a MurAc residue which, as is shown in Fig. 2.13, is esterified with a glycosyl phosphate end-group of the teichoic acid as the coupling unit.

■ **Lipopolysaccharides** are characteristic of the cell walls of gram-negative bacteria. While the specific antigen action typical of all bacteria depends, in the case of gram-positive germs, upon the oligosaccharide side-chain on teichoic acids and simple extractable polysaccharides in the cell wall, the lipopolysaccharides of gram-negative germs are plainly the antigens[17]. This is one of the reasons why the biochemistry of this class of compounds, above all in salmonella and *E. coli*, has been intensively investigated for many years. As Fig. 2.7 (p. 74) shows, the lipopolysaccharide, with protein and the usual membrane phospholipids, forms a second so-called outer membrane next to the cytoplasmic membrane. The inner region of this membrane, that is to say the lipid part and protein, is similar from

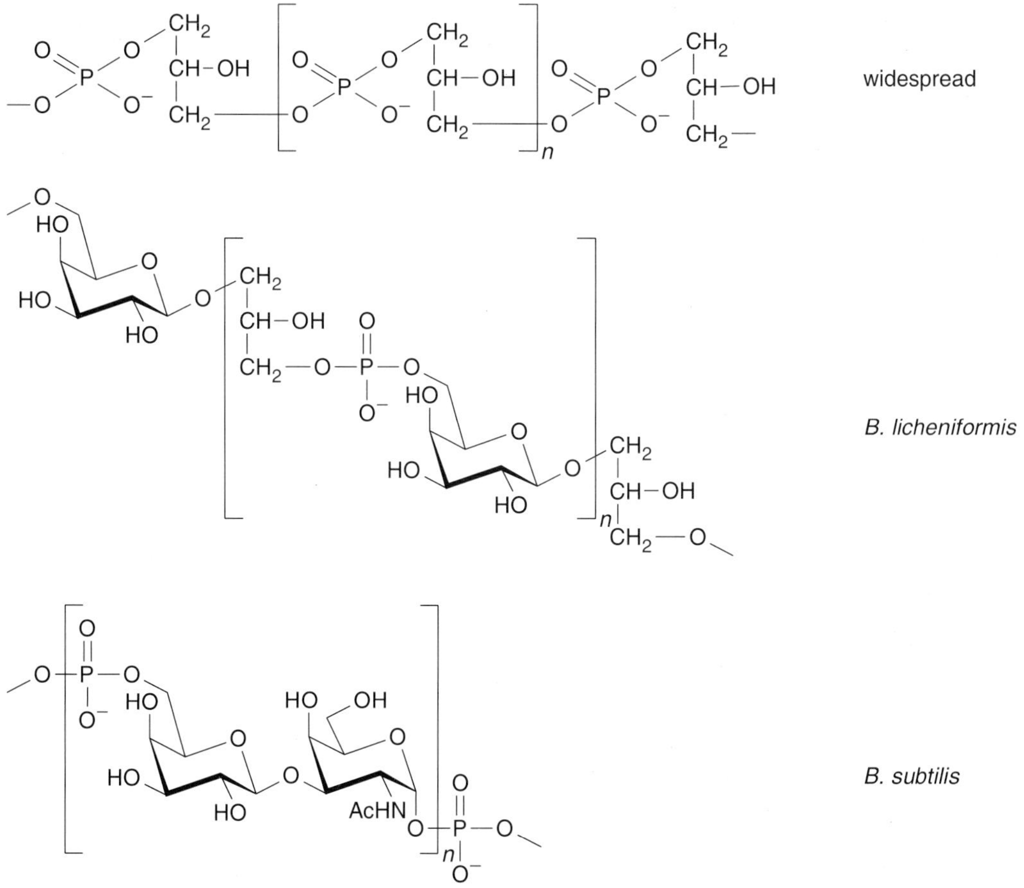

widespread

*B. licheniformis*

*B. subtilis*

**Fig. 2.11** Different structure types of *teichoic acids* of gram-positive bacteria.

organism to organism. On the other hand, the outer region, that is to say the polysaccharide chains, are of greater variety and form the so-called O-antigens of the gram-negative bacteria. In the case of salmonella, about 1000 different serologically ascertainable types can be distinguished. The possibilities in variation appear to be unlimited. The range of the structural variations is brought about by change of the type, number, and linkages of the monosaccharides in the repeating units but also through the incorporation of unusual monosaccharides which otherwise do not appear in the living world. Over 100 different monosaccharides have been identified already in bacterial polysaccharides[3]. One is tempted to look for a biological purpose in this unusual breadth of variation. A possible explanation is certainly the attempt by the microorganism to escape recognition by a hostile-minded environment with the help of camouflage of its surface. The lipopolysaccharides all have a very similar basic structure (Fig. 2.14).[18]

Lipid A (in contrast to the usual phospholipids of the membrane, which is denoted as lipid B) is the anchor with which the lipopolysaccharide is fastened in the outer membrane through hydrophobic interaction. Lipid A, the endotoxin, is responsible for the pyrogenic action of gram-negative bacteria[19]. Even in a concentration of 1 ng/kg the lipopolysaccharide produces high fever. The framework of lipid A is formed from two $\beta(1\rightarrow6)$-linked GlcN-residues, of which all of the functional groups, except for the 1, 3', and 4'-positions, are acylated with fatty acicds of $C_{12}$- $C_{16}$ chain length. The 3-hydroxymyristic acid ($C_{14}$), connected through nitrogen, can in turn again be acylated (Fig. 2.15).

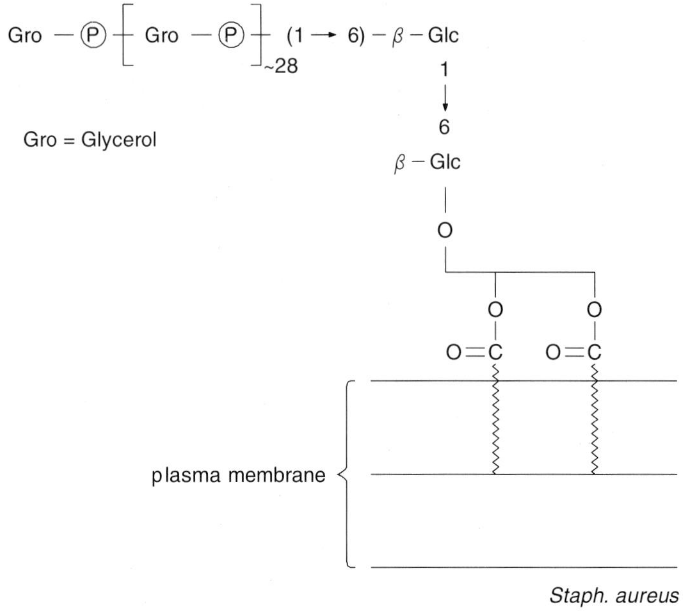
Gro — (P) [ Gro — (P) ] ~28 (1 → 6) — β — Glc
1
6
β — Glc
O
O
O
O=C
O=C
Gro = Glycerol
plasma membrane
Staph. aureus

**Fig. 2.12** *Lipoteichoic acids* are linked with a phosphate diester bond onto a glycolipid. In the example shown the link is to the 6-position of an *isocellobiosyl diglyceride*.

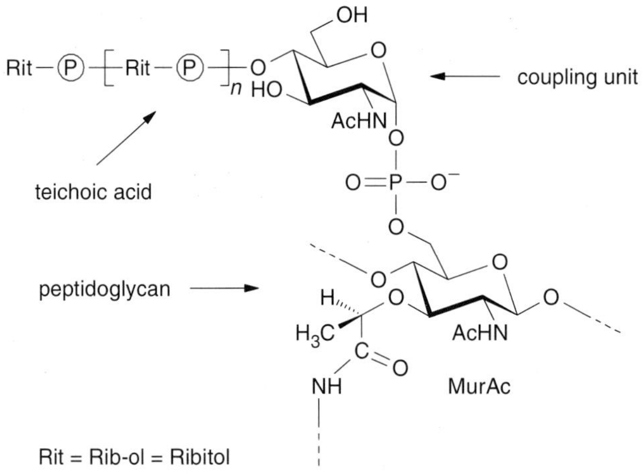
Rit—(P)[Rit—(P)]n O
OH
HO
AcHN
O
O=P—O⁻
coupling unit
teichoic acid
peptidoglycan
H
H3C
C=O
NH
AcHN
MurAc
Rit = Rib-ol = Ribitol

**Fig. 2.13** Teichoic acids of the cell wall of gram-positive bacteria are covalently linked with MurAc of the peptidoglycan.

The so-called core structure of the lipopolysaccharide is substituted at the 3'-position with Kdo as a glyconic end group. The very acid labile 3-deoxy-ketosyl bond facilitates considerably the hydrolytic separation of the lipid A portion from the rest of the polysaccharide.

The core structure of the bacteria examined so far is very similar to that in lipid A (Fig. 2.14, p. 79). The *O*-specific polysaccharide is joined onto the core. A great number of structures, above all from *E. coli* and salmonella, are now elucidated. Each variant of an original form possesses a different *O*-specific polysaccharide. The repeat units are in general very short, that is seldom longer than five mono-

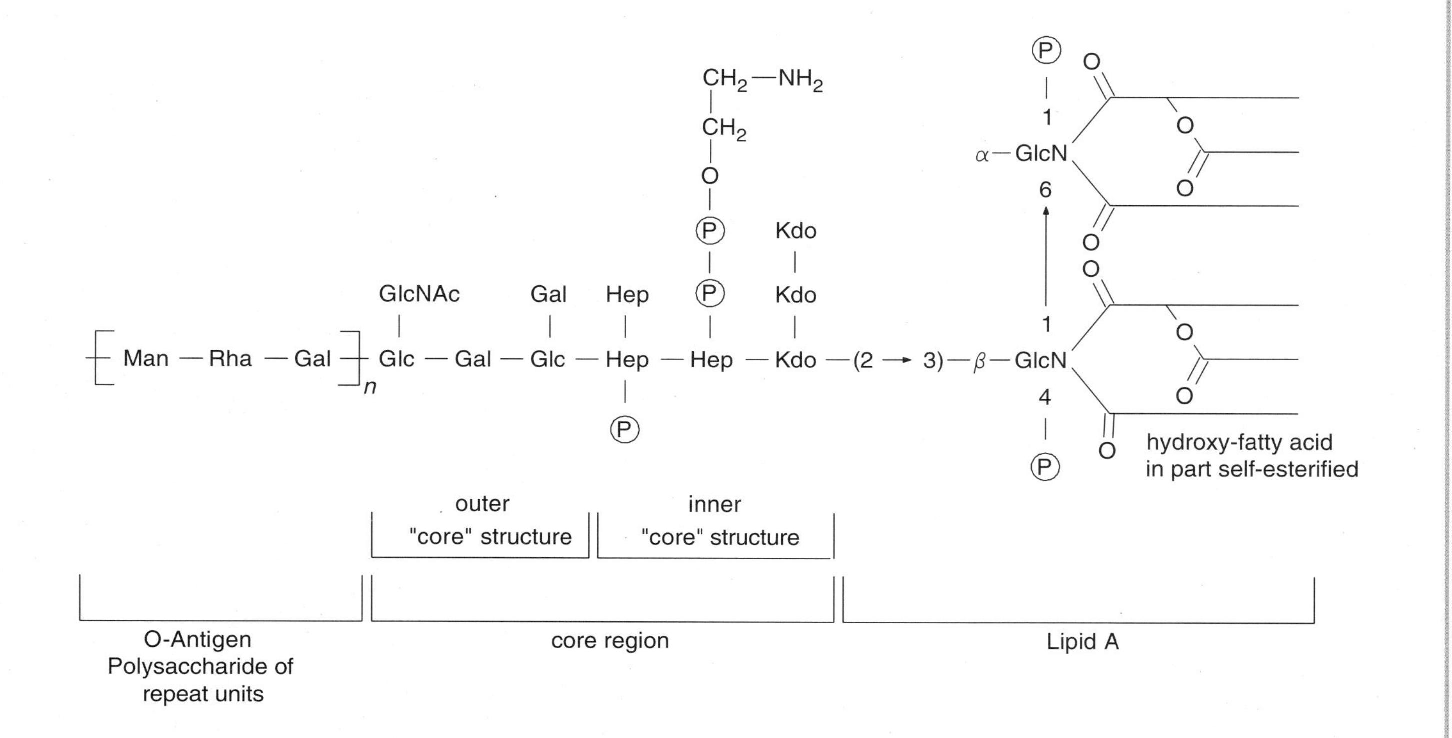


**Fig. 2.14** A typical *lipopolysaccharide* from the cell wall of a gram-negative bacterium.

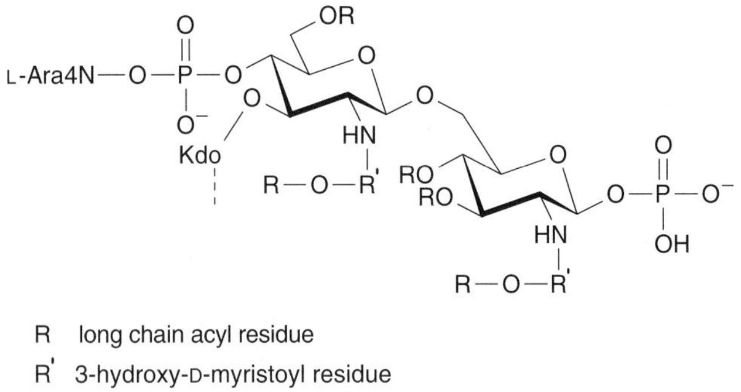

R long chain acyl residue
R' 3-hydroxy-D-myristoyl residue

Fig. 2.15 Lipid A form Salmonella.

saccharide units. A decasaccharide, up till now the longest proven repeating unit discovered, is surely a notable exception[20].

**Specific extracellular polysaccharides and capsular polysaccharides.** The variety of the hitherto described bacterial polysaccharides are further amplified through such structures which either, in the case of gram-negative bacteria, are not anchored in the outer membrane or, in the case of gram-positive bacteria, are not bound covalently to the teichoic acid. The polysaccharides appear as voluminous capsules (Fig. 2.7, p. 74) round the compact cell wall and are, if present, also responsible for the immunological properties of the organism. In respect to their importance to the organism, similarities can be drawn with the lipopolysaccharides of gram-negative bacteria. The variety can act as a kind of camouflage which serves as a protection for the organism.

Naturally, this is not the case for dextrans and fructans which are produced in great quantities by *Leuconostoc* and *Streptococcus* types, when a suitable source of low molecular weight carbohydrate is available, such as sucrose. It can be imagined that the organism makes these relatively insoluble extracellular polysaccharides in order to have a surrounding matrix for undisturbed growth, for example in dental plaque.

### 2.1.2.2 Cell walls of fungi and yeasts

The fungi (yeasts and lichen are included), as eukaryotes, occupy a middle position between the plant and animal kingdoms. Their presence is inconspicuous, yet not of inferior biological importance. They are the means by which the enormous amount of plant biomass, after dying, is returned again into the cycle of organic compounds, along also with the minerals, rather than forming a permanent growing rubbish dump covering the planet.

The cell walls of the fungi, as various as the appearances of the organism itself, are constructed very differently. This fact leads to the use of the cell wall polysaccharides, which form 80–90% of the cell wall mass, for taxonomic classification[21]. Also, the fungi evidently make use of the tried principle of imbedding fibrilous homopolysaccharides in a matrix of amorphous heteropolysaccharides. Chemically different categories form, for example, the following combinations:

| | |
|---|---|
| cellulose – mannan | cellulose – glycogen |
| mannan – glucan | chitin – mannan |
| chitin – glucan | chitin – chitosan* |
| poly-GalNAc – galactan | |

* de-*N*-acetylated chitin

Under the partly inaccurately and superficially described cell wall polysaccharides of the matrix, $\beta$-D-glucopyranans with many different types of linkage and of various degrees of branching, as well as the corresponding $\alpha$- and $\beta$-mannopyranans, appear to play the major role. It should be emphasised particularly that the mannans, and probably also the glucans, are present as glycoproteins[22]. In this context,

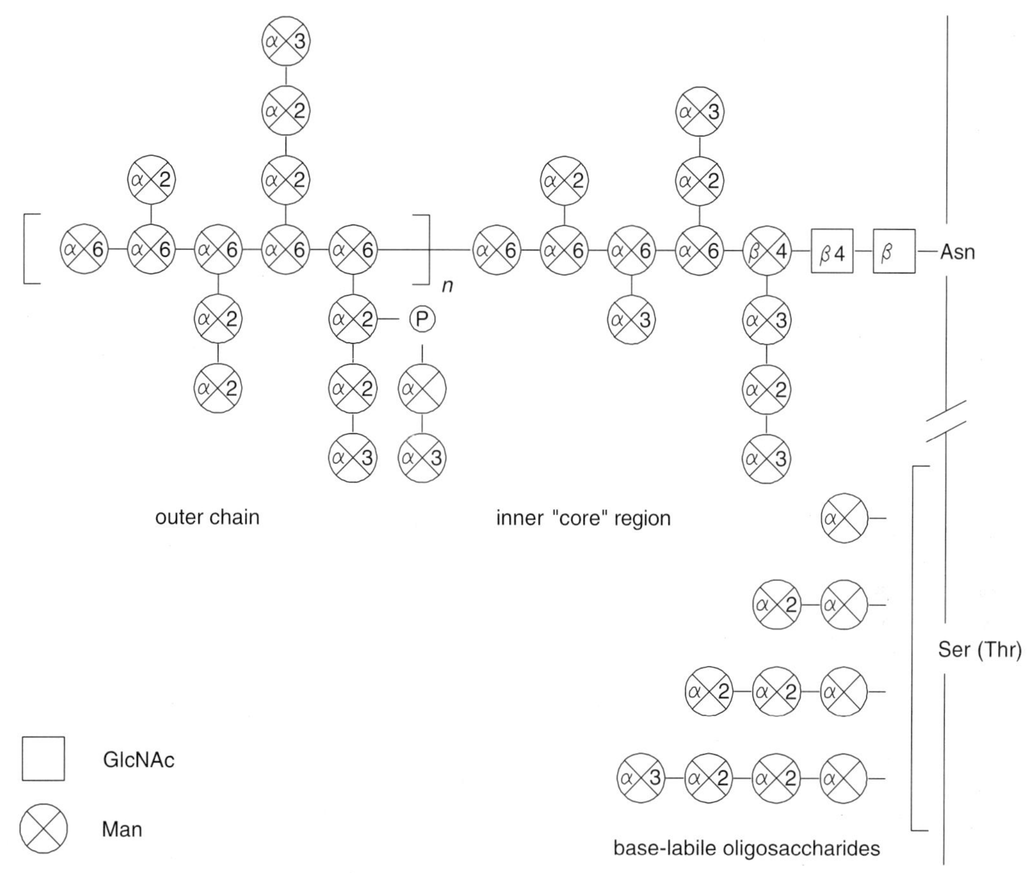


**Fig. 2.16** The *mannan* of *Saccharomyces* is a glycoprotein. The structure presented here is that given in the article by T. Nakajima and C. E. Ballou, *J. Biol. Chem.*, **1974**, *249*, 7685. There is some disagreement with the known high-mannose core structure accepted today.

the mannan from *Saccharomyces cerevisiae* has been well investigated. It is bound to the protein by an *N*-glycosidic link from the *core region*. On a long chain are fixed many shorter side branches (Fig. 2.16). Additionally there are shorter (up to tetrasaccharide), *O*-glycosidically linked manno-oligosaccharides distributed over the whole protein. The linkage of these chains is mostly $\alpha(1{\rightarrow}2)$.

The variation in structure of the cell wall polysaccharides of fungi is emphasised above all by the appearance of different fibrous polysaccharides. The unusual occurrence together of cellulose and chitin in *Ceratocystis ulmi*[23] appears to be quite astounding.

Chitin, like cellulose, is a typical thread-forming polysaccharide. In the case of fungi, the combination of chitin with all possible glucans as matrix polysaccharides means that they are the building material most frequently encountered for cell walls. With the use of chitin as a fibre-building component, the possibility arises of additional interaction through hydrogen bonding involving the amide groups. This is clearly shown in the detailed structure of α-chitin represented in Fig. 2.17[24].

It now seems certain that chitin in its natural environment occurs as a proteoglycan and that non-covalent interaction with structural proteins – for which the *N*-acetyl groups are especially suited – plays an important role in the construction of building elements of the cell wall in fungi. Unfortunately, it must be said that modern, accurate research results on the construction of the cell walls in fungi are very rare. This may be because of the difficulty of isolating chitin in the native state. Even so, it is certain that the multiplicity in types, with the already mentioned differences in the chemical composition of the cell wall, is a great hindrance to systematic investigation.

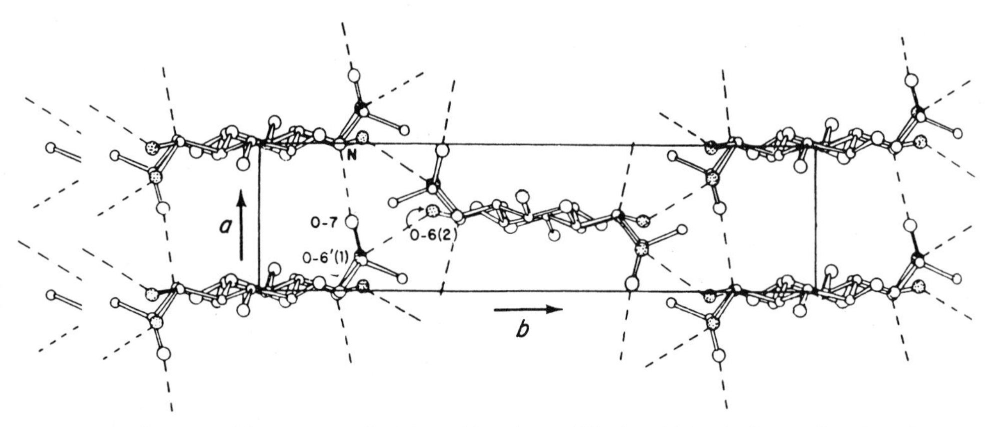


**Fig. 2.17** Side view of the monosaccharide residues in α-chitin, in which a hydrogen bond can be recognised between the acetamido carbonyl (O-7) and the N-H hydrogen in the neighbouring molecule.

### 2.1.2.3 Plant cell walls

This section must begin with the question: what is a plant? If one assumes that life on earth began in the oceans, then one can count blue algae together with photosynthesising bacteria, as the oldest known living objects at around $3 \times 10^9$ years old[25]. The border between the two groups is indistinct. Common to both is the capability of providing, photosynthetically, for their living needs and the fact that they are prokaryotes. The demarcation of plants from other living objects can be viewed in terms of physiology for nutrition: all plants are autotrophic, that is they can live with light energy and inorganic matter. This is also the main distinguishing difference between blue algae and normal bacteria. As regards the cell wall though, no great difference exists between bacteria and blue algae. Blue algae possess, as do gram-negative bacteria, a murein-sacculus which is strengthened through layers of pectin, hemicelluloses and possibly also cellulose. If one takes the cell wall as a measure, then one can subdivide all living objects into three fundamental kingdoms, eubacteria, archaebacteria, and eukaryotes (Fig. 2.18)[26].

Eubacteria comprise all organisms dealt with hitherto. Archaebacteria are only known as fossils. From eukaryotes, then, have developed what we know today alongside the prokaryotes as living objects: plants, fungi, and animals. This subdivision shall be retained in the following discussion.

Because eukaryotic algae are, in general, water plants, the significance and function of the cell walls are for them different from those of land plants. In the case of algae, there is no necessity for the whole organism to maintain and support itself against gravity but on the other hand the organism must withstand the water pressure and thereby maintain a stable, functional form. These requirements are met by the special composition of the cell walls. The typical cell wall polysaccharide of plants, cellulose, can be identified only with difficulty in red algae, possibly the oldest eukaryotic algae. It is certain that in the case of all red algae the content of cellulose does not lie, essentially, over 5% of the dry weight[27]. On the other hand, one finds in many cases *sulphated α-(1→3)-D-mannans* in up to 30% of the dry weight and above all *sulphated galactans,* which represent up to 40% of the major polysaccharide of red algae. The sulphated galactan that can be extracted with hot water from the dried algae is in general better known under the designation *agar* and *carrageenan.* These terms cover an abundance of structurally variant polysaccharides, which all have a more or less high capacity to form gels. This pronounced property may be very important for the mechanical stabilisation of the algae cells, comparable to the stabilisation in shape of an inflated rubber tyre. They best correspond to the needs of water plants. The gelling capability in the case of the sulphated galactans depends on two special structural elements: highly ordered helical regions which are loosened by hydrophilic, unordered sections of the chain[28] (Fig. 2.19).

In the wide-meshed cavities so formed, a large amount of water can be stored, firmly. As will be shown in the case of the alginic acids and pectins, the principle of gelation can be realised also through other structural features in polysaccharides. The galactans consist originally, that is to say as they are

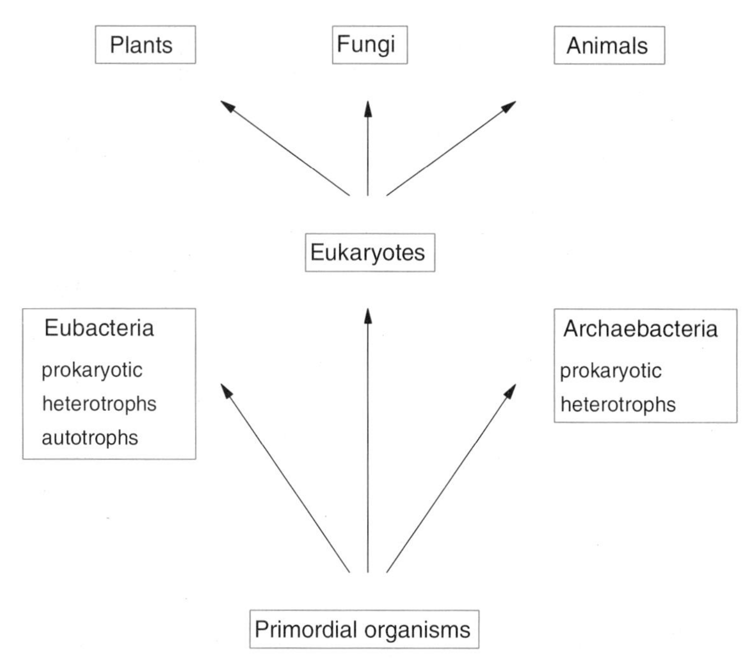


**Fig. 2.18** Living things evolved from primordial organisms in three main kingdoms. Plants are autotrophs. They exist both as prokaryotes and also as eukaryotes.

biosynthesised, of a repeating unit composed of two residues A and B. Residue A is D-Gal, $\beta$-(1→4)-linked, residue B is D- or L-Gal, $\alpha$-(1→3) linked (Fig. 2.20).

The chemical modification of both residue A and residue B can occur post-biosynthetically, as shown in Fig. 2.21.

In highly-ordered regions, repeat units consist, over long stretches, of the disaccharide *-(1→3)-β-D-Galp-(1→4)-3,6-anhydro-α-D-Galp-(1→3)-* (Fig. 2.22).

In the unordered regions, chemical modification through sulphation, methylation, and pyruvylation appears to be more or less arbitrary. The high charge density favours the binding of water.

Green algae are not very much younger than red algae in terms of their historical evolution. On many grounds they are considered the forerunners of the land plants. With respect to the glycans present, their exist great similarities between the two groups. Many green algae contain crystalline cellulose as a fibre-forming polysaccharide but in many it is replaced by fibre-forming mannans or xylans. Glucomannans and pectins are swelling, gel-forming matrix polysaccharides. Thanks to the enormous variety of types of green algae, which are represented by three classes with fourteen orders and more than 20,000 species (new species are discovered constantly ), a relatively high heterogeneity of the cell wall polysaccharides is understandable. It is sufficient here to point out the clear relationship with land plants whose cell wall polysaccharides are discussed at the end of this section.

The youngest group of algae, in terms of their evolutionary history, are the brown algae, *Phaeophyta*, as the rough phylogenetic derivation in Fig. 2.23 (p. 86) shows.

Many representatives, *Macrocystis*, can reach remarkable sizes with thallus lengths of over 50 m. Their size alone makes many brown algae conspicuous inhabitants of the coastal regions of the world seas. They have an economic value which should not be underestimated.

Cellulose plays a subordinate role in the structure of these giant plants with an average content based on dry weight of around 5%. Numerous polysaccharides of differing structure are found in various proportions depending on the type and species. In this regard, the terms *laminaran*, *lichenan*, and *fucoidan* are recognised for these heterodisperse glycans from brown algae. Many polysaccharides are covalently bound to protein.

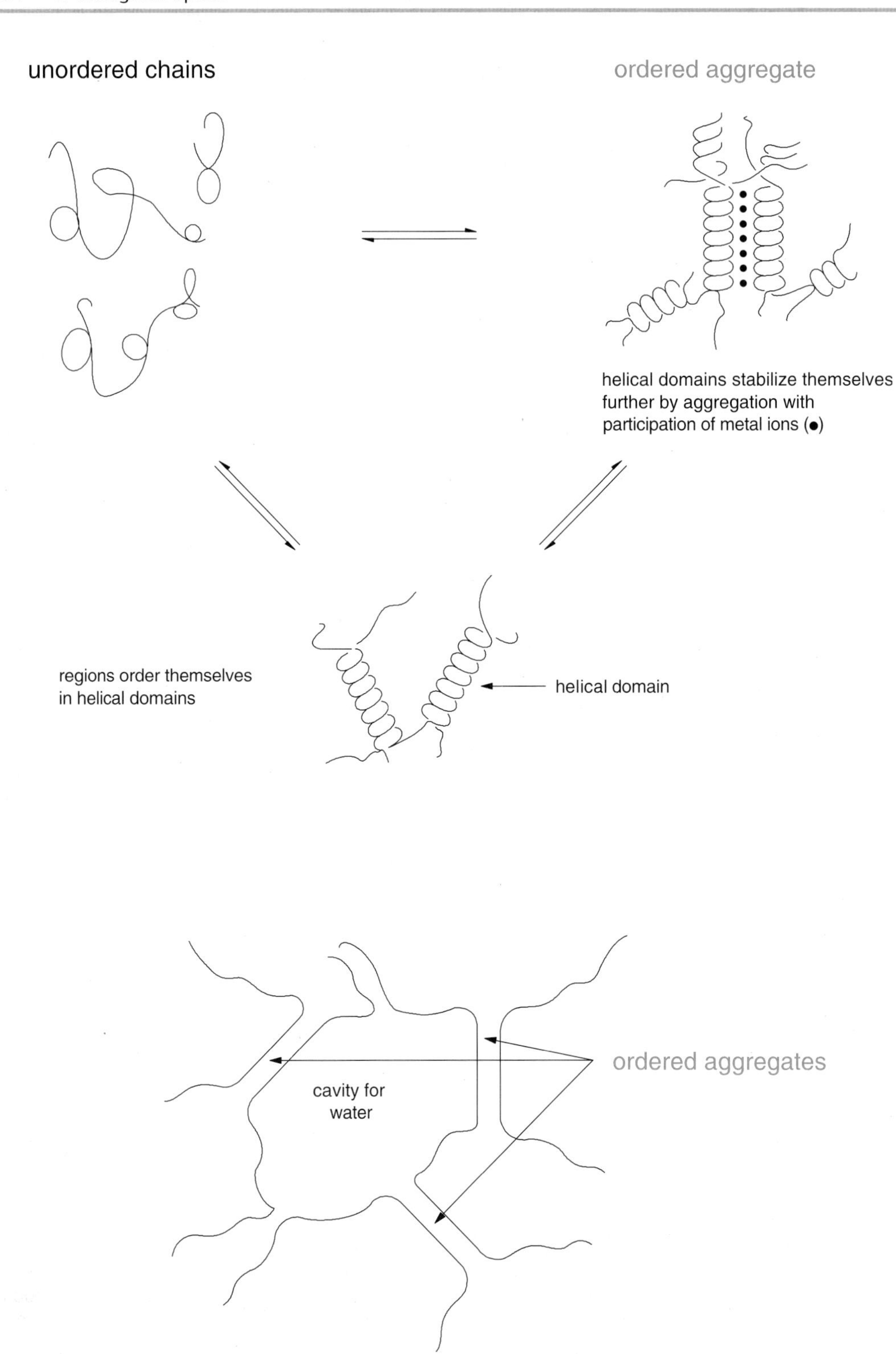


**Fig. 2.19** Representation of a model showing the origin of stable gels from algal polysaccharides.

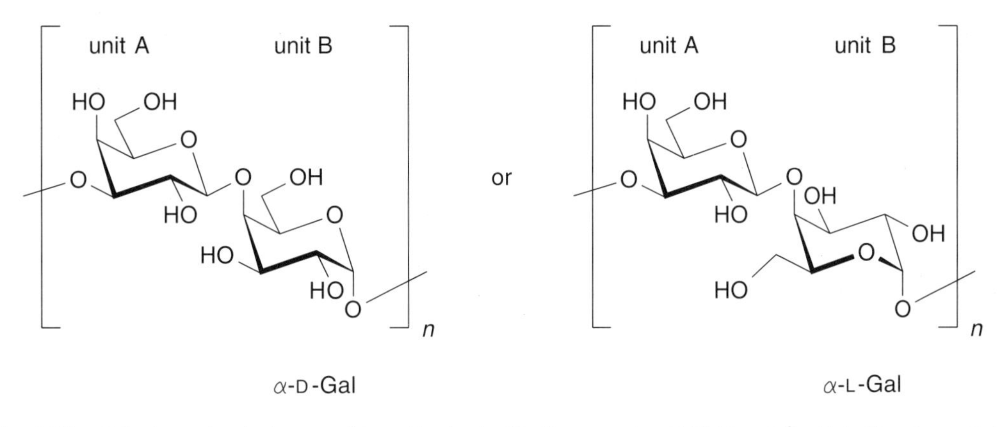


**Fig. 2.20** *Galactans* of *red algae* are biosynthesised with the repeat unit $[A\text{-}B]_n$ = $[\text{-}\beta\text{-D-Gal}p\text{-}(1\rightarrow4)\text{-}\alpha\text{-D-Gal}p\text{-}(1\rightarrow3]_n$ or $[\text{-}\beta\text{-D-Gal}p\text{-}(1\rightarrow4)\text{-}\alpha\text{-L-Gal}p\text{-}(1\rightarrow3]_n$.

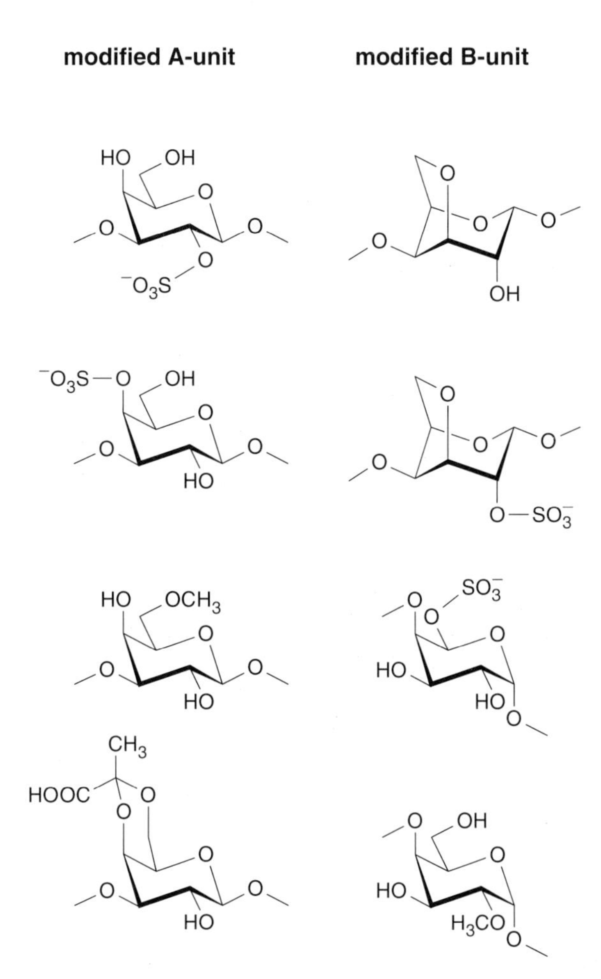


**Fig. 2.21** The units A and B of the *red algae galactan* can be modified post-biosynthetically in various ways.

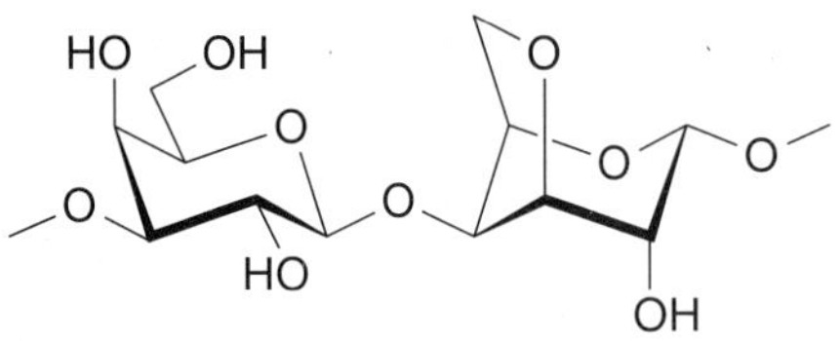

**Fig. 2.22** The repeat unit in *galactans* which, as longer chains, make possible highly ordered helical structures in polysaccharides.

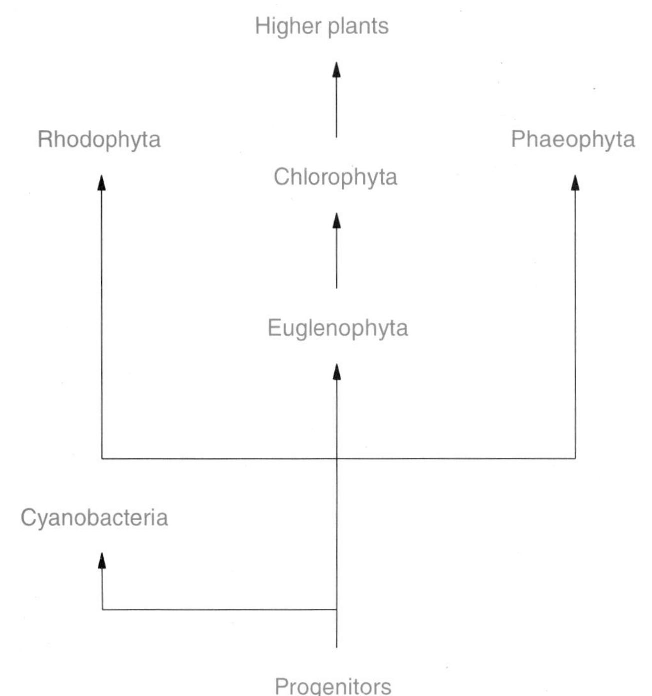


**Fig. 2.23** Phylogenetic relationship between the algae group and higher plants.

The most important group of polysaccharides from brown algae are the *alginates*. They make up about 40% of the dry-weight of the plant. It is clear that the alginates of brown algae confer morphological stability. If the alginates of the plants are removed by extraction, the tissue collapses to a brown amorphous mass. Alginates form not only part of the cell wall but are found principally in the intercellular space. As their name signifies, alginates are polyanionic compounds which in the native state occur as salts of cations which are in the surrounding sea water, namely $Na^+$, $Mg^{2+}$, and $Ca^{2+}$. $Ca^{2+}$ ions especially, but still more strongly $Sr^{2+}$, are very firmly bound. Cross linking by means of these ions is responsible for the especially high stability of the alginate gels[29]. Alginate may be extracted relatively easily from the crushed dry mass of brown algae by ion exchange with the help of alkali or ammonium salts in neutral solution or with dilute mineral acid[30]. The native, that is to say in this case the biosynthesised structure of the alginates is $\beta$-(1→4)-D-mannuronan. Post-biosynthetically, $\alpha$-(1→4)-L-guluronate residues are formed through epimerisation at the 5-position. Alginates of different species, from species of differing ages, and also from different organs in a plant differ in the ratio of the two building blocks. In very young tissue, sometimes almost pure mannuronan is found. The older and tougher the tissue, the higher, in general, is the proportion of L-GulA. Alginates are excellent gel-formers. Fundamentally, the formation of a stable gel depends upon the intermolecular association of regular sequences in the polysaccharide, as has already been shown for the agar gels (see Fig. 2.19, p. 84). The difference lies in

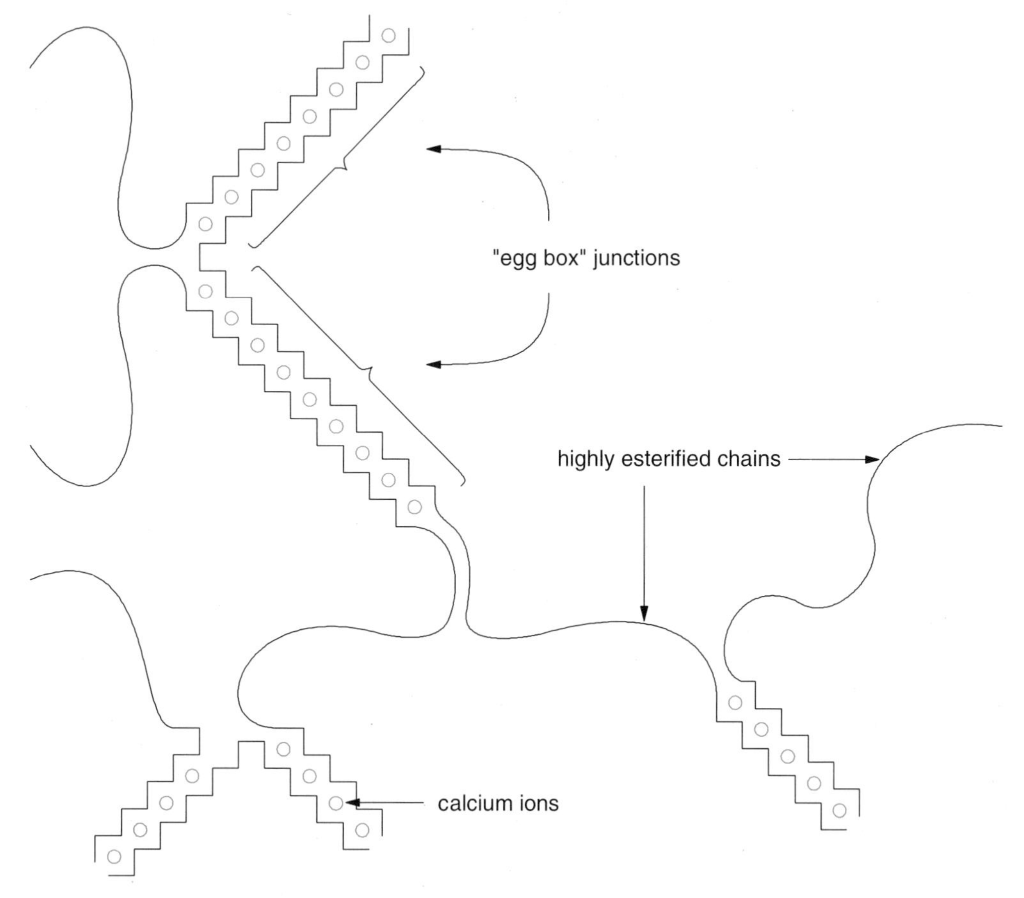


**Fig. 2.24** Gels from alginates of brown algae and pectinates from land plants are formed by a wide-mesh arrangement of highly ordered segments from intermolecular salt bridges (*egg box* junctions). In the case of the alginates with poly-L-GulA, it alternates with hetero-sequences comprising $\beta$-D-ManA and $\alpha$-L-GulA.

the type of intermolecular binding. It is now certain[31] that block segments of $\alpha$-L-GulA are responsible for very stable association over salt bridges with divalent cations such as $Ca^{2+}$ or $Sr^{2+}$. Intermolecular chelation of $Ca^{2+}$ with poly-$\alpha$-L-GulA is optimal with a block of about 15 monosaccharide units. The aggregated zones are very graphically described as "egg box junctions"[32]. For the formation of wide-meshed gels, however, either extended blocks of $\beta$-D-ManA must follow or, more or less statistically mixed, heterosequences with L-GulA and D-ManA (Fig. 2.24)[28]. This principle of gel formation is followed also in the gelation of pectinates from higher plants, only with different monosaccharide building blocks (see p. 89f).

Today, alginates have considerable technological importance. An example of their application is to produce immobilised microorganisms. Suspended in a sodium alginate solution, bacteria can be enclosed in a very stable gel by dropwise addition of calcium salt solution and may then be manipulated in a granular form without damage.

The evolutionary change of water plants into land plants must have proceeded inevitably with an enormous biochemical and morphological change. This change concerns, especially, the further evolution of the cell wall and the extracellular matrix, which also was to adapt, organ-specifically, to the biological functions in the new environment. For water plants, large surfaces for optimal exposure to radiation and optimal material exchange can be realized without difficulty. Buoyancy in the water allows full development also, without specific stabilising structural elements, such as those which are vital for the land plants, especially the trees. Plants strive for the light for photosynthesis in the leaves, which are spatially remote from water and mineral supplies obtained through the roots in the earth. Both supplying organs then needed to be connected in the most stable manner possible, by a self-supporting stem. That demanded the development of new, particularly firm material.

Green algae make use of cellulose as a fibre-building component in cell walls. This in many respects unique polysaccharide is finally the basis of all cell walls in land plants. The cellulose fibres are embedded in a matrix which, according to type, organ, and above all stage in growth, can vary considerably in its composition. Matrix materials are other polysaccharides, glycoproteins, and *lignin*. One distinguishes, according to the stage in growth, primary and secondary cell walls. The first arises during the period of cell growth. It is relatively thin, permeable and flexible. The primary cell walls of all land plants are very similar and a differentiation of the cell walls according to organ type has not yet set in.

After conclusion of cell growth, the secondary cell wall develops, above all by thickening and solidification. This is attained by storage of more cellulose and, above all in the case of wood formation, through lignin storage. Secondary cell walls of different organs can be assembled very differently.

Cellulose[33] is certainly the most abundant single polysaccharide, if not the most abundant polymeric compound. The literature on this, one of the most significant raw materials of humanity, is difficult to survey. It is sufficient here to point out the role of cellulose as an essential cell wall constituent of land plants. The main source of cellulose is wood, which depending on the type, contains between 40 and 50% of this polysaccharide. Sources of cellulose are also grasses in the widest sense, whose content of cellulose in the dry weight, however, never rises above 30%. Cellulose constitutes about a third part of typical organs for reserve polysaccharides, such as potatoes or sugar beet. Cotton wool consists of nearly pure cellulose (Fig. 2.25)[33].

| Biopolymer | Conifers | Deciduous |
|---|---|---|
| Cellulose | 51 | 47 |
| Hemicelluloses | 17 | 23 |
| Lignin | 28 | 22 |

**Fig. 2.25** Estimated composition in % of the most important biopolymers of wood.

Wood is the main constituent of the secondary wall and is a good example of the use of certain linear polysaccharides for the construction of high load-bearing parts of organ. Cellulose, again as an important component of wood, has a structure ideally suited to fufil this task. As a pure homopolysaccharide with exclusively $\beta$-(1→4)-linked Glc*p*-residues, – between about 3000 for primary cell walls, up to 15 000 for secondary cell walls – cellulose has an extended and ribbon-like structure. This structure is itself stabilised through intramolecular hydrogen bonds (Fig. 2.26)[33] and allows the tight packing of several, anti-parallel chains next to each other and mutual bonding through intermolecular hydrogen bonding (Fig. 2.27)[33].

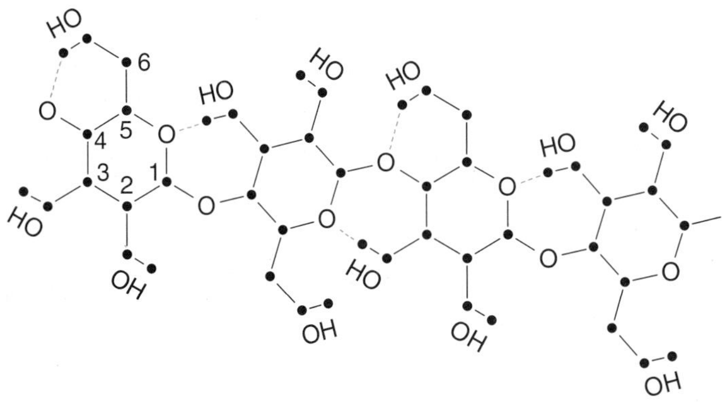

**Fig. 2.26** The ribbon structure of the cellulose molecule is stabilised by intramolecular hydrogen bonds.

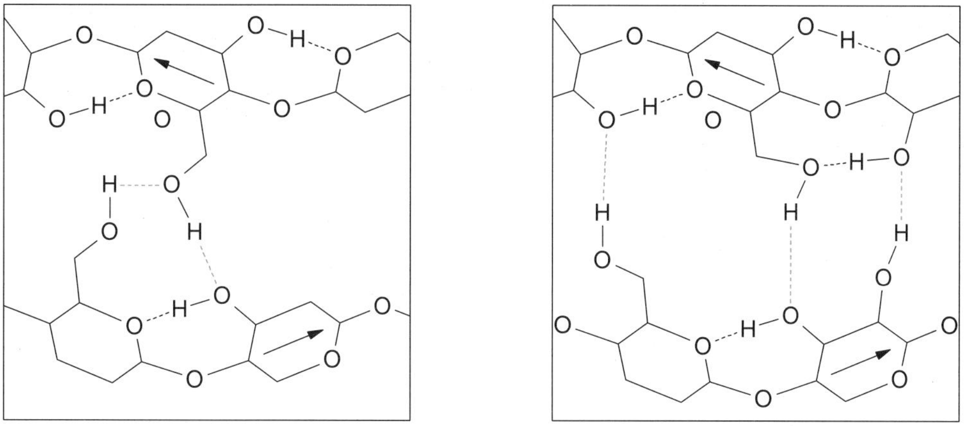

**Fig. 2.27** Two possibilities for the intermolecular stabilisation of antiparallel cellulose molecules by hydrogen bonding.

In this way microfibrils are formed with about 50 to 100 molecules which are distinguished by their water insolubility, chemical resistance, and extraordinary tensile strength. The cellulose fibres are embedded in an amorphous matrix of different polysaccharides, whose structures are very variable. The *matrix polysaccharides* can be classified roughly into two groups, the *hemicelluloses* and the *pectins*. Most of the compounds of these two groups are heteropolysaccharides, whose primary structures are, in part, still the object of current investigations. Very little can be said about the secondary or tertiary structure of such compounds *in vivo*, especially as isolation of native material from the secondary cell wall without chemical damage is almost impossible. Accordingly, the following thoughts refer to the primary cell wall of land plants which, as already indicated, allows certain generalisations to be made.

The primary cell wall[34] consists of up to 90% polysaccharides and 10% glycoprotein. In the primary cell wall cellulose is present to the extent of only 20–30%, the usual polysaccharides being markedly complex and in part only very inexactly defined. There is even a conviction that some polysaccharides of the primary cell wall have not yet been discovered. The *xyloglucans* are, after cellulose, the most abundant cell wall polysaccharides. In a (1→4)-β-D-glucan framework, xylose residues are found at irregular intervals, which for their part can again be glycosylated with D-Gal*p* and L-Fuc*p* units (Fig. 2.28)[34].

Xyloglucans bind directly to cellulose, and it is suggested that the content of xyloglucan controls the degree of aggregation of cellulose in cellulose fibrils through competitive binding. Arabinoxylans are the next group of hemicelluloses (Fig. 2.29)[34].

As side chains, they contain predominantly, alongside a little GlcA, L-Ara*f*. It has been suggested that these side chains are esterified with 3-carboxylic acid derivatives of phenylpropene which for their part allow a crosslinking through radical C,C-coupling (Fig. 2.30). (1→3)-β-Glucans, mannans, and *galactans* with variable side chains and various types of linkage complete the very different assembled mixture from one case to the next, of the known hemicelluloses.

A further group of cell wall polysaccharides of plant cells are the pectins, of which especially the galacturonans should be mentioned here. The product isolated from citrus fruits contains (1→4)-α-linked D-Gal*p*A residues. A continous chain of about 25 D-Gal*p*A residues is interrupted by a (1→2)-α-L-Rha*p* residue. Gal*p*A residues are present in block segments as their methyl ester. This, and the fact that pectins form very stable gels with calcium, leads to a description of the gel structure after the "egg-box" principle[28], already represented in the case of the alginates. It is plausible that also in land plants, at least in the non-woody parts, especially in fruits, gels contribute to the mechanical stabilisation of the cell wall.

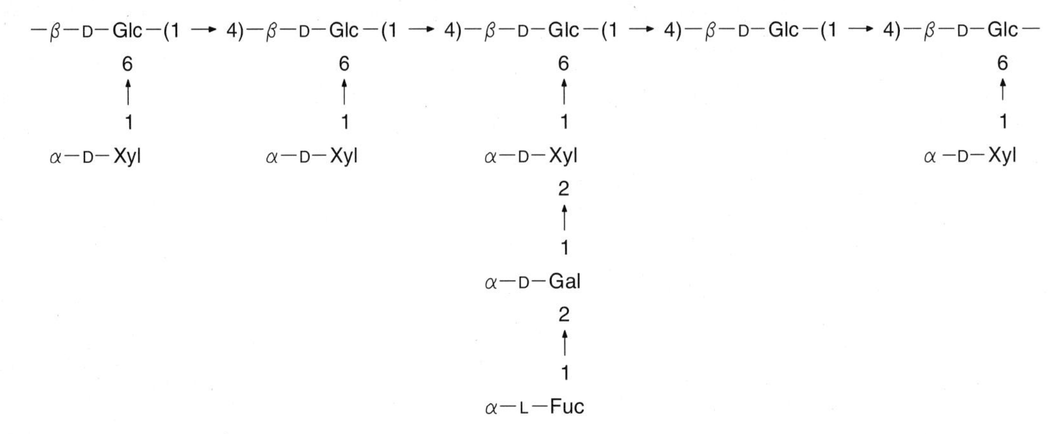


**Fig. 2.28** *Xyloglycans* belong to the class of *hemicelluloses* and are a major component of primary cell walls of higher plants. Their structure varies especially in the side chain. Shown here is a segment from the xyloglucan of the sycamore tree.

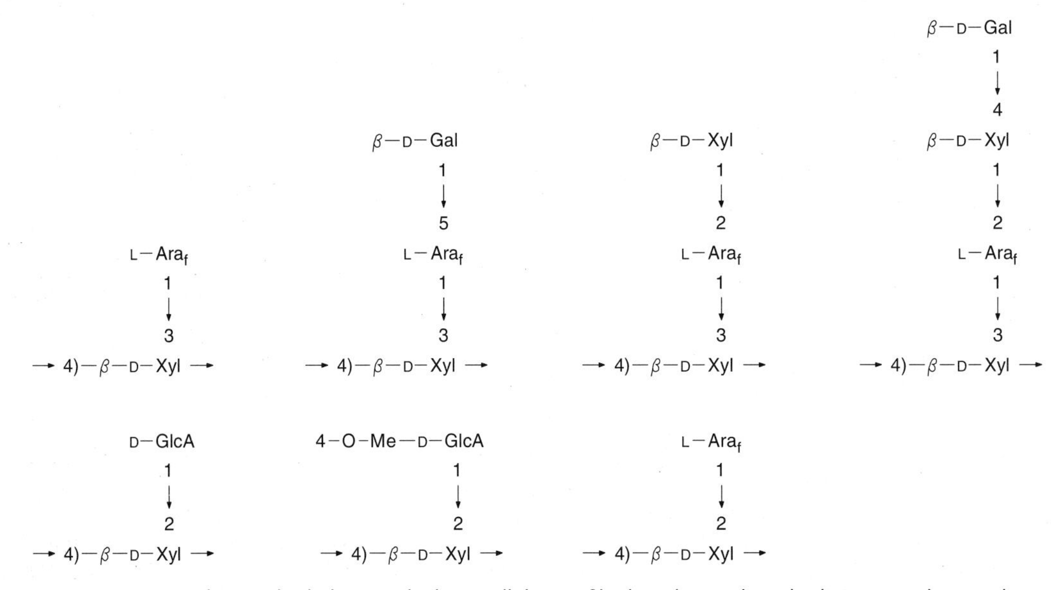


**Fig. 2.29** Heteroglycans also belong to the hemicelluloses of higher plants. The side chains are subject to large variation.

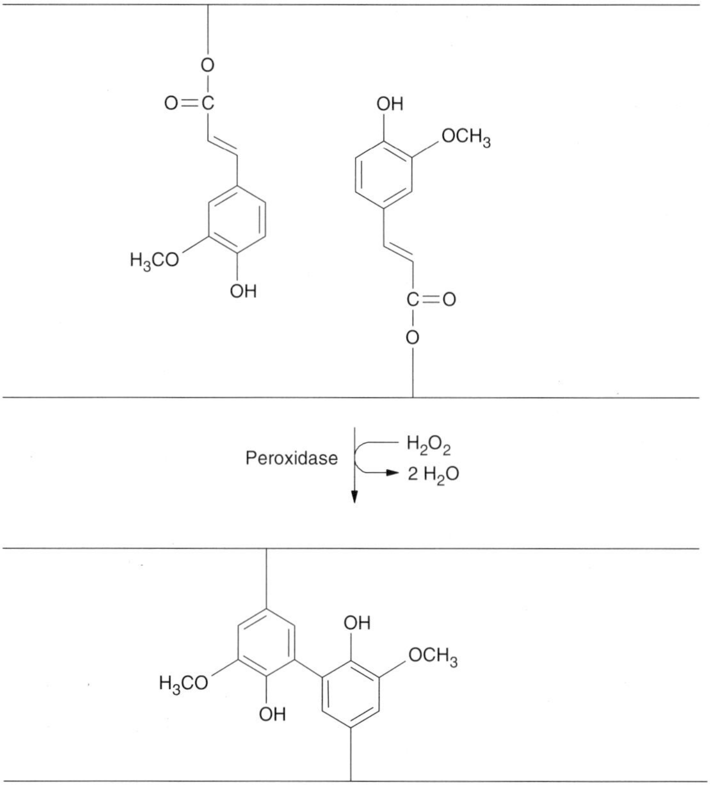

**Fig. 2.30** The side chains of *hemicelluloses* are partly esterified with *ferulic acid*. Oxidative coupling of the aromatic residues leads to crosslinking.

### 2.1.2.4 Extracellular matrix of animal organisms

The transition from the protected, firm cell wall of bacteria, of plants, and of fungi to the extracellular matrix of the animals is simplest to achieve morphologically in the case of invertebrates. The protection of the single cell occurs collectively in that a common solid outer shell surrounds the whole organism, not only to protect but also to give simultaneously support and form. The outer shell or the outer skeleton enables also the consolidation of muscle fibres necessary to produce motion, which in the animal world plays an essential role, and which certainly is a very important reason for changing the function of the material outside the cytoplasmic membrane. The dominant material for building the exoskeleton is *chitin*[35], that serves not only as building material for insect armour and insect wings, but also as the principal constituent of the outer skin of crustaceans, including the krill, and it also guarantees the firmness of mussels and oysters and the protection and mobility of cuttle fish.

As cellulose is combined in plants with matrix polymers in very different proportions, so chitin is found, according to origin, associated with varying amounts of foreign substances. However, in the case of chitin armour, the polymer matrix is set not in polysaccharide but in protein. Chitin arises in laminated form, very possibly as a proteoglycan, and is stored in layers between protein. The fibre direction changes in each new lamina so that the principle of construction is not to be compared with ferro-concrete (bacteria and plant cell walls) but resembles more a plywood sheet, tensile strength and compressibility being shown in all directions. In many chitin armours minerals are stored for additional hardening.

What in invertebrates is protecting and supporting exoskeleton, is in vertebrates bone, skin and connecting tissue. Higher development is accompanied by higher differentiation. The role of the solid framework for stabilisation of the organism and the consolidation of muscles is taken over by the bone-skeleton, whose constituents, principally, are collagen as fibrous protein, some glycoprotein and the mineral hydroxyapatite. Here, collagen displaces the polysaccharides completely as the fibre-forming material and it is interesting that collagen in bones is arranged in layers with mixed fibre-directions at right-angles, as is chitin in the chitin armour of the invertebrates.

Polysaccharides play an important role, though only in combination with proteins, in the construction of many different tissue-types (Fig. 2.31) as well as being constituents of secretions of all possible functions. General properties are required for large tensile or pressure loads with optimal flexibility and mobility, as one expects of sinews, cartilage or skin tissue, and of blood vessels. Polysaccharides also allow, through their special physico-chemical properties, the friction-free mobility of the joints, and they line and protect the digestive tract as mucus and are to be found overall as viscous secretions, where the organism requires *lubricating oil* for various purposes.

An inventory shows the importance of an especially important group of polysaccharides for higher animals, the *glycosaminoglycans* (Table 2.1)[36]. Especially intensively investigated, for obvious reasons, are the glycosaminoglycans of the mammals and of humans, to which, as the main polysaccharides of the extracellular matrix, the following section is devoted.

All glycosaminoglycans up to hyaluronic acid are proteoglycans of complex structure in which single polysaccharide chains may not be considered separately.

It is 70 years since the first efforts were made to develop methods for the isolation of native proteoglycans in order to understand, by exact analysis of these complicated molecules, the connection between the chemical composition of the extracellular material and their function. Today, this goal has still not been achieved, so that here it must be sufficient to give a rough insight into the construction of the extracellular matrix and to give information on the chemical structure, as far as it is known.

Tissue, in addition to cells, consists of the extracellular matrix, a mesh of polysaccharides and proteins, to which cells are fastened (epithelial cells, muscle cells) or into which cells are embedded (fibroblasts, osteoblasts). Connecting tissue forms the major part of the extracellular matrix, in which cells are only thinly distributed (Fig. 2.31). Connecting tissue is the major component in skin, cartilage, sinews, and bones and forms thereby not only the framework of the vertebrate organisms, but also the

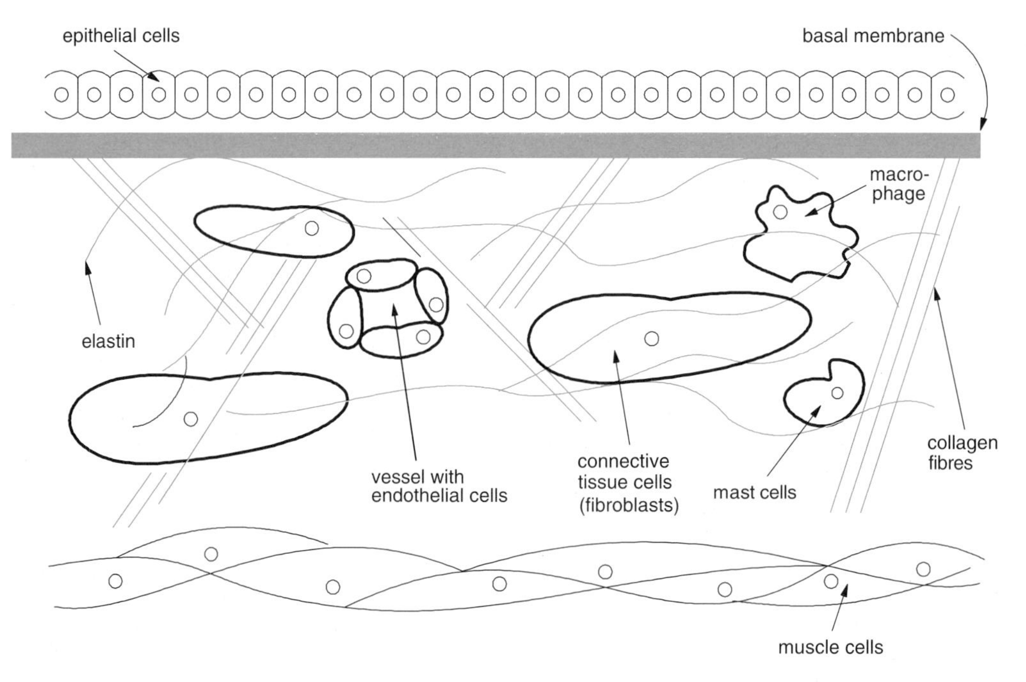


**Fig. 2.31** The section shows schematically essential components of vertebrate tissue, which are embedded in a matrix (not shown) of *glycosaminoglycans*.

**Table 2.1** *Glycosaminoglycans* of the mammals

| Glycos-amino-glycan | Molar mass (× 10³) | Repeat unit A–B monosaccharide A | monosaccharide B | Sulphate per DS unit | Bound protein | Other sugar components | Tissue distribution |
|---|---|---|---|---|---|---|---|
| **hyaluronic acid** | 4–8000 | –β–GlcA–(1→3)– | –β–GlcNAc–(1→4)– | 0 | – | | different tissues, skin, vitreous humor, cartilage, synovial fluid |
| **chondroitin sulphate** | 5–50 | –β–GlcA–(1→3)– | –β–GalNAc–(1→4)– | 0.2–2.3 | + | D-Galactose D-Xylose | cartilage, cornea, bone, skin, arteries |
| **dermatan sulphate** | 15–40 | –β–GlcA–(1→3)– or –α–L–IduA–(1→3)– | –β–GalNAc–(1→4)– | 1.0–2.0 | + | D-Galactose D-Xylose | skin, blood vessels, heart, heart valves |
| **keratan sulphate** | 4–19 | –β–Gal–(1→4)– | –β–GlcNAc–(1→4)– | 0.9–1.8 | + | *N*-Acetyl-D-galactosamine *N*-Acetyl-D-glucosamine D-Mannose | cartilage, cornea, invertebral disc |
| **heparan sulphate** | 5–12 | –β–GlcA–(1→4)– or –α–L–IduA–(1→4)– | –α–GlcNAc–(1→4)– | 0.2–2.0 | + | D-Galactose D-Xylose | lungs, arteries, cell surfaces, basal membranes |
| **heparin** | 6–25 | –β–GlcA–(1→4)– or –α–L–IduA–(1→4)– | –α–GlcNAc–(1→4)– | 2.0–3.0 | + | D-Galactose D-Xylose | lungs, liver, skin, mast cells |

protective layer against the outside world. The components – collagen fibres and glycosaminogylcans – are distributed in different ratios according to function of the tissue. It is suggested that the extracellular matrix, besides its function of mechanical stabilisation, also has an influence on cell development and cell motion.

Proteoglycans have a central *core protein* to which the various polysaccharides are covalently attached. The protein and number and type of the polysaccharide are different in each case. The proteoglycan from cartilage tissue has a molecular weight of $3 \times 10^6$ Da, in which case the weight of the protein amounts to about 300 kDa. The polysaccharide chains are mixed, a few keratan sulphate chains (about 5 kDa) and many chondroitin sulphate chains (15 to 20 kDa) alternate (Fig. 2.32).

The core protein has at its disposal a region which, additionally and still mediated by binding protein, can bind non-covalently to a further glycosaminoglycan, *hyaluronic acid* (Fig. 2.33). Hyaluronic acid is the largest glycosaminoglycan and is able to accumulate up to 100 proteoglycans to form a complex of gigantic size. The complex is readily seen under the electron microscope and has, approximately, the dimensions of a bacterial cell (Fig. 2.34)[36].

The *proteoglycan-hyaluronate complex* fills the greater part of the space in cartilage tissue. In it, embedded and associated with it, are collagen fibrils (Fig. 2.35)[36]. According to function and the mechanical stress, the composition between collagen and the various proteoglycans changes in the extracellular matrix. *Proteodermatan sulphate* isolated from blood vessels, sinews, and skin, has a relative small molecular weight of 100 – 200 kDa. The protein is only sparsely occupied with dermatan sulphate and chondroitin sulphate chains. Proteokeratan sulphate composed of about equal parts of polysaccharide and protein and with a molecular weight of 72 kDa is found in the cornea of the eye. Typical for the protein part are disulphide bridges and, for the polysaccharide part, the *N*-glycosidic linkage (see section 2.3.8.4, p. 217) to the protein (Fig. 2.36)

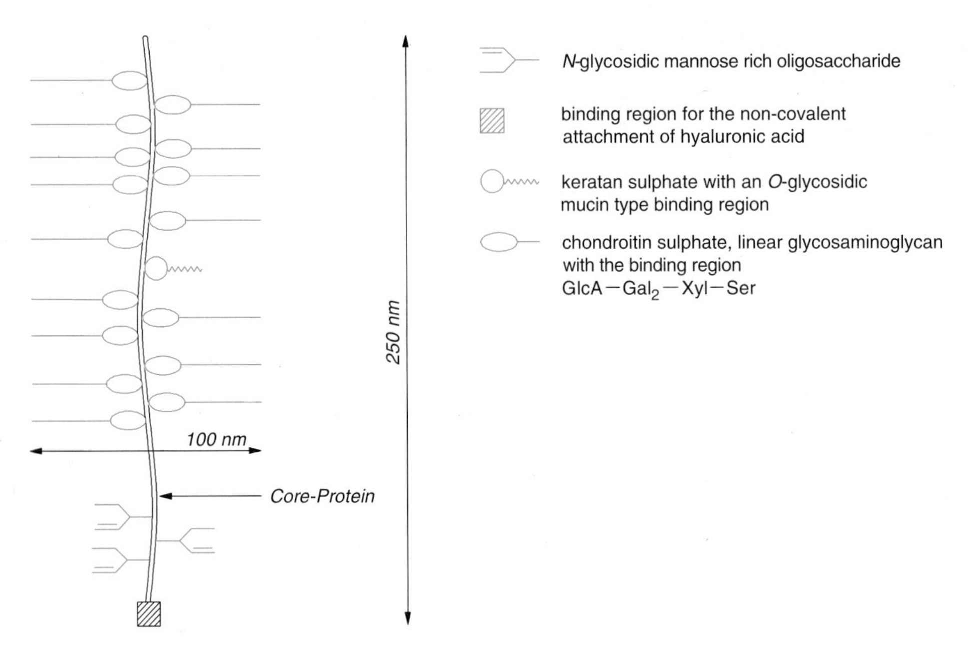


**Fig. 2.32** Schematic representation of a *proteoglycan* from cartilage tissue with a molecular weight of several million. The core potein carries different carbohydrate structures covalently bound.

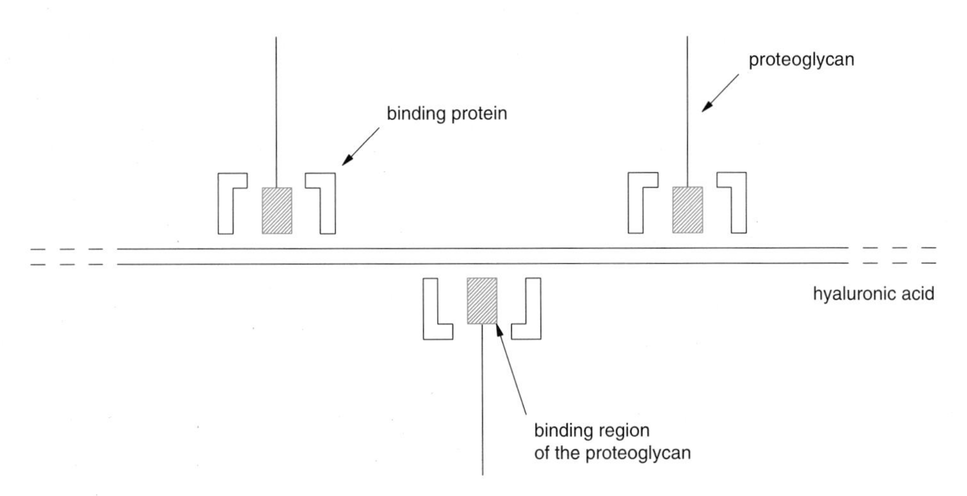


**Fig. 2.33** Hyaluronic acid is a glycosaminoglycan which contains no protein itself, but which has about 100 proteoglycans non-covalently bound by means of binding proteins

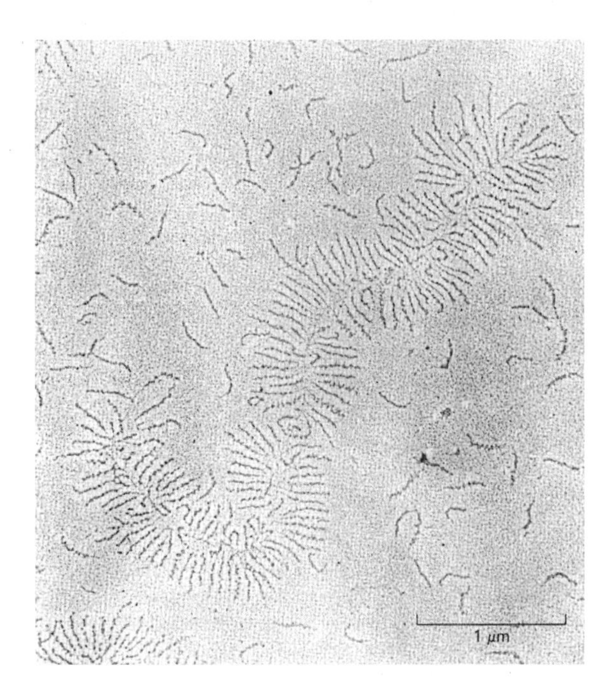


**Fig. 2.34** Electron micrograph of a *proteoglycan aggregate* from bovine cartilage.

**Fig. 2.35** Electron micrograph of *proteoglycan*s in the extracellular matrix of rat cartilage.

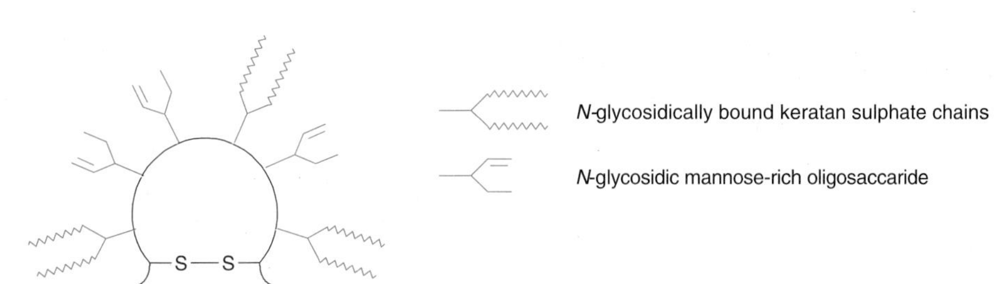


**Fig. 2.36** Schematic representation of the proteoglycan from the cornea.

Except for hyaluronic acid, which most probably has no peptide part, all glycosaminoglycans are linked either through a connecting trisaccharide that is *O*-glycosidically bound to serine or threonine in the core protein (Fig. 2.37) or exceptionally, as in the case of proteokeratan sulphate of the cornea, through an *N*-glycosidically linked core oligosaccharide[37].

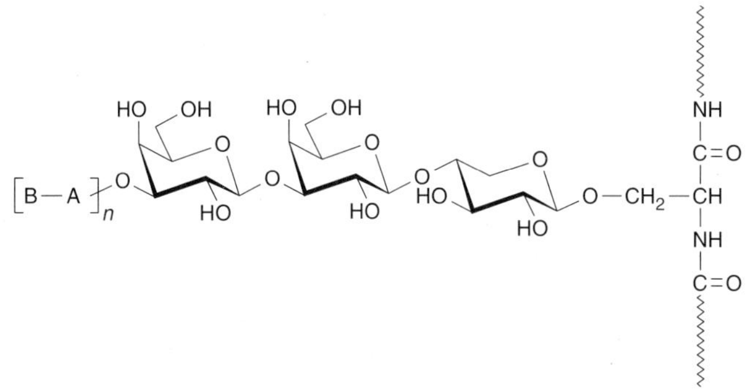

**Fig. 2.37** The core region of an *O*-glycosidically bound glycosaminoglycan (A-B is the repeat unit)

All glycosaminoglycans are composed of disaccharide repeat units (A-B) consisting of a hexose or hexuronic acid and a glycosamine[38]. With the exception of *heparin* and heparan sulphate, A- and B-units of all glycosaminoglycans are $\beta$-linked, but the $\alpha$-link from the glycosamine in the former leads to a conformationally unique situation (Fig. 2.38). This maybe connected with the fact that heparin and

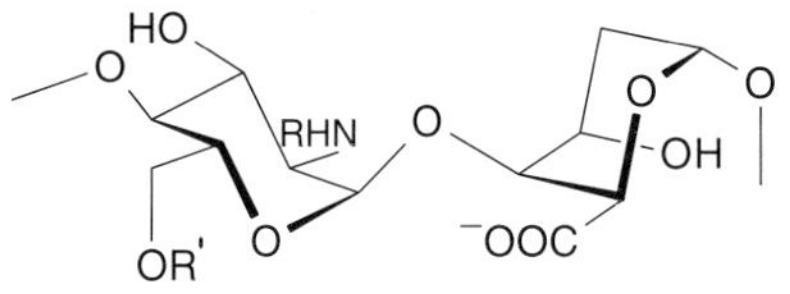

4)—α—D—GlcNAc/S—(1 → 4)—α—L—IduA—(1 →

heparin and heparin sulphate

R = $SO_3^-$ (S) or CO–$CH_3$

R' = $SO_3^-$ or H

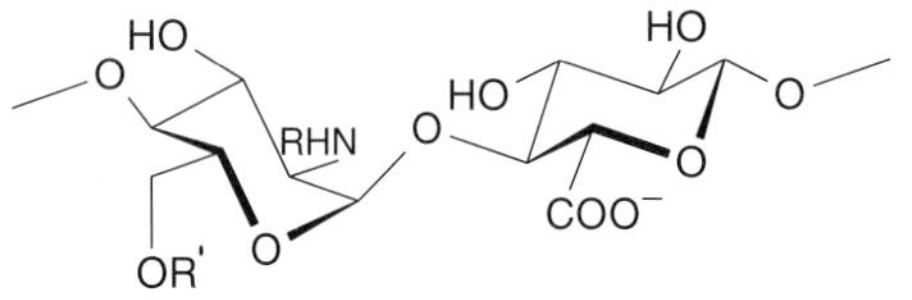

4)—α—D—GlcNAc/S—(1 → 4)—β—D—GlcA—(1 →

**Fig. 2.38** The repeat unit in heparin and heparan sulphate adopt a special conformation because of the $\alpha$-configuration of the B-unit (see Table 2.1, p. 93).

heparan sulphate are not typical glycosaminoglycans of the extracellular matrix, but associate with cell membranes, as perhaps proteoheparan sulphate with a hydrophobic region of the protein part, or heparin, which is to be found chiefly in the mast cells of the connecting tissue.

The most striking structural characteristic of the glycosaminoglycans is their high charge density caused by the sulphate residues (herewith hyaluronic acid is excepted) and by carboxy groups. This brings about the binding of osmotically active ions, as for example sodium, and thereby a very high tendency to take up water. The accumulation of charges along the chain, together with the relative rigid conformation of the $\beta$-linked pyranosyl residues, prevents back-folding and coiling of the glycosaminoglycans, which outstretched demand much space per mass. The tendency of the glycosaminoglycans to swell caused by conformation and charge is responsible for the high load-bearing capacity towards pressure of the cartilage tissue and other types of connecting tissue. The very wide-meshed, voluminous structure of the glycosaminoglycans is ideally suited as mechanical support of the tissue, whereby it allows simultaneously the free diffusion of dissolved molecules and the migration of cells.

As well as glycosaminoglycans together with collagen providing for mechanical solidity in the extracellular matrix of vertebrates, so their structural peculiarity also gives body liquids special properties. Joints are mechanically loaded with high masses. The synovial fluid, whose most important constituent besides water is *hyaluronate*, provides for frictionless functioning between the head of a bone (condoyle) and the glenoid cavity. Viscosity and shock absorbing properties can be ascribed to the structural characteristics already described.

The main proteoglycan of the follicular fluids has similar functions of protection from mechanical damage as is accomplished by the hyaluronic acid in joint fluids. However, although the former is comparable hydrodynamically with the proteoglycans, it has a very different chemical structure (Fig. 2.39). The core protein shows no binding region for hyaluronate and carries, besides a few relatively large dermatan sulphate chains, numerous *O*-glycosidically bound oligosaccharides of the so-called mucin type (see section 2.3.8.6, p. 226).

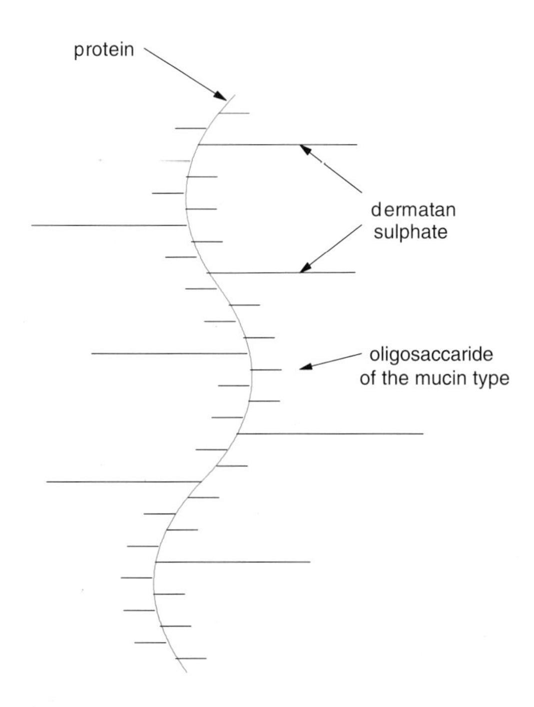


**Fig. 2.39** *Proteoglycan* of the *mucin type* from follicular fluid. The approximately 200–400 short oligosaccharides are *O*-glycosidically bound through GalNAc. In addition there are about 10–20 long dermatan sulphate chains uniformally distributed.

While the viscosity properties of this glycoprotein of mixed type are comparable with those of the the glycosaminoglycans, they lack in solution the load-bearing capacity towards pressure and the shock absorbing properties of the cartilage proteoglycan.

*Mucin type glycoproteins* carry *O*-glycosidically bound oligosaccharide chains of different sizes. They owe their name to the mucus secretions in which they were found originally. Today it is known that there are many *O*-glycosidic glycoproteins, which have nothing to do with the mucus secretions.

Mucins[39] are excreted from certain epithelial cells and serve the exclusive purpose of protecting organs from damage and, in the case of the gastrointestinal tract, they also provide for the smooth transport of food and its residue. They are found in the mucus of the bronchi, in saliva, and in the secretions of the genitalia.

Mucins contain about 50% of their weight in carbohydrates. The protein part consists of up to 50% of the amino acids serine and threonine, besides principally proline, glutamic acid, glycine and alanine.

The physicochemical behaviour is determined predominantly by the carbohydrate portion. The oligosaccharides of the mucins vary according to origin (Fig. 2.40).

Mucins appear to be present, for similar reasons as those discussed for the proteoglycans with glycosaminoglycan side chains, as rigid, outstretched sticks in solution and for that reason to be specially adapted for the formation of viscous, mucoidal solutions. In the submaxilla mucin of ruminants, with molecular weights of about 100 kDa, self aggregation is observed[40] to oligomers with more than a million complex-weight. The oligomerisation is controlled by ion content of the medium and through the density of NeuAc-residues. Although the oligosaccharide structure of the mucins in many cases resemble blood-group determinants or other antigenic determinants, no reason exists up to the present to ascribe to them a special function as carriers of information.

### 2.1.3 Polymeric carbohydrates as rapidly mobilised energy sources and reserves of raw materials

Low molecular weight carbohydrates, monosaccharides such as *glucose* or its isomers *mannose, galactose*, or *fructose* are transformed enzymatically into all other biomolecules. This, and the fact that monosaccharides arise directly out of inorganic materials by photosynthesis, gives carbohydrates a special position in the biosphere. The metabolism of all living things is based fundamentally on the transformation of glucose. All organisms, to a certain degree also microorganisms, lay down food reserves as soon as the food supply exceeds running requirements, so that at a later date they are able to fall back on them in case of shortage. In the case of animals, the build up of reserves occurs after feeding, in the case of plants after photosynthesis and, with many microorganisms, when other food ingredients limit the growth, for example minerals in the case of carbon-rich sources.

What could be more obvious than to deposit reserves in the form of carbohydrate? Formation of reserves also occurs under physiological conditions. The deposition does not occur in the form of low molecular weight, water soluble monosaccharides but rather as extensive, water insoluble polymeric condensation products. Large quantities of monosaccharides would increase unduly the osmotic pressure in the cell and cause a too strong, dangerous water-uptake. This danger does not exist in the case of the osmotically inactive, high molecular weight polysaccharides.

In contrast perhaps to the structural polysaccharides, reserve polysaccharides must be assembled quickly, that is when nourishment is still available, or during the assimilation of carbon dioxide. Even so, it must be possible for them to be broken down quickly to monosaccharides in times of need. These limiting conditions determine the molecular construction of typical reserve polysaccharides.

In the cell, reserve polysaccharides are mostly deposited in the form of globular aggregates, which allows very many starting-points for enzymatic access. Only in a few plants, and very rarely in lower animals, has the reserve polysaccharide a linear structure. Such compounds, as say the fructan *inulin* from Jerusalem artichoke and dahlia tubers or the *galactan* from many snails, reach only small molecular weights of some thousands. The vast majority of living things solve the problem in another way and much more effectively by being able to attack simultaneously their food reserve at many positions. The reserve polysaccharides of almost all plants and animals and many microorganisms are either *starch* with its high molecular weight component, *amylopectin*, and the low molecular weight component amylose or the likewise high molecular weight glycogen. Glycogen and amylopectin belong to the group comprising the largest, uniform biomolecules with molecular weights between $10^6$ and $10^9$. Both polymers are highly branched which results from the on average, linear chain length of 20 to 25 monosaccharide units with a weight of about 4000, in striking contrast to the total molecular weight.

The open, tree-like branched structure of amylopectin and glycogen offers catabolic as well as anabolic enzymes many starting points for simultaneous action. In many respects there exists great similarity between both polymers regarding structure and metabolism. Of necessity, however, the regulation of deposition and mobilisation is clearly different[41] but this is not considered within the framework of this book.

#### 2.1.3.1 Starch

*Amylose* and *amyopectin*, the two components of starch, are the carbohydrate reserve in almost all plants and, after cellulose, the main polysaccharides. Starch occurs in almost all tissue parts of green plants, predominantly however in organs provided for storage such as grain, root tubers, fruit and seeds

α–NeuAc–(2 → 6)–α–GalNAc–(1 →
sheep, bovine, armadillo
(upper-jaw salivary gland)

α–NeuAc–(2 ↘ 6)
α–GalNAc–(1 ↘ 3)
β–Gal–(1 ↗ 3) α–GalNAc–(1 →
α–Fuc–(1 ↗ 2)
pig
(upper-jaw salivary gland)

α–NeuAc–(2 ↘ 6)
α–GalNAc–(1 →
α–Fuc–(1 → 2)–β–Gal–(1 ↗ 3)
pig, dog
(upper-jaw salivary gland)

β–Gal–(1 → 4)–β–GlcNAc–(1 ↘ 6)
β–Gal–(1 → 4)–β–GlcNAc–(1 ↘ 6)
α–GalNAc–(1 →
β–Gal–(1 ↗ 3)
α–Fuc–(1 ↘ 4)
α–GalNAc–(1 ↗ 3)
β–Gal–(1 ↗ 3)
human (ovary)

β–Gal–(1 → 4)–β–GlcNAc–(1 ↘ 6)
α–GalNAc–(1 →
β–Gal–(1 → 4)–β–GlcNAc–(1 ↗ 3)
sheep
(tongue salivary gland,
stomach mucus)

α–Sia–(2 ↘ 6)
α–GalNAc–(1 →
α–Sia–(2 ↘ 6)
β–Gal–(1 → 4)–β–GlcNAc–(1 ↗ 3)
α–Sia–(2 ↘ 6)
β–Gal–(1 → 4)–β–GlcNAc–(1 ↗ 3)
α–Sia–(2 → 4)–β–GlcNAc–(1 ↗ 3)
rat
(colon mucus,
tongue salivary gland)

**Fig. 2.40** Oligosaccharide structures of the *glycoproteins of the mucin type* are clearly distinguished according to their source of organ origin and to the species of origin. The monosaccharide of attachment to the protein is, however, always α–GalNAc. *Sia* is acylated, mostly *N*-acetylated neuraminic acid., e. g. NeuAc.

of all types. In photosynthesis, starch arises directly in the chloroplasts of the green parts of plants and can be directly identified there after extraction of the chlorophyll with suitable solvents through the very sensitive deep-blue, iodine-iodide colour test. Photographs by electron microscopy show the temporary deposited starch grains in the stroma of the chloroplasts (Fig. 2.41)[36] as oval particles of about 1 $\mu$m diameter.

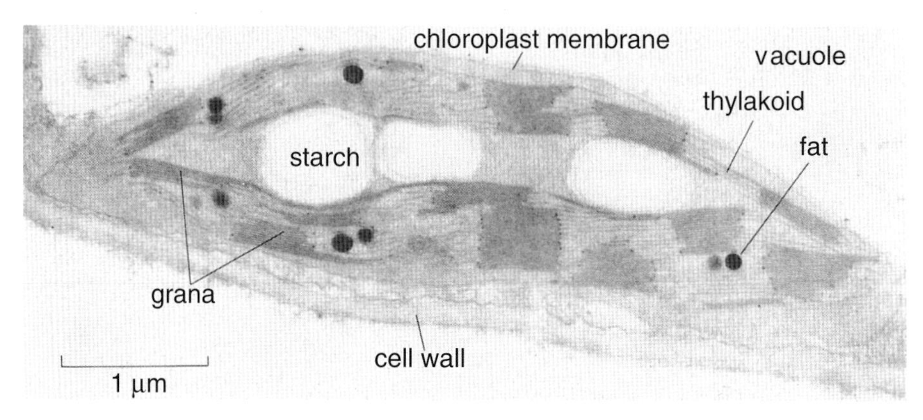


**Fig. 2.41** Electron micrograph of a *chloroplast*, the photosynthesising organelle in cells of the green components of plants. The starch grains are clearly seen as light particles of about 1 $\mu$m diameter.

With isolated chloroplasts starch can be synthesised *in vitro*. The amyloplasts in storage tissue contain no pigment and only sparingly developed and distributed membranes. Starch is stored there for a period. Starch granules, which are insoluble in cold water, can be isolated by destruction of the tissue by washing out, sieving, and filtration. They possess, depending on their origin, specific macroscopic forms (Fig. 2.42)[42] and a microscopic, clearly detectable radial order with alternately amorphous and crystalline regions (Fig. 2.43)[42]. By greater magnification, lamellae about 10 nm long and 5 nm thick are readily discernable in the growth-rings[43], which correspond well with the cluster-zones in the model of D. French and J. P. Robin et al.

By careful warming to not more than 60 °C in water, amylose may be dissolved out of the starch grains and, after centrifugation, can be precipitated by addition of ethanol. The amylose content in starch varies, according to origin, normally between 10 and 25 %, but can also assume extreme values. *Amylose* is a linearly constructed (1→4)-α-D-glucopyranan, in which bonding homogeneity can be proved by complete degradation with $\beta$-amylase. There is disagreement concerning the conformation of amylose chains. Currently, the stretched helical conformation and also the random coil with helical regions are discussed. The stretched helical form is the most likely on the grounds of the very narrow limits for the angles $\phi$ and $\varphi$[44]. In the presence of complexing agents, for example iodine-iodide, opinions are certainly united about the uniform helical conformation of amylose[45,46]. The inclusion complex contains the amylose helix with a periodicity of probably 6 glucose units.

Amylopectin is, as already discussed, a highly branched D-glucan with (1→4)-α-linked sequences and (1→6)-α-linked branching points. By the use of different catabolic enzymes, by partial hydrolysis, and also from the results of methylation analysis, it is possible to describe roughly the complex primary structure of amylopectin.

Two models have been discussed and both indicate something similar. The model of S. Peat[47] describes amylopectin with a single continuous main chain C, which carries, at the same time, the reducing end group (Fig. 2.44). On this are attached the B-chains of medium length through (1→6)-α-glycosidic linkages. On the periphery are found the short A-chains, likewise (1→6)-α-linked. The model of D. French[48] is somewhat more precise and indicates the average chain length. Accordingly the outer, short S-chains, are 15 – 20 glucosyl units long, the inner L-chains about 45 units and the main chain over 60 units long The main chain and L-chain are the skeleton of the molecule on which thick clusters of S-chains are attached. The periodic clusters explain in a certain way the radial, periodically changing

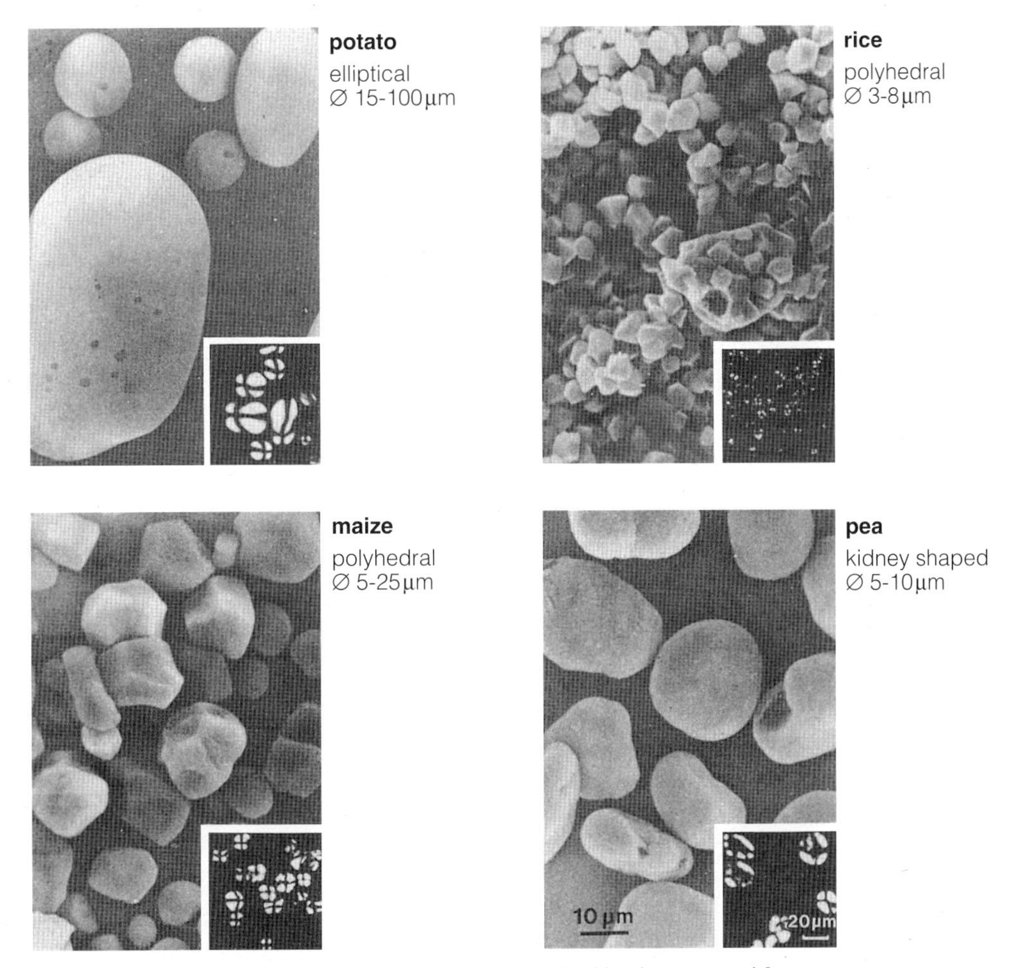


**Fig. 2.42** Starch grains of different origin may be distinguished by their size and form.

dense zones in the starch granules[49]. Numerous enzymes with different specificities can hydrolytically degrade amylopectin, completely or partially (Fig. 2.45).

$\beta$-Amylase is a typical plant enzyme which can break down amylopectin into maltose and the so-called limit-dextrin. Since the enzyme as an exoglycanase cleaves from the non-reducing ends and cannot cleave (1→6)-linkages, inevitably, a polymeric residue remains. The ratio of maltose/limit dextrin is a good measure of the degree of branching and the proportion of the different chain types.

### 2.1.3.2 Glycogen

The reserve polysaccharide of animals and fungi is *glycogen*. Since it has a large structural similarity with amylopectin of plants most of the research methods on structure and macroscopic organisation of starch are also used. As in that case, break down with $\beta$-amylase leads to maltose and a limit dextrin in certain relative proportions. Together with chemical methods for end group determination, numerous results of measurement yield an average chain length of (1→4)-$\alpha$-linked D-glucopyranosyl residues of 11–14, depending on the method used and the origin of the preparation. This means, in contrast to amylopectin, a clear denser branching. With the help of a coupled degradation, alternating with phosphorylase and isoamylase – phosphorylase cleaves off (1→4)-$\alpha$-linked D-glucopyranosyl residues up to about 4 units before a (1→6)-$\alpha$-branching point, isoamylase removes the remaining stub so that phos-

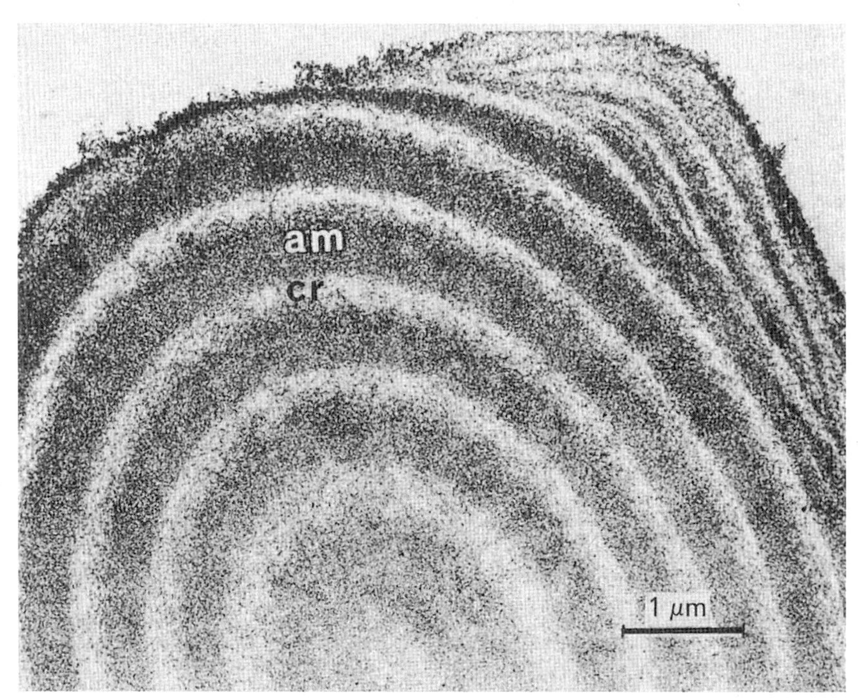


**Fig. 2.43** *Starch grain* from maize. Amorphous (am) and crystalline (cr) layers alternate.

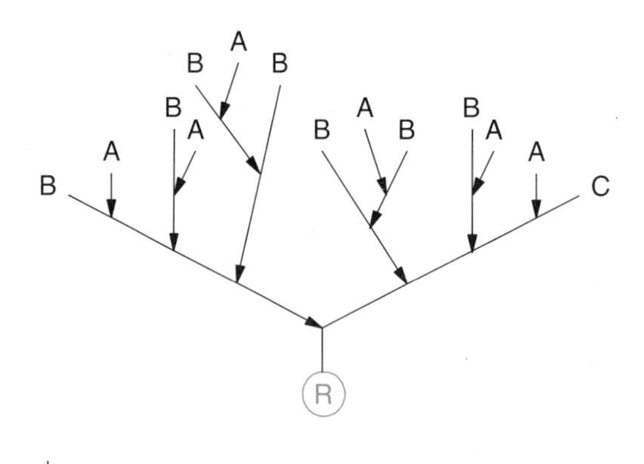


**Fig. 2.44** Schematic representation of an *amylopectin molecule*, according to a proposal of S. Peat.
**A** Short peripheral and
**B** mid-length inner chain;
**C** main chain which carries the reducing end R.

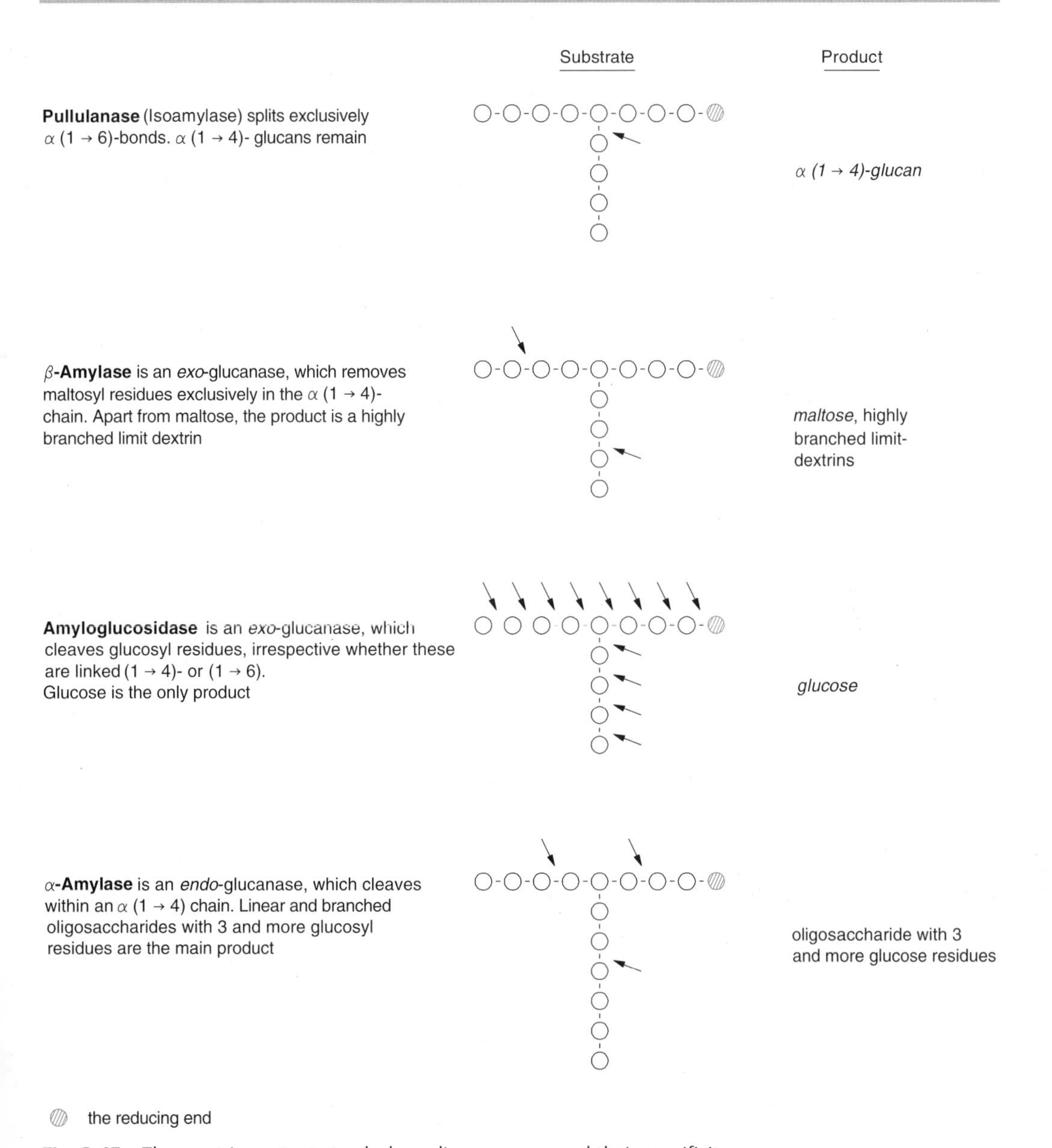


**Fig. 2.45** The most important starch-degrading enzymes and their specificity.

phorylase can attack again – it was possible to obtain three limit dextrins. From this a three-fold periodicity of accumulation of branching points in glycogen was derived, which led to the two suggested models of Meyer and Bernfeld[50] (**A**) and of Whelan[51] (**B**) (Fig. 2.46).

Model (**B**) gets closer to the truth because by the action of isoamylase glycogen can be broken down into a mixture of chains of (1→4)-α-linked D-glucopyranosyl residues, whose weight distribution is symmetrical and relatively narrow and corresponds approximately to the average chain length. Glycogen occurs as rosette-shaped granules deposited in the cytoplasm of almost all cells of animal tissue. Liver and muscle cells are especially rich in glycogen (Fig. 2.47). The small (100 – 150 nm) *glycogen granules* (α-particles of liver cells) differ from the *starch grains* (50 μm) of plants by the dimensions alone.

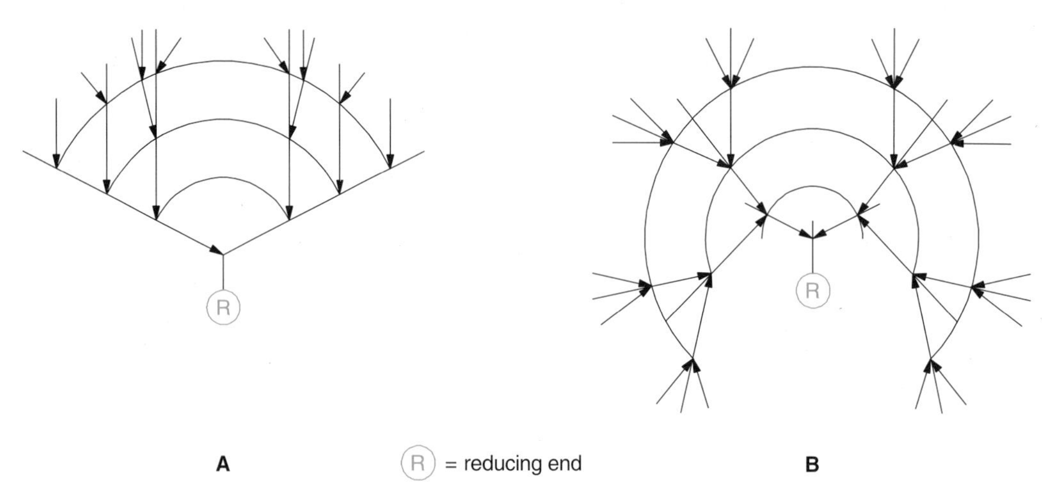


**Fig. 2.46** *Glycogen structure* of **A** Meyer and Bernfeld and **B** Whelan.

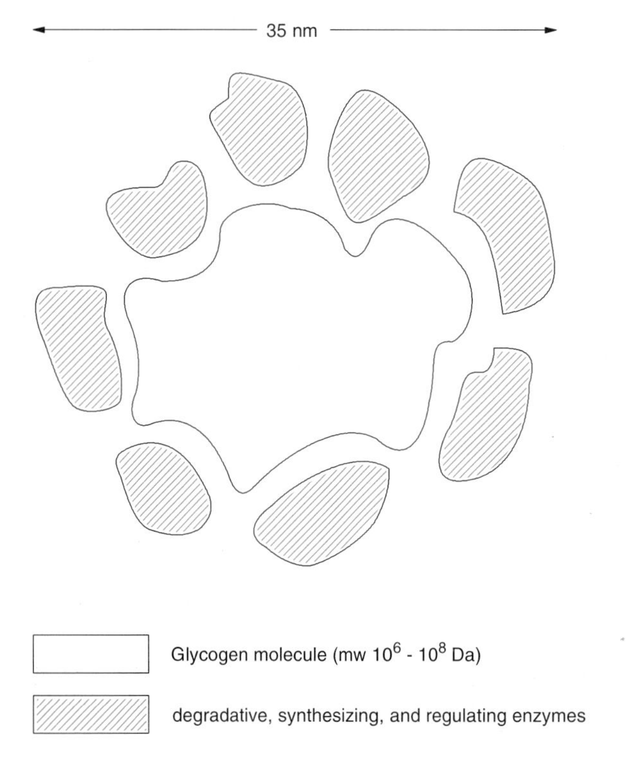


**Fig. 2.47** *Glycogen granules.* The highly ramified glycogen molecule, perhaps covalently bound as a proteoglycan, 2–3 fold in protein, is surrounded by a single layer of degradative, synthesising and regulating enzymes.

While starch grains – in regard to their size – are relatively free of protein, the $\beta$-particles (25–35 nm) of the glycogen granules are associated both with catabolic (phosphorylase), anabolic (glycogen synthetase) as well as with regulating (phosphoryrlase-kinase) enzymes, so that one can speak of a glycogen-enzyme complex. As a result of this non-covalent association with enzymes, it has been suspected for some time that glycogen is a proteoglycan. There are, currently, research results which suggest that the larger glycogen particles ($\alpha$-particles) consist of several lower-complexes formed from two to three glucan molecules covalently bound through protein ($\beta$-particles). The $\alpha$-particles, which usually occur in the liver, are then supposed to be formed from these lower-complexes by aggregation, through formation of intermolecular disulphide bridges[52].

#### 2.1.3.3 Fructans

Fructans are polymers of $\beta$-D-fructofuranose and belong either to the inulin type or to the phlein type. The former is $(2\rightarrow1)$-linked, the latter $(2\rightarrow6)$-linked (Fig. 2.48). In comparison to starch or glycogen, the reserve fructans are small molecules with about 20 to 30 fructose units. They form, as it were, a transition between the highly branched macromolecules amylopectin and glycogen on the one hand and, on the other, the non-reducing reserve oligosaccharides sucrose, raffinose, stachyose of many plants and the likewise non-reducing trehalose of the fungi and insects (see section 2.1.4). Inulin is found in dahlia and Jerusalem artichoke tubers and is suitable as a starch substitute for diabetics.

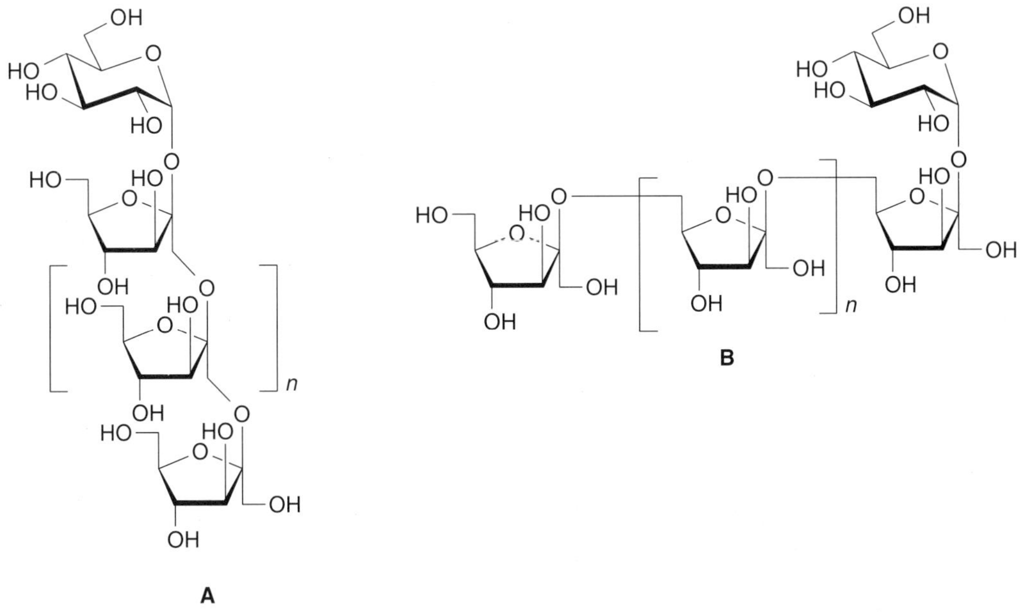

**Fig. 2.48** *Fructans* are reserve polysaccharides of some higher plants and algae. They belong either to the **A** *inulin* type or to the **B** *phlein* type. Both polymers carry a sucrose unit on the aglyconic end.

### 2.1.4 Low molecular weight carbohydrates in extra-cellular fluids

The body fluids of all living things are aqueous solutions. Cells and compartments in cells remain separate from one another, and many cells for their part are again delimited from the outside world. The physicochemical demarcation occurs through hydrophobic membranes which, as already discussed, are mechanically protected and secured by cell walls or the extra-cellular matrix. The membranes prevent the uncontrolled flow of dissolved cell components. In order that, nevertheless, an exchange of water soluble compounds and minerals can take place, specific transport systems in the membranes, which are discussed later, provide for the flow of materials. While microorganisms obtain their nourishment direct from the outside world and transport this in dissolved form through the cell wall and membrane in the cytoplasm and transport corresponding waste materials to the outside, higher organisms in contrast make use of special systems for the provision of cells of the different organs.

Carbohydrates, especially glucose, are not only key substances of general metabolism, but they are also ideally suited for the transport of energy and the equivalent of raw materials in body fluids. Most low molecular weight representatives have nearly unlimited solubility in water and possess no charge.

### 2.1.4.1 Sucrose and oligosaccharides of the raffinose family

The problem of transport of carbohydrates begins with photosynthesis in the chloroplasts of green plants, first inside the cell, then from cell to cell, and finally in the long-distance transport from organ to organ (Fig. 2.49).

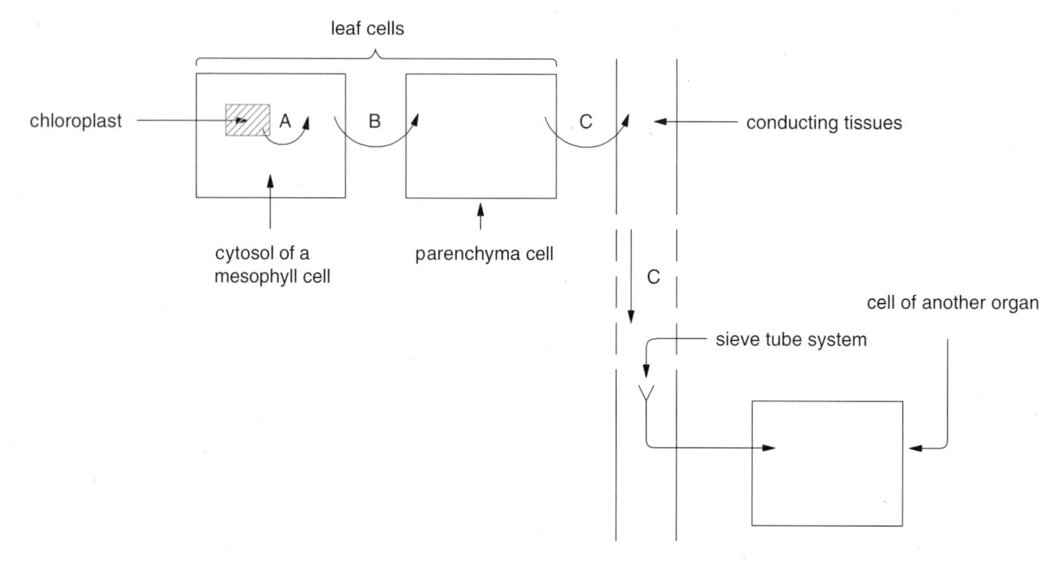


**Fig. 2.49** Transport of carbohydrates, **A** inside the cell, **B** from cell to cell and finally **C** from organ to organ takes place by means of low molecular weight, water soluble compounds.

The long-distance transport is achieved through the phloem of plants in the form of the phloem sap with which, especially, the assimilation products are transmitted to the non-photosynthesising cells. The phloem sap is a very concentrated solution of 50 – 300 g of solid material per litre. About 90% of this solid material is carbohydrate, principally sucrose but also derivatives of sucrose and various sugar alcohols. A complicated osmotic mechanism ensures transport of the solution in the phloem. The compounds transported in the sap are selected previously by the membrane transport mechanism which apparently keeps back all reducing sugars. *Sucrose* is formed in high concentration by the reaction sequence shown (Fig. 2.50).

The dephosphorylation step of sucrose-6-P especially determines the reaction gradient. Sucrose is the form of transport of carbohydrates in almost all higher plants, and in many is also the reserve carbohydrate, as for instance in sugar cane or sugar beet, whose pressed juice can contain up to 16% sucrose. The metabolisable monosaccharides glucose and fructose are liberated enzymatically from sucrose when required without expenditure of energy. The high potential of sucrose towards group transfer allows, even without expenditure of energy, the formation of activated glucose, UDP-Glc or ADP-Glc (uridine or adenosine diphosphoglucose) (Fig. 2.51) for the synthesis (see section 2.3.5, p. 189) of the reserve polysaccharide starch.

The question arises why nature has chosen sucrose as the transport carbohydrate in plants. While other forms of transport are supported, the decisive factor is possibly that sucrose is non-reducing and correspondingly insensitive towards chemical, especially oxidative influences. Dimerisation has, moreover, the important advantage of maintaining as low as possible the osmotic pressure of the phloem sap. A further saccharide-specific argument is possibly the high potential of sucrose, already mentioned, for group transfer. In the case of hydrolysis, the thermodynamic equilibrium, catalysed by the

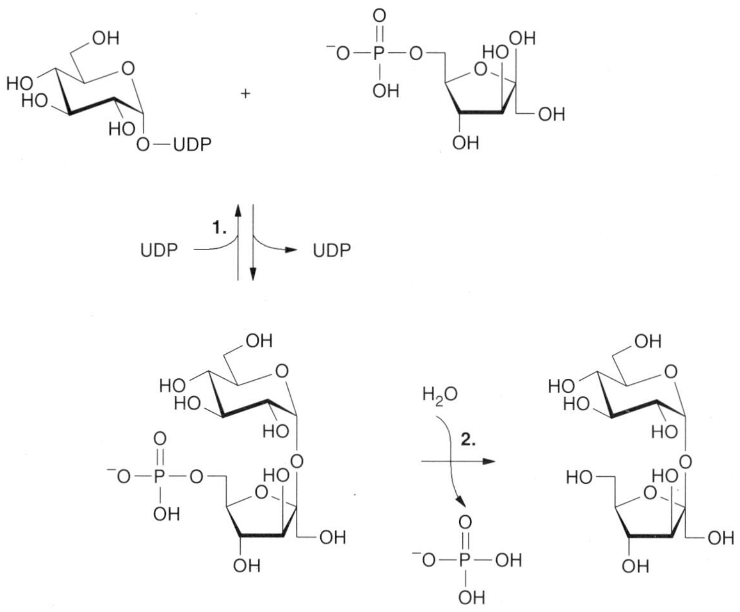

**Fig. 2.50** *Sucrose*, the most important transport carbohydrate of green plants, arises in an irreversible synthetic sequence by the dephosphorylation of sucrose 6-phosphate.
**1.** Sucrose phosphate synthase
**2.** Sucrose 6-phophatase

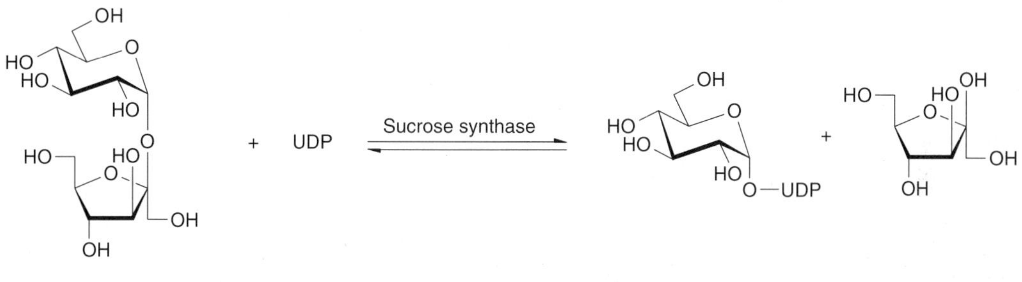

**Fig. 2.51** The high potential for group transfer from sucrose allows the synthesis of *activated glucose* by glucosyl transfer to UDP.

enzyme invertase, lies fully on the side of the components glucose and fructose, only by reason of the fact that out of one species 6 isomeric compounds can arise.

A further lowering of the molar concentration in phloem sap, and thereby the osmosis, can be achieved by oligomerisation of the sucrose. In many plants, beside sucrose, products of α-D-galactosylation, *raffinose*, *stachyose*, *verbascose*, and *ajugose* are found as forms of carbohydrate transport and storage (Fig. 2.52).

As with sucrose, these are non-reducing sugars. The oligosaccharides arise, surprisingly, not in the normal manner of galactosylation by a transferase and nucleoside diphosphogalactose (for example UDP-Gal), but through transfer from a *myo*-inositol derivative widely occurring in photosynthesising plants, *galactinol*, which for its part arises by galactosylation with UDP-Gal. Why this route is followed is still not clear. Galactinol is itself not used as a transport agent.

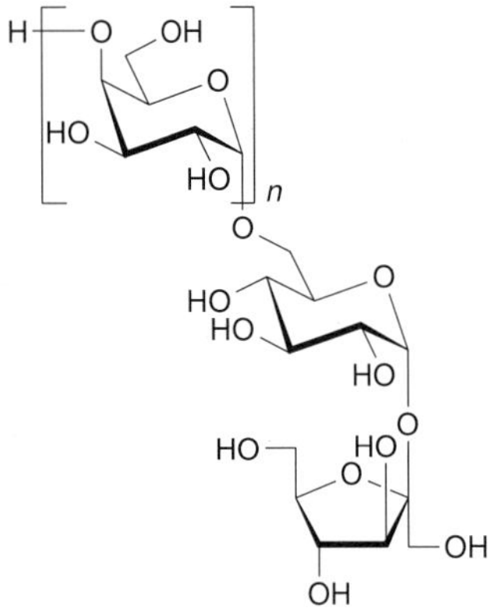

n = 0 Sucrose
n = 1 Raffinose
n = 2 Stachyose
n = 3 Verbascose
n = 4 Ajugose

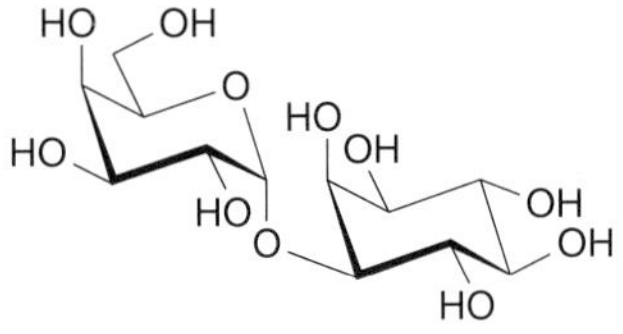

Galactinol

**Fig. 2.52** The *raffinose family* of non-reducing plant oligosaccharides. Raising the molecular weight lowers the osmotic value of the phloem sap. The galactosyl donor is *galactinol*, a galatosylated *myo*-inositol, which is not transported itself.

### 2.1.4.2 Trehalose

The non-reducing disaccharide α,α-trehalose (Fig. 2.53) fills an entirely similar function to the purely plant disaccharide sucrose as a low molecular weight transport-, and possibly also reserve-carbohydrate.

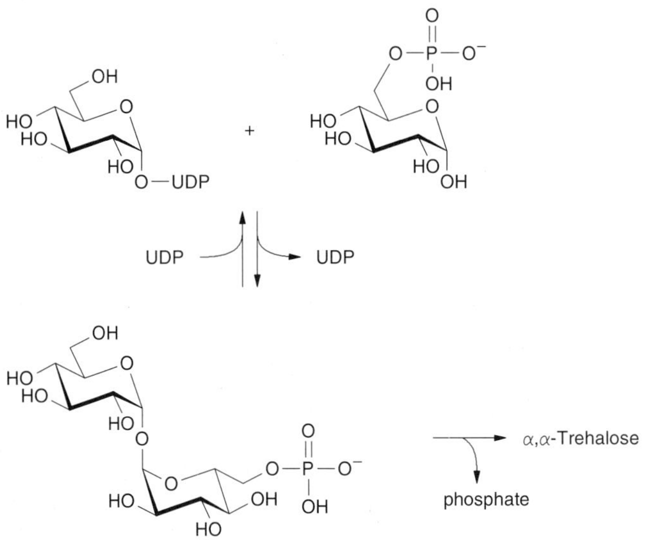

**Fig. 2.53** α,α-Trehalose is biosynthesised in an analogous manner to sucrose (see Fig. 2.50, p. 107).

Trehalose is found primarily in insects, but also in other invertebrates, in fungi and yeasts, and even in many plants. Its biosynthesis is comparable with that of sucrose (Fig. 2.53). In yeasts and insects, trehalose has a storage function as well as a transport function. In the case of a plentiful carbohydrate supply in yeasts, trehalose is synthesised, deposited in certain organelles, and mobilised on need. In insects and invertebrates, trehalose is the *main blood sugar* in concentrations of up to 2%, which exceeds the glucose content in the blood of vertebrates. This fact is surely an indication that similar selection criteria, as were described already for sucrose and the oligosaccharides of the raffinose family, were decisive also for the selection of trehalose.

### 2.1.4.3 Blood glucose

In vertebrates, and of course in humans, D-glucose is the main form of transport for carbohydrates. The relationships concerning this between storage organs (liver), transport (blood) and the organs of consumption (muscle, brain- and nerve tissue) are complicated (Fig. 2.54)

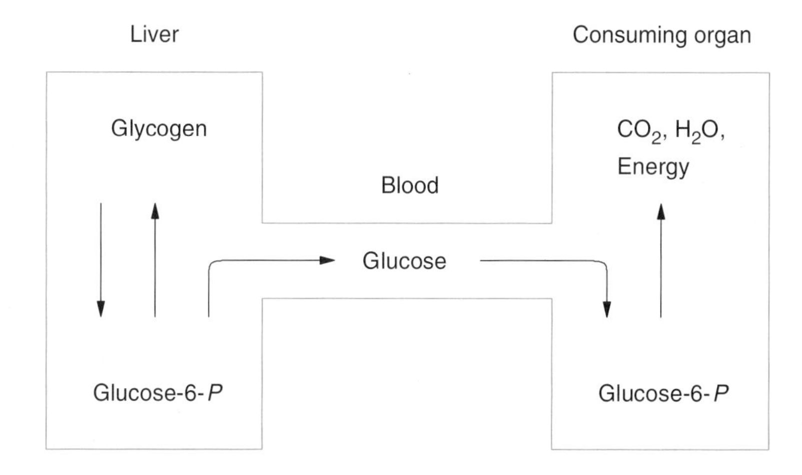


**Fig. 2.54** *Glucose* facilitates the transport of equivalents of energy from the liver through the circulatory system to the body's organs, as shown in the very simplified scheme.

The course of events is delicately regulated so that the blood always shows a constant glucose concentration. Even slight disturbances lead to irreversible damage, especially of the central nervous system. For the provision of energy of the nerve tissues and of erythrocytes, glucose is irreplaceable. Under normal conditions it also provides all other cells with energy. It is the starting material for the biosynthesis of pentose phosphate and thereby of RNA and DNA. Glycerol of the membrane lipids arises exclusively from glucose.

Under fasting conditions the glucose concentration in the blood of humans, depending on the individual, amounts to 0.5 – 1 g/L. This concentration can rise after a carbohydrate-rich meal to about 1.3 g/L, but after a short time it again attains the normal value. In one hour humans consume, depending on the body weight and activity, about 20 g of glucose. The brain and the nervous system are especially dependent on the permanent supply of glucose as an energy source. Transport processes in relation to glucose and the role of glucose in metabolism are considered in sections 2.2.1 (p. 130) and 2.3 (p. 173).

After having tried in the last section to rationalise the advantages of using non-reducing disaccharides such as sucrose and trehalose as transport forms for carbohydrates, it is difficult to find an explanation why the most highly developed living beings, of all things, do not make use of them. Perhaps the highly developed and therefore very sensitive brain makes it necessary to employ a small, directly usable molecule such as glucose. The blood vessels in the brain are especially impermeable to organic compounds of all types and form thereby, the so-called blood-brain barrier as a protection from disturbing influences. Only for a very limited number of compounds, amongst them glucose, is there a highly specific transport system by which the brain and the nerve cells are provided with the necessary nourishment. For a small molecule such as glucose, a very specific transport system may be simpler to develop than for a more complex molecule, in which case, naturally, the disadvantages discussed above, especially the high osmotic pressure, must be compensated.

### 2.1.4.4 Milk oligosaccharides

Lactose, as the main carbohydrate for the nourishment of new born mammals, is not necessarily a form of transport of carbohydrate in the sense of the functions discussed in the preceding sections. It is included in this chapter, nevertheless, for two good reasons. Lactose serves, as sucrose, trehalose, and glucose for the provision of the organism with energy and raw materials for biosynthesis, and it is the third of only three naturally occurring, biosynthesised disaccharides, directed for a definite use.

Lactose forms in the milk glands of female mammals only after their offspring has been born. A hormonal signal is necessary and with that lactose arises by galactosylation of free glucose, this biosynthesis differing fundamentally from that of both sucrose and trehalose. Lactose synthesis is maintained by the stimulus of suckling. If the demand for milk ceases and the stimulus of suckling is lost, then the synthesis of lactose is also switched off. Only in the presence of *α-lactalbumin* is glucose an acceptor substrate for galactosyl transferase. Without the milk protein α-lactalbumin, the Michaelis constant of galactosyl transferase for glucose lies at 1.4 mM, that is far above the physiological concentration of the monosaccharide. This type of regulation is biologically meaningful since the galactosyl transferase normally has the task to galactosylate GlcNAc end groups in glycoconjugates and is consequently an important enzyme of glycoprotein and glycolipid biosynthesis. It is distributed in the organism of all mammals and is "misused" on demand only momentarily for the production of lactose. For mammalian organisms it is manifestly advantageous to synthesise α-lactalbumin as the regulatory unit through hormonal impulse for galactosyl transferase. Two birds are therby killed with one stone, since α-lactalbumin, by no means present only in catalytic concentration, is required as a protein source, albeit inferior, for the young.

If the milk of various mammals is analysed, then differences in the composition of the carbohydrate portion are particularly noticeable. Cow's milk contains about 50 g/L of lactose and only traces of other oligosaccharides. Human milk on the other hand contains alongside about 85 g/L of lactose a multitude of complex oligosaccharides in a concentration of at least 5 – 17 g/L. While the colostrum of humans, the milk secreted immediately after birth, contains no greatly increased concentration of the so-called milk oligosaccharides[53], the colostrum of other mammals is, astonishingly, in contrast to the later delivered milk, considerably richer in the milk oligosaccharides. It may be conjectured that these in part fairly complex compounds have important protective functions for the new born. Evidently, in the case of slow growing human babies, in contrast to calves and lambs, the protection must be maintained longer.

Almost all oligosaccharides of human milk contain lactose as the reducing end group. Attachment of GlcNAc, (1→3)-*β*- or (1→6)-*β*- onto the galactosyl residue produces different core structures, which can be extended further through (1→3)-*β*- or (1→4)-*β*-D-galactosyl residues. The Gal-GlcNAc unit can occur repeatedly so that a multitude of the so-called extended core structures – 17 have been identified up to now – are possible. The attachment of sialyl- (by *sialic acid* is meant, in general, the *N*-acetylated neuraminic acids – NeuAc; various sialic acids differ in the type and position of the acyl group) and/or fucosyl-residues increases the number of the milk oligosaccharides considerably. Lists of proved structures and their possible biological importance are to be found in the literature[54]. Certain genetically determined glycosylation paths result in certain structural patterns of the milk oligosaccharides which can serve to classify them (Fig. 2.55)[53].

If it is remembered that there can be more than 20 core structures depending on the glycosylation rules, which on their part can be further elaborated by different fucosyl- and sialyl transferases, either in competition with one another or independently one after another, over a hundred potential milk oligosaccharides may be derived without difficulty. It is striking that many antigenic determinants of the specific blood group structures, for example H, are again found in the milk oligosaccharides (Fig. 2.56).

Obviously the same genetically determined glycosyl transferases in the gland cells responsible for the construction of the milk oligosaccharides and glycoproteins of the mucin type are involved with the specific blood group determinants. The production of the milk oligosaccharides may be accidental and therefore biologically meaningless. The presence in high concentration and the similar structure of the *false* acceptor lactose may deceive the glycosyl transferase of the milk gland cells which are normally focussed on *N-acetyllactosamine* as the acceptor. Accident or not, it seems certain today that the oligosaccharides of human milk play an important role in the prevention of bacterial infection in the new born. The attachment of germs to the epithelial cells of the middle ear, the urinary tract, and the digestive tract could be competitively inhibited by the oligosaccharides of human milk.

### 2.1.5 Chemical and physiochemical modulators of proteins and lipids

With improvements in analytical techniques, it is becoming ever more apparent that proteins with covalently attached oligosaccharides are not the exception, as could be asserted until not too long ago, but rather the rule. The glycosylation of proteins and of lipids plays a more and more noticeable role in all organisms including bacteria but above all in the most highly developed mammals. The

α−Fuc−(1 → 2)−β−Gal−(1 → 4)−Glc — 2-*O*-Fucosyllactose

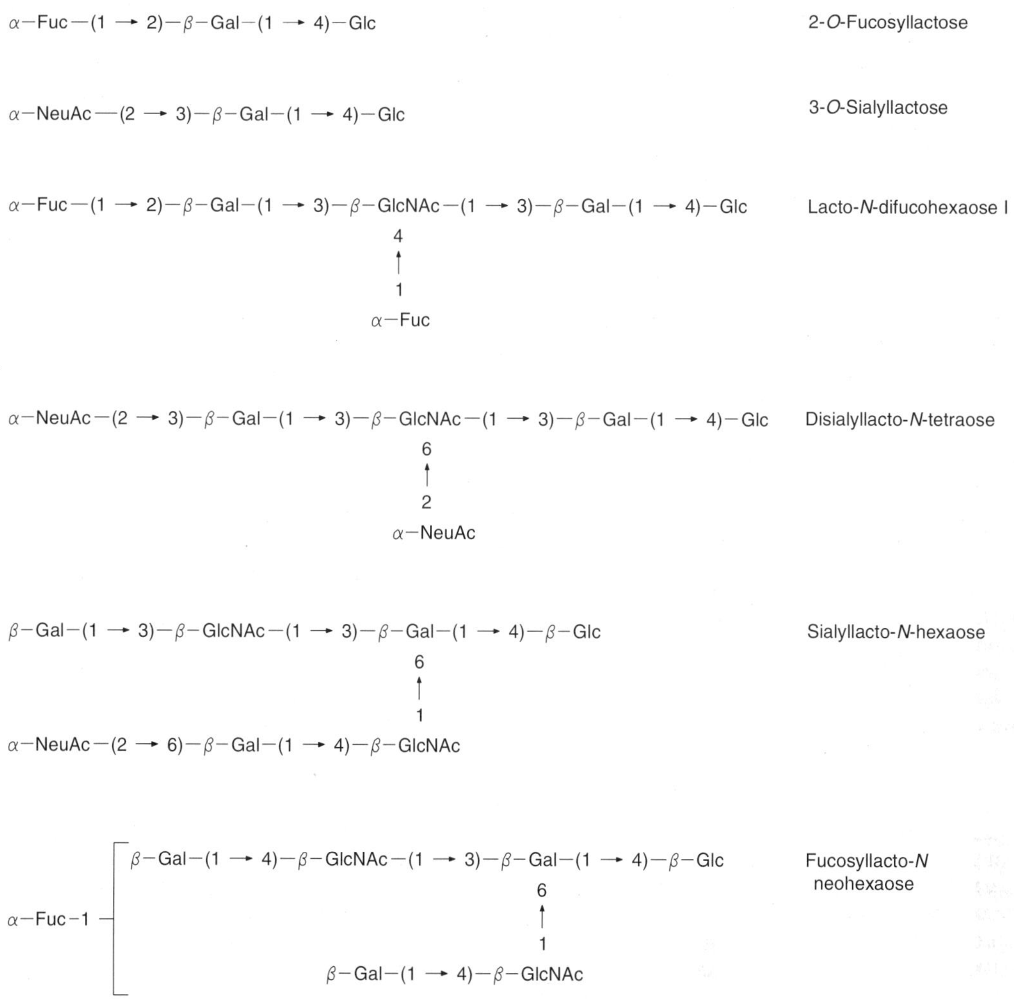

α−NeuAc−(2 → 3)−β−Gal−(1 → 4)−Glc — 3-*O*-Sialyllactose

α−Fuc−(1 → 2)−β−Gal−(1 → 3)−β−GlcNAc−(1 → 3)−β−Gal−(1 → 4)−Glc — Lacto-*N*-difucohexaose I
(GlcNAc 4 ← 1 α−Fuc)

α−NeuAc−(2 → 3)−β−Gal−(1 → 3)−β−GlcNAc−(1 → 3)−β−Gal−(1 → 4)−Glc — Disialyllacto-*N*-tetraose
(GlcNAc 6 ← 2 α−NeuAc)

β−Gal−(1 → 3)−β−GlcNAc−(1 → 3)−β−Gal−(1 → 4)−β−Glc — Sialyllacto-*N*-hexaose
(Gal 6 ← 1)
α−NeuAc−(2 → 6)−β−Gal−(1 → 4)−β−GlcNAc

α−Fuc−1 — [ β−Gal−(1 → 4)−β−GlcNAc−(1 → 3)−β−Gal−(1 → 4)−β−Glc — Fucosyllacto-*N* neohexaose
(Gal 6 ← 1)
β−Gal−(1 → 4)−β−GlcNAc ]

**Fig. 2.55** *Oligosaccharides of human milk* can be placed, according to A. Kobata, in seven structural series. Oligosaccharides of the lactose series, the lacto-*N*-tetraose series, the lacto-*N*-hexaose series and the lacto-*N*-neohexaose series are shown here.

covalent associations between oligosaccharides on the one hand and proteins and lipids on the other are combined today under the collective name glycoconjugates, to which belong, according to definition, the proteoglycans dealt with in section 2.1.2.4 (p. 91). As already intimated, with the carbohydrate structures it is a question of oligosaccharides of various sizes and various degrees of branching, but sometimes also only of monosaccharides. The carbohydrate portions in glycoproteins can fluctuate between a few percent and more than half of the total mass and in the case of glycolipids the carbohydrate portion is in preponderance. The most important difference between the carbohydrate structures of the glycolipids or glycoproteins (glycosylated proteins) – they shall be called glycoconjugates in the following (see section 2.1.1, p. 67) – and that in proteoglycans, for example of the extracellular matrix in vertebrates, or in lipopolysaccharides of the cell wall of gram-negative bacteria, is their biosynthesis which is ensured by highly specific transferases and is thereby genetically determined. The very economic principle of the *repeating unit* is not employed in the case of glycolipids and glycosylated proteins, and the principle of *post-biosynthetic modification* for the production of structural variety only very sparingly. A finished, built glycoconjugate always possesses the same conserved carbohydrate

α−Fuc−(1 → 2)−β−Gal−(1 → 3)−β−GlcNAc−(1 → 3)−β−Gal−(1 → 4)−Glc **A**
4
↑
1
α−Fuc

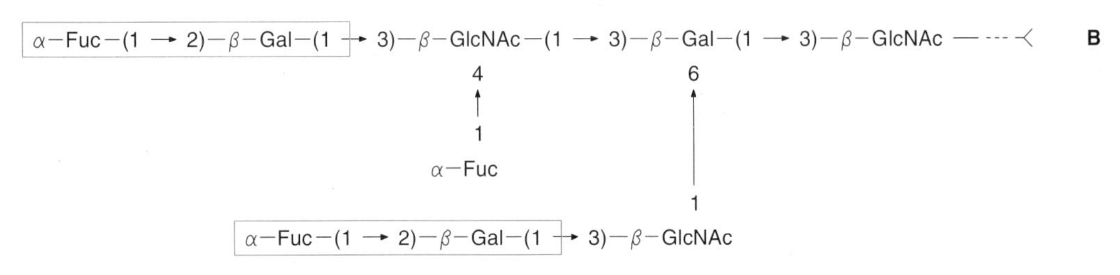


**Fig. 2.56** Structural similarity of the *blood group determinants* of *glycoproteins* of the *mucin type* with many oligosaccharides of human milk is attributable to the action of the same glycosyl transferases in the secretory glands. **A** Milk oligosaccharide *lacto-N-difucohexaose 1* is contrasted with the **B** H-determinant of glycoproteins of the blood group O.

structures. A certain microheterogeneity can be attributed to a definite content of incomplete molecules.

There are only a limited number (about 10) regularly occurring monosaccharides in glyconjugates of higher animals in comparison to the 20 proteinogenic amino acids (see section 2.1.1). The apparently unlimited, documented structural variation arises solely through different branching possibilities, change of anomeric configuration, and naturally also through the variation of the sequences.

There are some cases, proved beyond all doubt, where certain carbohydrate structures in glycoconjugates are definite, unambiguous recognition features in the widest possible sense. The number of these cases, some of which are dealt with in section 2.2.2 (p. 134), is comparatively small. The idea, that oligosaccharides possess in especially high measure the possibility of conserving and transmitting information is stimulating and inspires the research in the newly created area of glycobiology. One endeavours to seek a code behind certain structural elements, comparable to the DNA code or the biologically relevant tertiary structure in proteins determined by their primary structure. The future will show whether this was generally the purpose of evolution or whether, more likely, accidental, on hand oligosaccharide structures were enlisted for the release of certain signals, and that the function of the oligosaccharides in the preponderant number of glycoconjugates is trivial.

Carbohydrates possess in the first place the property of being hydrophilic. In this they are differentiated above all from the lipids but also from many proteins. Thus, nothing is more obvious, than to make a lipid or protein compatible with an aqueous medium through glycosylation. Oligosaccharides, above all highly branched ones, have relatively rigid conformations which are well adapted to place ionic groups and also lipophilic structural elements in certain positions on a protein surface or a membrane. In these cases the oligosaccharide serves as rigid framework for the installation of certain functional residues. Oligosaccharides can be very voluminous and when densely distributed on the surface of a protein protect against each type of denaturation as well as against attack of degrading enzymes. There are reasonable examples of all these functions. Some of these will be discussed in the following sections.

### 2.1.5.1 Glycoproteins of the body fluids

Almost all proteins found in the extracellular fluids are *glycoproteins*. The variation in respect to the relative proportion of protein to carbohydrate is large. *Ovalbumin* contains about 3% carbohydrate, *human serum-antithrombin III* contains 15% and salivary gland glycoprotein can consist of up to 50% carbohydrate. The carbohydrate part can be concentrated in a glycoprotein in only one oligosaccharide, as in ovalbumin, but it can be spread out also amongst about 800 simple disaccharide residues, as in the *submaxilla mucin* of sheep.

The chemical structure of glycoproteins allows recognition of two types of conjugates, the *O*-glycosidic and the *N*-glycosidic glycoproteins. The former carries the carbohydrate portion bound *O*-glycosidically to serine or threonine, the latter *N*-glycosidically to the amide nitrogen in asparagine. Both groups differ not only in the type of linkage but also in the type of the *bridgehead*-saccharide. In the case of *O*-glycosidic glycoproteins, with very few known exceptions (Fig. 2.57), an α-D-GalNAc is always bound to the protein (Fig. 2.58). The *N*-glycosidic glycoproteins all carry a core pentasaccharide

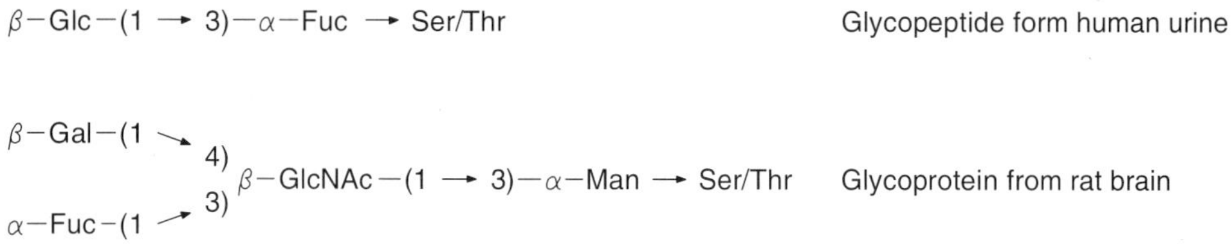

β−Glc−(1 → 3)−α−Fuc → Ser/Thr  Glycopeptide form human urine

β−Gal−(1 → 4)<br>α−Fuc−(1 → 3)<br>β−GlcNAc−(1 → 3)−α−Man → Ser/Thr  Glycoprotein from rat brain

**Fig. 2.57** Some unusual *O*-glycosidic glycoproteins or peptides have no *N*-acetyl-D-galactosamine as *bridgehead* monosaccharide.

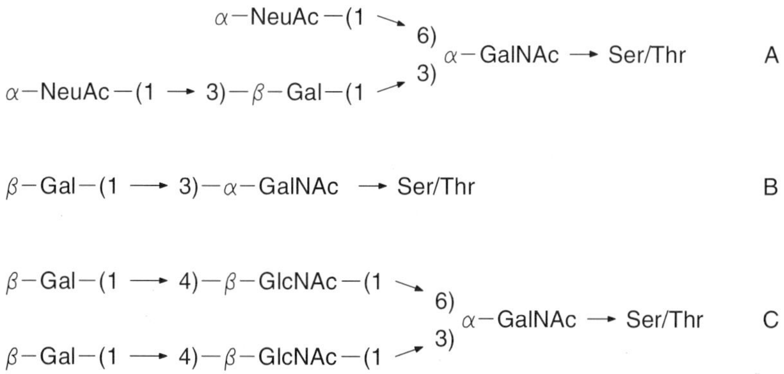

α−NeuAc−(1 → 6)<br>α−NeuAc−(1 → 3)−β−Gal−(1 → 3)<br>α−GalNAc → Ser/Thr  A

β−Gal−(1 → 3)−α−GalNAc → Ser/Thr  B

β−Gal−(1 → 4)−β−GlcNAc−(1 → 6)<br>β−Gal−(1 → 4)−β−GlcNAc−(1 → 3)<br>α−GalNAc → Ser/Thr  C

**Fig. 2.58** Typical *O*-glycosidic oligosaccharides in glycoproteins as for instance **A** in *glycophorin A* of human erythrocytes , **B** in the antifreeze protein of fish , or **C** in the mucin of gastric mucosa of sheep , are always bound to the protein through an α-GalNAc.

(Fig. 2.59) consisting of a chitobiose unit β-linked with the protein, on the 4-position of which is linked a β-Man*p* residue, which for its part carries two α-Man*p* residues, (1→3)- and (1→6)-linked, respectively. This so-called inner core is found in all *N*-glycosidic glycoproteins.

α−Man−(1 → 6)<br>α−Man−(1 → 3)<br>β−Man−(1 → 4)−β−GlcNAc−(1 → 4)−β−GlcNAc−(1 → Asn

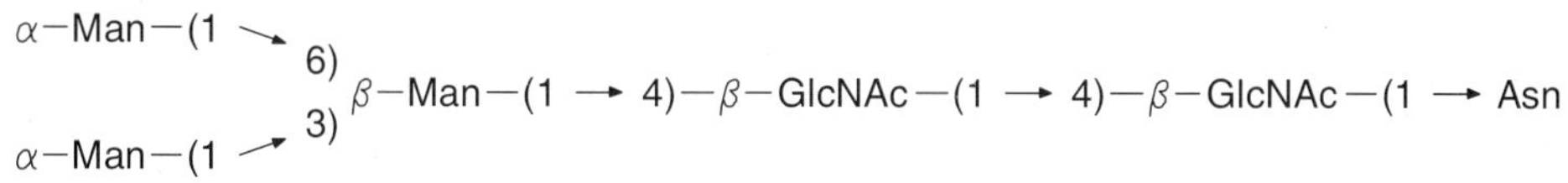

**Fig. 2.59** All *N*-glycosidic oligosaccharides in glycoproteins contain the core pentasaccharide shown, that is always bound onto an asparagine in the peptide chain.

Starting from the so-called bridgehead-structures, through further glycosylation, the in part function- and organ-specific, but also the species specific glycoproteins are formed . This apparently strict specificity is contradicted though by the fact that identical structures arise from sources which exhibit no connection whatsoever and that a certain glycoprotein can also show, beside complete oligosaccharides, *incomplete* oligosaccharide structures. The original assumption that a glycoprotein carries either only *O*- or only *N*-glycosidic oligosaccharides cannot be sustained. There are many known examples in which both types occur, as say human plasminogen, which carries an *N*-glycosidic oligosaccharide on Asn 288 and an *O*-glycosidic oligosaccharide on Thr 345[55]. However, as a rule, a glycoprotein is either of the *O*- or *N*-glycosidic type.

If one considers the oligosaccharide structures of some *O*-glycosidic glycoproteins, then a clear regularity stands out in respect of the monosaccharides utilised, the type of linkage, and the end group. One is therefore inclined to assign to the carbohydrate part a rather general chemical and physicochemical modulating function (see section 2.1.2.4, p. 91). Serum and salivary gland secretions are well researched media in respect of their glycoproteins (see Fig. 2.40, p. 99). Up to the present day, about 20 *O*-glycosidic glycoproteins have been isolated and identified from different species of vertebrates (Fig. 2.60).

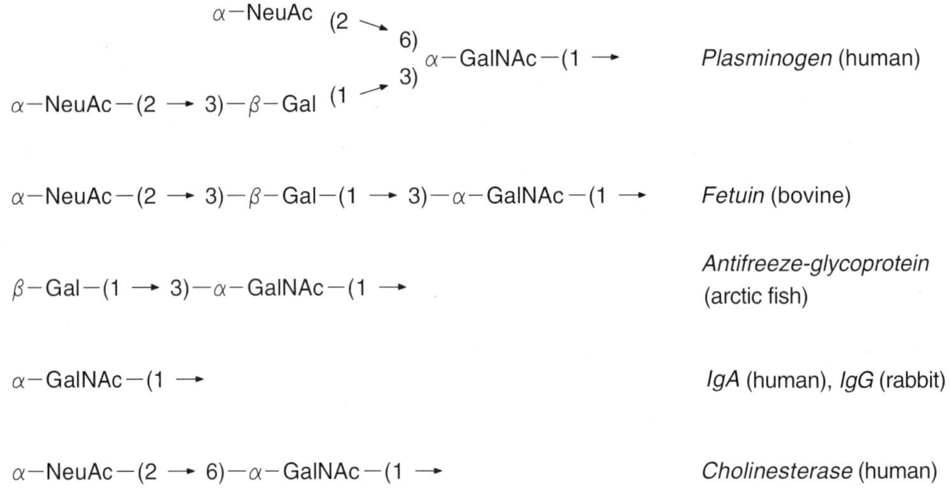

α−NeuAc (2 → 6)
α−GalNAc−(1 → *Plasminogen* (human)
α−NeuAc−(2 → 3)−β−Gal (1 → 3)

α−NeuAc−(2 → 3)−β−Gal−(1 → 3)−α−GalNAc−(1 → *Fetuin* (bovine)

β−Gal−(1 → 3)−α−GalNAc−(1 → *Antifreeze-glycoprotein* (arctic fish)

α−GalNAc−(1 → *IgA* (human), *IgG* (rabbit)

α−NeuAc−(2 → 6)−α−GalNAc−(1 → *Cholinesterase* (human)

**Fig. 2.60** The *O*-glycosidic oligosacchrides in plasma glycoproteins of different origins have great structural similarities (see Fig. 2.40, p. 99).

The necessity of certain oligosaccharide structures for biological function has been proved only in a few cases. An artificial alteration can be brought about by treatment with sialidase, that leads to a general decrease in the charge density on the protein surface and possibly also causes a functional modulation. For example, a certain glycoprotein (von Willebrand factor) is necessary for the attachment of blood platelets to injured vascular cells. With enzymatic removal of the NeuAc end groups, the glycoprotein loses this capability [56]. The function of the so-called ***anti-freeze*** glycoproteins in the serum of numerous polar fish types appears to be unambiguous. The glycoproteins, as a protection against cold, prevent the formation of ice because they are taken up by the seed crystals. The glycosylation is regular just as is the structure of the protein which is formed out of 17 to 50 repeat units of the tripeptide AlaAlaThr. The tripeptide carries on each threonine the disaccharide, typical for *O*-glycosidic glycoproteins, β-Gal-(1→3)-α-GalNAc[57]. It is certain that the biological function of glycoproteins protecting against frost is bound up with the existence of the oligosaccharide. Removal, chemical modification[58] or also enzymatic glycosylation of the 3-position of the terminal galactose with sialyl transferase[59] removes completely the effect of the glycoprotein in protecting against frost.

An example of an *N*-glycosidic glycoprotein is the well researched ***ovalbumin*** from chicken's eggs. The protein carries, as already intimated, only one oligosaccharide residue on Asn 292, although the repeatedly proved peptide Asn-X-Ser (Thr) as the signal sequence (***sequon***) for the *N*-glycosylation occurs twice in the primary structure of the protein. This alone speaks for the high specificity of the glycosylation reaction[60]. If the structures of the different isolated oligosaccharides are considered, then the strong fluctuation of the monosaccharide composition and branching pattern stand out (Fig. 2.61.).

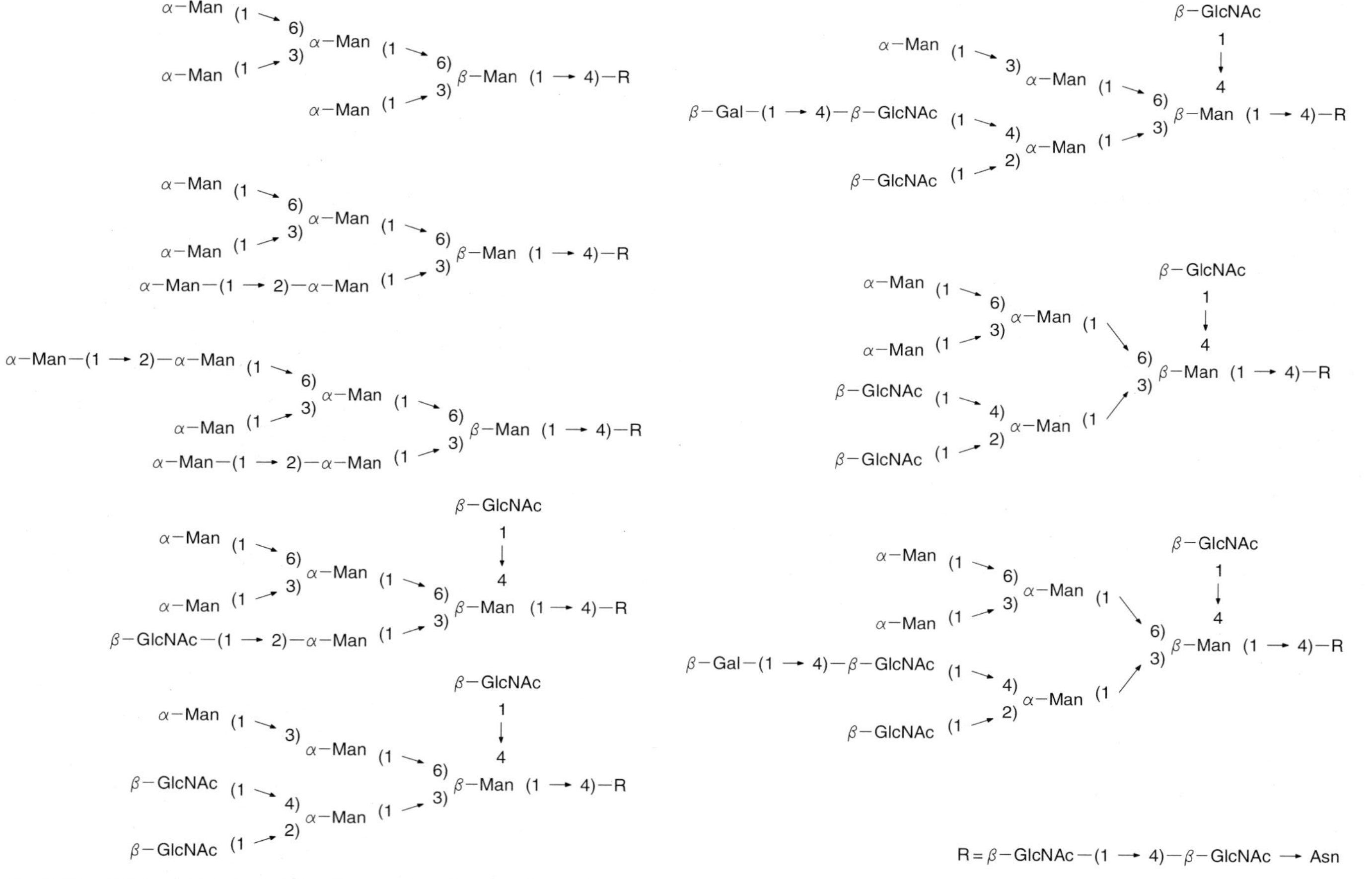


**Fig. 2.61** Different *N*-glycosidic oligosaccharides which are found in *ovalbumin*.

In this case it is very difficult to attribute a specific function to these oligosaccharides. Up till now, the most plausible explanation for the appearance of this so-called microheterogeneity is incomplete biosynthesis of the finished glycoconjugate. The construction of *N*-glycosidic glycoproteins is not only the result of glycosylation but also of glycoside cleavage, an occurrence which is designated as "processing" or "editing" (Fig. 2.62). This extraordinary complex biosynthetic process (see section 2.3.8.5, p. 218) accounts, in fact, for all the different oligosaccharide structures found in ovalbumin and in other gly-

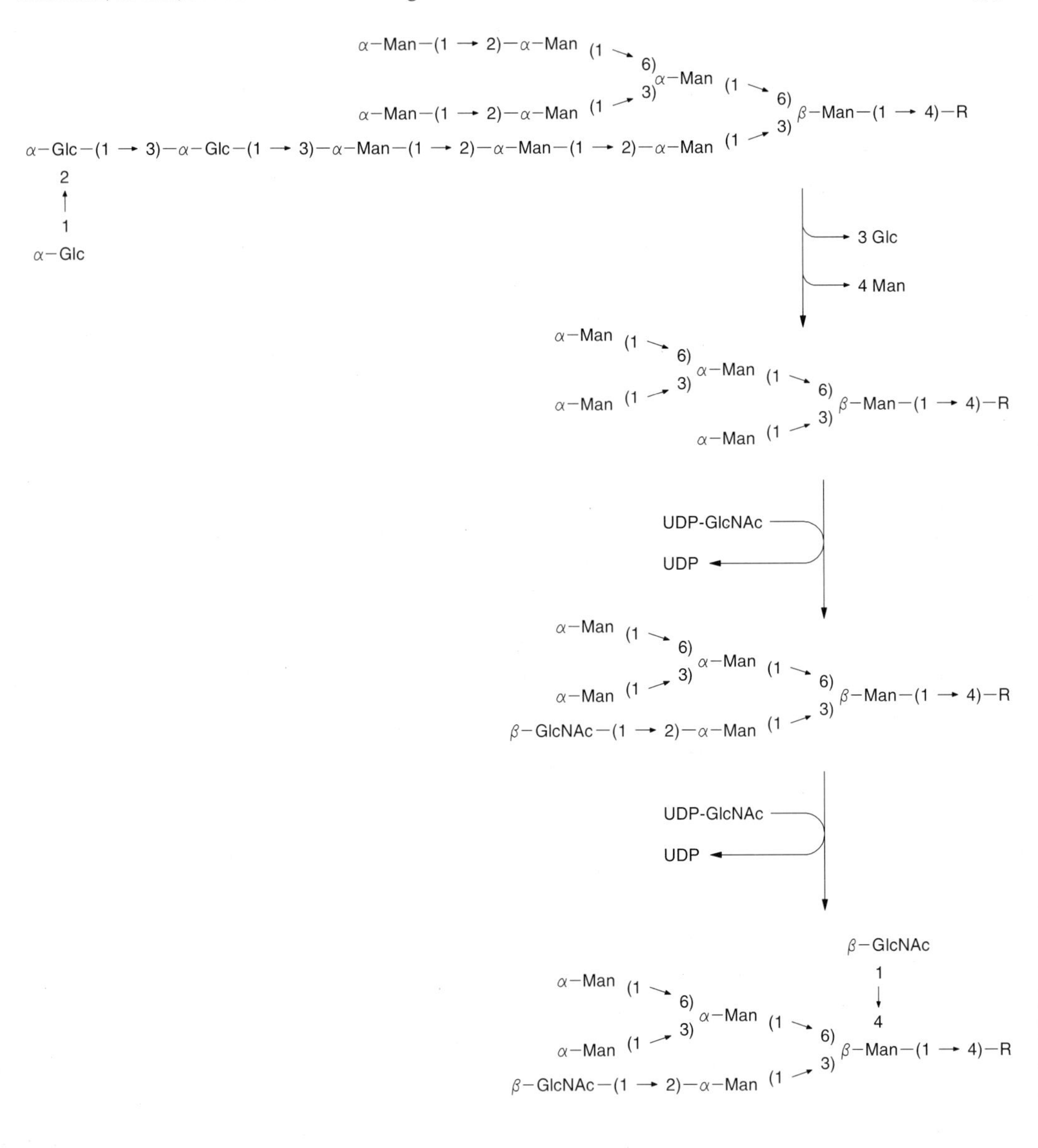


**Fig. 2.62** The phenomenon of microheterogeneity of glycoproteins can be explained by the complicated biosynthetic pathway of the oligosaccharide structures. Different structures represent different stages of this pathway. In contrast to the proteins, *immature* glycoproteins also leave their site of production. Given here is a part of the so-called "processing pathway" for *N*-glycosidic bound oligosaccharides which are found in *ovalbumin* (Fig. 2.61).

coproteins. Despite all the possible heterogeneity of the products, the biosynthesis of the *N*-glycosidic glycoproteins nevertheless is systematic.

From species to species, even from organ to organ, distinct differences can occur in the glycosylation pattern in *N*-glycosidic glycoproteins. In Fig. 2.63 the oligosaccharides from α1-acidic glycoproteins from rat and human serum are compared.

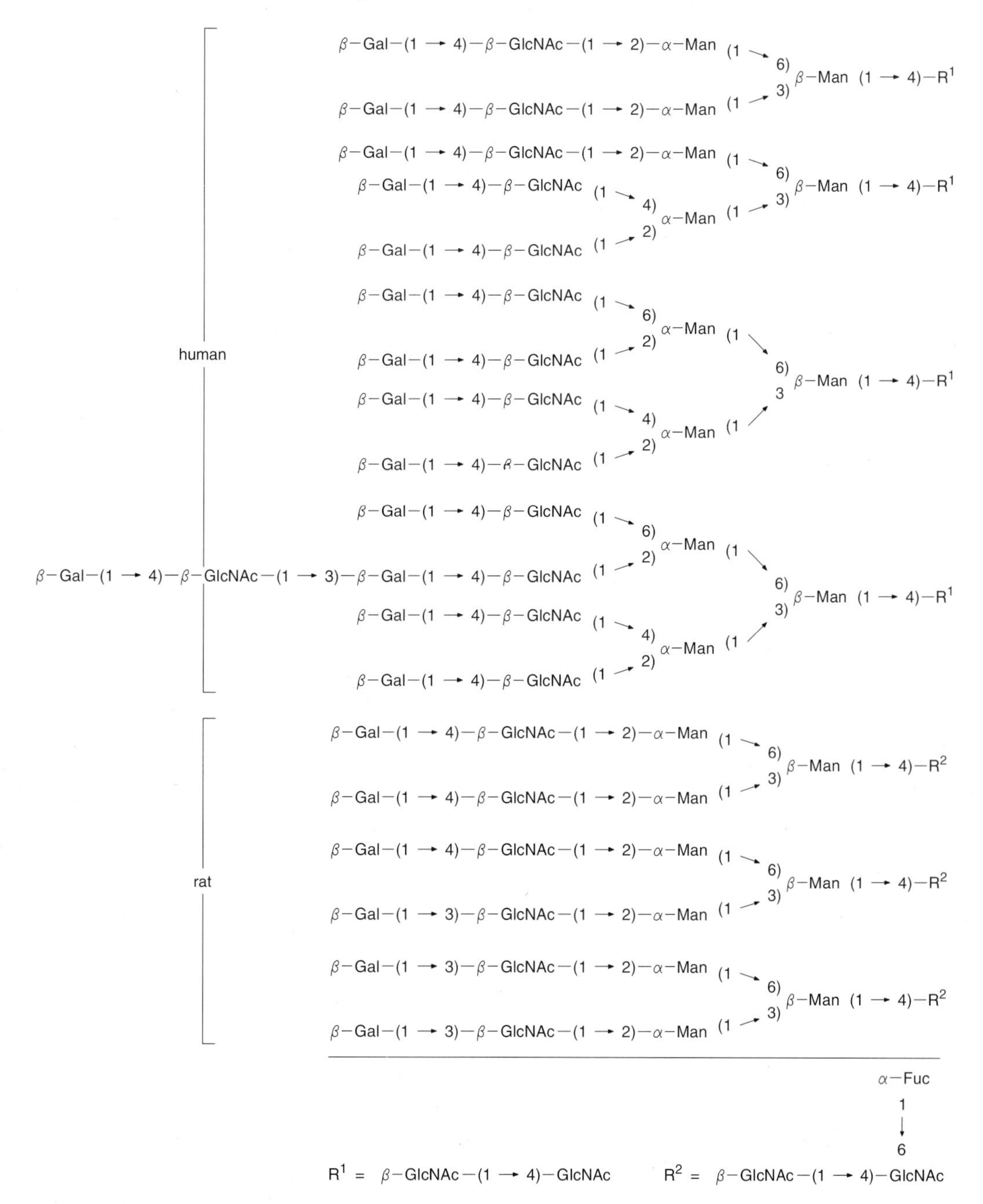


**Fig. 2.63** Comparison of some oligosaccharide structures from serum α-1 *acidic glycoproteins* of the rat and of humans. Especially striking is the lack of the α-fucosylation in the proximal region in the case of humans. Similarly striking is the higher percentage of branching of the oligosaccharides in human glycoproteins.

Despite general similarities, it is striking that none of the oligosaccharide structures of the human α1-acidic glycoproteins carries an α-fucosyl residue at the GlcNAc-residue adjoining the protein. Peripheral α-fucosyl residues are found throughout, but they are not shown in Fig. 2.63. The oligosaccharide structures of the glycoproteins from rat serum contain in part the peripheral β-Gal(1→3)β-GlcNAc group which is not found in humans. Also the degree of branching is distinctly different: about 80% of the *rat* oligosaccharides belongs to the biantennary, complex type, and more than 90% of those of humans are tri- and tetra-antennary.

This complicated method of synthesis of the *N*-glycosidic oligosaccharides found only in eukaryotes, supports the idea that there is a meaningful function for the glycoconjugates. It seems improbable that all higher organisms produced the genetic, extraordinary lavish apparatus of the highly specific glycosyl transferases and the almost equally specific glycosidases, and developed the apparently very involved and uneconomic reaction pathways just to surround proteins, lipids and membranes with a hydrophilic coat. Probably, the reason for the presence of these oligosaccharides is so that many different tasks may be accomplished. For one purpose, an exactly built oligosaccharide is required but for the others, any carbohydrate structure suffices. If the optimal synthetic apparatus exists then it can be used for the many purposes, including the trivial ones. Microbiological research in the coming decades will give an answer to these important questions.

### 2.1.5.2 Membrane glycoproteins

Cell membranes of mechanically stressed cells, for example the blood cells, would probably be more unstable and more vulnerable, were there not highly branched oligosaccharide structures on the surface which simultaneously also improve the hydrophilic contact with an aqueous medium, the blood plasma. As hydrophilic substances they must be linked covalently with lipophilic components of the membrane, consequently with lipids or also with proteins, which immerse lipophilic segments in the membrane or pass through the membrane and thereby are part of the membrane structure (Fig. 2.64).

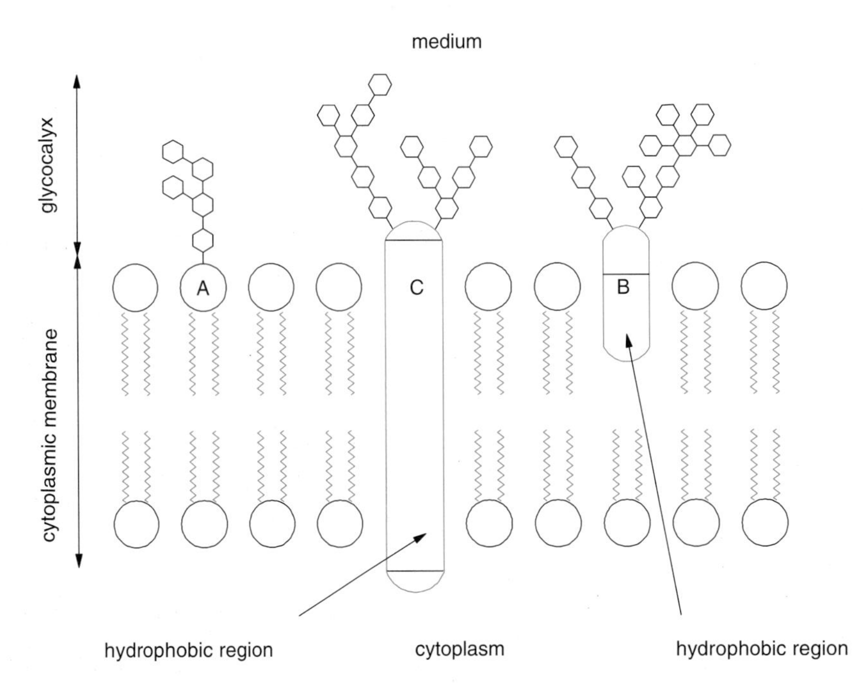


**Fig. 2.64** Schematic representation of a section of a membrane. The side facing the extracellular space is glycosylated. This zone is called the *glycocalyx*. **A** Lipids or **B** proteins carry the carbohydrates structures which are submerged in the membrane or which, in the case of many glycoproteins, penetrate through to the **C** cytoplasmic side.

For many years the human *erythrocyte membrane* has been one of the best researched objects which, in respect of its composition, certainly contains no more secrets. Two glycoproteins, *glycophorin* and the so-called *Band 3* (derived from its electrophoretic behaviour) are amongst the main components of the membranes of erythrocytes and project far out of the membrane with their oligosaccharide chain. The carbohydrate occurs exclusively on the outside, that is directed towards the medium, in this case to the serum. The sequence of the relatively small glycophorin comprising 131 amino acids was the first completely determined primary structure of a membrane protein. The some 16 oligosaccharide structures are all attached on the amino end of the peptide chain that lies on the outside of the membrane. The carboxy end occurs on the inside of the membrane. The number of glycophorin molecules per cell is estimated at about 600000. In spite of this surely not accidentally high concentration, it remains a mystery why humans who lack this glycoprotein are apparently entirely healthy[61].

Three glycophorins (A, B, and C) may be distinguished, each clearly distinct in the peptide sequence. The main glycophorin is A, that for its part is again polymorphic and occurs in the form of two peptides each of 131 amino acids, whose first and fifth amino acid from the *N*-terminal end differ (Fig. 2.65). The protein penetrates the membrane and projects on the outside with about 70 amino acids of the *N*-teminal domain. Here are situated also all *O*- (15) and *N*-glycosidic (1) oligosaccharide structures which constitute about 60% of the total glycoprotein[62]. The oligosaccharides are amazingly uniform (Fig. 2.65). Many of the *O*-glycosidic chains carry, depending on origin, occasional A, B, or H blood group determinants.

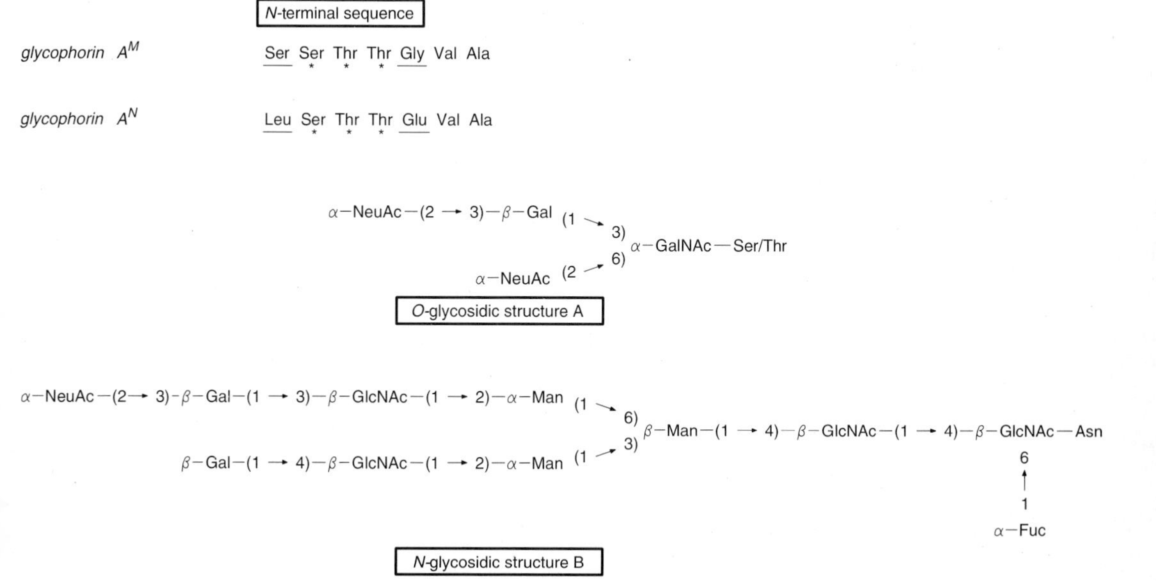


**Fig. 2.65** Protein part of human glycophorin A consisting of 131 amino acids occurs in two variants, which are differentiated in respect of the first (Ser; Leu) and fifth (Gly; Glu) amino acids counting from the *N*-terminus. Both proteins always carry 15 *O*-glycosidic oligosaccharides, almost exclusively the shown structure A, which are found on the amino acids occupying positions 2, 3, 4, 10–15, 22, 25, 37, 44, 47 and 50 from the *N*-terminus, and a single *N*-glycosidic chain on Asn 26, whose structure corresponds mainly to B, with possible variants.

Blood group substances are glycoproteins or also glycolipids which are anchored likewise in the erythrocyte membrane and by reason of their specific carbohydrate structures A, B, H, Le (the blood group Le with the sub-groups $Le^a$ and $Le^b$ is named after the blood donor Lewis in which this structure was first found) etc., show specific antigenic actions. Human erythrocytes may be classified according to these effects. As is apparent from Fig. 2.66, the peripheral monosaccharide residues form different typical determinants. The inner chain elements of type I and type II show an additional differentiation

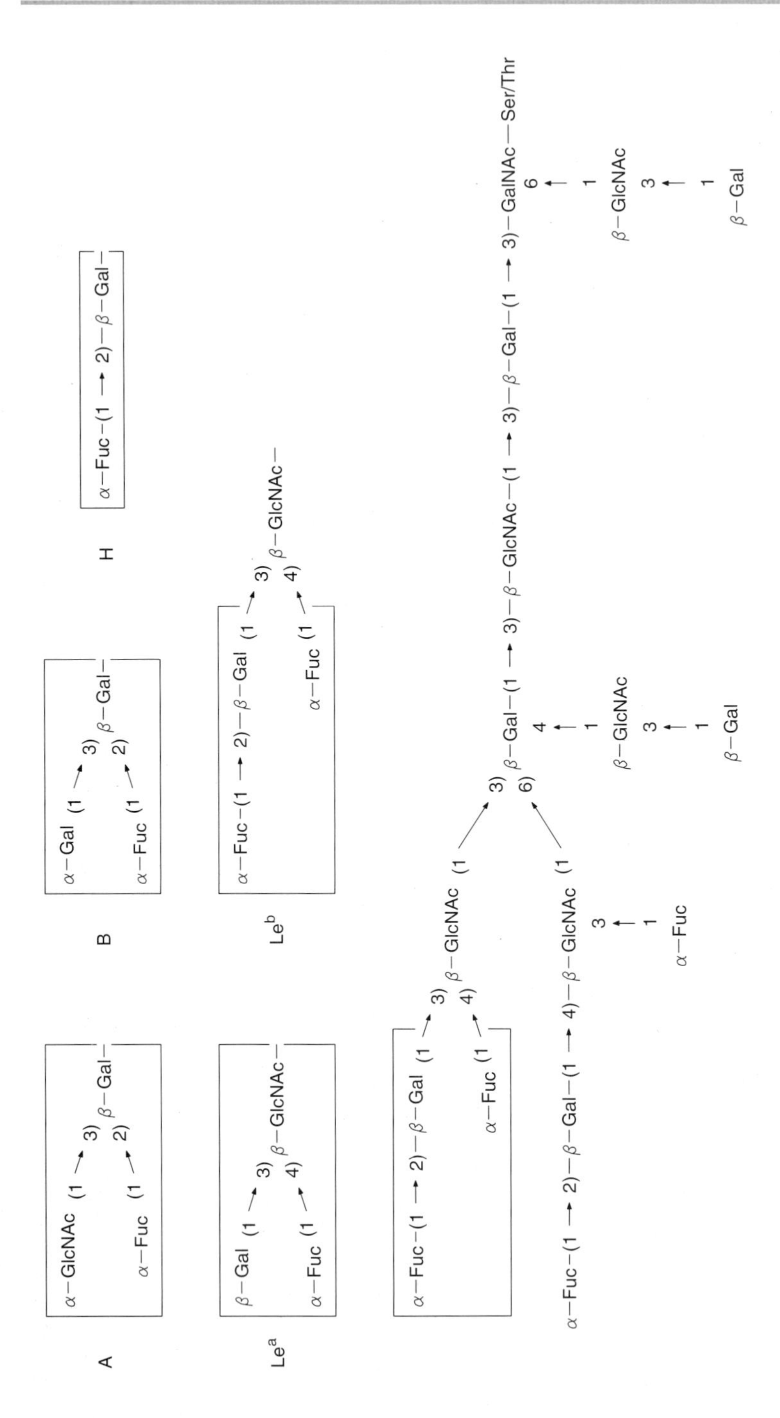


**Fig. 2.66** The peripheral monosaccharide units of the blood group determinants. The inner structure, as shown in the complete *Le^b-oligosaccharide* (bottom), can be very complex.

with −β-Gal-(1→3)-β-GlcNAc− (type I) or −β-Gal-(1→4)-β-GlcNAc− (type II) (see the structure of the complete Le^b-oligosaccharide in Fig. 2.66).

The branching points in the simplified biosynthetic scheme (Fig. 2.67) indicate that when there is an insufficient supply of nucleotides, too small a transfer activity and many other factors, *incomplete* carbohydrate chains can remain.

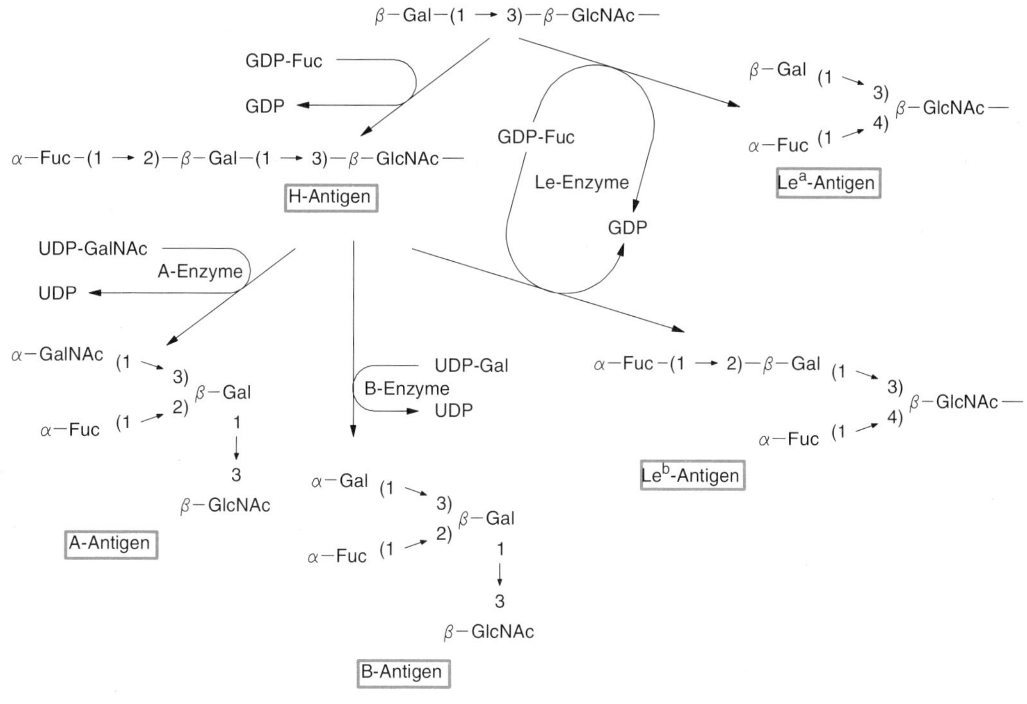


**Fig. 2.67** The blood group determinants arise from a preliminary structure. Mixed types form subgroups, as for example $A_1$ and $A_2$. $A_2$-Individuals also have H-determinants which are lacking in $A_1$-individuals.

Thus erythrocytes arise with the different determinants. Actually it can be proved that blood group A individuals, who can be separated once more serologically into sub-groups $A_1$ and $A_2$[63], differ through distinct microheterogeneity in their carbohydrate determinants. $A_2$ Individuals have, besides the normal A-determinants, still incomplete H-determinants. In $A_1$ individuals the H-determinants are almost completely missing[64].

In this connection, it should perhaps be noted that the blood group specific carbohydrates in many individuals occur not only on the surface of the erythrocytes but also as dissolved mucins in body fluids, for example saliva, seminal fluid, mother's milk, etc. It is decidely difficult to assign a special biological function to the blood group glycoproteins. Certainly, it cannot be assumed that with all specific structures there is an associated specific function, which can be of use to the individual. Of course, it is also difficult to understand here why the markedly complex carbohydrate structures should be biosynthesised without purpose.

The glycoproteins of the erythrocyte membrane introduced here are especially well researched since their isolation and purification is simple in comparison to other membrane glycoproteins. Fundamentally, it can be said that glycoproteins are to be found in all membranes of eukaryotic cells. They are oriented in the cytoplasmic membrane with their carbohydrate portion always to the extracellular side. The intracellular membrane systems, such as endoplasmic reticulum or the vesicles of the Golgi apparatus, also carry glycoproteins. These, however, are stored always with their carbohydrate to the inner side of the organelle. As far as investigations have revealed, the oligosaccharides of all membrane glycoproteins are – whether *N*-glycosidic or *O*-glycosidic – of the same general structure as those of the blood serum or other body fluids and of the erythrocyte membrane. The originally entertained supposition, that the carbohydrate portion in glycoproteins determines their clearly defined transport in and out of the cell, cannot be proved, since proteins, whose glycosylation has been hindered by specific inhibitors, find their way equally well. Consequently, there remains for the time being no better explanation for the existence of glycosyl residues than as a general aid to the stabilisation of protein structure and to increase their hydrophilicity. The carbohydrate residues of the coat-protein of most viruses probably have similar functions. As investigations on the *Semliki-Forest*-virus showed, the synthesis of the viral coat-protein passes through the same stages as that of the glycoprotein of the host cell[65,66].

In this connection of special actual interest are the highly glycosylated coat-proteins of the *HIV-1 retrovirus, gp 120* and *gp 41.* It was established that inhibition of *N*-glycosylation led to a distinct decrease in the pathogenicity of the virus. It was therefore deduced that the membrane glycoproteins have a function in the binding onto certain receptors of the T-lymphocyte[67]. This indeed may be valid generally for glycoproteins of virus membranes. The occurrence of many different oligosaccharide structures[68,69] make a targetted recognition mechanism rather unlikely.

Recently, a group of *membrane bound mucins* were reported that are, as in the case of *episialin*, highly glycosylated (50–80% carbohydrate portion) glycoproteins of over 300 kDa, and that have a special function on the surface of the epithelial cells facing the lumen. Their form, strengthened by the protein structure but above all by the innumerable *O*-glycosidic glycan chains, projects like a voluminous rod (Fig. 2.68) about 500 nm in length far above the glycocalyx into the lumen of the organ component and so prevents, it is thought, the adhesion of epithelial cells opposite one another in capillaries[70]. The principle of a structural element stiffened through carbohydrates[71] is found also in the case of receptors, which are raised on a *spacer* fastened onto the the cell surface and thereby project, easily accessible, into the extracellular space (see section 2.1.5.4, p. 124).

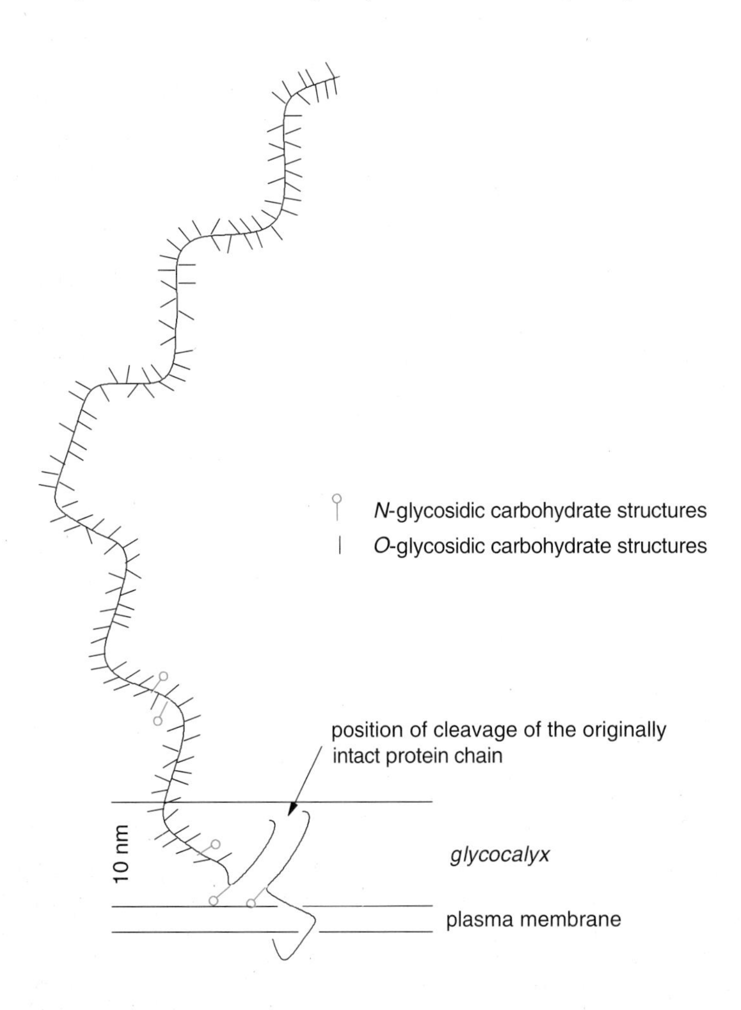


**Fig. 2.68** *Episialin* is a long membrane protein consisting of 30–90 repeat units, each of 20 amino acids, which is highly *O*-glycosylated and thereby projects like a voluminous rod beyond the *glycocalyx* out into the lumen of the epithelial layer. The 10 nm thickness of the *glycocalyx* shown in the diagram gives an impression of the size of the membrane protein. The originally continous protein chain is split at the indicated position, although the fragments remain together, but not covalently linked.

The high content of the *O*-glycans in sialic acids contributes, by a very large charge density, to the *anti*-adhesion properties of the membrane glycoprotein of the episialin type. Normally, that is in healthy cells, episialin is found only on the side of the epithilial cells facing the lumen. In the case of metastasising tumour cells, an enhanced expression of episialin is observed as is also the fact that in malignant cells the episialin is distributed over the whole surface of the cell and not only that of the lumen.

More recent investigations refer to membrane glycoproteins ("surface layer" glycoproteins) even in the case of archeabacteria. The biosynthesis of these S-layer glycoproteins shows remarkable agreement with that of the eukaryotic glycoproteins[72]. The glycan structures are built up from repeating units.

### 2.1.5.3 Glycosylated structural proteins

The extracellular matrix in animal tissue also contains besides proteoglycans, proteins which, as *collagen* and *elastin*, are responsible for the mechanical stability of the organism and which increase its load-bearing capacity. In addition there are also proteins such as *fibronectin*, which provide arrangement both of the extracellular matrix itself as well as also the cell in the tissue. The organisation of tissue is one of the still in large measure unsolved problems of molecular biology. There are many indications that the structure of the cell surface as well as the structure of the polymers involved in the construction of the extracellular matrix determine organisation and order through specific, non-covalent interactions. In the preceding section it was shown that the membranes of all eukaryotic cells consist of 2–10% carbohydrate, which constitute the outer layer of the cell, the glycocalyx. This layer merges more or less smoothly into the extracellular matrix, so that often, morphologically speaking, a boundary is ascertained only with difficulty. For the greater part the extracellular matrix consists of proteoglycan. Also, the fibrous proteins, except for elastin, are glycosylated. What could be more obvious than to think of a special role for the carbohydrate structures in the organisation of cells and extracellular matrix? At the same time, it is not appropriate to speak of a specific interaction mediated by carbohydrates. For that the structures in question are not sufficiently differentiated.

Procollagen is synthesised in the fibroblasts and many proline and lysine residues are also hydroxylated there. On selected hydroxylysine and hydroxyproline residues $\beta$-galactosylation first takes place, then the galactosyl residue is $\alpha$-glucosylated in the 2-position. The completed procollagen contains about 2% carbohydrate, apparently statistically distributed (Fig. 2.69). Collagen is released solely as a triple helix into the extracellular space, in which the disaccharide residues are directed to the outside.

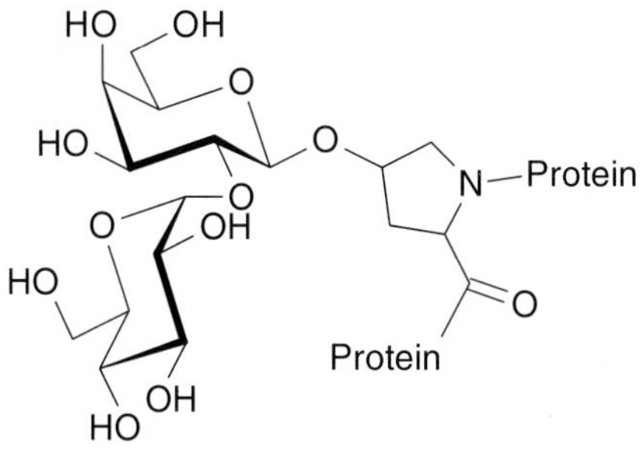

**Fig. 2.69** The *triple helix* of the *collagen* is glycosylated on many hydroxylysine and hydroxyproline residues with an $\alpha$-D-glucopyranosyl-(1→2)-D-galactopyranose unit.

It can not be said what purpose the glycosylation serves. Possibly an intermolecular interaction of the carbohydrate residues with one another is conceivable in the organisation of the collagen fibres.

A series of so-called adhesion-glycoproteins, of which *fibronectin*[73] and *laminin*[74] are especially well investigated, are specifically in contact both with collagen as well as with proteoglycans and with cells.

*Fibronectin* consists of two identical, 2500 amino acids long peptide chains which are linked together by two disulphide bridges at the carboxy ends (Fig. 2.70). Four different specific contact points are known. One can speculate about the participation of carbohydrate structures in the organisation of cells in the extracellular matrix but probably this originates rather from the glycoconjugates of the cell

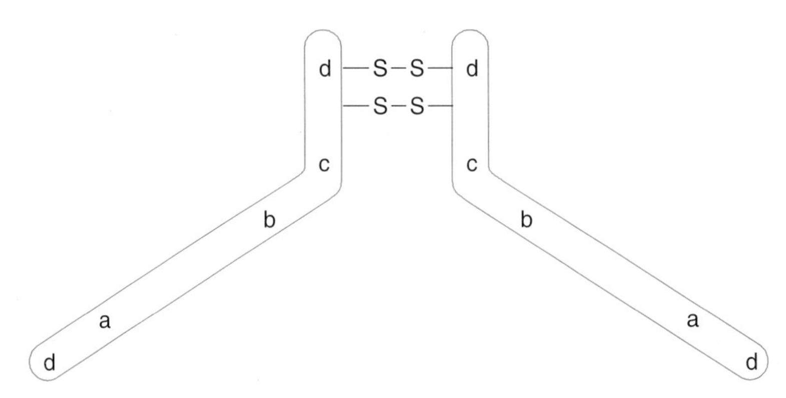


**Fig. 2.70** *Dimeric fibronectin* brings about through specific interaction of its binding domain with **a** collagen , **b** glycoproteins of the cell surface, the proteoglycan of the **c** extracellular matrix, and **d** fibrin the anchoring of the cells in the connective tissue.

membrane than from the carbohydrates of the fibronectin aggregate. In the case of metastasising cancer cells, the interaction between cells and fibronectin is disturbed.

The protein has three manifestations. The dimeric molecule is soluble and probably has a function in the blood plasma in blood clotting. Fibronectin is found as oligomers on the cell surface and finally, insoluble fibronectin-fibrils form a part of the extracellular matrix. In oligomers and fibrils additional disulphide bridges are responsible for the aggregation. It may be surmised that fibronectin is involved in the morphogenesis of living things in that it steers cell migration and cell arrangement. How this mechanism functions in detail is not known. It would be paradoxical if a single defined macromolecule in different positions of the organism assumed different binding functions. Therefore it must provide besides a common base-structure sequence which differs from fibronectin to fibronectin and which is responsible for the necessary specificity of cell adhesion. Certainly the construction of fibronectin has similarity to the construction of the antibodies with preponderant constant domains and variable regions. It must also not be forgotten that fibronectin is only one of many, still little investigated adhesion glycoproteins.

The glycoprotein *laminin*, which together with collagen and heparan sulphate is one of the main components of the basal membrane, has similar functions as fibronectin. Epithelial cells are in contact with certain domains of laminin via specific receptors. Relatively little is known about the manner of glycosylation of the known adhesion proteins. Correspondingly, also no statement can be made about the function, whether specific or unspecific, of the oligosaccharides side chains.

### 2.1.5.4 Glycosylated lipids

Without membranes there would be no cells and no compartmentalisation in cells. The cytoplasmic membrane and also the intracellular membrane system are asymmetrically constructed. The outside has a different composition to the inside. Chiefly responsible for this asymmetry are the *glycolipids*, which are found exclusively on the side of the cytoplasmic membrane facing away from the inside of the cell. The membrane consists of about 50% protein and up to 50% lipids, of which, according to the cell, 1 – 30% are glycosylated. In respect of their amphiphilic nature, the main membrane lipids, phosphatidylethanolamine, phosphatidylserine, phosphatidylcholine, and sphingomyelin do not differ greatly from the glycolipids. Both possess a hydrophilic head group and a hydrophobic tail. This distribution of physicochemical properties causes the membrane lipids to form, spontaneously, lipid double layers, that is to say membranes.

The glycolipids of bacteria and of plants are almost all derived from diacylglycerol, whose free hydroxyl group is glycosylated. The glycolipids[75,76] of animal cells are glycosylated derivatives of ceramide. Plant cell membranes contain as glycolipid principally galactosyl- (structure **A** in Fig. 2.71) and digalactosyl-diglycerides as well as the unusual sulphoquinovosyldiglyceride, represented only in the chloroplast membrane of green plants (structure **B** in Fig. 2.71). Glycolipids in bacterial membranes,

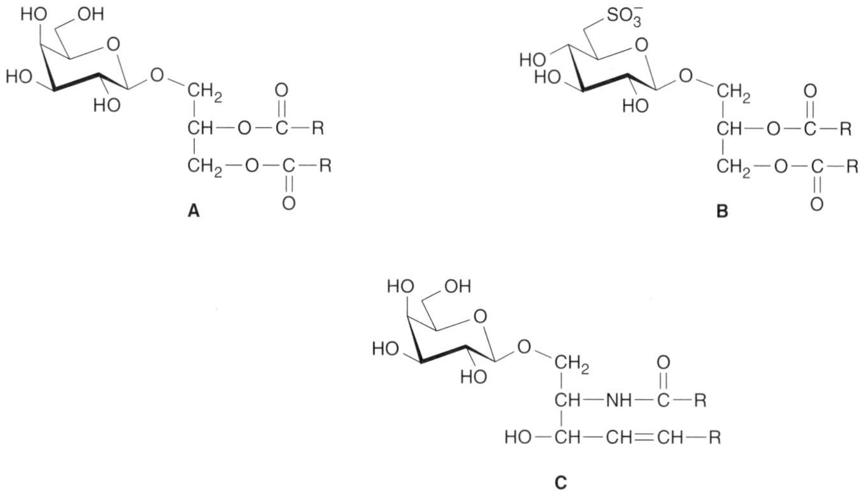

**Fig. 2.71** Glycolipids of plants are **A, B** glycosylated diacylglycerols , whereas those of animals are **C** glycosylated ceramides.

if present at all, are largely unimportant as contact points with the outer world since they are masked by the outer layer.

The glycosylation pattern of the cell membrane in animals is very diverse and varies in part a great deal from one type to another but also, in the case of higher organisms, within an organism. For example, *galactosyl ceramide* (*galactocerebroside*), the main glycolipid in the myelin of white brain matter, (structure **C** in Fig. 2.71), that the nerve cells build up in multilayers, is only to be found there. Galactosyl ceramide is missing, as far as one can detect, in the membranes of all other cells. In nerve tissue, galactosyl ceramide sulphated in the 3-position (sulphatide) (Fig. 2.72) is found in relatively small concentration.

Galactose is replaced by glucose as the bridgehead monosaccharide in the glycolipids of other cells and in complex glycolipids. The most interesting group in the membrane glycolipids are the *acid* gangliosides, whose oligosaccharide structures are characterised by the peripheral NeuAc-residue (Fig. 2.73, p. 127). In this regard, the gangliosides differ from the *neutral* glycolipids (Fig. 2.73).

The frequently observed agreement of peripheral oligosaccharide structures of membrane glycoproteins and membrane glycolipids is astonishing. Blood group antigens of both classes of compound have similar determinants but different core-oligosaccharide structures. These facts are easily explained from the course of the biosynthesis. Proteins, as lipids, are subject to different initial glycosylation in the endoplasmic reticulum, which depends on the acceptor structure – here ceramide, there acceptor sequence in the protein. If the substance-specific core-region exists, then the type specific peripheral glycosylation can take place. Thus, it is understandable why many soluble glycoproteins produce oligosaccharide structures similar to those of the membrane glycoproteins.

Glycosphingolipids together with glycoproteins form the glycocalyx of the cell surface, and one can ascribe their role to cell differentiation and morphogenesis. Regarding the carbohydrate part, the structures are very variable and often have a typical cell pattern that can alter with the differentiation in a definite manner. Disturbance in the catabolism of these glycolipids leads to the so-called storage diseases principally of the nervous system which is especially rich in glycosphingolipids[77]. More than a hundred different glycosphingolipids are known[78]. These are subdivided according to their inner carbohydrate structure (core structure) into the ganglio- globo-, and lacto-series, etc. (Fig. 2.73, p. 127).

All *glycosphingolipids* have a characteristic shape like a large L. The hydrophobic leg sticks into the membrane and the other hydrophilic part stands virtually at a right angle to it and accordingly clings to the membrane surface. Although today, as already mentioned, functions for all of the complex glycosphingolipids are ascribed as recognition markers of cells in the widest sense[79], the general biological function above all must depend on the pronounced amphiphilic property of this glycolipid. It can very

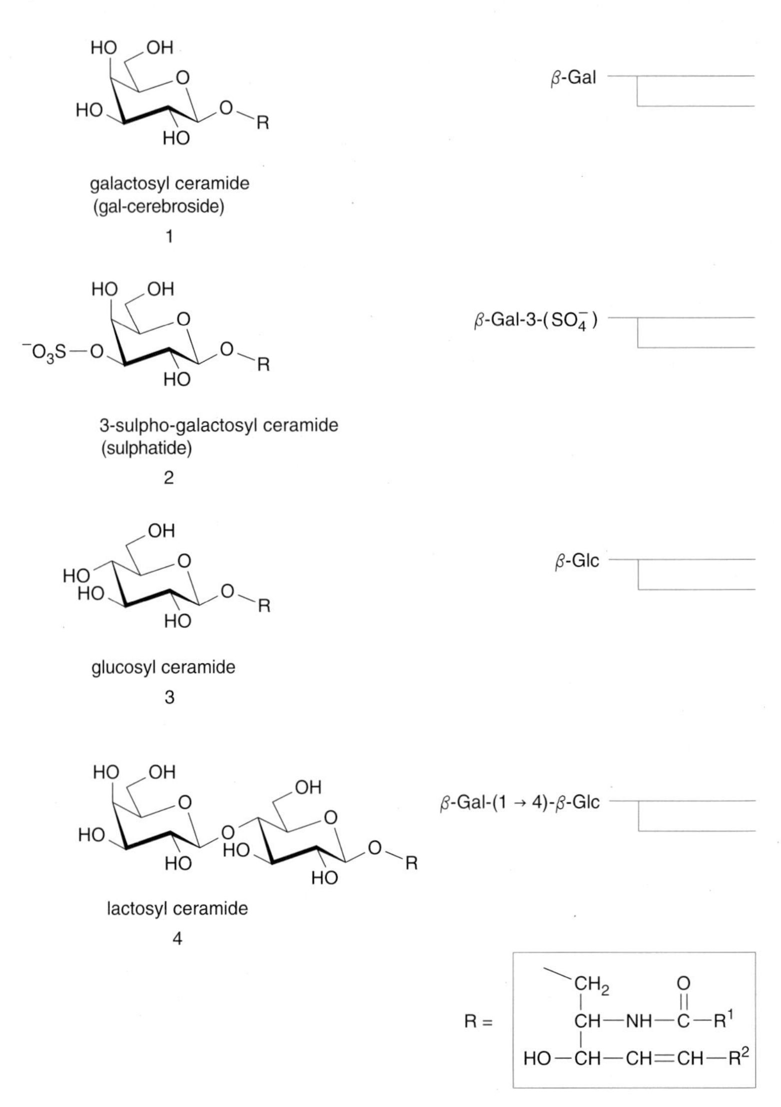


**Fig. 2.72** Some simple *glycosyl ceramides*. The galactosyl derivatives are found in white brain matter in especially high concentration – compound 1 forms about 170 mMol/g of the dry weight.

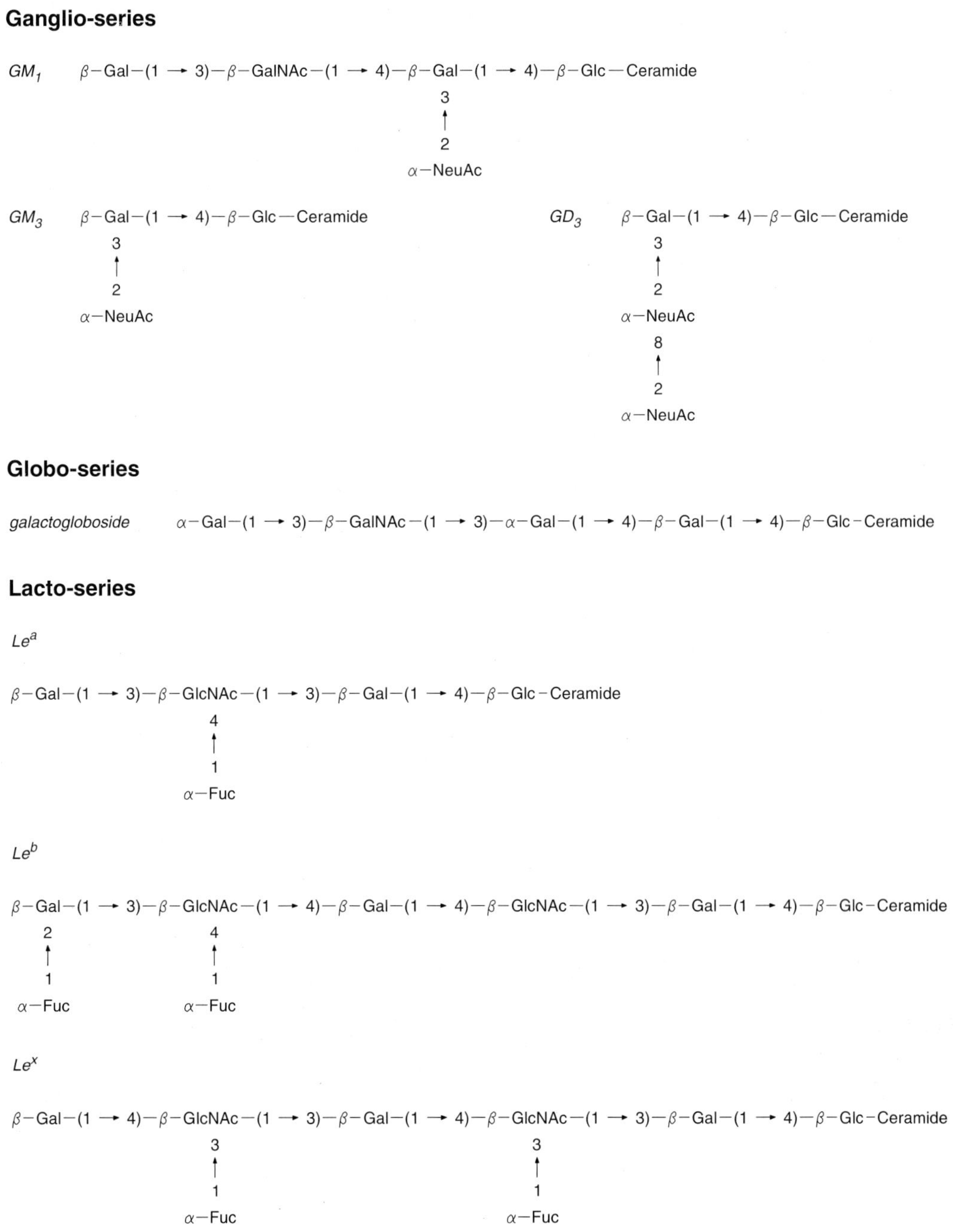

## Ganglio-series

*GM$_1$* β−Gal−(1 → 3)−β−GalNAc−(1 → 4)−β−Gal−(1 → 4)−β−Glc—Ceramide

3
↑
2
α−NeuAc

*GM$_3$* β−Gal−(1 → 4)−β−Glc—Ceramide

3
↑
2
α−NeuAc

*GD$_3$* β−Gal−(1 → 4)−β−Glc—Ceramide

3
↑
2
α−NeuAc
8
↑
2
α−NeuAc

## Globo-series

*galactogloboside* α−Gal−(1 → 3)−β−GalNAc−(1 → 3)−α−Gal−(1 → 4)−β−Gal−(1 → 4)−β−Glc-Ceramide

## Lacto-series

*Le$^a$*

β−Gal−(1 → 3)−β−GlcNAc−(1 → 3)−β−Gal−(1 → 4)−β−Glc-Ceramide

4
↑
1
α−Fuc

*Le$^b$*

β−Gal−(1 → 3)−β−GlcNAc−(1 → 4)−β−Gal−(1 → 4)−β−GlcNAc−(1 → 3)−β−Gal−(1 → 4)−β−Glc-Ceramide

2 4
↑ ↑
1 1
α−Fuc α−Fuc

*Le$^x$*

β−Gal−(1 → 4)−β−GlcNAc−(1 → 3)−β−Gal−(1 → 4)−β−GlcNAc−(1 → 3)−β−Gal−(1 → 4)−β−Glc-Ceramide

3 3
↑ ↑
1 1
α−Fuc α−Fuc

**Fig. 2.73** Structures of some *glycosphingolipids* from the cell membranes of humans. The assignment to a certain series is based on the different core structures.

well be imagined that the glycosphingolipids, with their typical compact molecular form which allows a dovetailing of the carbohydrate residues of different molecules on the cell surface, contribute considerably to the mechanical stabilisation of the plasma membrane. By their presence solely on the outer side of the cell they are obviously, just as the other components of the glycocalyx, the structural markers which can be recognised specifically by other cells, by viruses, bacteria, and antibodies, and by the biopolymers of their own extracellular matrix (see section 2.1.5.2, p. 118).

While the biological function of most of the glycolipids considered here still can be explained through the trivial alteration of physicochemical properties, it is very hard to do so in the case of the relatively recently discovered lipid constituents of the endoplasmic reticulum[80]. This area of research concerns the so-called *GPI anchor* which is a type of carrier for proteins, above all glycoproteins, which must be deposited on the outside of the membrane through vesicle transport[81].

The linkage of the protein to the GPI anchor takes place in the lumen of the endoplasmic reticulum (ER) (Fig. 2.74). The protein anchored in the ER-membrane by the C-terminus is cleaved directly on the

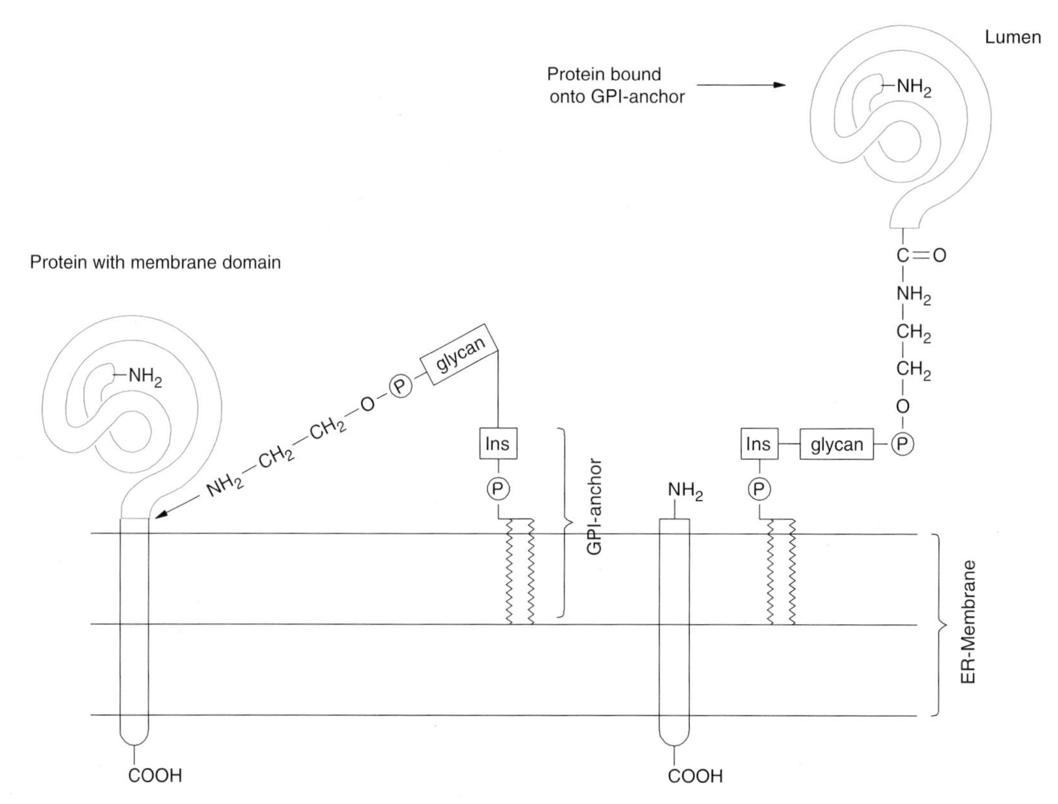


**Fig. 2.74** Transfer of a protein to a *GPI-anchor* on the lumen side of the ER-membrane. The GPI anchor holds proteins on a long arm *at a distance*. This may be significant primarily for the final placement of the proteins in the plasma membrane on the cell surface.

membrane and transferred onto the amino group of a terminal ethanolamine residue in the GPI anchor. An increasing number of membrane proteins are being found which are bound in this way to the cytoplasmic membrane. Especially well researched is the GPI-anchor of the variable surface glycoprotein of *Trypanosoma brucei*. The structure of this GPI anchor shown in Fig. 2.75 is found also, with relatively slight variation in the glycan part, in higher organisms and this means of attachment appears to be a general principal of the membrane anchoring of glycoproteins and also proteoglycans[82].

The question arises as to what advantage it can have that glycoproteins, which have in themselves the possibility of anchoring in membranes by reason of hydrophobic domains, prefer consolidation in

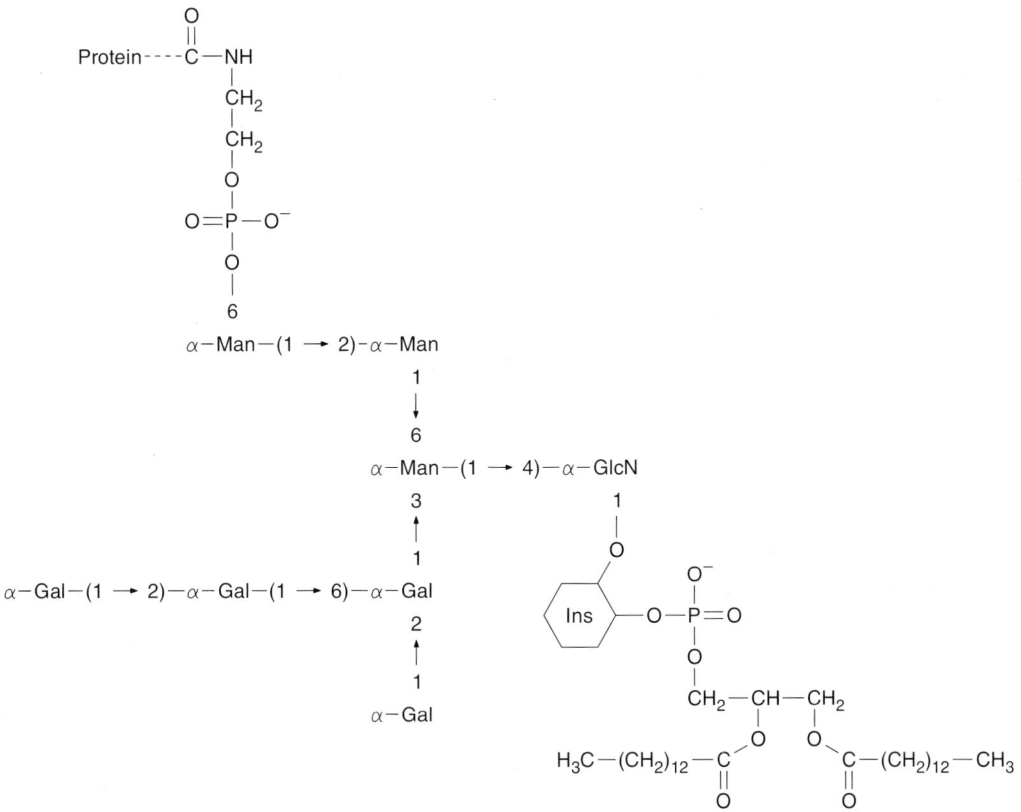

**Fig. 2.75** All of the *GPI anchors* investigated so far have a similar general structure. Variations can occur in the glycan structure. It should be noted that the glucosamine glycosidically bound to the inositol carries a free amino group. In addition, phosphate residues can occur on the glycan structure. The GPI anchor from *Trypanosoma Brucei* is shown here.

the membrane through a GPI anchor. Speculation that this allows an easier detatchment of the protein from the membrane into the extracellular space through phospholipases are difficult to prove and perhaps only meaningful if it is associated with the quick liberation of the variable glycoprotein coat of parasitic protozoa. In fact, endogenous phospholipase C activity was proved (the phospholipases C sever the hydrophilic from the hydrophobic part of the membrane lipids) in trypanosomes, which leads to the conversion of the normally insoluble glycoprotein coat into a soluble form. Surely, however, this very special function is not applicable generally for the GPI-glycoproteins of higher organisms. It must be assumed for the present that the GPI-anchor is a type of hydrophilic spacer, which allows the bound protein to project into the extracellular space, that can be advantageous, for example, for its enzymatic action or receptor function (see Section 2.1.5.2, p. 118 and Fig. 2.68, p. 122).

# 2.2 Specific biological processes

The foregoing sections have introduced carbohydrates in their abundance as building materials for the construction of living things of all types, and the biological role of the carbohydrates was denoted as almost totally static and in the widest sense as unspecific. In this section 2.2 and also in 2.3, the dynamics of change will be described, to which the carbohydrates themselves are subjected and which in certain cases they are able to initiate thanks to their presence in biological systems. In this connection the functions are very specific, that is to say only definite compounds are changed or initiate a change. They are not exchangeable and their absence has negative consequences for the organism.

## 2.2.1 Central role in general metabolism

It could be imagined that bacterial cell walls might be built out of different materials or that extracellular matrices might vary in their composition with regard to biopolymers and yet both could be equally functional, but it surely cannot be imagined that the monosaccharides glucose and fructose are replaceable in their central function.

Carbohydrates are not only quantitatively, as already discussed, but also qualitatively an important constituent of living systems. They deliver the necessary energy in order to maintain the functions of the cell, and they are the raw material out of which all other organic components of a living thing are built, either directly or indirectly. The assimilation of carbon dioxide in green plants delivers in fractions of a second, via 3-phosphoglyceric acid, 3-phosphoglyceraldehyde which already can be looked upon as a carbohydrate derivative. Complex biochemical transformations allow pentose phosphates and hexose phosphates to be produced in the Calvin cycle (see section 2.3.6, p. 193) from these starting materials and therefrom the multitude of the naturally occurring carbohydrates and the abundance of compounds of other classes of substances. These interrelationships are investigated in some detail in the following sections. Fig. 2.76 shows in a very simplified form the central role which carbohydrates play in metabolic events.

Carbohydrate metabolism itself will be the subject of section 2.3 (p. 173), so a discussion is not presented here of the dynamic interrelationships amongst carbohydrates as a class of natural products. It is moreover surely understandable that within the framework of a book of this nature, a detailed treatment of general biological metabolism is not included. The reader is referred to a relevant text of biochemistry from where also the terminology of classical biochemistry can be obtained.

### 2.2.1.1 Relation to lipid metabolism

Carbohydrate and lipid metabolism are closely connected and are well coordinated , for both supply the *acetyl-S-CoA* necessary for the extraction of energy in the citric acid cycle (Fig. 2.77). An important principle of general metabolism is the fundamental irreversibility of synthetic (anabolic) and degradative (catabolic) metabolic pathways. Without this principle, metabolism could not be regulated and could not be fitted to the requirements of an organism. Good examples are the two pairs of contrary metabolic pathways, the so-called $\beta$-oxidation of fatty acids and fatty acid biosynthesis as well as glycolysis and gluconeogenesis.

The common product of carbohydrate degradation and lipid or fatty acid breakdown, acetyl-*S*-CoA, is the *fuel* for the cell's energy needs. Both carbon atoms of the *activated acetic acid* are converted into carbon dioxide in the citric acid cycle, also called the tricarboxylic acid cycle, with formation of chemical energy.

If the organism has the problem of an excessive supply of acetyl-*S*-CoA, be it for reasons of a too high carbohydrate or fat uptake or for reasons of a too low energy requirement, or both, then there is only one possibility to utilise acetyl-*S*-CoA, namely to transform it into fat, that is adipose fat – a transformation of acetyl-*S*-CoA into carbohydrate is not possible. This biologically very significant event serves to store carbon and hydrogen in the most concentrated and easily mobilised form possible for times of high energy needs or for periods of hunger. Non-physiological obesity can only be treated by raising the energy needs or through self-imposed hunger periods. Naturally, not only are the physiologically relative unimportant triglycerides built with acetyl-*S*-CoA but also the phosphatides and

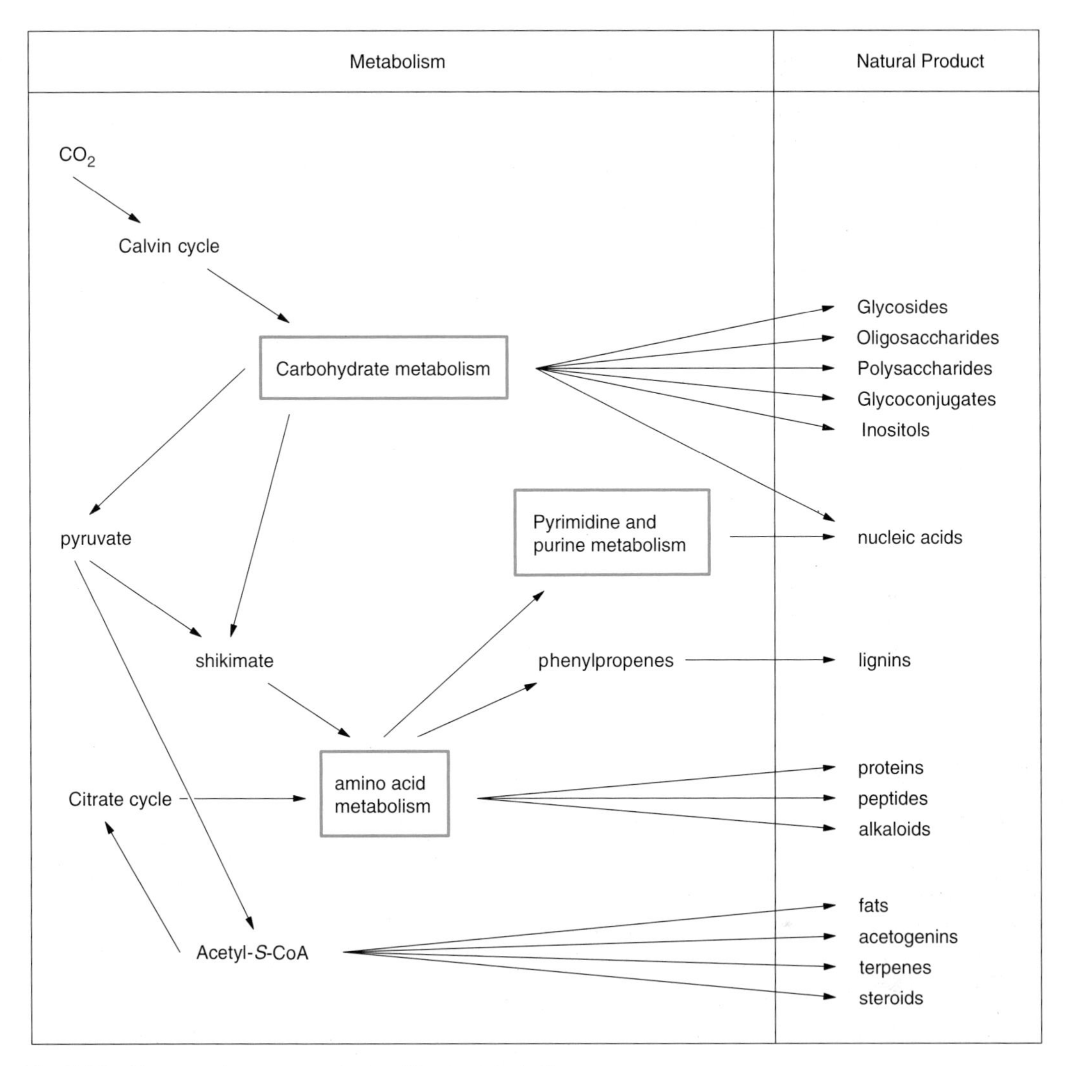


**Fig. 2.76** The most important relationships in metabolic processes.

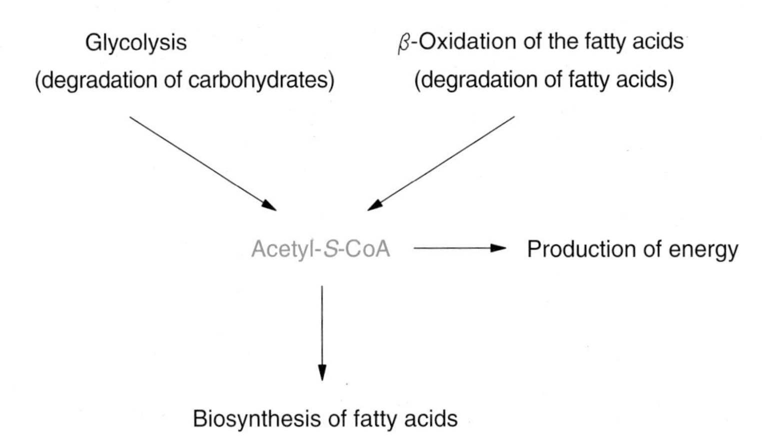


**Fig. 2.77** Production of energy from acetyl-*S*-CoA, formed as a result of the breakdown of carbohydrates and fatty acids. Excess acetyl-*S*-CoA can be used for the synthesis of fatty acids, not however for the synthesis of carbohydrates.

ceramides necessary for the construction of the essential membranes. The biosyntheses of the steroids and terpenes and other *nonsaponifiable* lipophilic compounds start from acetyl-S-CoA and, because of that, from the carbohydrates.

### 2.2.1.2 Relation to amino acid metabolism

Carbohydrate metabolism is not so directly bound with amino acid metabolism as with lipid metabolism. Direct connections between monosaccharide decomposition and biosynthesis of amino acids are known only in a few cases. For example, alanine is produced generally from pyruvate arising from glycolysis. Glutamate and, less generally, aspartate arises from the ketodicarboxylic acids 2-ketoglutarate and oxaloacetate involved in the citric acid cycle. Otherwise the pathways for the biological production of amino acids are very complex and, correspondingly, the connections with carbohydrate metabolism are also complex and detailed. Under normal conditions, the requirements in amino acids can be met from proteins in the food. Only plants and bacteria are autotrophic with respect to amino acid acid production.

### 2.2.1.3 Carbocyclic compounds derived from carbohydrates

Both the aromatic system of the benzene ring and also the saturated system of the cyclohexanes are widespread in nature. The two carbocycles arise from entirely different pathways, yet each directly from carbohydrate metabolism. Not all living things are able to make benzene derivatives. This synthetic ability is reserved for green plants and microorganisms. The two components out of which finally the carbocycles are built are *phosphoenolpyruvate* (*PEP*) and *D-erythrose 4-phosphate.* PEP arises through glycolysis and D-erythrose 4-phosphate is involved in the monosaccharide inteconversions of the pentose phosphate pathway. From both, through an aldol type reaction, *3-deoxy-D-arabino-heptulosonate 7-phosphate* is formed (Fig. 2.78).

A cyclisation with a complex mechanism yields 3-dehydroquinate from which, after further transformations, is produced shikimate that has given the name *shikimate pathway* to this typical metabolic pathway for the biosynthesis of benzene derivatives. Shikimate is the precursor of all aromatic amino acids, and not only these but also, starting from phenylpyruvate, of the phenylpropenes as starting materials for lignin biosynthesis (see section 2.1.2.3, p. 82) (Fig. 2.76 and 2.78).

Cyclohexane is the basic skeleton of the inositols and similar compounds. The inositols, as polyhydroxycyclohexanes, are closely related to monosaccharides and to the alcohols derived from monosaccharides, the alditols. Phosphorylated and combined with lipids, they possess (especially *myo*-inositol) great significance as components of biological membranes (see section 2.1.5.4, p. 124). *L-myo-Inositol 1-phosphate* arises by cyclisation of an intermediate produced directly from D-glucose 6-phosphate by activation with $NAD^+$, and is the precursor of all cyclitols (Fig. 2.79).

As the hexaphosphoric acid ester, *phytic acid*, *myo*-inositol serves in plant seeds and in bird erythrocytes as a phosphate reservoir. While *myo*-inositol occurs in the cell membranes of possibly all living things, of the in total 9 theoretically possible isomers, only a further 3 have been discovered in plants until now (Fig. 2.80, p. 135). Monomethyl ethers of some inositols are also known as plant constituents[83]. All compounds of this type are derived from *myo*-inositol which for its part is to be found as a component of membrane lipids (D-1-phosphatidylinositol) in the cell membranes of all living things (Fig. 2.81, p. 135).

L-*myo-Inositol 1-phosphate* arises, as already indicated above, from D-glucose 6-phosphate through a cyclisation reaction activated by $NAD^+$. The inositol biosynthesis is a fine example of how in nature activation is attained through the temporary introduction of a carbonyl group. Dehydrogenation of D-glucose 6-phosphate at the 5-position makes the neighbouring C-atom 6 into a nucleophilic enol-C-atom, which under stereochemical control of the enzyme D-glucose-6-phosphate-cyclase enters into an intramolecular aldol reaction with the carbonyl-C-atom 1. The cyclisation is concluded through reduction of the keto group with the NADH arising at the start of the sequence. In this cyclisation, $NAD^+$ has a catalytic function (Fig. 2.79).

The decomposition of *myo*-inositol occurs typically not by reversal of its formation but with molecular oxygen with participation of *myo*-inositol oxygenase. *D-Glucuronate* is formed which can be decomposed through general carbohydrate metabolism. The chemical resistance, not only of the inositols, but also of the open chain sugar alcohols (alditols), especially against oxidising agents, is probably also

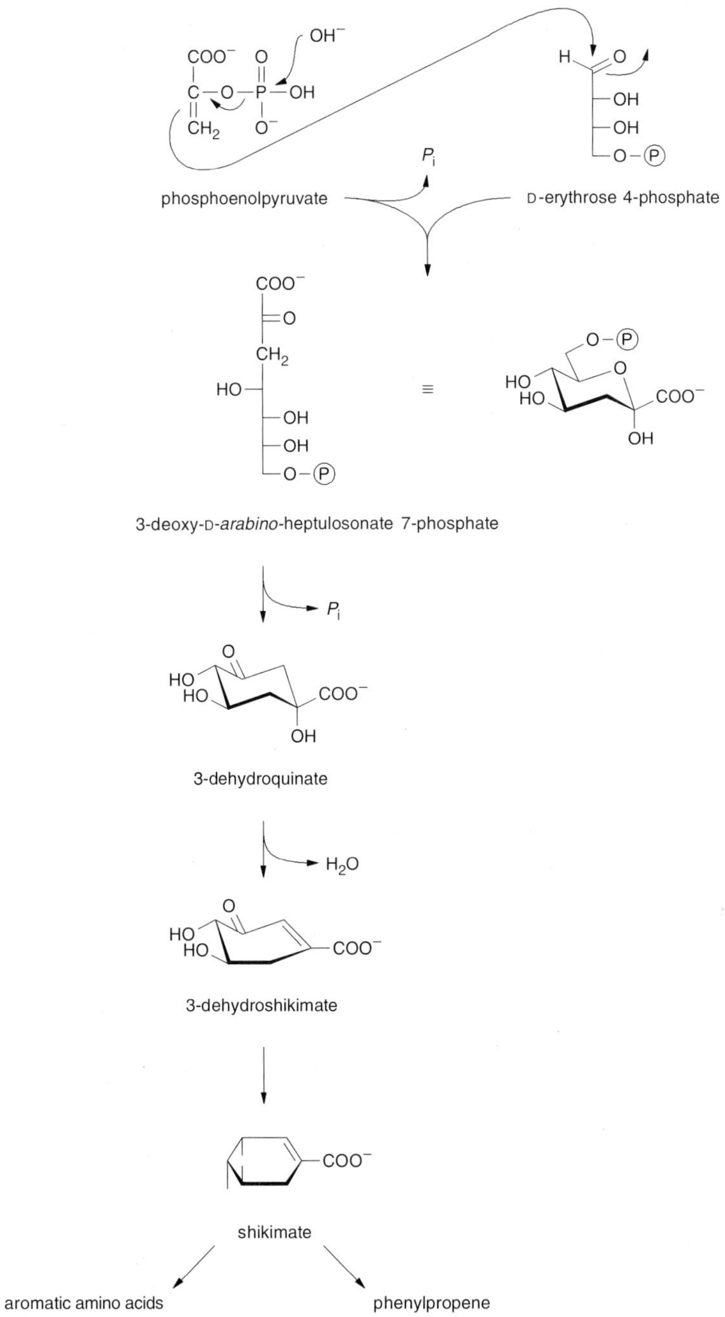

**Fig. 2.78** Simplified schematic representation of the *shikimate pathway* for the biosynthesis of aromatic, six-membered carbocycles (benzene derivatives). Both *phosphoenolpyruvate* and D-*erythrose 4-phosphate* arise from carbohydrate metabolism.

the reason why *myo*-inositol in the form of an α-D-galactoside (*galactinol*) is found in most green plants (see section 2.1.4.1, p. 106).

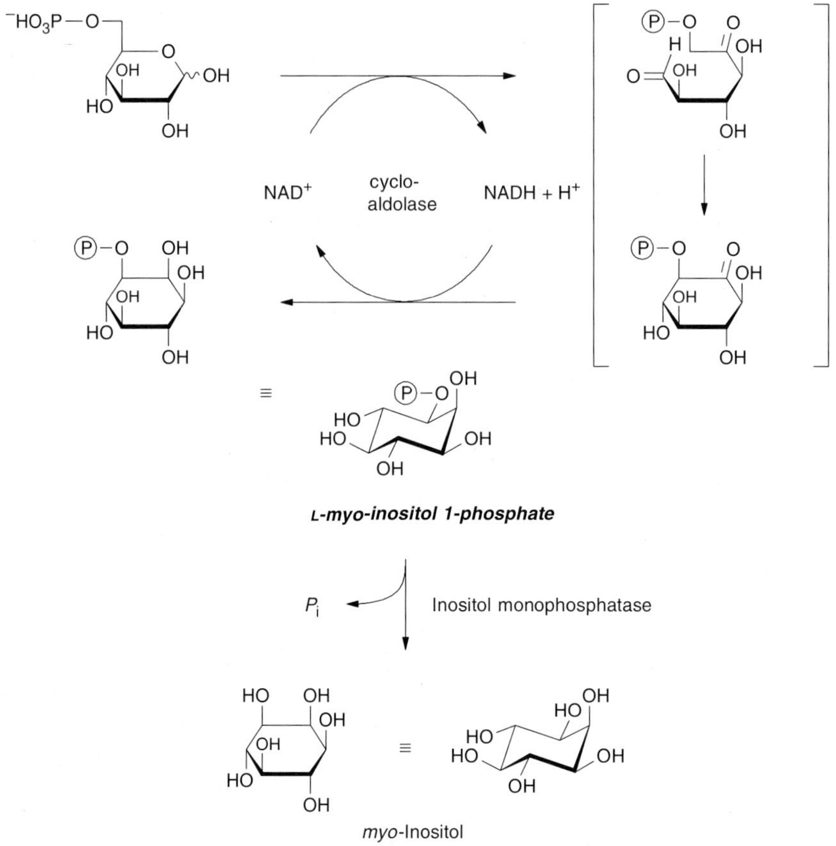

**Fig. 2.79** The *cycloaldolase reaction* converts D-glucose 6-phosphate into 1-L-*myo*-inositol 1-phosphate and therewith provides access to all naturally occurring cyclohexane derivatives.

## 2.2.2 Biological recognition[84]

The discovery of complex carbohydrate structures in glycoproteins and in glycolipids of all eukaryotic cell types, and the genetically costly biosynthesis of numerous, highly specific glycosyl transferases and glycosidases necessary for the synthesis of the complex oligosaccharide, have aroused general suspicions that the carbohydrate structures must be information carriers of some sort. This has been demonstrated in some cases, though also disproved in many, but it would be more accurate to say that the situation is still completely uncertain.

That proteins are able to bind oligosaccharides specifically with great affinity, if the structures of the individual monosaccharide residues, the positions of linkage and the anomeric configurations are correct, is well known. Antibodies are a good example of this phenomenon. The possibility exists therefore for proteins to read the exceedingly complex *language* of the oligosaccharide structures built from a relative simple *alphabet*. The question is only for what purpose?

Some attractive ideas were related to the possibility of influencing the activity of enzymes or the possibility of *addressing* enzymes with the help of carbohydrate structures[85]. In a careful study on 30 different pancreatic ribonucleases, it was shown, however, that neither enzymatic activity depended on whether and to what extent the protein is glycosylated nor did the excretion in the intestines suffer if an enzyme lacked covalently bound carbohydrate structures.

It seems certain that the way in which information is transferred by carbohydrates is not similar to that involving the genetic code, requiring its duplication, copying, and translation. For that, the mole-

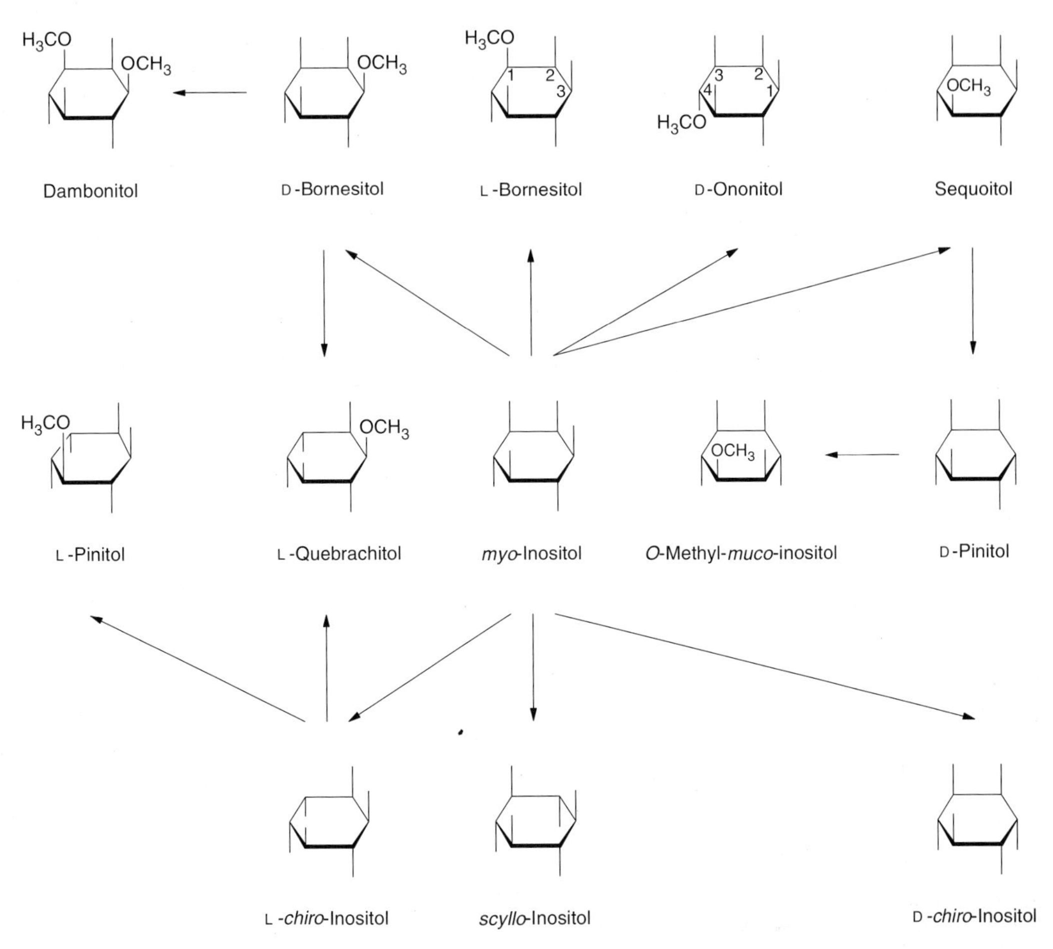


**Fig. 2.80** Some of the naturally occurring cyclitols and their derivatives. The arrows indicate their biogenesis from *myo*-inositol.

**Fig. 2.81** *D-1-Phosphatidylinositol*, a component of membrane lipids.

cular diversity or microheterogeneity of the complex oligosaccharides in a glycoprotein or amongst the glycolipids of a cell membrane is too large. The question arises therefore, what is recognised and what is released in the biological process? Also, it cannot be imagined that the entire structure of an oligosaccharide composed of about 10 monosaccharides means something specific; how should it be read? However, it could well be imagined that proteins, as receptors in the widest sense, can recognise small elements in the structure of an oligosaccharide such as different anomeric configurations, or different linkage positions, and naturally also, different monosaccharide units. A series of examples of this type are known and well investigated. In most cases the recognition concerns small, mostly peripheral structural elements. Thus, the question remains why large, complex, structures are biosynthesised when only a small part of it suffices for the signal transmission. This and many other questions on the biological role of the glycoconjugates are still unanswered and need further and more intensive investigation. For the moment, it is sufficient to comment that there is a specific recognition process in which carbohydrates are involved, beyond doubt. Some of these recognition processes are introduced in this section.

#### 2.2.2.1 Lectins as specific carbohydrate receptors

Should information of some type be inherent in carbohydrate structures then there must be complementary receptors which are able to specifically bind these structures and to recognise them in a differential manner. In general, the dissociation constant between ligand (carbohydrate structure) and receptor (according to definition always proteins, although in the meantime specific carbohydrate-carbohydrate recognitions are also discussed) is a valid measure of the binding, that is the precision of recognition. In fact, for about 100 years plant proteins have been known, today denoted as *lectins*, which have the ability to make red blood cells (erythrocytes) stick together, that is to agglutinate them[86], a property which led to them originally being called *phytoagglutinins*. A further clue was the extraordinary toxicity of rizinus seeds, whose aqueous extract contains the toxic principle *ricin*[87], described by its discoverer Hermann Stillmark (1888). It later became apparent that the protein ricin – one molecule kills a cell, that is to say $10^{-9}$ g would kill a grown man – is not identical with the agglutining carbohydrate-binding protein but is only associated with this and also has large structural similarities with it.

Over several decades agglutinins were found in almost all plant families, but they are especially abundant in legumes. The main sources are the plant seeds with in part very high concentrations of a specific lectin. The jack bean (*Canavalia ensiformis*), for example, contains in its seeds, in up to 3% of the total weight, the agglutinin *concanavalin A*. This was crystallised by Sumner as early as 1919. In the 1950's, it was ascertained that haemagglutination depended on the binding of carbohydrates on the erythrocyte surface by the agglutinin, and that it is possible to distinguish different blood groups with different agglutinins. Because of their selective properties, the American W. C. Boyd suggested in 1954 that these specific, carbohydrate-binding plant proteins be named lectins (from the Latin *legere*, to choose). The term has gained acceptance and is applied today very generally to carbohydrate binding proteins, even to those not of plant origin.

Different plant lectins agglutinate erythrocytes in a biologically meaningless process by binding to blood group specific carbohydrate structures (Fig. 2.82)[87]. In order that plant lectins can effect agglutination, they must possess at least two separate binding points for the specific carbohydrate structure. This should be one of the criteria for the use of the term lectin. Indeed, as already indicated, there is today a very widely held view to denote as lectins those carbohydrate binding proteins which are not antibodies and which show no enzymatic activity. With this definition, the circle of the proteins to be included is greatly widened.

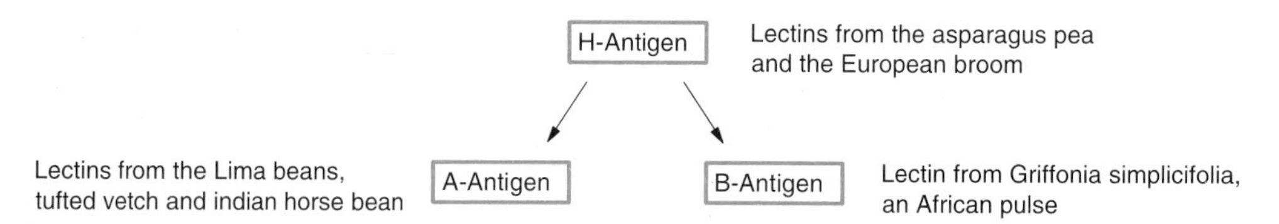


**Fig. 2.82** *Plant lectins* are able to distinguish the different blood group antigens of the ABO system.

The structure and binding specificity of very many plant lectins are known today. The biological significance of these proteins remains an open question. Contradictions concern, especially, the conspicuous properties of the plant lectins to recognise and to bind certain monosaccharides and oligosaccharides. GalNAc is specifically bound by many plant lectins but does not generally occur in plants. There has been no shortage of suppositions as to the biological function of the plant lectins but they cannot be reduced to a common denominator. The toxic lectins may have a protective function for their organism against plant-eating pests and many lectins may, since they are fungicides, protect against infection. The lectins of the legumes could be especially concerned with the colonisation of nitrogen-fixing bacteria. In point of fact it is possible in some cases to prove the specific binding of such nodule bacteria with lectins of the host plant (see section 2.2.2.9, p. 159).

Generally, however, no common biological function can be found for all the plant lectins. It must be asked, therefore, whether the striking carbohydrate binding quality of the plant lectins has distracted from the search for a hitherto unknown biological function of these proteins. The uncertainty as to the importance of the plant lectins continues to pose a problem, as they are not enzymatically active and are not carbohydrate receptors originating in the immune system of another living thing. Up to the present their biological function is based on the recognition of carbohydrate structures, according to the usually accepted definition of lectins. Actually, such a definition is not fully permissible, since the animal lectins are mostly membrane proteins and their structure differs very much from each of the plant lectins.

#### 2.2.2.2 Lectins in liver-cell membranes and in sera

As explained in the foregoing section, proteins from plants which have the special property of recognising and binding with high affinity to certain monosaccharides, and also oligosaccharides, were originally described as lectins.

In 1970, Ashwell, Morell, and their co-workers[88] made the observation from experiments carried out in rats and rabbits that serum glycoproteins have a distinct lifetime and after constant times are filtered out and broken down by the liver. This appears to be a natural purification process of the organism. If the glycoprotein is treated with sialidase, an enzyme that splits the end NeuAc residues from the antennae of the glycoprotein oligosaccharide, and the so treated and, additionally, radio-marked glycoprotein is placed again in the blood circulation, then an accelerated filter process is observed. For example, human *asialo-ceruloplasmin* that has been radio-marked on the end group galactosyl residue by treatment with galactose oxidase followed by reduction with tritiated sodium borohydride, $[^3H]NaBH_4$, is removed in a few minutes from the circulation of rabbits[89]. Untreated, natural ceruloplasmin has a half-life in the serum of t/2 = 56 hours. Normally the next monosaccharide in the oligosaccharide chain of the *N*-glycosidic glycoprotein is galactose, which, after separation of the *N*-acetylneuraminic acid stands on the periphery. If the galactose is also removed, this time with $\beta$-galactosidase, then the next monosaccharide, GlcNAc is exposed. Astonishingly, this glycoprotein is retained even longer in the circulation than the intact original glycoprotein. Not only glycoproteins behave in this manner, but also whole blood cells, whose membrane glycoprotein has been subject to the same treatment with sialidase and $\beta$-galactosidase.

It seems reasonable to conclude from these experiments that the liver cells (hepatocytes) contain a receptor protein in the membrane that apparently recognises the peripheral galactose of the desialylated glycoprotein and binds with greater affinity. As is now known, binding is followed by infolding of the membrane in question (Fig. 2.83), the building of a vesicle and with it transport into the cell, an occurrence which is named *endocytosis.*

The principle of endocytosis is general for all cells, in particular for liver cells, and is employed in order to take up macromolecules or also whole cells so that they may be digested in the lysosomes (see section 2.2.2.3, p. 140). The reverse occurrence, exocytosis, serves for the transport of freshly synthesised macromolecules outward through the cytoplasmic membrane.

The Gal receptor of the liver – better called the Gal/GalNAc receptor, since GalNAc is bound even better to the receptor than Gal – occurs in all mammals. In birds a similar system is found, only here, apart from NeuAc, Gal must also be removed since the receptor of birds recognises only the peripheral GlcNAc. There is no question that the receptors, which are also called animal lectins, quite specifically recognise and bind the end of chain monosaccharide residue, in the case of mammals Gal, in the case of birds GlcNAc, and thereby give the signal for endocytosis.

The receptors from rat liver[90] and from chicken liver[91] have been isolated and in the past years have been especially intensively investigated[92]. In spite of the different ligand specificities, both *liver lectins* have surprising structural similarities that probably is connected with their common function as endocytosis receptors.

The primary structure of the lectin from chicken liver has been fully elucidated, as have those of a good many other lectins. Most structural investigations have been conducted on chicken lectin because of its relative small molecular weight of 26000 Da. Large homologies in the primary structures allow

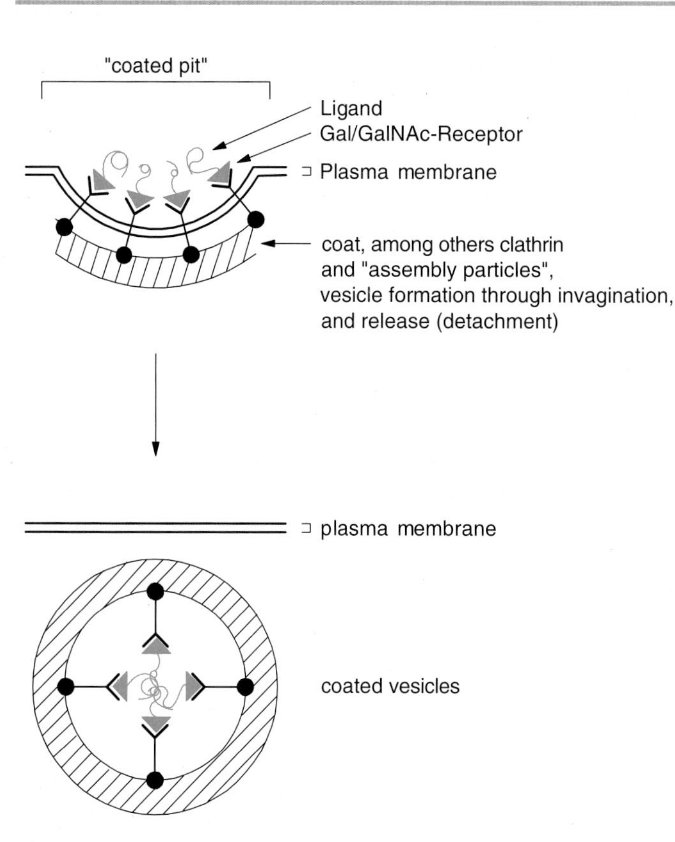


**Fig. 2.83** Schematic representation of receptor mediated *endocytosis*. Initially, the ligand-receptor complex migrates into the "coated pits".

conclusions to be made about higher order structures of other animal lectins. The chicken lectin consists of four domains[92] (Fig. 2.84).

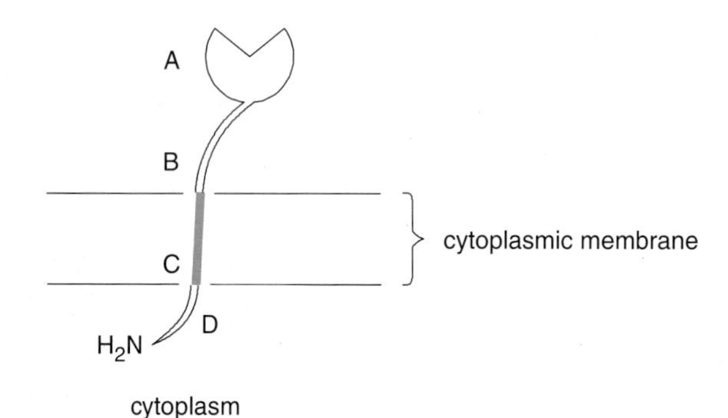


**Fig. 2.84** The *GlcNAc-specific chicken lectin* embedded in the *hepatocytic membrane*. **A** is the GlcNAc binding domain, **B** a long stem comprising 22 amino acids. **A** and **B** together are159 amino acids and an *N*-glycosidic carbohydrate outside the plasma membrane. **C** is the *trans*-membrane domain comprising 25 of the most hydrophobic residues and **D** the *N*-terminus built up from 23 hydrophilic residues.

The *N*-terminus with 23 amino acids is found on the cytoplasmic side of the membrane. The seventh residue, serine, carries a phosphate group whose function is still unknown. The trans-membrane domain consists of 25 mostly hydrophobic amino acids. The major part of the lectin with the *C*-terminus and in total 159 amino acids is found on the outside of the membrane. Of this sequence a globular part comprising 136 amino acids can be separated with subtilisin. This part is the actual GlcNAc receptor. Its affinity does not differ between each of the intact lectins. This so-called carbohydrate recognition domain (CRD) sits on a short stem of 22 amino acids, which in position 67 carries the only *N*-glycosylated asparagine of the lectin. The carbohydrate part in this position might possibly contribute to a stiffening of the stem in the style of a ruff. It is very interesting that chicken lectin is anchored in the membrane as a hexamer. Correspondingly, one multi-attenary ligand presents six binding sites[92]. This so-called "clustering effect" leads to a multiplication of the affinity by many orders of magnitude[93] (Fig. 2.85).

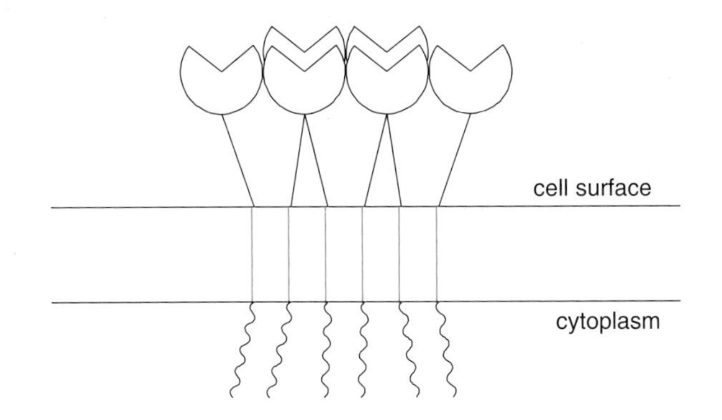


**Fig. 2.85** *Chicken lectin* is bundled as a hexamer in the membrane. This arrangement, designated as "clustering", makes possible the binding of many glycosyl end groups of a multi-antennary glycoconjugate. The so-called "clustering effect" leads to an exponential strengthening of the affinity.

As already indicated, chicken lectin and other animal lectins, for example the Gal receptors from rat liver, possess large structural similarities. By genetic manipulation, the CRD of the chicken lectins may be combined with the membrane anchor of the rat lectin. Incorporated into rat liver cells, only GlcNAc end groups were recognised but not Gal end groups. Two of the three rat lectins – it is not yet known why three are necessary – differ only in the carbohydrate portion, but the third is distinct from the first two in also having a different protein structure. Altogether, the rat lectins are almost twice as large, which expresses itself above all in a lengthening of the stem and of the cytoplasmic end. In spite of the different binding specificity, the CRD's of the chicken lectins and the rat lectins show 40–50% agreement in their amino acid sequence.

Animal lectins appear to be ordered in families whereby the family membership depends on particular structural properties. The *C-type lectins*, to which belong also the receptors of the hepatocytes discussed in the preceding section, possess as a common structural feature a binding position for $Ca^{2+}$, hence the designation C-type[94]. These lectins are receptors initiating endocytosis, cell adhesion molecules (CAM) (see section 2.2.2.6, p. 148)[95] or secreted proteins of the extracellular matrix and of the serum. To the latter belong the very much studied mannose binding proteins in the serum of mammals[96]. They are defence proteins, which independent of the immune system, attack foreign germs, fungi, and bacteria, because they are taken up by Man-oligosaccharides of the cell surface and initiate the complement mediated lysis of the intruder. The carbohydrate recognition domain (CRD) of the mannose binding proteins of the rat could be crystallised with an asparaginyl-octasaccharide and the three dimensional structure determined with a 1.7 Å (0.17 nm) resolution. In these investigations the role of $Ca^{2+}$ as a structural stabilising factor became very clear, the ion itself entering into binding with the carbohydrate. Schematically represented in Fig. 2.86[97] is how in the crystal both CRD's lie *back to back* and use their $Ca^{2+}$ side for binding to a mannosyl residue.

The specificity of binding is not too large and is extended to D-GlcNAc, D-Glc and, interestingly, also to L-Fuc. D-Gal is not recognised. The importance of the 3- and 4-positions in the ligands is underlined. By suitable juxtaposition, the conformation of the 3- and 4-hydroxy groups in D-Man appears just as that of the 2- and 3-equatorial hydroxy groups in L-fucose (Fig. 2.87, p. 141).

Here are minimal structural elements for effective binding, that is sufficient for recognition of a carbohydrate ligand. In these special cases, where ligand-receptor interactions are known exactly, it is also possible to manipulate the specificity of a receptor, in that certain amino acids in the binding region can be exchanged through "protein engineering". In the case of the D-mannose binding proteins, two glutamic acid-asparagine pairs are responsible for the binding of the D-Man 3- and 4-hydroxy groups. It is known that in D-Gal binding domains (hepatocyte Gal/GalNAc receptors) one of these pairs is exchanged for the pair aspartic acid-glutamine. Actually, through this simple exchange of functionality in the receptor, it is possible to convert its specificity from D-Man into D-Gal[97]. These investigations, especially those of K. Drickamer, contribute to a high degree to our understanding of the specific recognition processes between carbohydrates and proteins.

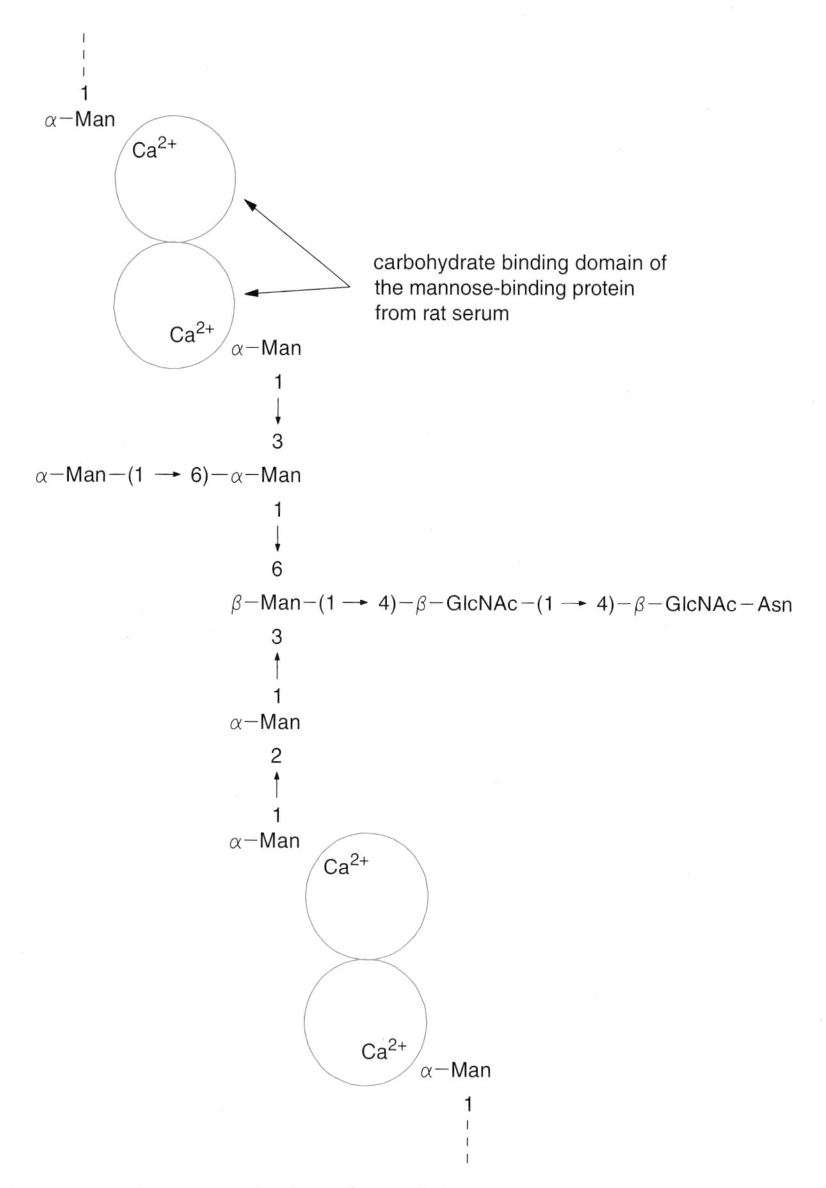


**Fig. 2.86** In the mannose binding protein of the rat, a ternary complex is formed between the carbohydrate binding domain, a $Ca^{2+}$, and a terminal mannosyl residue. This model is based on the X-ray structure analysis of the crystalline ligand-receptor complex.

### 2.2.2.3 Man-P receptors in the recognition of lysosomal proteins

All high molecular weight constituents of the cell and extracellular matrix have only a limited lifetime. At the end of this average time they are broken down through hydrolytic cleavage and the components are then available for reuse in biosynthesis. Degradation takes place in the cell in an organelle specifically designed for this purpose, the lysosome. Lysosomes are separated from the cell interior with a membrane and differ most importantly in the acidic environment of about pH 5 of their lumen compared to pH 7.2 in the cytoplasm. *Lysosomes* are endowed with very aggressive hydrolases whose pH

**Fig. 2.87** Superposition of a D-glucosyl, D-mannosyl, and an L-fucosyl residue. The structural conformity of the 3- and 4-positions in D-mannose or D-glucose with the 2- and 3- positions in L-fucose is apparent when the latter is drawn in an appropriate manner.

optimum lies in acidic media. These *acid* hydrolases are relatively inactive in the neutral pH of the cytoplasm which provides a certain protection for the cell against lysosomal enzymes which eventually escape.

The lysosomal enzymes, of which about 40 of all specificities are known, arrive from the place of their origin in the endoplasmic reticulum at the Golgi apparatus, where they receive a marker at an early stage of their glycosylation (see section 2.3.8.5, p. 218). This marker allows them to be sorted out later from all other proteins and specifically, with the help of vesicle transport, to be conveyed into the lysosomes (Fig. 2.88). The lysosomes serve for the breakdown and reutilisation of extracellular polymers, bacteria and its own cell constituents (Fig. 2.89).

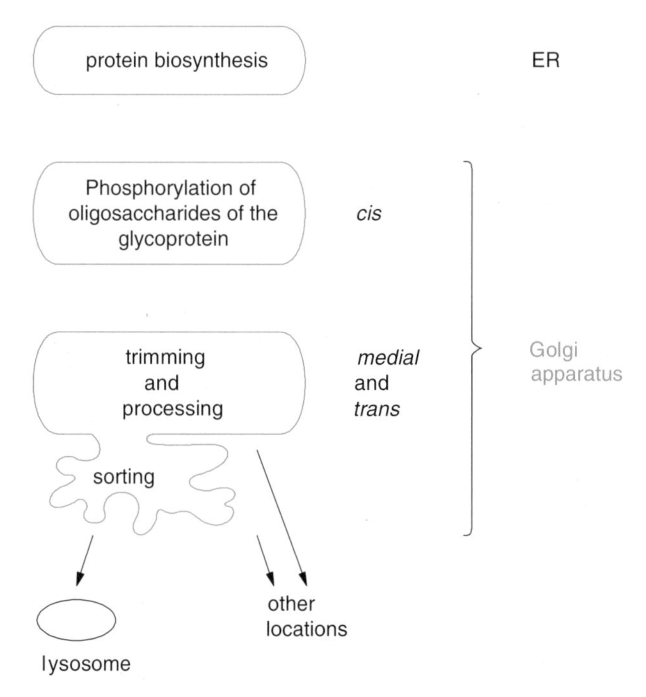


**Fig. 2.88** The lyosomal hydrolases are marked in the *Golgi apparatus* by phosphorylation of oligosaccharide residues and, with the *address* supplied, are sorted and sent to the lysosomes.

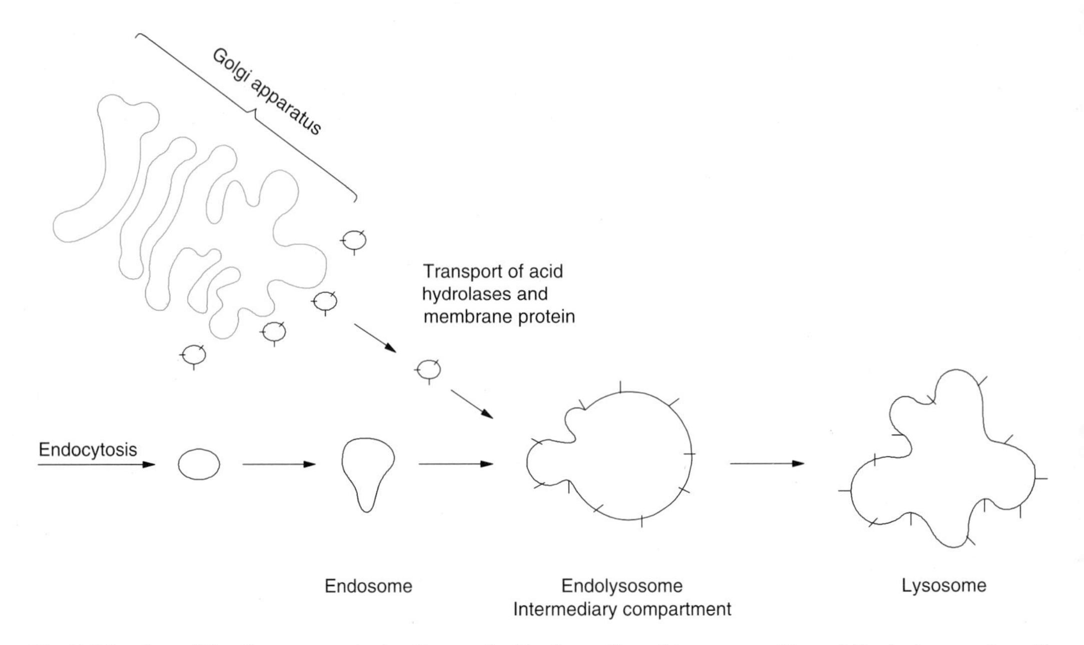


**Fig. 2.89** One of the three suggested pathways for the formation of lysosomes. The *acid* hydrolases unite with vesicles of endocytosis (endosomes), which contain materials capable of bringing about degradation, to give *endolysosomes*, which after maturing become true *lysosomes*. Endolysosomes can also absorb phagocytosed bacteria (*phagolysosome*) or individual cell organelles *(autophagolysosome).*

The first phase of glycosylation of *N*-glycosidic glycoproteins in the endoplasmic reticulum (ER) finishes with a so-called highly branched "high mannose" oligosaccharide. After transfer to the *cis*-Golgi system (Fig. 2.89), one or more of the peripheral mannose residues are phosphorylated in the 6-position. The mannose residues receive their phosphate group in an unusual form and manner (Fig. 2.90). Apparently, the later lysosomal enzymes have a currently unknown signal-region at their disposal in the protein, which is recognised by a GlcNAc-P-transferase (*N*-acetylglucosamine-phospho-transferase). UDP-GlcNAc serves as donor for GLcNAc-P, which is transferred to the 6-position of a terminal mannosyl residue.

The phosphate group remains in this manner temporarily masked until an unspecific phosphoglycosidase removes the GlcNAc residue and exposes Man-6-P residues. The specific phosphorylation signals that these corresponding proteins are lysosomal proteins. In the *trans*-Golgi network specific receptors wait in the form of transmembrane proteins for the Man-6-P (Fig. 2.91).

Currently, two different receptors are known, which differ considerably in their molecular weights (about 220 kDa and 45 kDa, respectively), but only insignificantly in their binding specificity towards phosphorylated mannose oligosaccharides. It is suggested that the receptors recognise not only a Man-P residue but also extended oligosaccharide structures. Inhibition studies with one and multi-antennary Man-P oligosaccharides make it very probable that there are at least two binding sites at a definite distance from each other for the Man-P residues (Fig. 2.92)[98,99].

*Vesicles*, formed by a budding process and which contain only marked lysosomal proteins, travel to the endolysosomes or lysosomes, fuse with their membrane, release the protein as result of the pH change, and travel, after renewed budding off, with the unoccupied receptor, back to the Golgi apparatus (Fig. 2.91)[36]. The immediate hydrolytic separation of the phosphate which follows prevents the possible transport of the enzyme back again to the Golgi apparatus. The Man-P-receptors are also found in small quantities in plasma membranes. There they evidently take care of lysosomal proteins, mistakenly lost during the process of endocytosis, to ensure that they arrive nevertheless at their destination.

The principle demonstrated in this special case of targetted vesicle transport for macromolecules may be universal for eukaryotic cells. However, it is questionable whether the principle of oligosaccha-

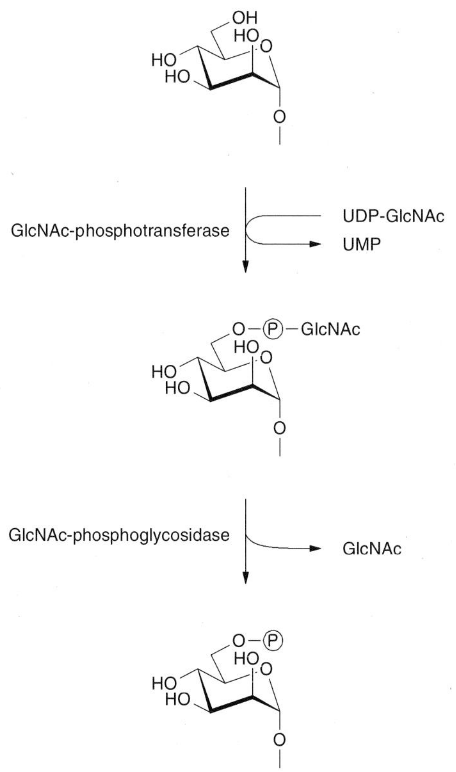

**Fig. 2.90** Phosphorylation of the mannosyl residue in the *cis*-Golgi system. The enzyme GlcNAc-phosphotransferase must be able to bind onto an hitherto unknown recognition site of the lysosomal enzyme.

GlcNAc-phosphotransferase: UDP-GlcNAc → UMP

O–Ⓟ–GlcNAc

GlcNAc-phosphoglycosidase: → GlcNAc

O–Ⓟ

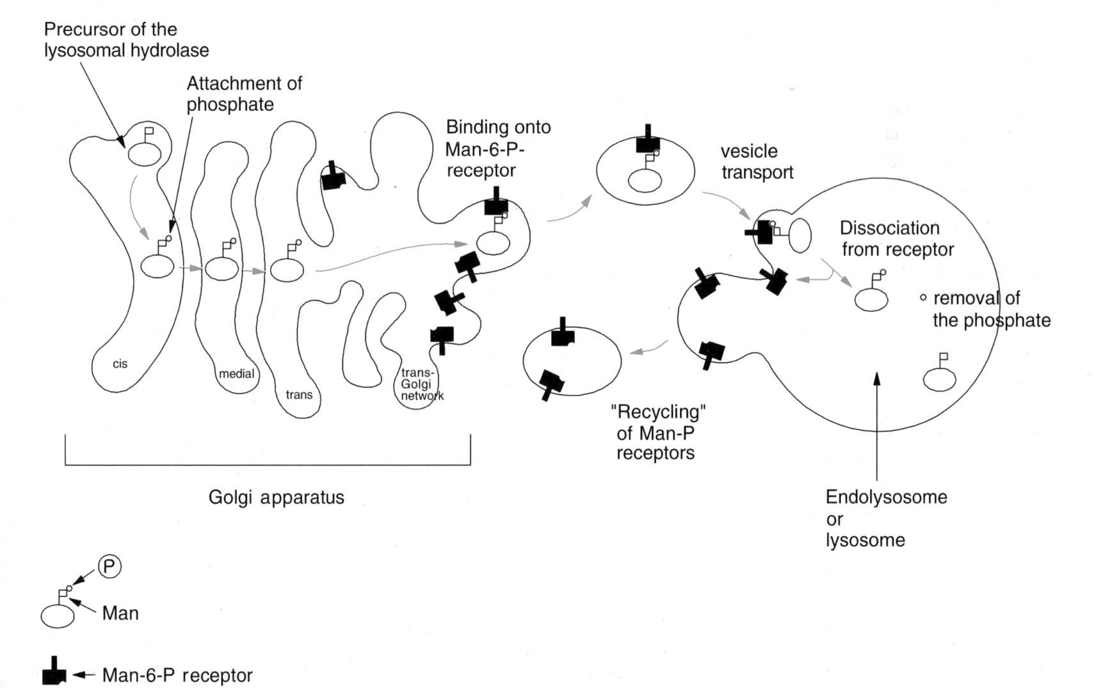


**Fig. 2.91** *The mannose-6-P receptor*, after completion of its work, is conveyed again to the Golgi apparatus.

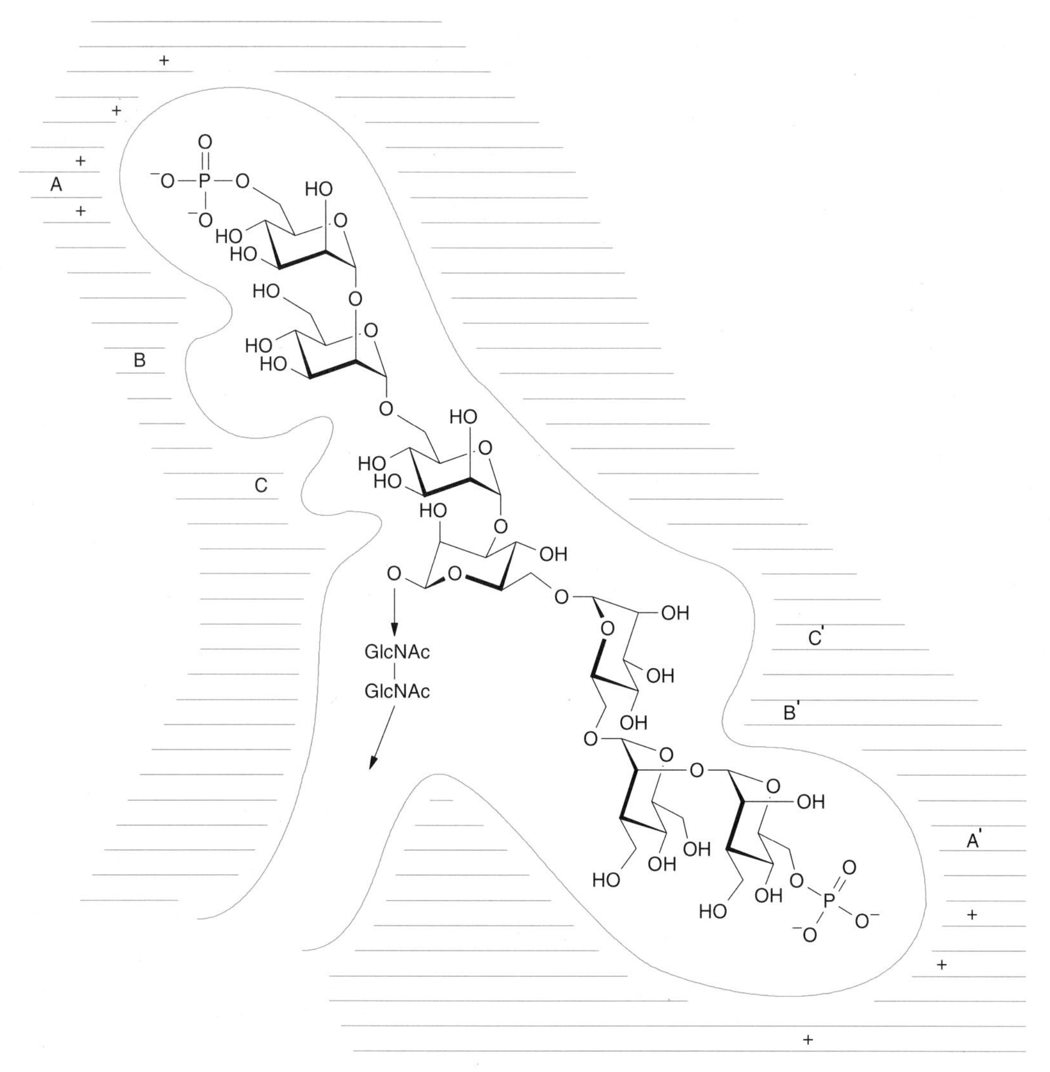


**Fig. 2.92** Proposed model for an extended binding region of the mannose-6-P receptor from alveolar macrophages of the rabbit.

ride marking, for example through phosphorylaytion, is universal. Since various Man-P residues of the type shown (Fig. 2.90) can arise on a multiantennary oligosaccharide, very high affinity in the nanomolar region is produced through the so-called "clustering effect" (multiplication of the binding constant through various contact points). The Man-P-signal is one of the few unambiguously proved recognition features which is based on a carbohydrate structure.

### 2.2.2.4 Heparin-antithrombin and blood coagulation

*Heparin* is a highly sulphated polysaccharide whose influence on the process of blood coagulation has long been recognised[100]. Knowledge leading to clarification of the mechanism of action is relatively new[101]. Blood coagulation is very finely regulated through two mechanisms. One enzyme cascade consisting of many steps catalyses the formation of fibrin, responsible for wound closure, from fibrinogen (Fig. 2.93). Inhibition in the cascade can occur in various places. One of these is the conversion of *pro-*

*thrombin* into *thrombin*, which for its part catalyses the conversion of *fibrinogen* into *fibrin*. The highly active inhibitor is a complex between *antithrombin III* and *heparin*, whose action is many times stronger and above all begins very much quicker than that of antithrombin III alone. The complex inhibits not only the conversion of prothrombin into thrombin, but also deactivates already existing thrombin, as it forms an inactive complex with the latter.

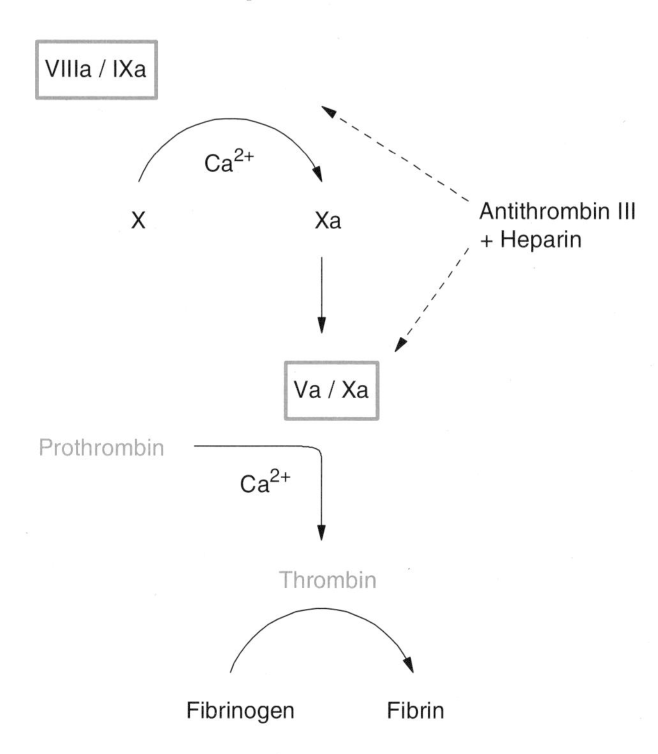


**Fig. 2.93** Schematic representation of a small part of the *blood coagulation* cascade. The interaction of *antithrombin III-heparin* occupies a key position both in the inhibition of the conversion of factor X into factor Xa and also in the transformation of *prothrombin* into *thrombin*.

Experiments in various laboratories showed in the 70's that commercial heparin preparations can have very different activities according to their origin. Careful separation of the heterogeneous mixtures led eventually to the knowledge that only a relatively small part showed high affinity to antithrombin III and, with that, biological activity. Degradation of such fractions and isolation of smaller oligosaccharide sequences narrowed down the structure of the effective sequence to the pentasaccharide shown (Fig. 2.94)[102].

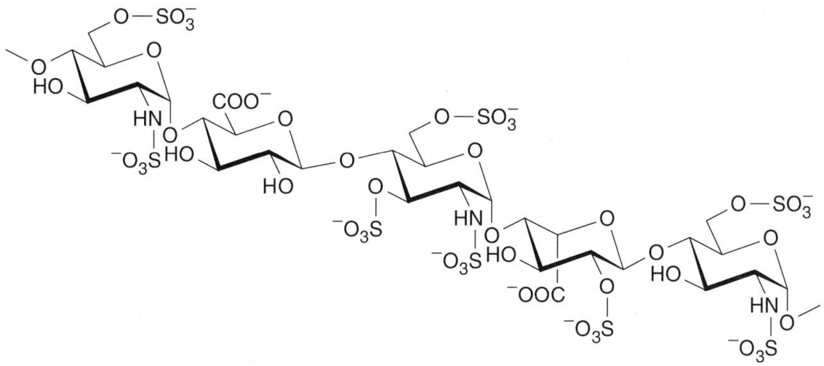

**Fig. 2.94** The structure of the active *pentasaccharide segment* in *heparin* which binds to *antithrombin III*.

It is now certain, since the chemical synthesis of this compound has also been accomplished, that a pentasaccharide sequence in native heparin is responsible for the specific binding to antithrombin. From these in-depth investigations, above all with structural variants of the pentasaccharide, it becomes clear that it is not so much the chemical peculiarity of the basic carbohydrate framework

-α-D-GlcN-(1→4)-β-D-GlcA-(1→4)-α-D-GlcN-(1→4)-αL-IduA-(1→4)-α-D-GlcN-

that is responsible for the specific action but the exact steric arrangement of anions in the form of the *N*- and *O*-sulphate groups as well as of the carboxylate groups. Recently, the complementary binding region of antithrombin III was elucidated through X-ray studies[103] whereby the association can be clearly attributed to a salt bridge between the anions of heparin and the corresponding ordered cationic lysine residues in the binding region of antithrombin III. The binding of the heparin leads to a fast conformation change in the antithrombin III, which is the cause of the high affinity towards thrombin.

It may be supposed that *heparan sulphate*, which differs from heparin only by a somewhat smaller average molecular weight and a lower sulphate content, also has similar functions as this. For example, the presence of heparan sulphate on the surface of vascular endothelial cells is surely responsible for protection against spontaneous blood coagulation. On the basis of very recent investigations, heparan sulphate on the cell surface and in the basal membrane is supposed to play a special role in the binding and thereby the control of growth factors.

### 2.2.2.5 Recognition in the fertilisation process[104,105]

With plants and animals, the fertilisation between female and male organisms of the same species is a fundamental principle of their reproduction. Species specificity is an important requirement and apparently depends upon mutual molecular recognition between female and male germ cells. The species specificity prevents the unallowed combination of different kinds of genetic material. The fertilisation processes of the sea urchin and the mouse have been especially well investigated. The surface of the sea urchin egg cells (diameter about 100 μm) – on the outside of the plasma membrane – consist of a vitelline membrane about 1 μm thick and a jelly coat about 30 μm thick made from proteoglycan (Fig. 2.95).

Approach of the sperm cells to the egg cells is triggered by *chemotaxis*, which in the case of the sea urchin depends on the secretion of a small peptide by the egg cell. The *vitelline sheath* is a thick, stable network of glycoproteins and contains the specific receptor for a surface protein (*bindin*) of the sperm cells. The uniting of egg and sperm cell begins with the so-called acrosomal reaction (the acrosomal vesicle empties its contents – hydrolases – through exocytosis and the gelatinous layer around the egg is degraded enzymatically) after first establishment of contact between sperm and the outermost layer of the egg cell. Fundamentally, these processes are very similar also in mammals and are especially well researched in the case of the mouse with respect to contact between sperm head and the *zona pellucida* (ZP, about 8 μm thick) surrounding the plasma membrane of the egg cell. The latter may be compared with the very much thicker jelly coat of the sea urchin (Fig. 2.95).

In general, fertilisation (fusion of the cells) does not occur *in vitro* between egg cells of one species and sperms of another. Similarly, a sperm cell does not bind to an already fertilised egg cell of the same species. However, if the zona pellucida of the egg cell is removed, then it can be fertilised but the embryo between dissimilar species is not viable and therewith not capable of life. The specific surface contact has consequently two important functions; avoidance of crossing between different species as well as avoidance of multiple fertilisation (polyspermy).

Of the three components of the mouse zona pellucida: *ZP1*, *ZP2*, and *ZP3* – all three are glycoproteins – with molecular weights of 200, 120, and 83 kDa, it can be established by *in vitro* experiments that only ZP3 has affinity for the surface of the sperm head. The inhibition of fertilisation by isolated ZP3-glycoprotein, but not by ZP1 or ZP2 is unambiguous. After fertilisation, ZP3 loses its ability to bind sperms.

ZP3 contains a serine and threonine rich polypeptide of 424 amino acids with three to four *N*-glycosidic oligosaccharide structures of complex type and in addition an undetermined number of *O*-glycosidic oligosaccharides. The glycoprotein appears to be decidely heterogeneous in denaturing polyacrylamide gel electrophoresis (SDS-PAGE). The ability to bind spermatoza and to initiate the acrosomal reaction is associated also with the oligosaccharide and not the protein component alone, since neither long heating of ZP3 to 100 °C nor treatment with denaturing reagents such as urea or SDS has any effect.

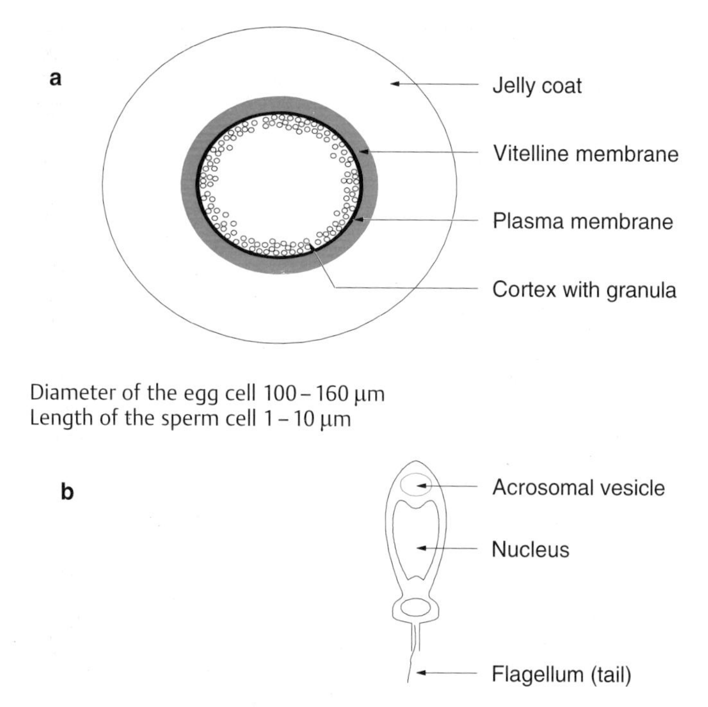


**Fig. 2.95** **a** Egg and **b** sperm cells of the sea urchin (relative sizes in the illustration are not to scale).

The glycopeptide arising from ZP3 by proteolytic decomposition does not lose the specific properties of the intact glycoprotein to inhibit the binding of spermatoza to egg cells. Removal of the *N*-linked oligosaccharide from ZP3 is without influence on the inhibition. However, with removal of the *O*-glycosidically linked oligosaccharide from ZP3 the inhibitory effect is lost. Through selective reductive cleavage of the *O*-linked oligosaccharides from ZP1, ZP2, and ZP3 one obtains each time oligosaccharide mixtures of which only the compounds obtained from ZP3 inhibit the take-up of spermatoza into egg cells[106]. However, one has to take into consideration that the acrosomal reaction is not triggered by these oligosaccharides and the glycopeptides mentioned above but only by the intact glycoprotein.

It can be ascertained by gel chromatography that of the broad spectrum of different sized ZP–3 *O*-oligosaccharides, only the fraction with a molecular weight between 3400 and 4600 shows inhibitory action against receptors in spermatoza. The isolation of a chemically homogeneous oligosaccharide with specific action has not yet been achieved. Athough this is so, treatment with six different exoglycosidases makes it seem very probable that α-galactosyl- and α-fucosyl end groups are recognised specifically by the spermatoza receptors, because the oligosaccharide fraction loses its inhibitory action both through incubation with α-galactosidase from coffee and also through incubation with α-fucosidase from bovine kidney. The inhibitory action disappears also through treatment with galactose oxidase, whereby the 6-hydroxymethyl group of a terminal galactosyl residue is oxidised to an aldehyde group.

It is thought today that the inability of the zona pellucida to bind sperm after successful fusion can be traced to the effect of exoglycosidases, which the ovum delivers through exocytosis in its mucus shell. Experiments on ZP3 from pig oocytes and boar sperm yielded fundamentally similar results. In this case, however, *O*-[107] and *N*-glycosidically[108] linked oligosaccharides appear to be responsible for the specific binding of the sperms. Of course, the effect with the carbohydrate structures still bound covalently to the protein is greater by far than that of the isolated oligosaccharide mixture.

Even though in the case of mouse ZP3, α-D-galactosyl residues and perhaps also α-L-fucosyl residues are proved necessary for the recognition between oocytes ZP and sperm receptors, it is still very difficult to imagine that these structural elements alone in a large population of many different oligosaccharides are responsible as epitopes for the specific interaction between egg cells and sperms. An explanation, perhaps at this moment somewhat premature and risky, may be that a special pattern with respect to number and local distribution of fully *normal* oligosaccharides and not a special epitope is responsible for the species specific binding of sperm cells and egg cells. As may be inferred from the foregoing discussion, the envelope of the egg cell was examined almost exclusively, for obvious reasons, on its adhesive role. Very little is known about the carbohydrate binding receptors of the sperm head. Thanks to the fact that (1–4)-$\beta$-D-galactosyl transferase was found on the tip of the head of all mammalian sperm, this receptor finds a place in the present discussion[109].

### 2.2.2.6 Cell adhesion

Multi-cell organisms consist of associations of different cells in very different surroundings. An organism arises because the cells organise themselves and their surroundings in precise order after a predetermined plan. In the final analysis, it is the molecules of the cell surface, which effect specific cell adhesion and so morphogenesis, which control formation of the organism. Today, the complexity of these events involving steps in the development of an organism can only be guessed at, and a discussion must be confined to relatively well researched examples abstracted from this enormous subject. In some probability, the principles of the morphogenesis of multi-celled organisms may be very similar and depend fundamentally on the interaction of glycoconjugates of the cell surface amongst themselves and with the extra-cellular matrix as usually disposed. The extent to which carbohydrates play herein a clear, differentiating role can only be guessed in most cases.

■ **Organisation of a unicellular organism**. It is understandable that primitive organisms are better suited for the investigation of the principles of cell adhesion than highly developed ones because the analysis is facilitated by the absence of many complicating factors.

The slime mould *Dictyostelium discoideum* is one of the simplest models for the organisation of cell aggregates. *Dictyostelium discoideum* exists under normal living conditions as a unicellular organism that feeds like an amoeba, and grows and reproduces by cell division. If the unicellular organism is deprived of nourishment, generally bacteria, a chemotatic signal is given off by the emission of cAMP into the surroundings, by which the unicellular organisms are caused to move towards one another. Simultaneously, a programme is switched on that leads to a distinct change in protein production, that on its part results again in a change of disposition of the cell. The cells are now able to aggregate and to form organised cell clusters, apparently small slime particles, which are capable of movement on a surface. Apparently, in a hostile environment, the cell cluster grows better than a single cell. The slime cluster undergoes a further organisation in that three different types of cell become differentiated, for the footplate, for the stalk, and for the fruiting body (Fig. 2.96) and together they build a structured organism[110]. With this the final stage is reached. The fruiting body gives off hardy spores from which themselves, as long as the conditions are favourable, again develop single celled amoebae.

The slime mould model is a relatively easy system in which to investigate *in vitro* the development of molecular surface structures for cell aggregation[111,112]. Experiments with tunicamycin and with different glycosidases gave a good indication of the significance of carbohydrate structures in cell adhesion in the case of *Dictyostelium discoideum*. *Tunicamycin* is an inhibitor of *N*-glycoprotein biosynthesis (see section 2.3.8.8, p. 230). Glycosidases can degrade already biosynthesised glycoconjugates of the cell surface and so modify the surface structure. Whilst tunicamycin blocks the aggregation capability of single cells in many experiments completely and in others, however, only partially, treatment of aggregated cells with $\alpha$-mannosidase results in the complete elimination of cell organisation[113].

An attempt to obtain direct proof for the function of glycoconjugates in cell adhesion was made with help of proteolytic fragments of isolated membrane-glycoproteins. Of five isolated glycopeptides, two were able to prevent the aggregation of the cells. Apparently, the glycopeptide blocked the receptors in the membranes of intact cells. Such receptors are called generally "cell adhesion molecules" (CAM). They need not necessarily be lectins (see section 2.2.2.1, p. 136), that is carbohydrate binding molecules.

Independent research groups succeded finally in isolating a homogeneous glycoprotein from aggregated cells which because of its molecular weight is named *gp80*, signifying a glycoprotein with a molecular weight of 80 kD – a customary if confusing type of nomenclature. Antibodies both against the protein part and the carbohydrate portion of gp80 can be obtained and with them the aggregation of single cells can be inhibited (Fig. 2.97). The gp80 is biosynthesised only in the first phase of cell aggregation and only in this phase are the mentioned antibodies effective. It may be conjectured that the different organisation phases are regulated by different surface molecules.

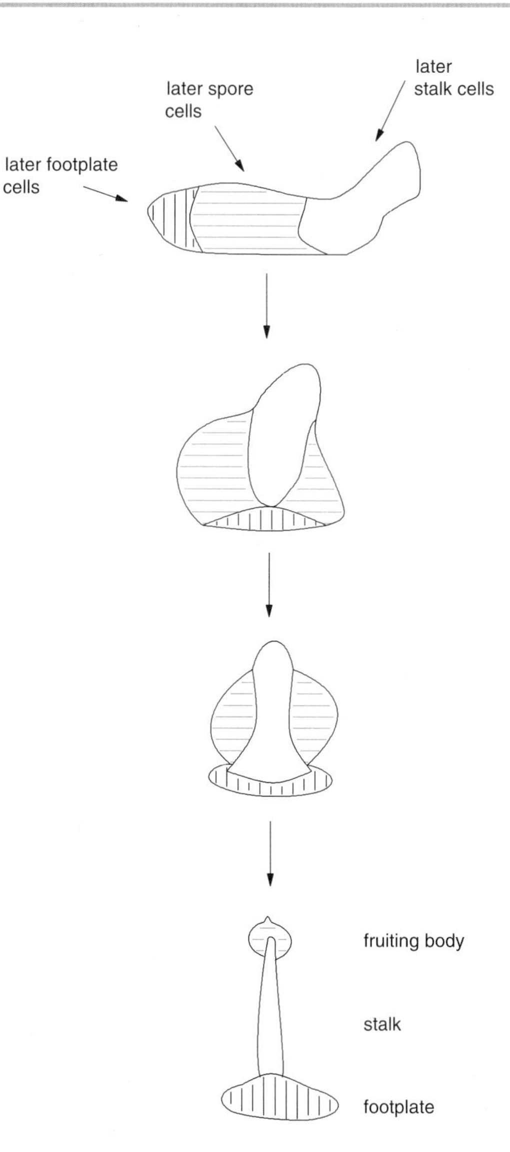


**Fig. 2.96** Footplate cells, stalk cells and spore cells locate themselves as a result of surface contact in the multi-cell organism of the slime mould *Dictyostelium discoideum*.

Glycoprotein gp80 contains two different types of oligosaccharide, of which only one is involved in cell adhesion. It behaves as an *N*-glycosidically bound oligosaccharide, which is bound and recognised by *concanavalin A*. This specificity for concanavalin A points with great probability to α-D-mannosyl end groups, whose existence was indeed verified through the effect of treatment with α-D-mannosidase (Fig. 2.97).

From the influence of the antibody against the protein part of gp80 as well as against its oligosaccharide portion, it may be concluded that gp80 behaves both as the receptor (protein as CAM) and also the ligand (oligosaccharide of the cell adhesion system). The double role of a cell surface glycoprotein, being at the same time receptor and ligand (Fig. 2.98), is one of, in principle, three possibilities for the contact adhesion between cells. In this case it is described as homophilic binding.

In addition, there is the possibility of heterophilic binding, if one cell provides the receptor and the other the ligand. In a third possibility, the cell surface structures on two similar or perhaps different cells are recognised and bound by a soluble lectin. In the course of this section examples of all types are considered.

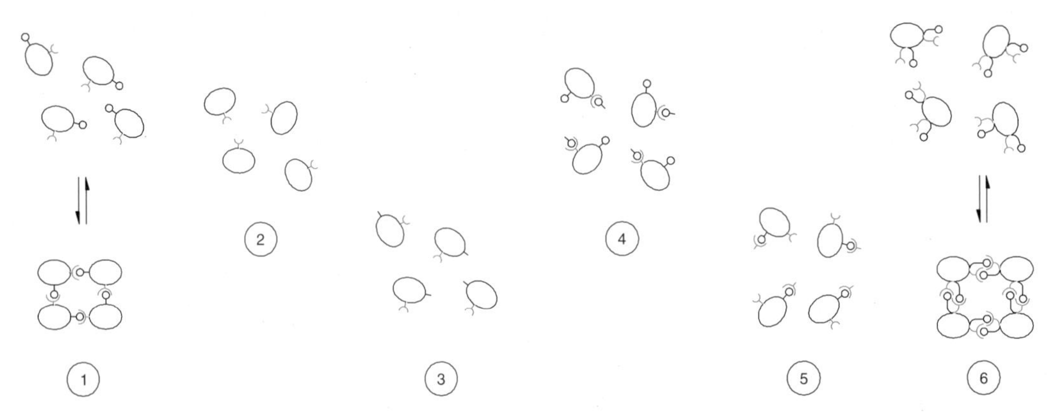


**Fig. 2.97** The aggregation of cells can be facilitated through glycoproteins of the cell membrane and receptors, which are themselves glycoproteins.
① The aggregation may be prevented in several ways
② The *N*-glycosylation of proteins is fully blocked with tunicamycin (section 2.3.8.8, p. 230), and the cells remain without *N*-glycosidic surface proteins
③ Glycosidase cleavage of end groups important for recognition
④ Glycopeptides, whose oligosaccharide part corresponds to the adhesion-mediating glycoprotein, inhibit the aggregation as low molecular weight ligands
⑤ Antibodies against the adhesion mediating glycoproteins prevent the aggregation
⑥ In the case of *Dictyostelium discoideum,* the adhesion-mediating glycoprotein *gp 80* may be both ligand and receptor, since antibodies against the carbohydrate part and antibodies against the protein part of gp 80 both act as adhesion inhibitors

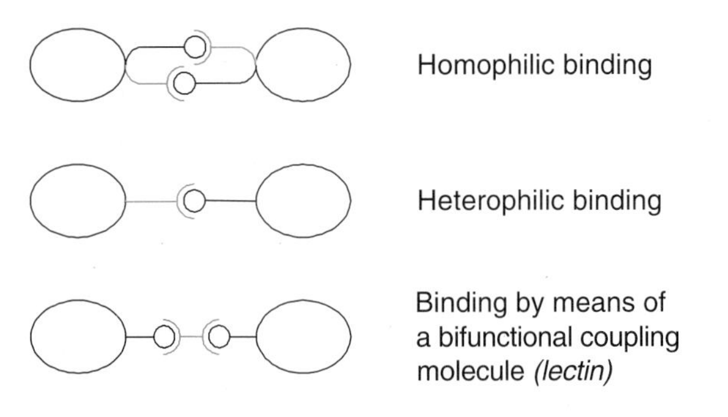


**Fig. 2.98** There are fundamentally three types of *adhesion mediation* between cells.

Careful interpretation of the experimental results regarding the biological function of gp80 in the first aggregation phase of *Dictyostelium discoideum* goes too far, on the whole. A cell aggregation can be traced only indirectly to carbohydrate structures, which are not even completely defined chemically and whose receptor is also still unknown with respect to its structure. Contradictory statements regarding the participating carbohydrate structures have not contributed to a clarification of the facts[114]. This shows the difficult position in which science may find itself even with simple models. It should also be considered that in the case of gp80, as well as with other glycoconjugates of oligosaccharides, the oligosaccharide may only have the function of stabilising the essential protein in a required conformation for cell adhesion. The experimental results can be interpreted satisfactorily with such a secondary function.

Beside the possible homophilic adhesion through gp80, there is with *Dictyostelium discoideum* the possibility of still other types of binding of cells, both with one another and also onto the support. With the help of a secreted lectin, *discoidin-1,* which is synthesised directly after withdrawal of nourishment, the amoebae migrate in streams to an aggregation centre, moving as it were, hand over hand on the self-produced lectin, which then also serves for fixing the cells together. This form of aggregation is the third of the aforementioned possibilities (Fig. 2.98). There is no information about the structure of the carbohydrate ligands of *discoidin-1.*

■ **Fibronectin, laminin, and integrin help in the organisation of the mammalian organism.** Naturally, that which is relatively easily observed and investigated in the Petri dish in the case of *Dictyostelium discoideum*, provides greater experimental difficulties in the case of higher organisms. Tissue cultures of mammalian cells – fibroblasts, cells of the connective tissue are especially well suited – are also prepared, today albeit at some expense, and used for investigation of cell-cell- and cell-matrix-binding. A cell separated from its associated state with other similar cells also retains its capability to aggregate specifically or to bind specifically to an isolated component of the extracellular matrix.

Information about the molecular structure of the self-binding components of cell membranes or the extracellular matrix is obtained generally through the use of antibodies which can inhibit measurable aggregation of cells or their attachment to the matrix. The immunisation of rabbits with cell membranes or matrix material produces an antibody mixture against the molecules of the cell surface or of the matrix. The naturally divalent antibodies are transformed proteolytically, thus artificially, into monovalent antibodies, which still possess their specific binding capability, yet can no longer accomplish aggregation (Fig. 2.97). After cleavage of these antibody fragments, the fractions are tested for their capability to inhibit cell aggregation or cell-matrix binding. Likewise, the membrane components or matrix components are separated and a search is made for the fraction which can reverse again the inhibition of cell aggregation or matrix binding.

Thus, in principle it is possible to search specifically for components which are responsible for cell-cell adhesion or cell-matrix adhesion and to isolate these compounds preparatively. Most *cell adhesion molecules* tracked down by the immunological method are glycoproteins. This fact alone makes the compounds intensively researched objects of glycobiology and such research is further driven by the partially established suspicion that a deviation from the normal biosynthesis of the adhesion molecules results in the unregulated proliferation of cancer cells and above all in the so dangerous metastasis. In this respect, the relationships between connective tissue cells and their surrounding matrix are of special interest.

*Fibronectin* (see section 2.1.5.3, p. 123) is a glycoprotein of the cell surface of fibroblasts, which in degenerate cells occurs to only a very diminished extent or also not at all[115]. An earlier experiment from the 1970's showed that fibronectin dissolved in the medium causes transformed fibroblasts to behave as normal cells, that is on contact to suspend growth and to cease proliferation. This control of the growth occurs through the so-called contact inhibition. The transformed spherical cells, which hardly bind to the wall of the container, again assume the normal flattened form of normal fibroblasts and adhere firmly to their underlying base (Fig. 2.99)[116].

*Fibronectin* is found not only on the cell surface but also in plasma and, regarding its origin, differs in the degree of aggregation and in solubility. The monomer has a molecular weight of about 220 kDa and today is one of the best researched glycoproteins of the extracellular matrix. Fibronectin is one of the most important regulation factors for cell-cell aggregation, cell matrix attachment, cell movement and cell morphology. It binds an astoundingly large number of macromolecules, amongst others the glycosaminoglycans heparin, heparan sulphate and hyaluronic acid, and naturally also whole cells. Low molecular glycoconjugates, such as gangliosides, are also bound by fibronectin.

Through proteolytic degradation of fibronectin it is possible to isolate fragments which show specific binding properties. In this way it is possible to correlate certain domains of the intact protein with specific binding regions (Fig. 2.100). Especially prominent are binding regions for heparan sulphate and collagen as well as binding locations for the cell membrane components. They are likely to have special significance for the well-ordered deposition of cells in the extracellular matrix. Fibronectin is a glycoprotein and it has long been a puzzle to what extent the carbohydrate structures are important for its function.

It is known that the oligosaccharides – of which there are five per fibronectin monomer – are found in the region of the collagen and cell binding locations. The oligosaccharides all belong to the complex *N*-glycosidic type and contain the monosaccharides GlcNAc, Man, Gal, NeuAc, and Fuc in differing amounts. In the presence of the inhibitor of *N*-glycosylation tunicamycin, it is possible to produce in cell culture carbohydrate free fibronectin. Such fibronectin, regarding its biological binding properties, is indistinguishable from normal glycosylated fibronectin. This result is one of the most important pieces of evidence against an over interpretation of glycoforms.

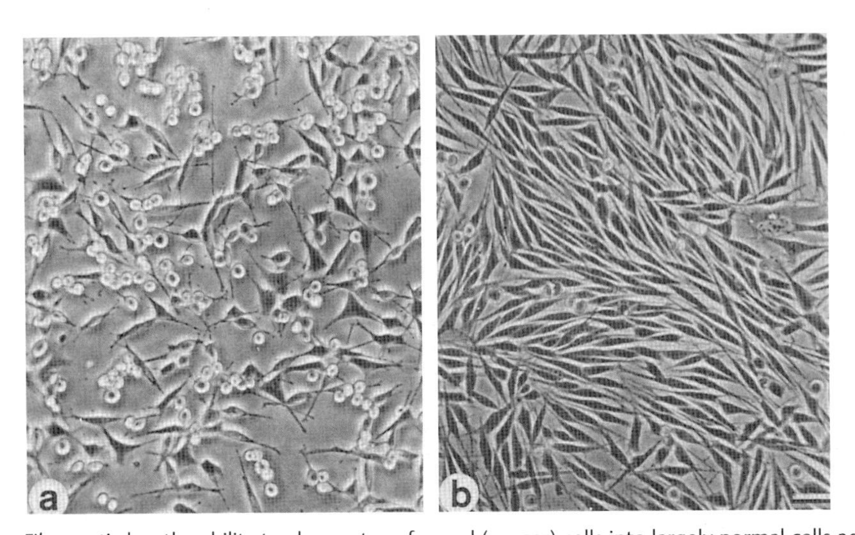


**Fig. 2.99** *Fibronectin* has the ability to change transformed (cancer) cells into largely normal cells again.
**a** Deformed, round fibroblasts of the rat kidney, which have arisen through treatment with sarcoma virus
**b** After the addition of fibronectin, the cells again assume their normal spindle-shaped habit and arrange themselves parallel to one another

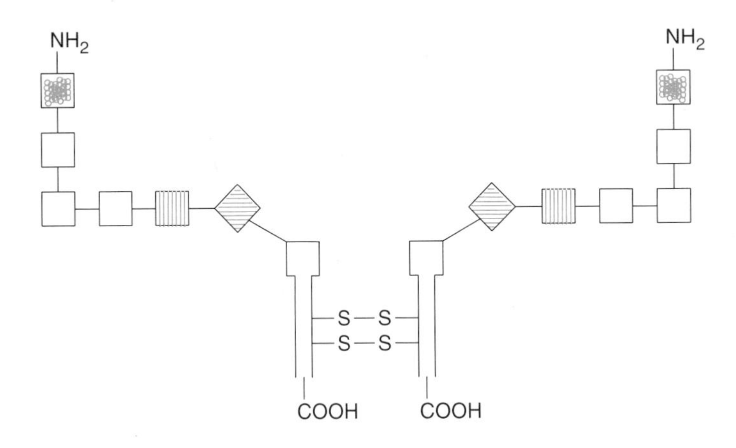


**Fig. 2.100** *Fibronectin* is a dimeric glycoprotein, whose both similar but not identical halves are linked through two disulphide bridges on the C-terminus. Three important binding domains are almost symmetrically arranged.

The significance of the general stabilisation of membrane- and plasma-proteins by glycosylation was already referred to in section 2.1.5 (p. 110 ff.). Also with the glycosylated fibronectin, a distinctly higher resistance towards proteolytic degradation and thereby a longer lifetime can be demonstrated in comparison to the non-glycosylated protein.

Even though the carbohydrate structures of fibronectin itself are not obviously involved in the process of specific adhesion there are, nevertheless, firm results which indicate that the adhesion of fibronectin to the extracellular matrix, to collagen and to heparan sulphate, is brought about by specific binding of the oligosaccharide structures of the cell surfaces. It has been suggested that different glycosylation patterns of the cell surface are responsible for the very different affinities of different cell types to fibronectin. Especially, the very strong binding of blood platelets differs from the relative loose binding of fibroblasts to fibronectin. These differences are biolgically meaningful because fibroblasts require adequate freedom of movement for their migration through the connective tissue, while blood platelets must become fixed at the site of injury of the tissue, in order to be able to assist in the repair of the wound and in blood coagulation.

Fibronectin is only one of the many glycoproteins which mediate an anchoring of the cells in the matrix and the binding of the cells to one another. An important constituent for all basal membranes is the glycoprotein *laminin*. The basal membrane is a special tissue which, for example, surrounds muscle fibres, and which delimits epithelial cells from the connective tissue or endothelial cells of the blood vessel wall from the underlying epithelial cells. Similarly to fibronectin, *laminin* mediates the linkage of cells to one another and of cells to the macromolecules of the extracellular matrix, the main fibre protein of the basal membrane, *collagen IV*, as well as to *heparan sulphate*. Typical of laminin is its cross-shape, which is revealed with an electron microscope thanks to its large molecular weight of about 900 kDa. The protein consists, as shown in Fig. 2.101, of three polypeptides of different sizes[117].

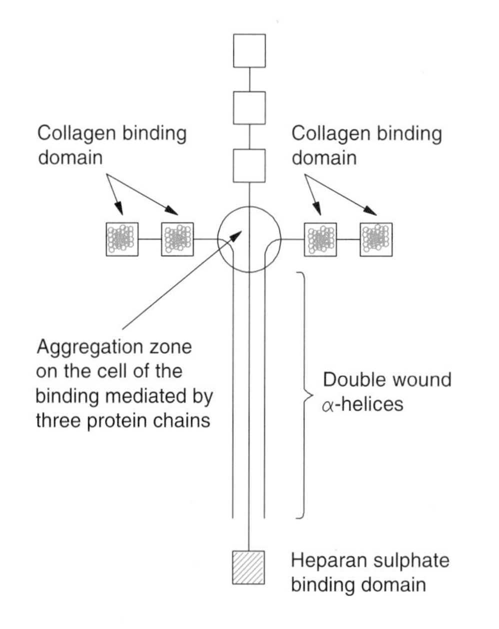


**Fig. 2.101** *Laminin structure.* The trimer is highly glycosylated.

*Laminin* has been given special attention in the recent times since its participation was suggested in the fixation of tumour cells before their passage through the basal membrane[118]. The penetration of the basal membrane is the critical step for the formation of tumours by metastasis[119]. Not until after the passage through the basal membrane out of the blood circulation into tissue are tumour cells safe from the immune system of the organism and able to begin to build a new tumour.

The carbohydrate of laminin constitutes about 15% of the total mass and is found distributed very differently in the various protein domains. On the basis of the foregoing results, a role is attributed to them in the mediation of binding between cells, also tumour cells, and the basal membrane. There is no lack of experiments to prove such a connection, yet merely the size of the molecules makes an unequivocal proof of the carbohydrate-mediated binding interaction between cells and laminin extraordinarily difficult. Especially remarkable are the unusually many different structures of the oligosaccharides which are probably only of the *N*-glycosidic type. Different length polylactosamine chains can carry at their end and as branches α-Gal and NeuAc residues. The relatively unusual α-Gal end group for glycoconjugates is alloted a special signal function which has not yet been unambigously demonstrated (Fig. 2.102)[117].

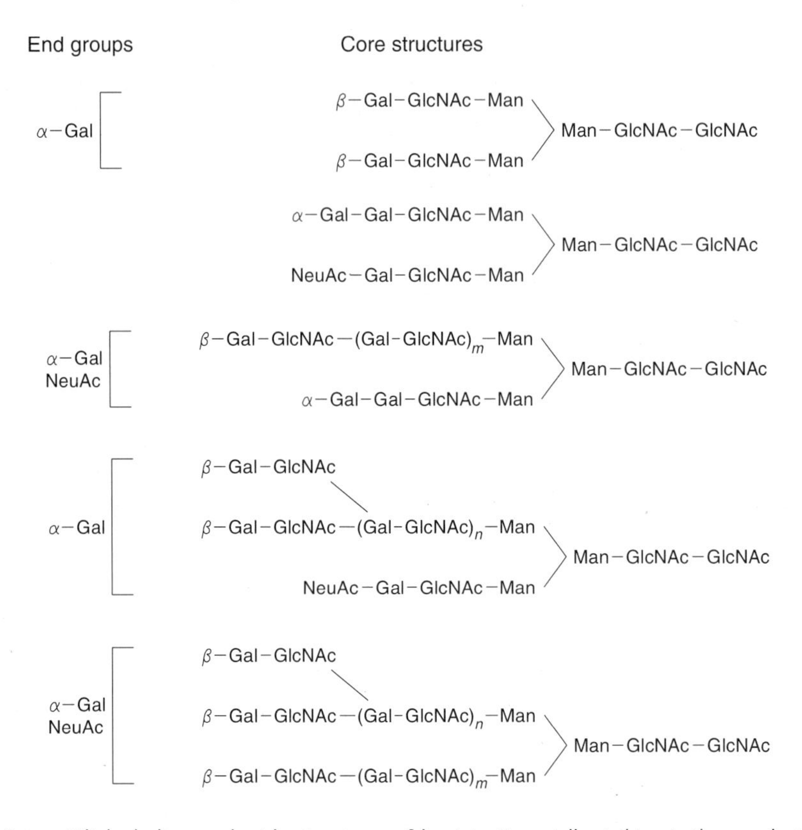


**Fig. 2.102** Some *N*-linked oligosaccharide structures of *laminin*. Especially striking is the α-galactosyl residue appearing as the end group and the *N*-acetyllactosamine occurring partly as a polymer chain.

There are also preliminary indications that laminin preparations without carbohydrate ligands can no longer induce certain types of cell behaviour, for example the ability to attach themselves flat and spread on a laminin base. Complementary receptors on the cell have not yet been found. The many open questions about laminin are sufficiently interesting to make it a main object for further glycobiological research, and especially also of cancer research.

As the third type of glycoprotein, integrin mediates the binding of cells together and of cells to the extracellular matrix. Unlike fibronectin and laminin, ***integrin*** remains anchored in the cell membrane with a short piece of the peptide chain at the C-terminus (Fig. 2.103)[120].

Integrin consists of two independent peptide chains α and β with molecular weights which vary beween 120–180 kDa. In each pair the α-chain is the larger. Up to the present more than 20 intergin dimers have been identified, which arise through combination of several α-chains with three different β-chains. The $\beta_1$-integrin is found in most cell types and brings about the main binding to the extracellular matrix. The $\beta_1$-chain contains a relatively extended glycan structure of the *N*-glycosidic type, which as demonstrated, is essential for the binding of cells to fibronectin with high affinity. The difference between the molecular weight of intracellular undeveloped $\beta_1$-polypeptide and mature polypeptide anchored in the membrane amounts to at least 20 000 and is attributed to the transformation of "high mannose" precursor structures into large "complex type" structures (see section 2.3.8.5, p. 218).

Three separable glycopeptides arise by treament with pronase. The weight difference of the heaviest peptide is surprising, according to whether the original integrin arose with or without addition of inhibitor (mannonojirimycin, MNJ). It is supposed that the large glycan structure, similar to laminin, is polylactosamine, which is modified on the periphery.

MNJ inhibits specifically the ER α-mannosidase which leads to the first step of the necessary degradation of the "high mannose" oligosaccharide (see section 2.3.8.5, p. 218). The degradation of the

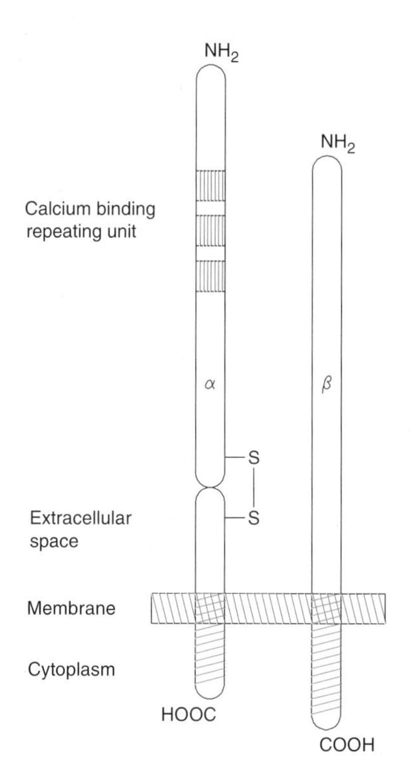


**Fig. 2.103** *Integrins* (about 20 dimers are known) are anchored next to one another in the cell membrane. The $\alpha_5\beta_1$-integrin shown here schematically specifically mediates the binding of the cell onto *fibronectin*. The *N*-glycosylated binding region for fibronectin is found on the *N*-terminus of the $\beta$-chain.

mannosyl-oligosaccharide is the prerequisite for the synthesis of the "complex type" oligosaccharides and glycan structures. It can be shown that *integrin-*$\alpha_5\beta_1$ from MNJ-treated fibroblasts does not bind to immobilised fibronectin. Likewise, it can be demonstrated that MNJ-treated fibroblasts, in contrast to normal fibroblasts, have lost their capability to about 70% for adhesion to a fibronectin base (Fig. 2.104)[120].

The oligosaccaride structures appear to be equally significant for the attachment of another integrin-$\alpha_6\beta_1$ from mouse melanoma cells to laminin. If the cells are treated with periodate solution and the oligosaccharide structure thereby destroyed, or if the cells are grown in the presence of the transglycolase inhibitor tunicamycin, then they lose partly their capability to bind to laminin. That we are dealing with damage to the oligosacccharide of the $\alpha_6\beta_1$-integrin can be proved by *in vitro* experiments with isolated integrin. Removal of the carbohydrate structure with *N*-glycanase resulted in the loss of the binding capability of integrin-$\alpha_6\beta_1$ to laminin[121]. Unfortunately, it has not yet been possible in this last case to identify the definitive carbohydrate structures responsible for the binding. The experiments have proved, however, that without the carbohydrate of the integrin neither binding of normal cells to the fibronectin of the extracellular matrix nor of tumour cells to the laminin of the basal membrane is possible.

■ **Selectins.** The invasion of tissue and formation of a new tumour begins with binding of the integrins of a tumour cell to the laminin of the basal membrane. The leukocytes in the bloodstream behave in an entirely similar way to the tumour cells if it is required in the natural course of events. The retardation and adhesion of leukocytes to the endothelial cells of the vessel wall is in preparation of an answer to inflammation. By signals from a focus of inflammation after infection or injury, both the surface of the endothelial cells and also that of the leukocytes are altered. The change has the consequence that the cells, which swim in the blood stream with high velocity, suddenly roll along on the vessel wall relatively slowly, as if the surface had become sticky. The process may be followed with a film camera and a light microscope. Again and again, the leukocytes are separated from the plasma stream and roll further until they finally cling fast, force themselves between the endothelial cells, penetrate the basal mem-

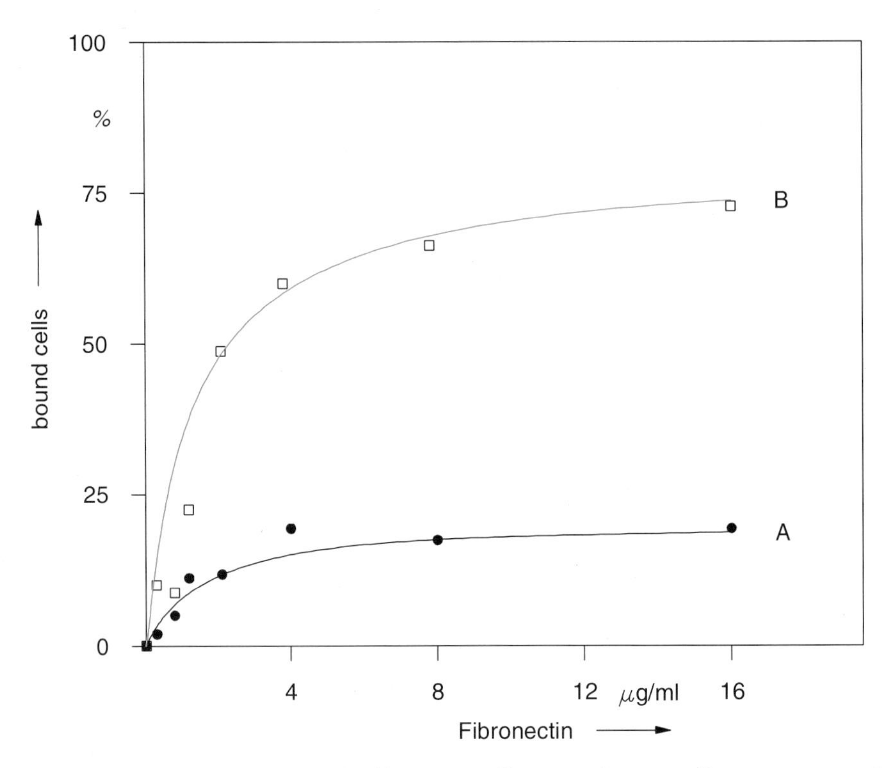


**Fig. 2.104** Binding experiment of human fibroblasts onto a fibronectin base. **A** Cells were pre-treated with the α-mannosidase inhibitor *mannonojirimycin* (MNJ); **B** experiment with normal cells.

brane, and gather themselves in the tissue about the focus of inflammation to fulfil their task (Fig. 2.105).

From the numerous indications as to a possible role of carbohydrates in the case of specific cell-cell adhesion, only very few have been proved through an unequivocal biochemical investigation. For the most part they are speculative ideas, substantiated to a greater or lesser extent. In certain respects this also applies to the very recently discovered group of carbohydrate binding proteins which have been given the general name *selectins*. The selectins are thought to be responsible for the specific binding interaction between different cell types as well as for the previously described adhesion between leukocytes and endothelial cells of blood vessels. Endothelial cells can be induced, through various stimulating factors such as those arising in the case of inflammation, to express a carbohydrate binding protein on the cell surface. The protein is named *E-selectin* or *ELAM-1* (endothelial leukocyte adhesion molecule). It binds specifically in the first place peripheral sialic acid (NeuAc), but necessary in addition for higher affinity is a fucosyl residue which, likewise peripheral, should be located on a polylactosamine chain. Obviously, there are binding sites available for both peripheral monosaccharides. The second binding site for L-Fuc is only released, however, when the first, for NeuAc, is already occupied. It seems clear today, on the grounds of inhibition experiments with synthetic sialyl Lewis$^x$ oligosaccharides (see section 2.1.1, p. 67ff), that the ligand on the surface of the leukocytes responsible for the attachment to endothelial cells carries a sialic acid bound (2→3)- on galactose and an α-L-fucosyl residue situated at the correct distance therefrom[122]. The strength of binding between the leukocytes and the endothelial cells, that is the transition from *rolling* to *adherence*, is regulated by duplication of the adhesion possibilities.

With regard to the action of the selectins, the *leukocyte-endothelial cell system* is today, very well investigated[123–125]. It can be expected that research activity in this area, which also has economic potential, will increase further. Syntheses of therapeutically applicable inhibitors of selectin-mediated leukocyte adhesion are worth special efforts. Currently, the publication rate in this highly topical re-

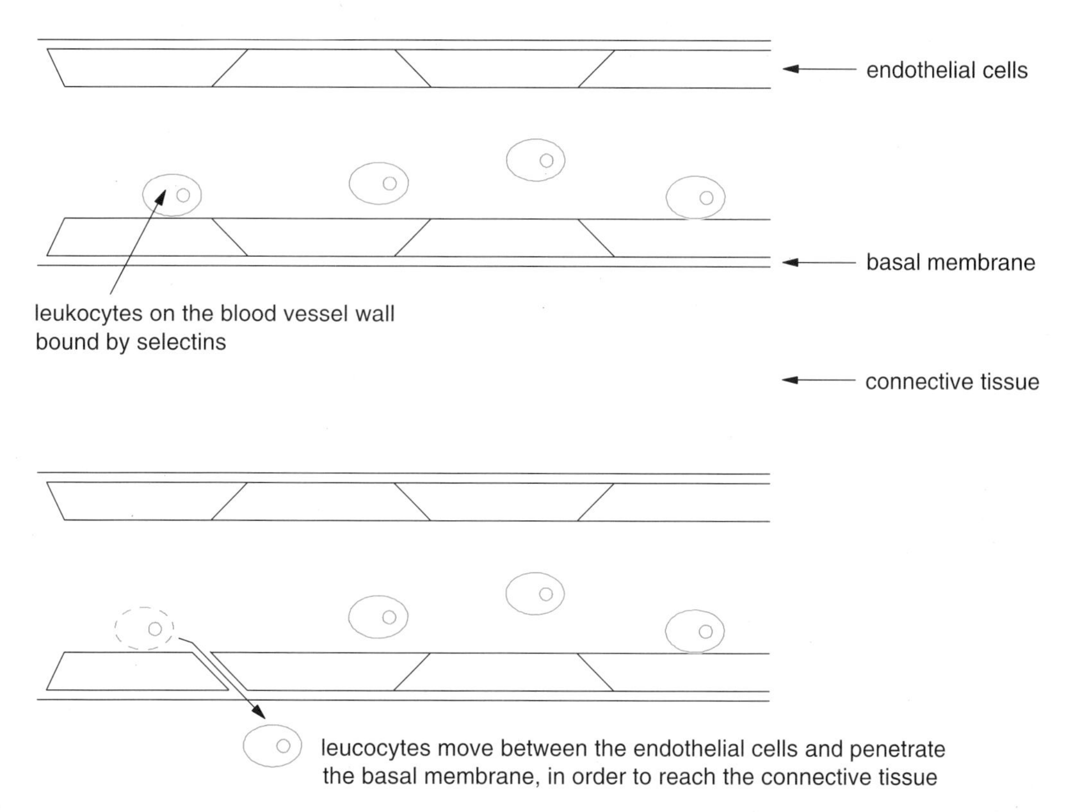


**Fig. 2.105** In inflammatory processes, leukocytes bind to the endothelial cells of the blood vessel wall through *selectin* mediated binding. It is now known that the peripheral oligosaccharide structure sialyl $Le^x$ of the leukocyte glycoconjugate is responsible for the binding to selectins.

search area is enormous, making it difficult to assess the often contradicting results, in order to obtain a clear picture of the actual molecular interrelationships.

### 2.2.2.7 Bacteria and viruses

Although bacteria really do not pose a general threat for mammals any more since the discovery and development of effective antibiotics, interest in the mechanism of bacterial infection is, nevertheless, still very much alive. Practical reasons still remaining are the possible development of resistance by bacteria, facilitated by plasmid transfer, as well their fast antigenic modification, both of which make a new therapy of the blocking of bacterial-cell contacts a worthwhile goal.

Modern experimental techniques facilitate laboratory investigations regarding the interaction of bacterial cells and also of viruses* with host cells and have led in preceding years to numerous impor-

* Viruses mostly produce oligosaccharides of the host cell[126]. With the help of the host's own glycosylation apparatus, the specific carbohydrate structures are attached to the designated sites in the viral protein and therefore, in contrast to bacteria or parasites, viruses are not easily controlled by the immune system of the host. Correspondingly, viruses also make use of the receptors of the host which frequently results in fusion with the host membrane, leading to infection. The glycoproteins of the viral coat are especially suited for contact absorption[127–129]. There are, as for example in the case of the *influenza virus*, also virus-based receptors, *haemagglutinins*, which bind interactively[130] with the oligosaccharide structures of the host cell which have NeuAc end groups. There are also virus-based sialidases which enable degradation of protecting layers of mucus on the epithelial cells of the respiratory tract in the attacked organism[131].

tant discoveries[132]. As with all cell-cell contacts, the major interest is in those molecules which interact first. They are the prerequisite for each infection and therewith also of the virulence. Many infections may be attributed to unspecific contacts, such as perhaps hydrophobic or ionic interactions, though numerous infections also exhibit distinct tropism and rightly suggest specific ligand-receptor interactions. The preponderant part of the ligands on epithelial cells recognised by the receptors of bacteria must be glycoconjugates. It may be supposed that the abundance alone of these surface molecules has led to the development of specific receptors directed specifically against them. The binding of the bacteria on glycoconjugates concerns, first of all, intact tissue. In the case of injured tissue, in contrast, the typical wound proteins such as fibrin, laminin, or fibronectin, as well as the main matrix protein collagen, are attacked.

Epithelial cells offer three types of glycoconjugates: *membrane anchored glycoprotein*, *proteoglycans* and *glycolipids*. Chiefly practical reasons have led to investigations of bacteria-ligand interactions being made almost exclusively on glycolipids, which have the advantage of being relatively easy to isolate and, what is more important, of exhibiting only one carbohydrate structure per molecule. However, there are also reasons to consider glycolipids as the preferred ligands for bacteria. Bacteria are forced to come close to the cell membrane of the attacked cell, in order to initiate the cell fusion necessary for the infection. Carbohydrate structures in the immediate vicinity of the cell membrane are hence especially favoured as ligands and such, above all, are to be found in glycolipids.

To determine the affinity between bacteria and certain glycolipids, use is made today almost exclusively of the so-called *solid-phase method*[133]. The glycolipids are separated on thin-layer plates and then, after isolation, are identified by mass and NMR spectrometry. Binding is demonstrated with radioactively marked, or antibody-labelled bacteria. With the help of a series of dilutions, quantitative data can also be obtained on the affinity. In Table 2.2 are shown some of the carbohydrate structures in glycosphingolipids specifically recognized by bacterial receptors.

It is noticeable that the lactosyl ceramide structure is one of the most frequently recognised groups, and this is especially so if the oligosaccharide chain stretches beyond the periphery. Biologically it is very significant if the bacteria acquires a receptor which can specifically recognise and bind a core region frequently occurring in glycolipids and also in glycoproteins. The idea of a cell-cell interaction assumes that the surfaces touch. This plausible mechanical picture can be applied however only with difficulty to the binding of a bacterial cell to a *hidden* carbohydrate structure of the host cell in the vicinity of the membrane.

An initial, relatively weak binding may be imagined on the periphery of the structure followed by specific and stronger binding on structures near the membrane. It may also be imagined, however, that oligosaccharide conformations of the glycolipids exist or can be induced in which the core region is not completely hidden but is directly accessible to the bacterial receptors. Certainly the accessibility is determined by many factors, both binding and also repulsive, in the immediate neighbourhood of the epitope, but it can be analysed only with difficulty with the experimental methods employed hitherto.

### 2.2.2.8 Mediation of chemotaxis by binding proteins

Sensory perception by organs provided for this purpose is to be found in general only in higher organisms. Of course, in the case of lower organisms such as bacteria, fungi, yeast, and algae, systems are available which fulfil similar functions as the sense organs. Perception can be traced to a relative simple process which is called *chemotaxis*. An external stimulus, in this case a certain chemical molecule, is bound by a receptor and the binding so transposed that the organism changes its behaviour. The process serves in the widest sense for the absorption of nutrients, which are necessary for the survival of the cell. Without discussing the physiology and biochemistry of the action as such, the well investigated system of the absorption of maltodextrins in *E. coli* and the chemotaxis triggered therewith, that is the motion of the cell through flagellating rotation towards the nutrient source, should be considered in the framework of this section. There is a clear case here of the biological recognition of carbohydrates.

Of course, *chemotaxis* is not only triggered by maltodextrins, that is the degradation product of starch, especially maltose, but also by other sugars, amino acids, and inorganic compounds. The bacterial cell (see section 2.1.2.1, p. 73) possess in the outer membrane a layer which displays permeable pores for a large number of water soluble components in the medium (Fig. 2.106)[134]. Between this membrane and the cytoplasmic membrane is found the periplasmatic space or the periplasma.

**Table 2.2** Some peripheral and non-peripheral (underlined) oligosaccharide strucures from glycosyl ceramides of certain tissues, which are recognised by bacteria and toxins

| Microorganism | Tissue attacked | Structural element recognised |
|---|---|---|
| *E. coli* | urinary tract | α-Gal-(1→4)-β-Gal |
| *Streptoc. pneumon.* | respiritory tract | β-GlcNAc-(1→3)-Gal |
| *Pseudomonas* | respiritory tract | β-GlcNAc-(1→4)-Gal |
| *Neisseria gonorrh.* | genital tract | β-Gal-(1→4)-β-Glc |
| | | α-NeuAc-(2→3)-β-Gal-(1→4)-β-GlcNAc |
| *E. coli* | – | β-GalNAc-(1→3)-*α-Gal-(1→4)-β-Gal*-(1→4)-β-Glc-Cer |
| *Pseudomonas* | – | β-Gal-(1→3)-β-GalNAc-β(1→4)-*β-Gal*-(1→4)-β-Glc-Cer |

| Ligand | Isoreceptor |
|---|---|
| Cholera toxin | β-GalNAc-(1→4)-**β-Gal-(1→3)-β-GalNAc-(1→4)-β-Gal-(1→4)-β-Glc**-Cer<br>3<br>↑<br>2<br>**α-NeuAc** |
| Tetanus toxin | α-NeuAc-(2→3)-**β-Gal-(1→3)-β-GalNAc-(1→4)-β-Gal-(1→4)-β-Glc**-Cer<br>3<br>↑<br>2<br>**α-NeuAc-(2→8)-α-NeuAc** |

The binding proteins bind in the unloaded state on the inner side of the pore and wait for the signal – in this case maltose or a short chain maltodextrin. The binding constant of the maltose-binding protein (MBP) to maltose is about 1 μm (at this concentration of maltose the protein is half-occupied with the ligand). The MBP changes its conformation after building the complex with the ligand, because of that becomes detached from the outer membrane, traverses the periplasmatic space and binds on the basis of the conformational change to the membrane receptor (carrier or *transducer*), which through this binding releases the signal with the help of a series of chemical modifications in the inside of the cell and passes it on further to the motor apparatus, the flagellum and their anchorage[135,136]. The high affinity of the MBP for maltose and also for the longer maltodextrins is responsible for release of chemotaxis. In addition, the MBP-ligand complex also mediates the transport of the short maltodextrins to the inside of the cell. The three dimensional structure of the MBP-maltose complex has been elucidated[137]. From this it is evident that the maltodextrin attaches in the binding region as in a terminus and inner contact exists only at the reducing end with the two glucose residues of a maltosyl residue. With this, though, the binding is optimal. Besides van der Waals interactions, 11 direct hydrogen bonds between ligand and receptor as well as 5 indirect such bonds mediated through water molecules can be detected.

### 2.2.2.9 Symbiosis between bacteria and plant cells

Interactions between bacteria and host cells are not always harmful for the latter, and this applies both for animal cells and for plant cells[138]. The symbiotic interrelationships between legumes, peas, beans and clover on the one hand and gram negative bacteria of the genus *Rhizobium* on the other are especially well known. Infection by the bacteria concerns exclusively the root hairs of the host plant, which protrude as extensions from specialised epidermal cells[139]. The binding is not only highly specific regarding these cells, as Fig. 2.107[138] (p. 160) shows, but also regarding the attacked plant.

Thus, *Rhizobium trifolii* infects only clover and *Rhizobium meliloti* only alfalfa. After infection the bacteria penetrate the epidermal cells of the host plant and induce formation of root nodules, which comprise about half of multiplied, modified bacteria and, for the other half, material of plant origin. The bacteria convert molecular nitrogen into ammonia which is fixed immediately in the form of L-glutamine and is given up into the plant.

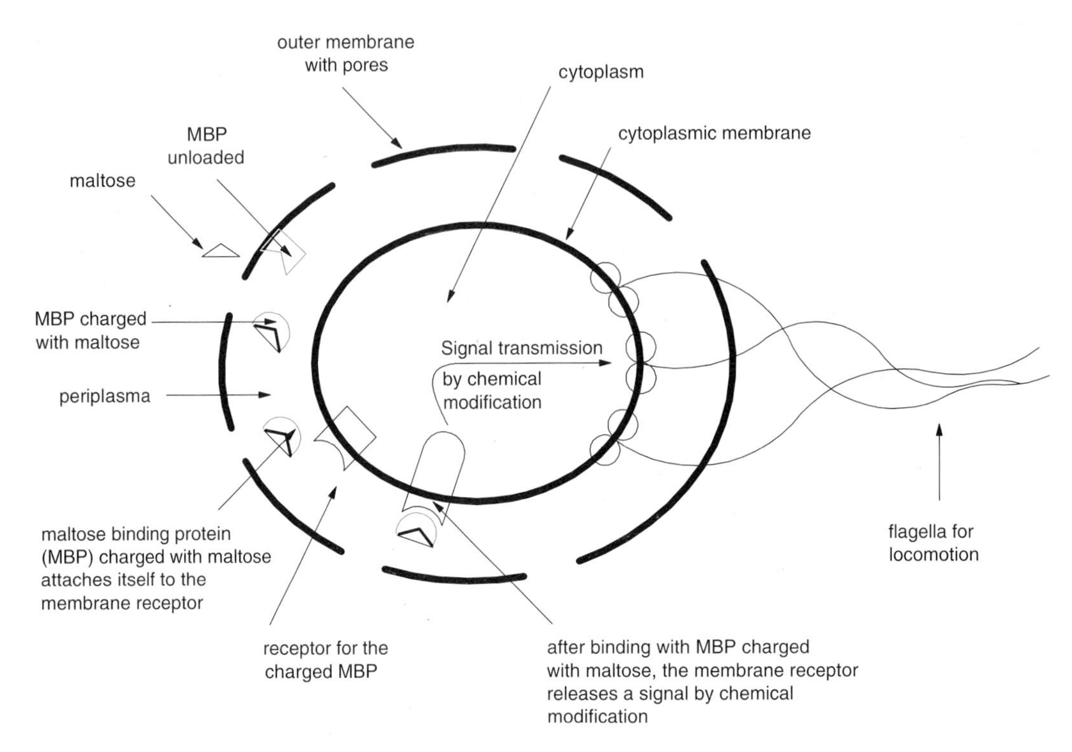


**Fig. 2.106** *Signal transmission* in bacteria.

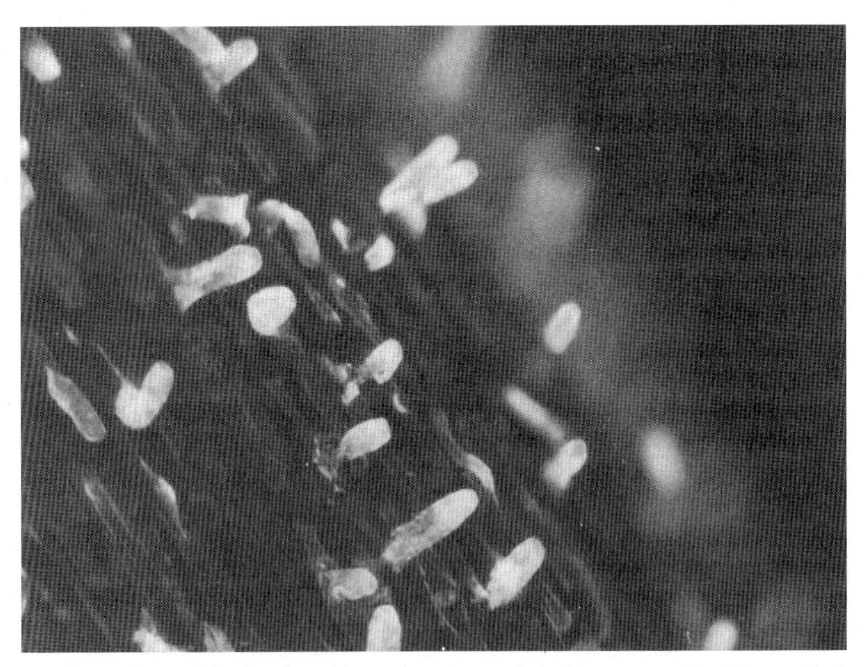

**Fig. 2.107** High specificity of the binding of bacteria to host cells is demonstrated in this case by the covering of clover root hair-tips with fluorescence marked *Rhizobium trifolii*.

The first, unspecific phase of the contact between bacteria and host cells appears to be mechanical and only leads to the binding of a relatively small number of bacteria. This process, however, induces the second specific phase of the binding with a ten-fold increase in the covering with bacteria. A first indication of the molecular mechanism of the infection was the cross-reactivity of surface antigens of the bacterial cells and the root epidermal cells. This astonishing result leads one to assume that both cell types have related molecular surface structures. The isolation of a protein *trifoliin A* both from clover seeds and also from the roots of clover provided the solution of the at first mysterious binding between two structurally similar surface antigens. Trifoliin A is found to be a multifunctional lectin, which both agglutinates the bacteria and also binds onto the root hairs. In each case the binding can be inhibited with 2-deoxy-D-glucose. Likewise, the infection of clover by *Rhizobium trifolii* can be inhibited. The mutual recogniton and binding between host cell and bacterium takes place correspondingly according to one of the three mechanisms possible by combination of similar epitopes with a soluble lectin, which in this case the host organism contributes (Fig. 2.108). A certain modification of the situation was found recently by investigation of the *Bradyrhizobium japonicum* – soya bean system whereby the receptor, the lectin, is thought to be a constituent of the bacterial surface and the plant cells contribute the carbohydrate ligands for binding[140].

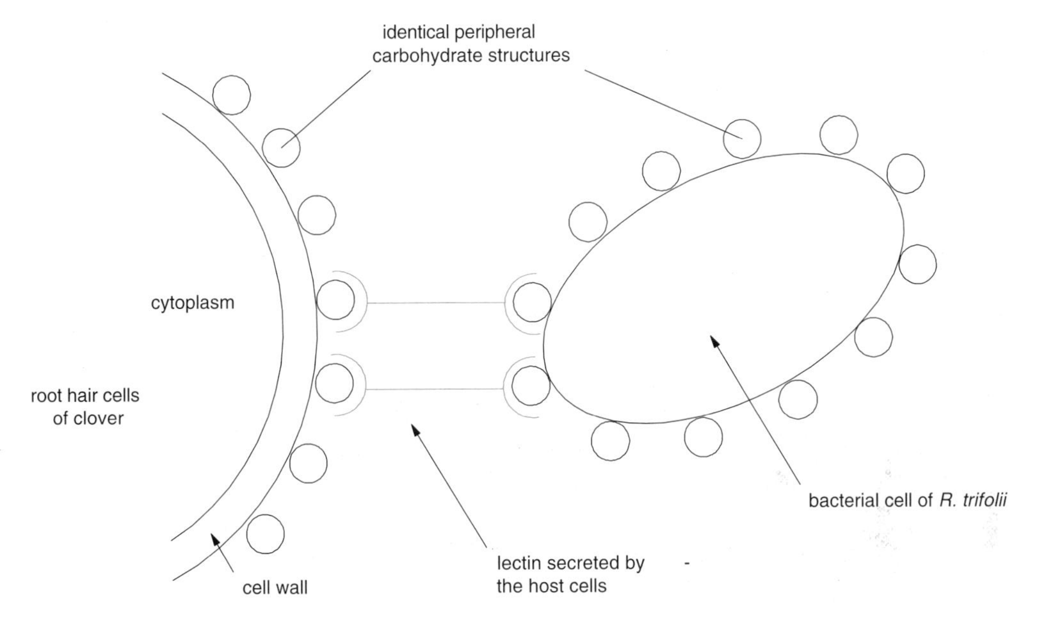


**Fig. 2.108** Binding of nodule bacteria onto root hairs takes place in the second, tight phase of binding through a bifunctional lectin secreted by the host cell.

The high specificity of the trifoliin binding is confirmed by the fact that neither 2-deoxy-D-galactose nor D-glucose are inhibitors, yet 2-deoxy-D-glucose is able to inhibit other symbiosis combinations. The suspicion that 2-deoxy-D-glucose might be an epitope of the lectin was originally aroused through the detection of this rare sugar in the capsular polysaccharide of *Rhizobium trifolii*. It could be shown histochemically with antibodies against trifoliin A that the lectin occurs exclusively in the tips of the root hairs (Fig. 2.107). Of course, trifoliin occurs only under certain conditions of growth. If the plant is supplied with sufficient utilisable nitrogen then no trifoliin A is liberated, even after infection with *Rhizobium trifolii*. This is a persuasive biological result for a specific, useful cell recognition for the plant cell under definite conditions. It is very probable that the lectin mediated binding of certain nodule bacteria on the root hairs of other legumes follows a very similar mechanism.

### 2.2.2.10 Defence and control mechanisms of plant cells

Plants must protect themselves, just as animals, against hostile organisms in their environment. In doing so, they make use of many different mechanisms. The most direct type, potentially, to prevent so-called plant-eating enemies, is for them to be unpalatable or poisonous. However, this defence mechanism is not sufficiently flexible against bacteria or fungi and so higher plants have developed more complex alternatives in which they react against the microbial pathogens only after infection.

It is mostly the *pathogen* itself that supplies the signal for defence to the plant. The defence consists first in the raised production of cell wall material for encapsulation of the seat of infection, but then also in the biosynthesis of antibiotically active, low molecular weight compounds, the *phytoalexins*[141]. In general, the reaction of the plant is restricted in this case to the site of attack and its nearest immediate surroundings. In addition, it is thought to give a signal by an as yet unknown, further-reaching mechanism, which induces in distant parts of the plant the synthesis of inhibitors for bacterial or fungal proteases, in order to neutralise the pathogen.

As signals, which are called in general *elicitors*[142], both chemical and physical stimuli are considered. They trigger off first the release of the endogenous elicitors, which switch on the phytoalexin biosynthesis. In general colloquial usage, the term elicitors is understood to mean only endogenous elicitors. Mostly, these are compounds which arise either from the cell wall of the attacked plant or that of the parasite and are liberated by endoglycanases. The effectiveness of such cell wall fragments is remarkable. In concentrations of under one nanomolar, the biosynthesis of *phytoalexins* is still induced in cell culture.

One of the first effective compounds, a heptasaccharide that stimulated the phytoalexin biosynthesis in soya bean seedlings, was discovered in, and isolated from an aqueous extract from the cell wall fragments of the fungus *Phytophtora megasperma*. The compound is a branched oligosaccharide consisting of seven *β*-linked glucopyranosyl residues (Fig. 2.109). The structure of the oligosaccharide was confirmed by chemical synthesis.

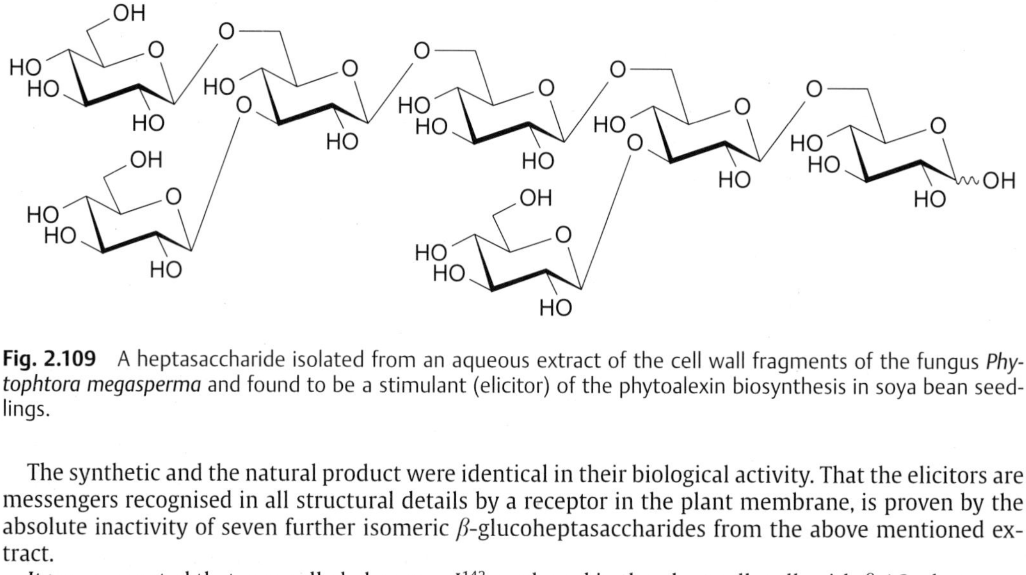

**Fig. 2.109** A heptasaccharide isolated from an aqueous extract of the cell wall fragments of the fungus *Phytophtora megasperma* and found to be a stimulant (elicitor) of the phytoalexin biosynthesis in soya bean seedlings.

The synthetic and the natural product were identical in their biological activity. That the elicitors are messengers recognised in all structural details by a receptor in the plant membrane, is proven by the absolute inactivity of seven further isomeric *β*-glucoheptasaccharides from the above mentioned extract.

It was suggested that a so-called glycanase I[143], anchored in the plant cell wall, with *β*-1,3-glucanase, *β*-glucosidase, and glucosyl transferase activity, is responsible for decomposition of the fungal glucan and liberation of the specific elicitor. This enzyme is, of course, also capable of degrading the elicitor. In the presence of isolated fractions of the soyabean membrane, the heptasaccharide is, however, safe from enzymatic degradation. Obviously, high affinity binding to the receptor suffices for its protection.

Numerous cases of similar elicitor effects are recognised today. In such instances it is also possible, as already noted above, that the effective degradation products originate from the cell walls of the host cells themselves and are liberated there either by enzymes of the pathogen or also – in the case of mechanical injury to the plant – by its own enzymes. The interrelationships between liberation of an elicitor and the production of a phytoalexin are presented in a simplified manner in Fig. 2.110.

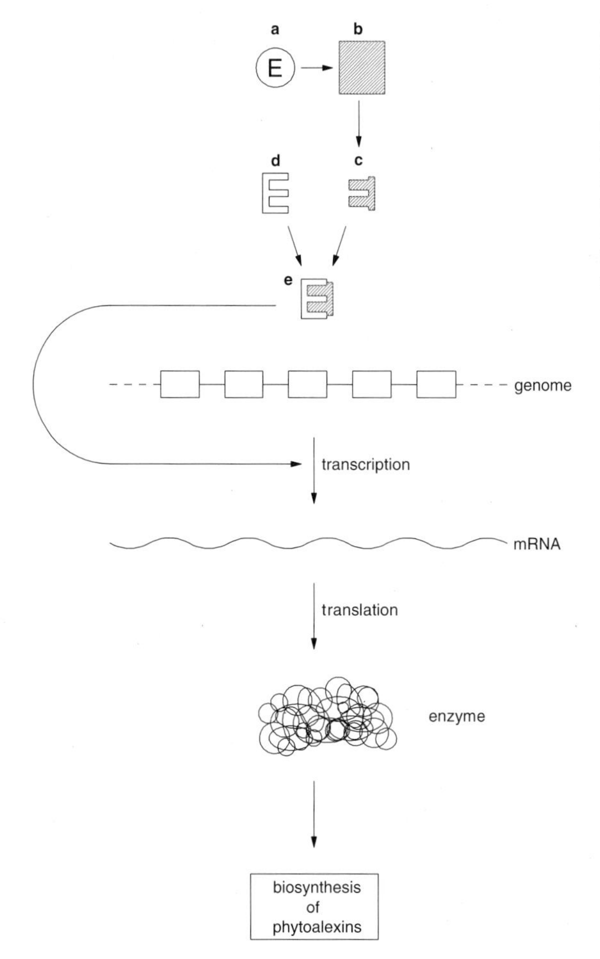


**Fig. 2.110** One of the three possible ways to protect plants by *phytoalexins* begins with the liberation of a cell wall fragment (heptasaccharide in Fig. 2.109) **c** from the cell wall of the fungus **b** through an enzyme of the infected plant cells **a**. *Elicitor* **c** binds to a receptor **d** of the plant. The signal complex **e** induces the transcription of certain plant genes into mRNA. Enzymes arise through translation, which are required for the biosynthesis of the defence substances.

It may be supposed that the plant cell wall in its chemical complexity not only has the mechanical function described in section 2.1.2.3 (p. 82) but also serves as an inexhaustible resevoir for signal compounds, which arise by degradation from the polysaccharides. To underline their specific effectiveness and to distinguish them from *normal* oligosaccharides, the elicitors are called oligosaccharins[144,145]. Such oligosaccharins cut out of the cell wall can also bring about astonishing effects under physiological conditions.

Evidently oligosaccharins of different origin are able to promote favourably, under identical growth conditions, the growth of different tissue types (roots, leaves, flowers). Such preliminary, isolated experimental results act as a stimulus to further invesigations.

The starting point of the new directions of research was the discovery of oligosaccharides as elicitors of the defence substances of plants. Following the definition given above, such elicitors are none other than oligosaccharins. The knowledge that carbohydrate structures not only release defence mechanisms but can also have quite general regulatory functions, such as control of growth, development and reproduction, opens up completely new areas of glycobiological research with plants[146].

### 2.2.2.11 Inositol triphosphate as a signal transmitter[147]

Signal transfer through the cell membrane always follows the same scheme. The outside of the cell membrane has at its disposal specific receptors which are able to bind, with high affinity, an agonist or antagonist (commonly a hormone, neurotransmitter, growth factor, toxin, or drug). The binding as an extracellular reaction releases a cascade of intracellular molecular events which finally lead to a macroscopic observable effect. The signal molecule which approaches the membrane from outside must be designated as *first messenger*, if one calls the intracellular signal molecule, as is usual in general colloquial usage, the *second messenger*. Signal transfer mediated by a second messenger gives, amongst other things, the possibility of more direct action. Only three such systems operating via a second messenger are known today: the adenylate cyclase-, the guanylate cyclase-, and the inositol triphosphate-diacylglycerol system. The latter constitutes about 30% of all hitherto known second messenger systems. In contrast to the extracellular signal molecules, which must themselves search out their object cells in order to find the specific receptor, the second messengers are universal messenger substances for all cells. The intramolecular reaction is correspondingly very uniform in contrast to the variety of the possible extracellular stimuli. Since inositol is a carbohydrate derivative, and its triphosphate as a second messenger has a central importance for signal transmission through membranes, these systems are discussed here in connection with general biological recognition.

Of the three systems mentioned, the inositol triphosphate system was first recognised at the end of the 1980's[148]. Similar to the sugars themselves, the sugar-like inositols moved into the spotlight of general scientific interest with the discovery of their biological significance. Most scientists drawn into the new area of research were relatively unfamiliar with this class of substances with their extremely rich variation in stereochemistry, and through their publications brought new perplexities into the already complicated system of inositol nomenclature. As mentioned earlier, of all the possible hexahydroxycyclohexanes, *myo*-inositol is by far the most abundant, although some isomers and derivatives also occur in nature. However, only *myo*-inositol is found in lipids. In the course of this section, when the term inositol is used then it is understood to mean *myo-inositol*.

Inositol possesses a plane of symmetry, which passes through the carbon atom with the single axial hydroxy group and the group lying opposite (Fig. 2.111). With the exception of these two hydroxy groups, alkylation or acylation and also phosphorylation at a hydroxy group leads to a chiral derivative.

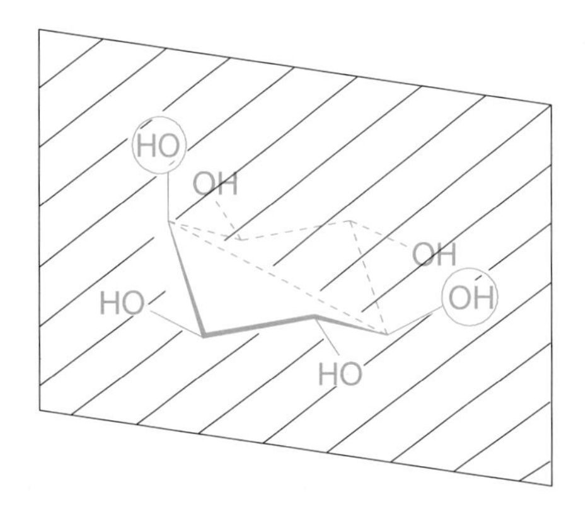


**Fig. 2.111** *myo-Inositol* has a symmetry plane. The compound is achiral. Modification on one hydroxy group, with the exception of those lying in the symmetry plane, affords a chiral molecule.

Chiral inositol derivatives obtained through enzymatic reactions are all pure enantiomers and therefore optically active. Of 66 possible isomers of the phosphorylated inositols[149], up to the present, 16 have been found in many different organisms. The expectation of further derivatives makes the introduction of a simple nomenclature necessary, to replace the current very complicated one, at least for biologically relevant molecules[150].

Confusion over the stereochemical complexity can be countered if the clear symbolism introduced by B. W. Agranoff is adopted. The inositol framework is symbolised by a turtle, in which the raised head is the axial hydroxy group in the 2-position. The other positions are distributed then as shown (Fig. 2.112).

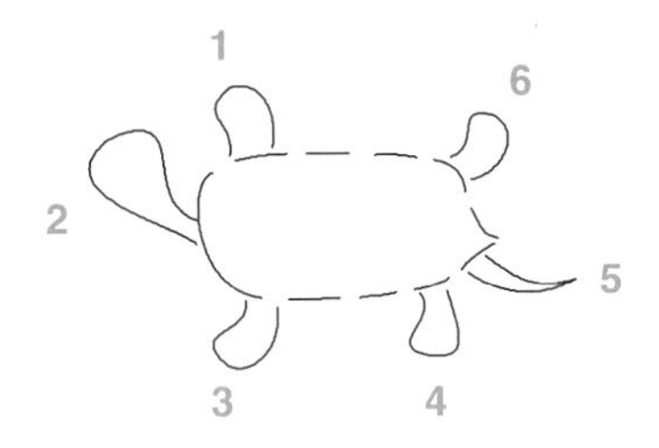


**Fig. 2.112** The **inositol turtle**, as a memory aid, facilitates remembering the stereochemical arrangement. If the *right fore-leg* is modified then that it is the **1-position** in the **D-form**. The *head* is always the **2-position** with the axial hydroxy group. If the *left foreleg* is modified, the **1-L-form** is present (3-D- and 1-L-forms are identical); in the **D-form**, the back leg **6** is to the right, **4** to the left, and **5** is the tail.

With the help of this picture, manipulation of formulae is greatly simplified. As already discussed in section 2.2.1.3 (p. 132), inositol is formed as a monophosphate from D-glucose 6-phosphate. This compound carries the phosphate group in the 3-position (left foreleg) and has a negative rotation. In abbreviated form, the compound is designated as Ins(3)P (Fig. 2.113). It can be formed from *myo*-inositol by direct phosphorylation with a kinase found only in plants.

Ins(3)P is hydrolysed by a specific phosphatase (Fig. 2.113). The so-formed inositol, together with that taken up through food, is the sole source for the biosynthesis of *phosphatidylinositol (PtdIns)*, one of the main components of lipid membranes. A phospholipase C can split the lipid into diacylglycerol and inositol monophosphate. This phosphorylated inositol is the enantiomer of that arising from D-glucose phosphate, that is Ins(1)P, according to the turtle picture. Ins(1)P can be hydrolytically cleaved by a phosphatase which can be inhibited with $Li^{+}$. It is important to know these inter relationships because they also determine the liberation of the $InsP_3$ as second messenger. Besides the usual membrane component, *1-phosphatidyl inositol (PtdIns)*, its monophosphate [PtdIns(4)P] and diphosphate [PtdIns(4,5)$P_2$] occur in the cell membrane, albeit in a very low concentration (Fig. 2.114). It was suggested that the higher phosphorylated lipids occupy preferentially the inner side of the membrane.

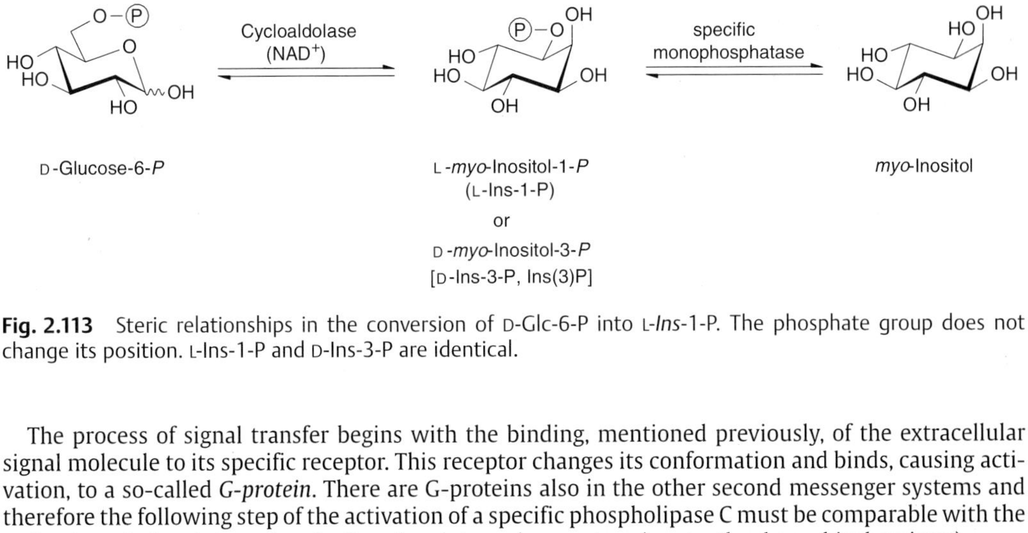

D-Glucose-6-*P* — Cycloaldolase ($NAD^{+}$) → L-*myo*-Inositol-1-*P* (L-Ins-1-P) or D-*myo*-Inositol-3-*P* [D-Ins-3-P, Ins(3)P] — specific monophosphatase → *myo*-Inositol

**Fig. 2.113** Steric relationships in the conversion of D-Glc-6-P into L-*Ins*-1-P. The phosphate group does not change its position. L-Ins-1-P and D-Ins-3-P are identical.

The process of signal transfer begins with the binding, mentioned previously, of the extracellular signal molecule to its specific receptor. This receptor changes its conformation and binds, causing activation, to a so-called *G-protein*. There are G-proteins also in the other second messenger systems and therefore the following step of the activation of a specific phospholipase C must be comparable with the activation of adenylate cyclase in the adenylate cyclase system (see textbooks on biochemistry).

Phospholipase C splits in fractions of a second PtdIns(4,5)$P_2$ into diacylglycerol, which remains in the membrane and for its part activates a protein kinase C as second messenger, and *Ins(1,4,5)$P_3$*, which as a water soluble component migrates into the cytosol (Fig. 2.115). Ins(1,4,5)$P_3$ binds intracellularly to specific receptors of the $Ca^{2+}$-reservoir and induces these ions to pour out. Then follow the usual changes caused by $Ca^{2+}$-ions.

The binding of Ins(1,4,5)$P_3$ is highly specific. Other isomers or less phosphorylated analogous compounds are not capable of liberating $Ca^{2+}$. There is an inclination here to make a comparison with the ManP-receptors. The excitation caused by Ins(1,4,5)$P_3$ is either reversed with a specific phosphatase or, through a renewed phosphorylation step whereby Ins(1,3,4,5)$P_4$ arises, is prolonged and modulated.

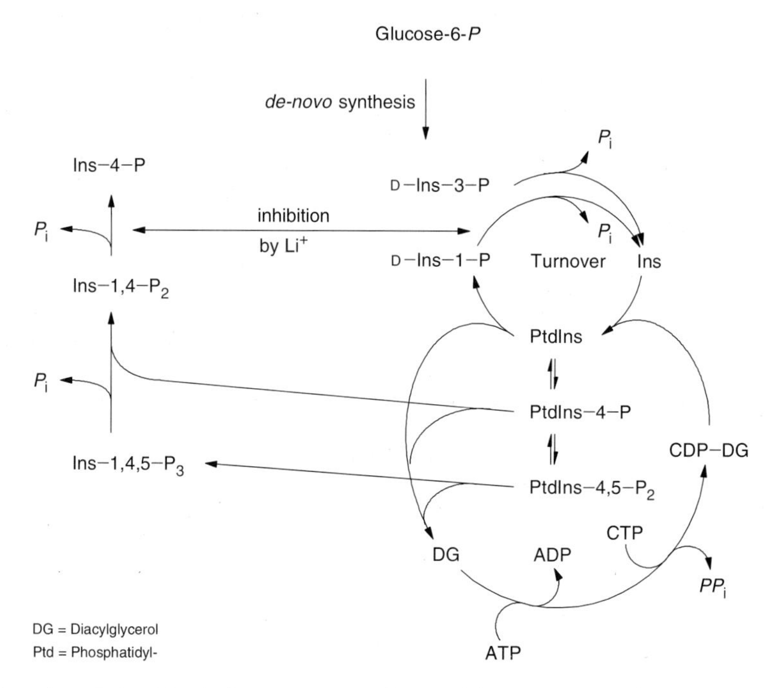


**Fig. 2.114** Complex relationships in the conversions of inositol derivatives. The phosphatidyl inositols (PtdIns) are involved.

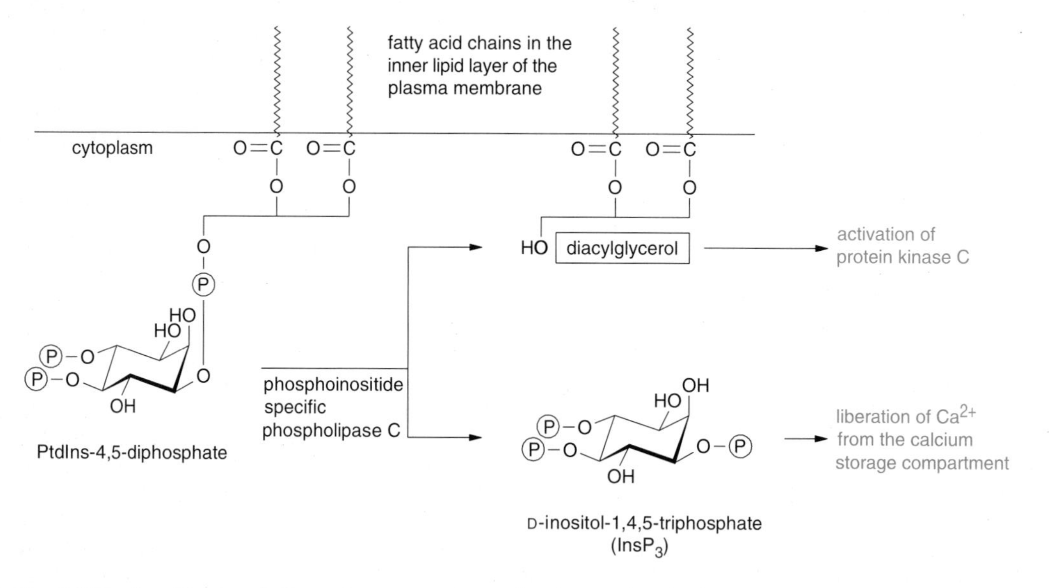


**Fig. 2.115** Specific *phospholipase C* corresponds to adenylate cyclase in the cAMP system.

There is still no information about the binding of Ins(1,4,5)$P_3$ to the receptor of the $Ca^{2+}$ reservoir and just as little about the receptor itself. It must be certain though that signal mediation through Ins(1,4,5)$P_3$ is a universal process. Here also, as in other already better and longer researched cases, as for instance the heparin-antithrombin III interaction, one can speculate that the carbohydrate part of the ligand serves as a rigid framework, in order to position charges in an appropriate stereochemical manner.

### 2.2.3 Chemically effective partners in biological processes

The role of carbohydrates described in the preceding sections was, in general, to take part passively in the synthesis or degradation of complex structures as substrate of enzymes, to alter the physiochemical properties of proteins or lipids, or to serve as signals for the transmission of biological information. In none of these functions do carbohydrates act as chemical agents in the events and thereby themselves become chemically altered. In fact, spontaneous chemical reactions of carbohydrates in biological systems without the participation of enzymes are exceedingly rare. In the following discussion, two biologically significant reaction types proceeding from carbohydrates are considered of which the first, the redox-reactions of ascorbic acid, is of greater physiological importance. The second, the condensation of hexoses with serum proteins, can lead to pathological changes in human organisms and, because of that, is undesirable.

#### 2.2.3.1 Ascorbic acid

*Vitamin C* or *L-ascorbic acid*[151] (hereafter the configurational descriptor is not used) is essential for all vertebrates and possibly for most types of animals, the insects included. Many animals can synthesise ascorbic acid in sufficient quantity for their own use. For humans and some animals such as apes, guinea pigs, bats and many birds, the carbohydrate derivative ascorbic acid is a vitamin, since a step in the biosynthesis, the oxidation of L-gulono-1,4-lactone to L-*xylo*-hex-2-ulosono-1,4-lactone, which spontaneously enolises, cannot be accomplished (Fig. 2.116).

Permanent withdrawal from ascorbic acid leads to severe symptoms of disease – spontaneous contusions, loss of teeth, poor wound healing – and finally death. Humans survive only about 3 to 4 months after their own reserves, predominantly in the adrenal glands and liver, are used up. With guinea pigs death occurs merely after about 20 days. It can well be imagined that scurvy, the deficiency disease of vitamin C, was greatly feared on sailing ships, whose crew had to give up for months fresh fruit and vegetables. The causes of the symptoms have been partly explained only recently. They are based on a disturbance in the building of the connective tissue.

In a precursor of collagen, the pro-$\alpha$-chain, proline and lysine residues are converted, to a great extent, into hydroxyproline and hydroxylysine. After that, three of these oxidised chains come together to give a three-stranded helix, which is called procollagen. This complex is transported out of the cell and transformed further extracellularly into collagen. It seems certain today that vitamin C is concerned substantially in the hydroxylation of lysine, but above all proline. If the hydroxylation of proline ceases in the case of a vitamin C deficiency, then the faulty pro-$\alpha$-chains are not capable of forming the triple helix of procollagen. Probably the formation of the helix is first facilitated through hydrogen bonds between hydroxyproline residues. The faulty pro-$\alpha$-chains are not further processed but are degraded again. Since collagen in the extracellular matrix is subject anyway to a constant degradation, the connective tissue becomes impoverished in this essential structural protein. The blood vessels then become brittle and the teeth lose their support because apparently the collagen transformation there, in contrast to other tissues, is comparatively quick. In contrast to hydroxyproline, hydroxylysine has, apparently, nothing to do with the intracellular building of the triple helix, but is necessary above all for glycosylation. To what extent the absence of this glycosylation can have consequences for the organism has not been clarified. Of course, lack of hydroxylysine has consequences for the extracellular cross-linking of the collagen molecules, an important prerequisite for the mechanical stability of collagen.

Besides the known specific function of ascorbic acid as a cofactor in the hydroxylation of proline and lysine, the details of which are still vague, the vitamin is assigned a very important role as a general anti-oxidant. Daily doses of a gram and more for humans – 50 mg per day is considered as the minimal dose – should have a favourable effect for the organism. Even larger quantities of the vitamin are harmless. Excess ascorbic acid is either excreted in the urine, or can also, after oxidation to dehydro-

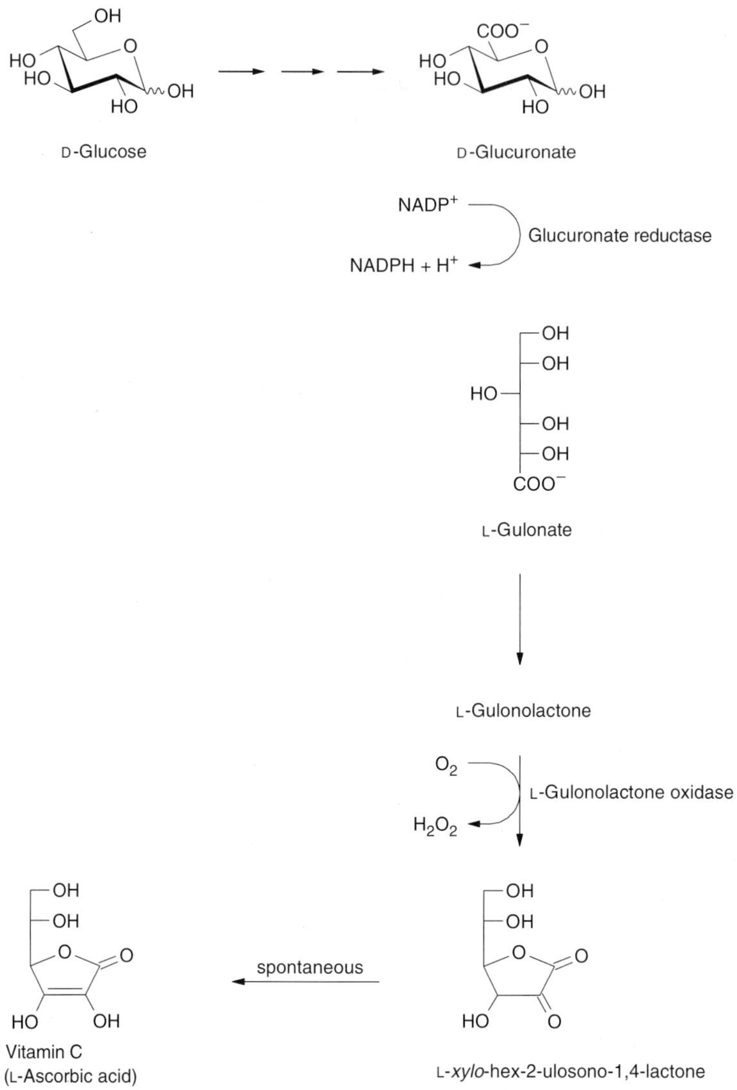

D-Glucose → → → D-Glucuronate

$NADP^+$ → $NADPH + H^+$ Glucuronate reductase

L-Gulonate

L-Gulonolactone

$O_2$ → $H_2O_2$ L-Gulonolactone oxidase

L-*xylo*-hex-2-ulosono-1,4-lactone

spontaneous

Vitamin C
(L-Ascorbic acid)

**Fig. 2.116** *Biosynthesis* of vitamin C (L-ascorbic acid) from D-glucose. The human organism has no L-gulono-lactone-oxidase and cannot carry out the corresponding step.

ascorbic acid, hydrolysis into *L-threo-hex-2,3-diulosonate*, reduction to *L-xylo-hex-3-ulosonate*, and decarboxylation of the latter, be conveyed as L-xylulose to the pentose transformations and therewith to the pentose-phosphate pathway (Fig. 2.117).

Especially, ascorbic acid provides an effective protection against free radicals in connection with molecular oxygen. Thus, ascorbic acid protects not only oxidation-sensitive redox-systems of general metabolism but also tissue which is exposed to constant high concentrations of oxygen and therewith unavoidably to the superoxide radical anion, known to be very toxic, which is formed through the one-electron reduction of molecular oxygen with a series of reducing agents *in vivo*. This intermediate abstracts hydrogen from other compounds and thereby forms radicals and, in turn, harmful peroxides which for their part deactivate vital transition metal ions. This vicious circle can be broken by antioxidants such as ascorbic acid (see below).

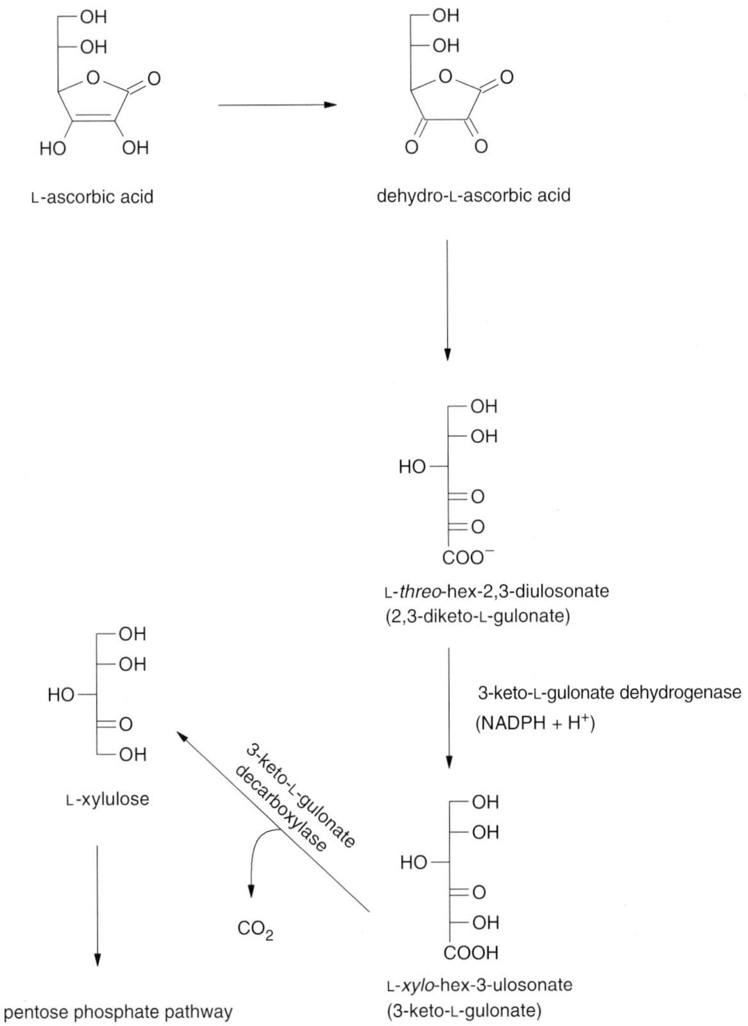

**Fig. 2.117** *Degradation* of L-ascorbic acid to *L-xylulose*, which can re-enter the pentose meabolism.

Viewed chemically, ascorbic acid is a glyconolactone, which arises from L-gulonolactone through further oxidation. The conjugation bestows the hydroxyl group in the 3-position a high acidity through resonance stabilisation of the corresponding anion (Fig. 2.118).

Ascorbic acid is a strong reducing agent and is very easily oxidised with a large number of oxidising agents and also with atmospheric oxygen in the presence of heavy metal ions and naturally also with peroxide radicals. The oxidation takes place by one-electron transfer through monodehydroascorbic acid as a reactive intermediate to dehydroascorbic acid. Both monodehydroascorbic acid and dehydroascorbic acid can be reconverted through NADH-dependent reductases again into ascorbic acid. The property of acting as an outstanding *biological antioxidant* lies in its ability to donate electrons very easily (Fig. 2.119).

Ascorbic acid is in general a more effective electron donor towards an organic peroxy radical than one of the other biological substrates R-H and is mostly in the situation of breaking the autooxidation cycle and so neutralising the injurious effects of the combination of free radicals and oxygen. This injurious action, amongst others, is responsible also for the ageing of tissue.

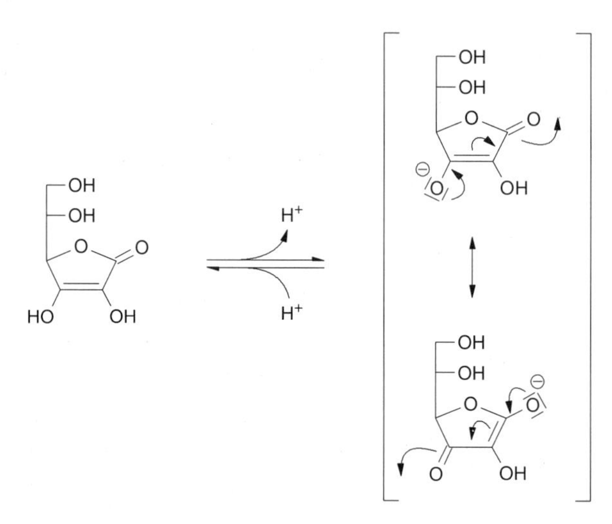


**Fig. 2.118** The 3-OH group is acidic. The corresponding anion is resonance stabilised.

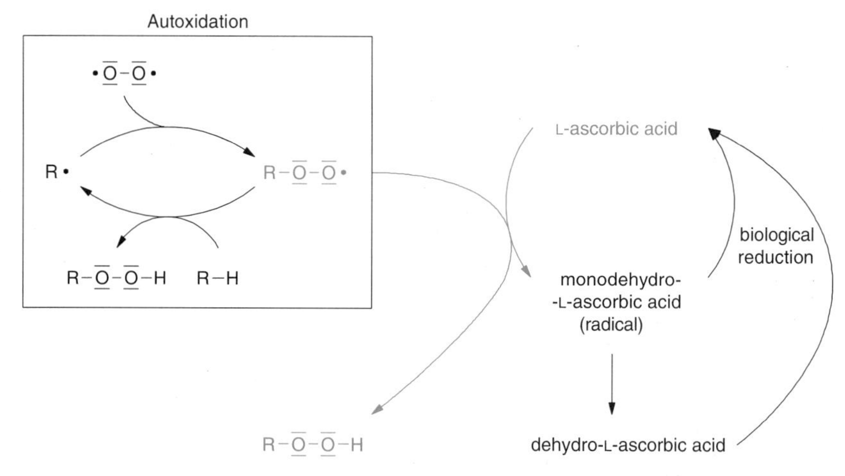


**Fig. 2.119** Damage by radical induced autooxidation of a substrate R-H can be avoided by strong anti-oxidants such as L-ascorbic acid. The high reactivity of the peroxide radical formed is nullified by its conversion to hydroperoxide. Both monodehydro-L-ascorbic acid (a radical) and also dehydro-L-ascorbic acid formed by further oxidation may be converted again into L-ascorbic acid by biological reduction.

### 2.2.3.2 Hexoses and serum protein

With about 90 mg/100 mL in blood serum, D-glucose, besides proteins, lipoproteins and lipids, is one of the main components of organic material and, expressed in molar concentrations, even the main component. If the molar concentration of glucose climbs above a value of 180 mg/100 mL then the organism begins to eliminate the excess through the kidneys in the urine. Hyperglycemia is a typical manifestation of diabetes, the sugar illness, which in general can be attributed to insulin deficiency. Insulin deficiency can depend on the genetically conditioned weakness of function of the insulin producing cells in the pancreas. It can, however, also be a result of general functional weakness of the organs in old age.

It has been known for some time, mainly through investigation of hexose transport* in erythrocytes with the help of radiolabelled D-glucose, that D-glucose combines covalently with protein, especially with the haemoglobin of erythrocytes. The reaction is not enzyme catalysed but occurs spontaneously and to a very large extent is irreversible. Since the glucosylation of haemoglobin is concentration dependent, more of the so-called *Hb A1$_c$*, the covalently modified haemoglobin, is found in the blood of diabetic patients. The quantity is a measure of the prevailing blood-sugar concentration in the preceding 4–6 weeks and so is used for diagnostic purposes. Under normal, physiological conditions this protein glycosylation is completely harmless, since the continual renewal of the blood protein taking place provides a replacement.

The situation is quite different if the blood sugar content, not only in the serum but also in tissues, remains unnaturally and permanently high and if, above all in old age, the renewal particularly of tissue protein, perhaps of collagen, occurs very slowly.

In the recent past it was suspected that an unnaturally high glycosylation rate of the body's own proteins could have something to do with pathological changes in the circulatory system and even with the accelerated ageing of the individual. On the basis of *in vitro* and *in vivo* experiments, the glycosylation of proteins is not thought to represent the final stage of the process, but to lead further to irreversible cross-linking between proteins. In the case of serum proteins, this could result in formation of insoluble aggregates which block the capillaries, and in the case of fibrous proteins of the extracelluar matrix it stiffens the cross-network of the tissue, thereby making it less elastic and more brittle. Actually, one of the remarkable features of old tissue is a higher degree of cross-linking of the fibrous proteins[152].

To a certain degree, the chemical reactions which are the basis for these changes should be explained and discussed separately. The reaction of hexoses with amino acids, peptides, or proteins has been known for a long time under the term *Maillard* or *browning* reaction[153]; it is, amongst other things, the cause of the brown colouration of beer to a greater or lesser extent. As carbonyl compounds, a hexose reacts with a free amine and forms an imine (Fig. 2.120).

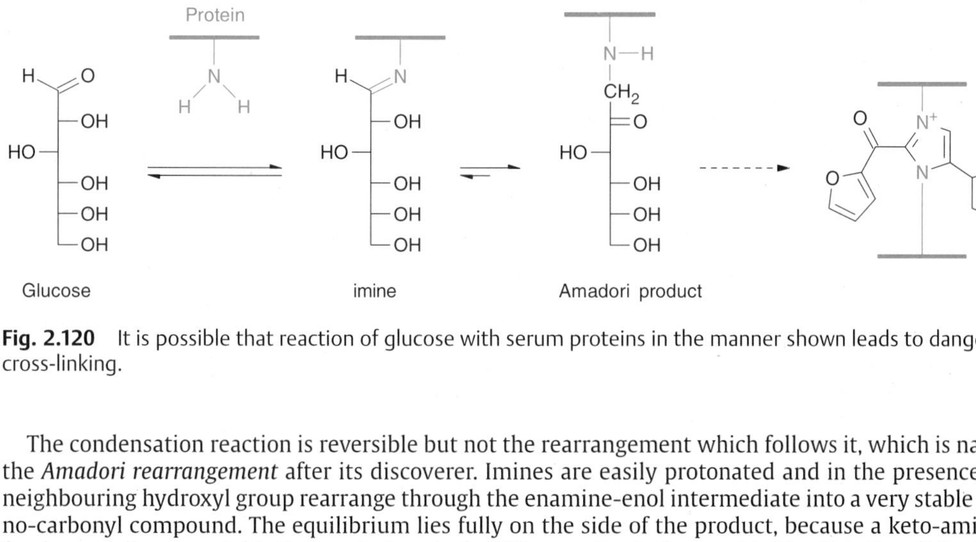

Protein

Glucose imine Amadori product

**Fig. 2.120** It is possible that reaction of glucose with serum proteins in the manner shown leads to dangerous cross-linking.

The condensation reaction is reversible but not the rearrangement which follows it, which is named the *Amadori rearrangement* after its discoverer. Imines are easily protonated and in the presence of a neighbouring hydroxyl group rearrange through the enamine-enol intermediate into a very stable amino-carbonyl compound. The equilibrium lies fully on the side of the product, because a keto-amine is fundamentally more stable than an imino alcohol. The reaction is favoured in the direction of the keto-amine because on the one hand the enamine is more stable than the imine, and on the other the enol is less stable than the ketone. Up to this point, the glycosylation of serum or tissue proteins is still harmless and has been demonstrated repeatedly. Opinions differ about the following reactions to this day[154]. Nevertheless, suggestions which are chemically truly plausible should be presented.

* Besides D-glucose, about 10% *D-fructose* and *D-galactose* are found in the serum.

With the *Amadori product*, there is again present a system capable of condensation reactions, that reacts further to the "advanced glycosylation end products" (AGE) with particular optical properties (brown colouration and fluorescence)[152] indicative of an aromatic system. The spectroscopic properties of the condensation products between glucose and proteins obtained *in vitro* agree in some degree with those of isolated collagen or lens protein from old human beings. A key compound appears to be *2-(2-furoyl)-4(5)-(2-furanyl)-1H-imidazole (FFI)*, which could be isolated from acid hydrolysate of AGE from glucose and bovine albumin (Fig. 2.120). It appears reasonable to explain the irreversible cross-linking of proteins through the imidazole derivative, arising by a series of condensation reactions.

Doubt about the correctness of this idea was raised by experiments with antibodies against artificially produced protein-coupled FFI. The antibodies could not bind to AGE-bovine albumin. It is now thought that the compound first arises as an artefact through the treatment of *AGE-bovine serum albumin* with acid and afterwards with ammonia and probably was not already present in the form shown as a cross-linking bridge-element[154]. No doubt exists, though, on the formation of cross-linkages between proteins through non-enzymatic glycosylation. The Maillard product that arises thereby, with the appropriate optical properties, is equally incontestable. Convincing experiments are those in which erythrocytes, which have been coupled with AGE-protein occurring naturally in the body, are looked upon as foreign and destroyed by the macrophages of the organism[152].

# 2.3 Metabolism

In a textbook concerning the biology of carbohydrates it is necessary to consider their metabolism, though only to a level which affords a general account of the area. An in depth and wider discussion of the theme, above all from the viewpoint of metabolism in general, is more appropiate in textbooks of biochemistry, of which there are many good examples[155].

Carbohydrates are starting materials for the production of all other natural products. This central position in general metabolism of all living things was considered in section 2.2.1 (p. 130). In this section 2.3, the very active and diverse transformations of carbohydrates amongst one another will be considered.

Research into the breakdown of glucose by glycolysis and in alcoholic fermentation at the start of this century was the first scientific high-point of modern biochemistry. The elucidation of the complex reactions of the photosynthetic assimilation of carbon dioxide followed in about the middle of the century, as well as that of the pentose-hexose interconversion. The last large research area of carbohydrate metabolism, the investigation of the synthesis of polymeric carbohydrates and glycoconjugates, was examined successfully more recently, so that today metabolism can be counted amongst the largely closed chapters of carbohydrate research and in each textbook of general biochemistry it is treated in detail. In the scope of this book, therefore, as already mentioned, only a few, important aspects of carbohydrate metabolism will be considered, which are necessary for the understanding of modern glycobiology. In this manner, the foundations are laid in general for a knowledge of the biochemical relationships in the mammalian organism.

## 2.3.1 Carbohydrate absorption

Carbohydrates are normally the main component of human nutrition. They are taken up in the digestive tract as polysaccharides, oligosacharides, or also as monosaccharides. The breakdown of polymers and oligomers into monosaccharides takes place here, since only these can get into the blood stream through the intestinal wall with specific transport systems. Digestion is only entirely completed in the small intestine where specially equipped intestinal cells (brush-border cells) absorb the monosaccharides.

Most frequently, the polysacccahrides absorbed in food are starch and cellulose, which arise from plants, and glycogen which is of animal origin. Even in the mouth amylase in saliva begins to degrade *starch* and *glycogen* to a mixture of the so-called maltodextrins as well as to their component units maltose and glucose, a process which is interrupted in the stomach, and is continued and completed in the small intestine through the action of the pancreatic amylase. Cellulose can be utilised only indirectly, if the mammal is equipped with a rumen, as is the case with ruminants, and the corresponding flora of the rumen bacteria. The bacteria of the rumen have at their disposal a *cellulase-β-glucosidase system* and finally degrade the glucose resulting from enzymatic hydrolysis anaerobically to acetate, lactate, and propionate as utilisable products as well as methane and carbon dioxide as non-utilisable products.

Both saliva and pancreatic amylase are alpha-amylases. The designation ***alpha*** is somewhat misleading. Retention of the anomeric configuration during hydrolysis is meant thereby, ***alpha*** being used in distinction to α. From an α-glycosidic compound is liberated an α-hemiacetal end group. In contrast, ***beta-amylases*** hydrolyse an α-glycosidic compound with inversion of configuration. beta-Amylases are found mainly in plants and microorganisms. They are very specific *exo*-(1→4)-α-D-glucanases and release exclusively the disaccharide maltose. The alpha-amylases are, in contrast, *endo*-(1→4)-α-D-glucanases and hydrolyse a maltodextrin chain almost randomly in its interior. For high activity, the substrate should possess at least five D-glucopyranosyl residues. Maltotriose is hydrolysed extraordinarily slowly and maltose almost not at all. A dextrin-(1→6)-α-glucosidase provides for hydrolysis of the (1→6)-α-linkages in the branching points of amylopectin and glycogen. The low molecular weight maltodextrins, especially maltose, as well as the other oligosaccharides sucrose and lactose taken up in food, are first converted on the outer wall of the epithelial cells of the small intestine by the specific, firmly anchored hydrolases, α-glucosidase (maltase), sucrose-α-glucohydrolase (*sucrase*) and *β-galactosidase*, into the free monosaccharides D-glucose, D-fructose, and D-galactose, which are immediately reabsorbed (Table 2.3).

**Table 2.3** Digestion and absorption of carbohydrates in mammals

| Organ | Enzyme | Substrate | Product |
|---|---|---|---|
| Mouth | Salivary amylase | Starch, Glycogen | Maltodextrins, Maltose |
| Duodenum (Lumen) | Pancreatic amylase, Dextrin-(1→6)-glucosidase | Amylopectin, Glycogen, Maltodextrine | Maltose, Glucose |
| Duodenum (Epithel) | Maltase, Lactase, Sucrase | Maltose, Lactose, Sucrose | Glucose, Galactose, Fructose |
| Small intestine – total length (brush border) | *Hexose transport through specific transport proteins* | | |

A resorption of carbohydrates takes place neither in the mouth nor in the stomach. With the complete hydrolytic decomposition of all polymeric and oligomeric carbohydrates to the monosaccharide stage – their exist, in additon to the named disaccharidases, hydrolases for glucuronides, aminoglycosides, and mannosides – the digestive process ends. The absorption of the hexoses and also other monosaccharides which are available then begins through the brush-border cells of the small intestine into the bloodstream.

The blood transports glucose and other monosaccharides for breakdown in muscle tissue, the heart, brain, and the inner organs. Under normal conditions, blood contains about 90 mg/100 mL (5 mM) of glucose. The constant nature of this concentration (homeostasis) is especially necessary, above all for the optimal upkeep of heart, brain and kidneys. If a carbohydrate-rich meal has not just been taken, then the concentration of monosaccharides in the intestinal lumen lies considerably below the blood-sugar concentration. It is necessary, therefore, to actively transport the monosaccharides, especially D-glucose, against a concentration gradient in the blood. The active transport system of the brush-border cells has been well researched. It favours D-glucose and D-galactose and is coupled with the transport of $Na^+$. Other monosaccharides, such as D-fructose, D-mannose, D-xylose, and L-arabinose are transported only with clearly less efficiency. The relative rates lie between 100 for D-Glc and D-Gal, 40 for D-Fru, 20 for D-Man, and 15 and 10, respectively, for D-Xyl and L-Ara.

The brush-border cells of the intestinal epithelium are suited for their particular function. Because of a concentration gradient, the membrane of the brush-border side transports $Na^+$ from the intestinal lumen into the cell (Fig. 2.121). This $Na^+$ influx, assisted additionally by a membrane potential, is the driving force for the glucose transport against a concentration increase. The D-glucose-$Na^+$ co-transport is strictly stoichiometric and produces a D-glucose concentration in the brush-border cell which lies above that in the blood.

The passage of D-glucose into the bloodstream through the membrane on the ***blood side*** of the brush-border cells is in fact also facilitated by a transport system but this runs spontaneously thanks to the concentration gradient. The active transport of D-glucose, which in the balance is energy consuming, is driven indirectly through an ATP-dependent *sodium pump*, which actively transports $Na^+$ against a concentration gradient through the membrane to the blood side in the circulation. This system keeps the concentration of $Na^+$ low in the brush-border cells and thereby maintains the co-transport of D-glucose-$Na^+$ out of the lumen in the cells (Fig. 2.121).

Active transport is required only for D-glucose, not for other monosacharides, whose concentration in the blood is normally so low that with the corresponding absorption of nutrient there always exists a driving concentration gradient.

### 2.3.2 Glycolysis and gluconeogenesis

Just as glucose gets from the intestine into the blood through active transport, it must be transported out of the blood into the cells of the organism. The transport systems of cells are similar and differ mostly only in their regulation. After entry of glucose into the cell it is converted immediately through hexokinase into *D-glucose 6-phosphate (GlC-6-P)* and is subjected to glycolytic degradation. Liver cells have at their disposal, additionally, the glucokinase specific for D-glucose. Utilisation of the monosaccharides taken up with nutrients takes place mainly in the liver. All hexoses are transformed into the key metabolite of carbohydrate metabolism, GlC-6-P, to which all carbohydrates lead, and from which then all further reactions proceed (Fig. 2.122).

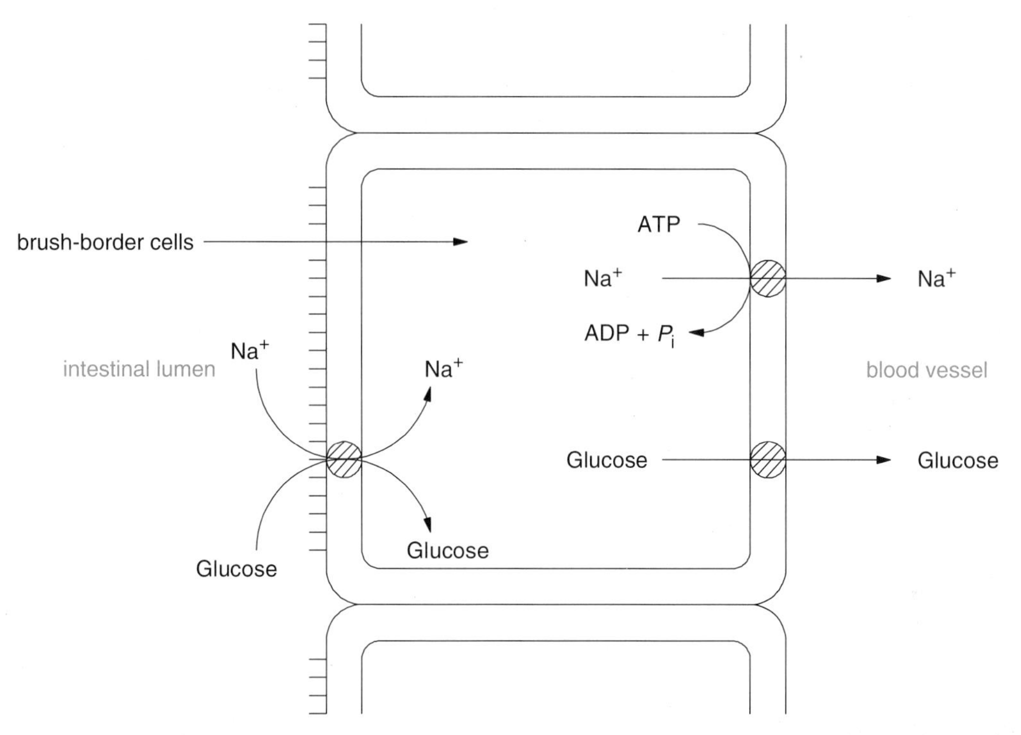


**Fig. 2.121** Glucose, as the major monosaccharide, is taken up in the small intestine by the brush-border cells and released into the circulation.
*Intestine side*: Active $Na^+$-co-transport of glucose against the concentration gradient of the intestinal lumen in the brush border cells.
*Blood side*: Mediated, yet not active transport of the glucose with a concentration gradient from the brush-border cells into the blood stream

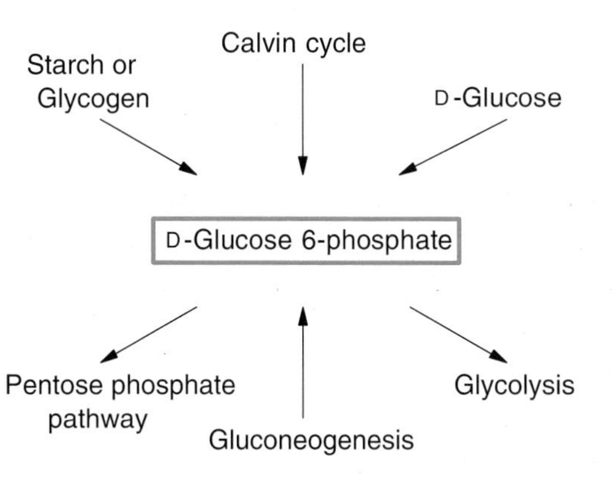


**Fig. 2.122** Central position of D-glucose 6-phosphate in carbohydrate metabolism.

Glycolysis is an important metabolic pathway for all organisms. GlC-6-P is degraded to two molecules of pyruvate, whereby in the balance, as energy-rich compounds, two molecules of NADH and two molecules of ATP are formed. Starting from D-glucose, ATP is required for two reactions (first and second activation reaction), but it is doubly regenerated again by two further reactions (first and second ATP-forming reactions) (Fig. 2.123).

The name *glycolysis* refers to the cleavage of a hexose into two C3 units, namely pyruvate, whereby an energy-rich compound is converted into two molecules of a less energy-rich compound. Glycolysis

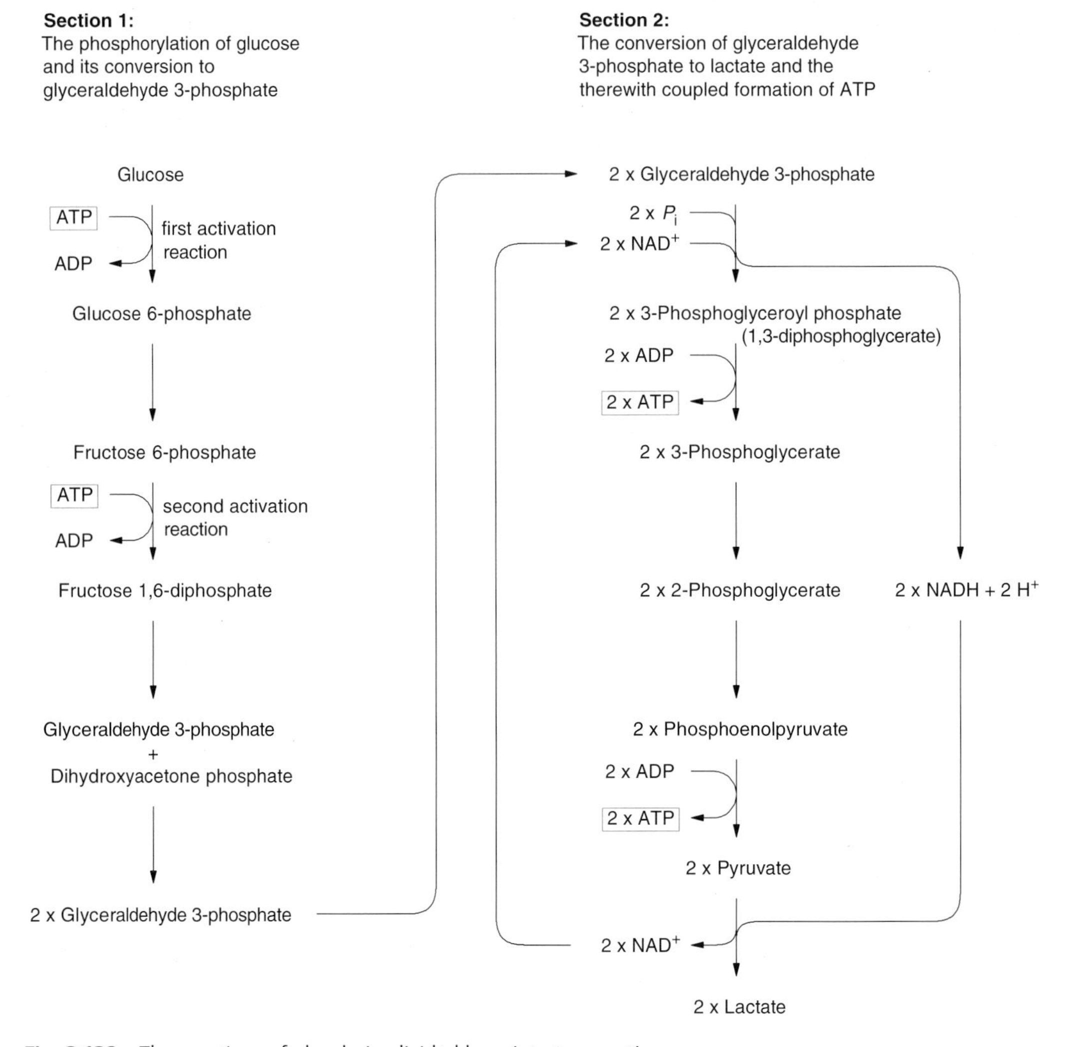


**Fig. 2.123** The reactions of glycolysis, divided here into two sections.

serves, consequently, first of all for the rapid production of energy, above all under conditions of oxygen deficiency when the end product of glycolysis, pyruvate, cannot be further converted. The necessary regeneration of $NAD^+$ for the dehydrogenation of D-glyceraldehyde 3-phosphate in step 6 can occur throug.. preliminary reduction of pyruvate to *lactate* in step 11. Therefore, under conditions of oxygen deficiency, there arises a temporary accumulation of lactate. The true *combustion* of the carbon and hydrogen of glucose to carbon dioxide and water takes place namely only through the complete oxidative degradation of pyruvate in the citric acid cycle, which is fully independent of glycolysis (see section 2.2.1, p. 130).

Alongside ATP, NADH is formed in glycolysis, and this must be continously regenerated so that the glycolysis does not falter through a lack of $NAD^+$ (step 6). Essentially, there are two ways by which NADH can deliver up its hydrogen. In the first, the hydrogen of the NADH is oxidised by oxygen to water eventually, through the electron transport system of the so-called respiratory chain with its numerous enzymes and co-enzymes, in which a part of the energy thereby set free gets lost not as heat but is utilised for the building of ATP (respiratory chain phosphorylation). In the second, it serves, as already mentioned above, for the reduction of pyruvate into lactate. The relationships between pyruvate combustion to carbon dioxide and water in the citric acid cycle and in the respiratory chain as well as the coupling of this in the production of ATP as utilisable chemical energy is schematically represented in Fig. 2.124.

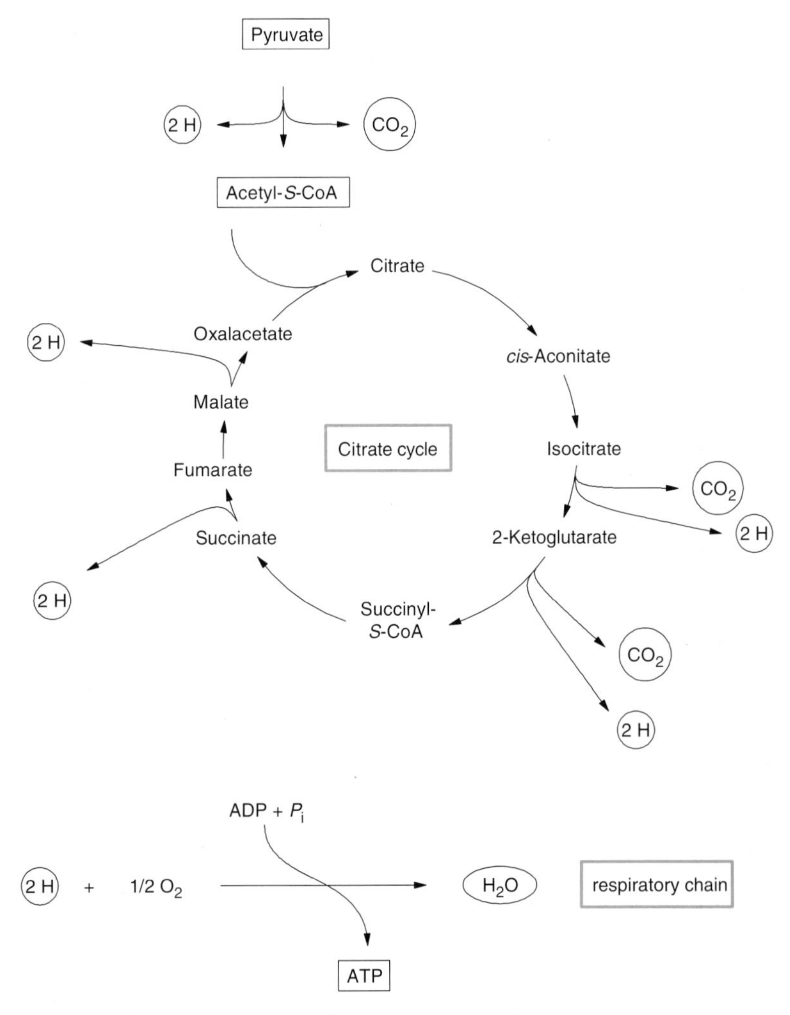


**Fig. 2.124** *Combustion* of pyruvate to carbon dioxide and water takes place in the citrate cycle and the linked respiratory chain. Utilisable energy arises in the form of ATP.

The formation of lactate makes glycolysis a closed process for energy production, which can even proceed in the case of an oxygen deficiency. In skeletal muscle, the glycolytic formation of lactate is a completely physiological process, when quick energy is needed and the oxygen supply from the bloodstream is not sufficient. In mammals, lactate finds its way back from the muscles to the liver, where it serves principally for gluconeogenesis.

Microorganisms, for example yeast, also naturally have at their disposal the enzymes of glycolysis, though here pyruvate undergoes a decarboxylation to acetaldehyde, which through reaction with NADH is transformed into alcohol and therefore ensures that $NAD^+$ is again ready for glycolysis. This process, familiar as alcoholic fermentation, served already in the last century for the proof of cell-free enzymic reactions (Fig. 2.125).

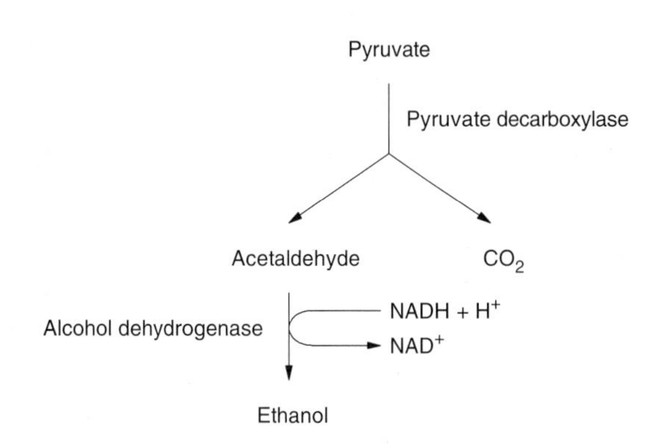


**Fig. 2.125** In the *alcoholic fermentation* pyruvate can be converted also to ethanol.

Glycolysis is not simply reversible (Fig. 2.126). Two steps of glycolysis, the conversion of phosphoenolpyruvate into pyruvate and of D-fructose 6-phosphate into D-fructose 1,6-diphosphate, must be reversed in a round about way and with other enzymes. If the first step, the phosphorylation of D-glucose to D-glucose 6-phosphate, is included then this step also must be reversed. The reversal of glycolysis, with the three named variations, is called gluconeogenesis.

Gluconeogenesis is the only possibility for the synthesis of glucose from non-carbohydrates, for example from lactate, propionate and the so-called glucogenic amino acids. The steps which are different for glycolysis and gluconeogenesis (Fig. 2.126) and which are catalysed by different enzymes, serve first of all for the independent regulation of glucose degradation (*glycolysis*) and glucose synthesis (*gluconeogenesis*). In principal, neither glycolysis nor gluconeogenesis are reversible.

Not only glucose is taken up in food, but the isomeric hexoses galactose, mannose, and fructose can also be taken into the bloodstream. There are different possibilities for their conversion into glucose or glucose derivatives. The conversion of mannose and fructose corresponds to the classical *Lobry de Bruyn – van Ekenstein transformation*. The isomerisation of galactose is a special case and is considered in the following section.

The Lobry de Bruyn – van Ekenstein chemical isomerisation also plays a role in metabolism. Certainly, the free monosacaharides are not converted into one another, although in microorganisms such interconversions are possible, but rather the corresponding 6-phosphates are involved. A conversion of D-mannose into D-glucose never occurs directly but always through D-fructose (Fig. 2.127, p. 180).

For a reversible transformation of D-glucose into D-fructose, there is available in mammals, moreover, the relatively unimportant possibility of reduction and reoxidation through the sugar alcohol D-glucitol (sorbitol) (Fig. 2.127, p. 180). It will seen in the next section that this pathway of epimerisation is always adopted if the isomerisation does not take place at the secondary alcohol group neighbouring the anomeric centre. It should also be noted that the biosynthesis of *N*-acetyl-D-mannosamine, very important for mammals, takes place from *N-acetyl-D-glucosamine* by way of its UDP-derivative (Fig. 2.128, p. 180).

### 2.3.3 Biosynthesis of oligosaccharides

■ **Nucleoside diphosphate sugars** are a special class of carbohydrate derivatives[156] whose discovery and investigation by L. F. Leloir[157] and his co-workers was of inestimable value. The anomeric phosphate ester group of the nucleoside diphosphate saccharide activates the glycosyl residue for enzyme-catalysed transfer reactions and the nucleoside residue serves glycosyl transferases or isomerases as an additional recognition factor for achieving specific syntheses. These so-called activated sugars fufil two extraordinarily important functions for synthetic metabolism:

1. They serve as substrates for the biosynthesis of all oligosaccharides and for the greater part of the polysaccharides; they are concerned in the synthesis of all glycoconjugates (see section 2.3.8, p. 206).
2. The biosynthesis of deoxysugars, of amino deoxysugars (with the exception of 2-amino-2-deoxysugars), of branched-chain sugars, of uronic acids and many other variants, which arise through modification of the normal hexopyranosyl framework, proceed from activated D-mannose or from activated D-glucose. The 4-epimerisation of D-glucose to give D-galactose should also be included here.

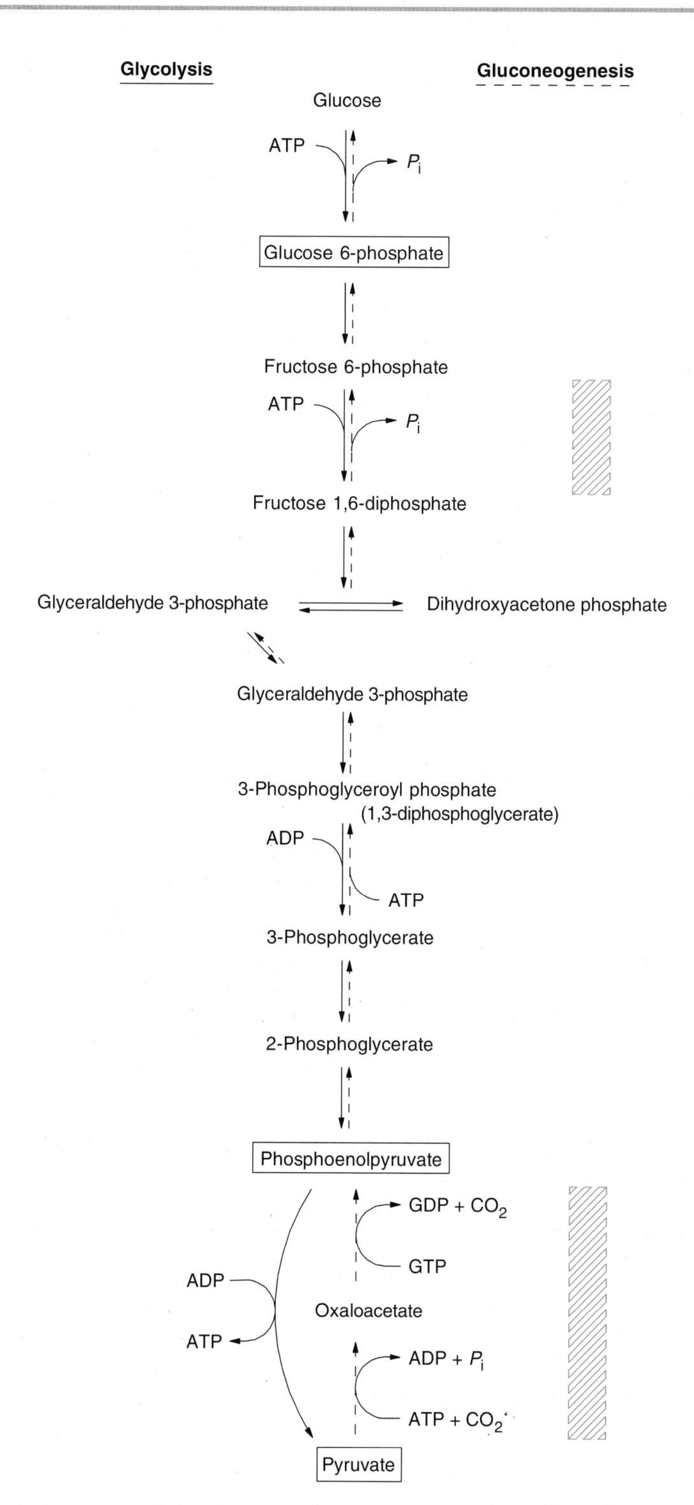


**Fig. 2.126** Glycolysis ⟶ and gluconeogenesis - - ► can run alongside each other, because certain reaction sequences ▨ are not reversible.

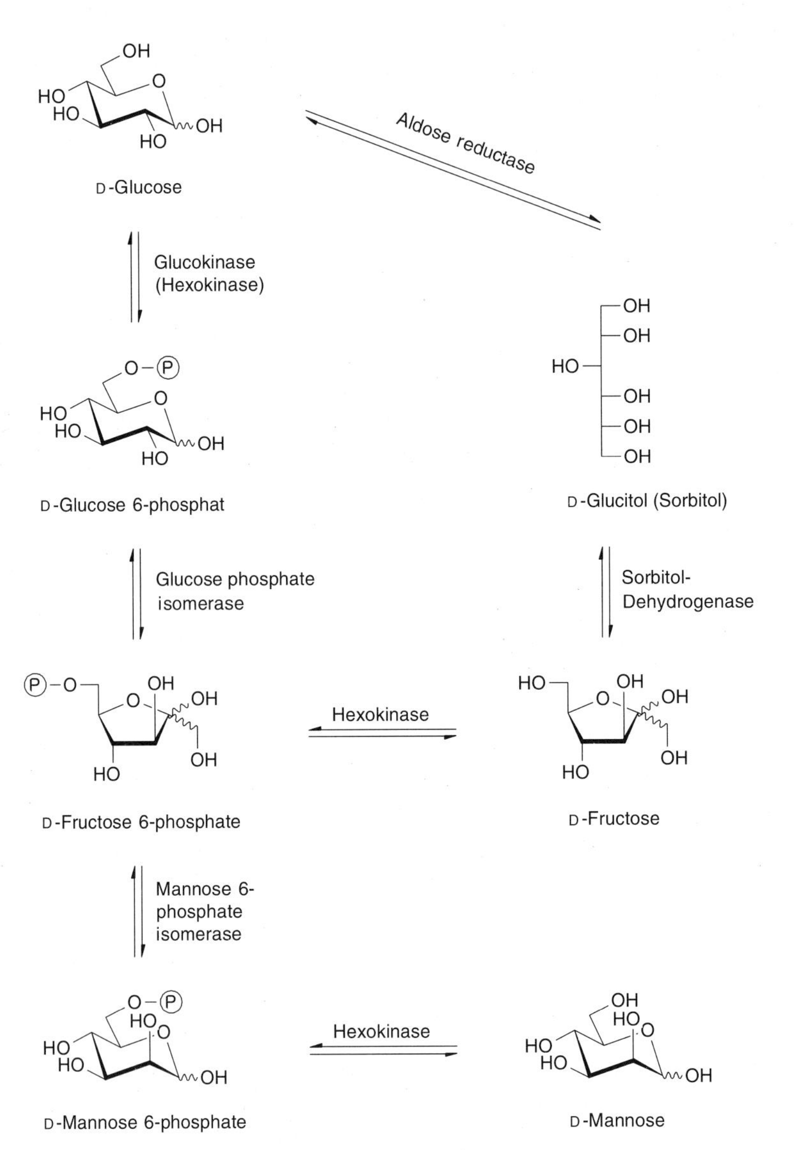


**Fig. 2.127** Biological *interconversion* of D-glucose, D-mannose, and D-fructose. A possible alternative pathway can take place by way of *D-glucitol (sorbitol)*.

UTP $PP_i$
UDP-glucosamine pyrophosphorylase
UDP
UDP-*N*-acetyl-D-glucosamine 2-epimerase

**Fig. 2.128** The conversion of GlcNAc into ManNAc is especially important for mammals, since NeuAc is formed from ManNAc.

What is clear today was still largely unrecognised about forty years ago in the case of many metabolic processes, not only those of the carbohydrates: namely the fact that degradative metabolic pathways cannot simply be the reversal of synthetic metabolic pathways if chaotic conditions in a cell are to be avoided.

It was especially the experience with *transglucosylation* starting from D-glucose 1-phosphate or from sucrose with use of phosphorolytic and also hydrolysing enzymes that revealed the general reversibility of most enzyme catalysed reactions, including hydrolysis, as a possible basis both for the synthesis and also the degradation of oligosaccharides and even of polysaccharides. The fact that with phosphorylase and glucose 1-phosphate the *in vitro* synthesis of glycogen and starch can be realised, as long as a "primer" is present, led in the 1940's to the general conviction that, also in higher organisms, phoshorylases are responsible not only for the degradation but also the synthesis of oligosaccharides and polysaccharides *in vivo*. The principle of transglycosylation from oligosaccharides and polysaccharides appeared to be a further possibility for biosynthesis, representing a type of disproportionation, that is a rearrangement of glycosyl residues.

Many bacterial enzymes are known – the dextran sucrases from *Leuconostoc* or the cyclodextrinases from *Bacillus macerans* especially have become well recognised – which are able to achieve great synthetic results at the expense of sucrose or also of maltodextrin.

The general principles of the biosynthesis of glycosides, oligosaccharides, and polysaccharides were first recognised in the mid 1950's. Almost all metabolic reactions which lead to the formation of a glycosidic bond proceed from nucleoside diphosphate sugars (NuDP-glycoses) and are catalysed by glycosyl transferases, which are very specific both with respect to the transferring sugar (monosaccharide or oligosaccharide) and also the nucleoside, as well as the acceptor which receives the glycose. The reaction follows the general principle:

NuDP-glycose + acceptor-OH —> Glycose-*O*-acceptor + NuDP

The *nucleoside* component in the activated sugar is predominantly uridine, but guanosine, adenine, thymidine and cytidine also occur. Many activated sugars, for example activated ketoses, have a phosphodiester bridge in place of a pyrophosphate group. Many nucleosides in activated sugars contain 2-deoxy-D-ribose instead of D-ribose (Fig. 2.129).

The extravagance of nature in using structurally complex activated monosaccharides as glycosyl donors is justified when it is remembered that the biosynthesis of specific glyconjugates of the cell surface depends on the capability of the appropriate glycosyl transferases to differentiate between subtle sterochemical alternatives. Both donors and acceptors appear to present to the enzyme adequate features for recognition.

All nucleoside diphosphate monosaccharides are biosynthesised by transfer of a nucleotide residue from a nucleoside triphosphate to an α-glycosyl phosphate with the help of a specific pyrophosphorylase (hexose 1-phosphate nucleotidyl transferase). Through the ensuing enzymatic hydrolysis of the liberated pyrophosphate, the otherwise unfavourable equilibrium for the synthesis can be displaced in the direction of the activated sugar (Fig. 2.130, p. 183).

The activation of the ketose neuraminic acid occurs, in contrast to that of the aldoses, through the transfer of the nucleotidyl residue, in this case of the cytidyl residue, directly onto the hemiacetal hydroxy group (Fig. 2.131, p. 183).

Transformations of the monosaccharide skeleton, so long as it does not involve positions 1 or 2, are in general carried out on activated sugars, whose ring form and anomeric configuration are fixed. One very important conversion for the metabolism of all living things is the reversible epimerisation at C-4 of D-glucose (Fig. 2.132, p. 184). The catalysing enzyme, UDP-D-glucose-4-epimerase, is widely distributed. It needs as a cofactor $NAD^+$, which has the task of effecting a dehydrogenation at the 4-position of the glycopyranosyl residue (glucopyranosyl or galactopyranosyl residue) with formation of a non-isolable intermediate. Reduction at the intermediate stage with the resulting NADH again produces the secondary hydroxy group in one or other configuration. Most 4-epimerases contain $NAD^+$ strongly bound in the active site.

The equilibrium of the 4-epimerase reaction is determined by the relative stability of the glucopyranosyl residue compared to the galactopyranosyl residue. Other similar epimerisations, for example L-arabinose to D-xylose, D-glucuronate to D-galacturonate, or *N*-acetyl-D-glucosamine to *N*-acetyl-D-galactosamine are carried out according to the same principle.

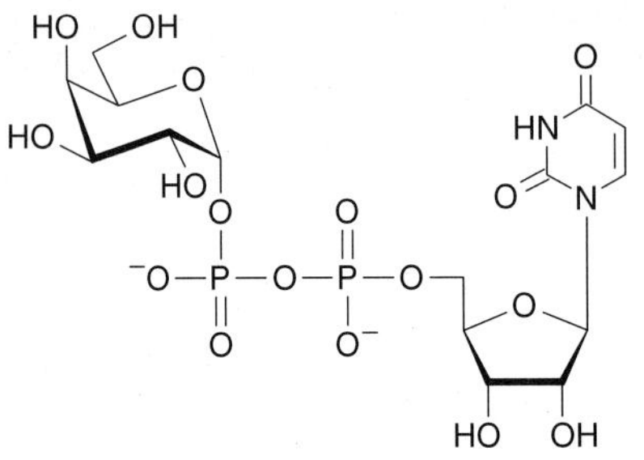

Uridine diphosphate D-galactose
(UDP-Gal)

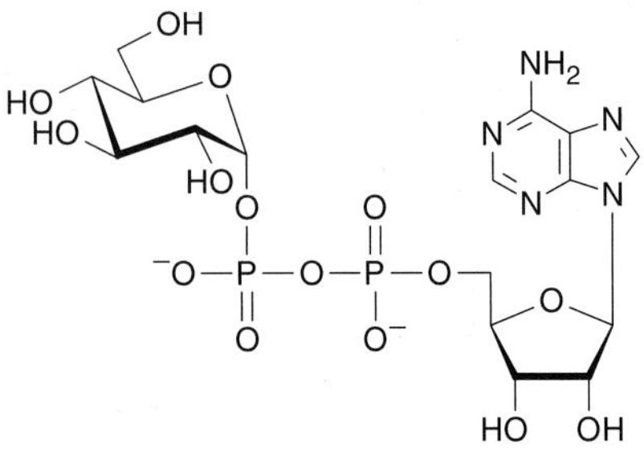

Adenosine diphosphate D-glucose
(ADP-Glc)

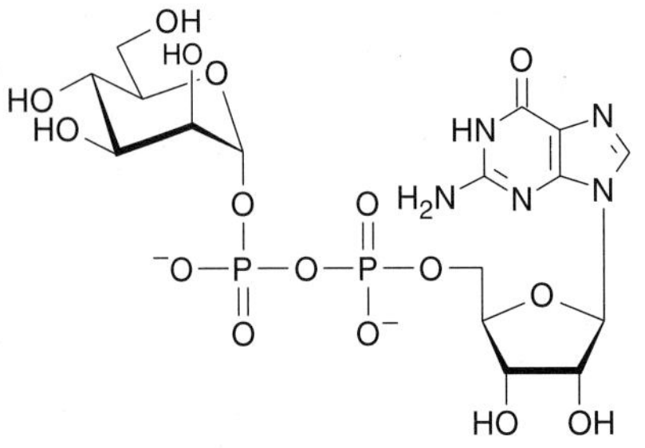

Guanosine diphosphate D-mannose
(GDP-Man)

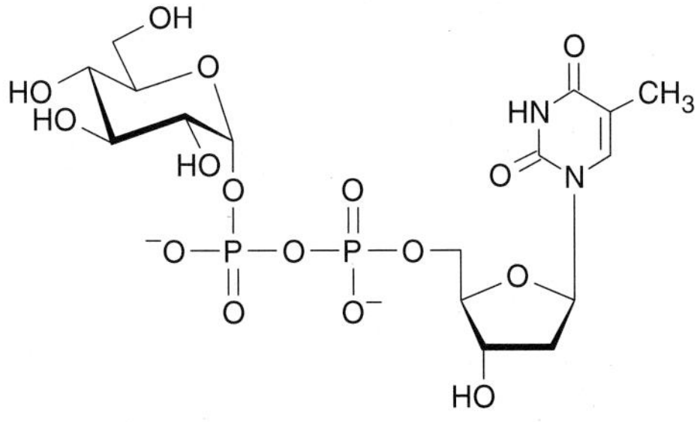

Thymidine diphosphate D-glucose
(TDP-Glc)

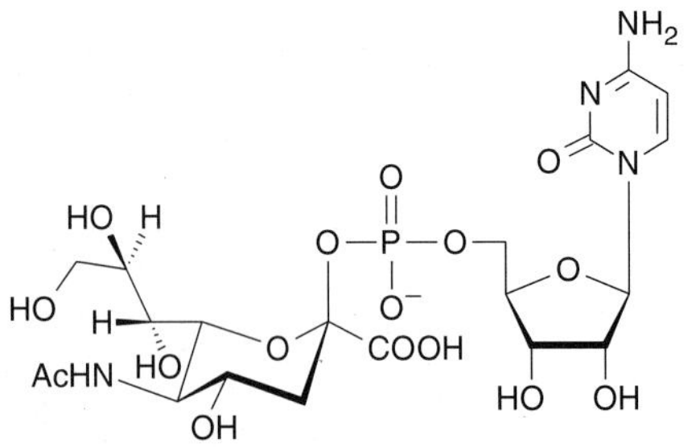

Cytidine monophosphate *N*-acetylneuraminic acid
(CMP-NeuAc)

**Fig. 2.129** Some monosaccharides activated as nucleoside diphosphates or nucleoside monophosphates.

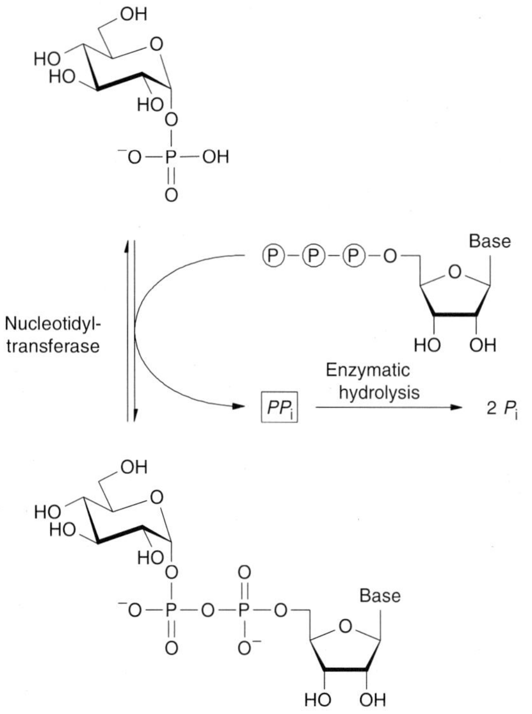

**Fig. 2.130** Biosynthesis of a nucleoside diphosphate glucose through transfer of a nucleotidyl residue onto glucose 1-phosphate.

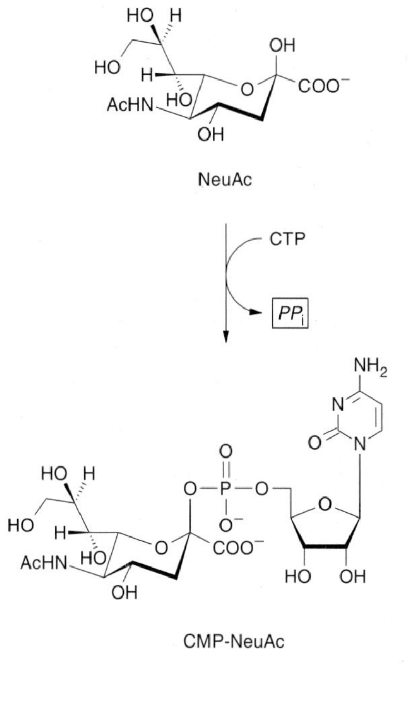

Nucleotidyl-transferase

Enzymatic hydrolysis

$PP_i$

$2\ P_i$

Base

**Fig. 2.131** *Activation* of *N-acetylneuraminic acid* (NeuAc) is typical for ketoses.

NeuAc

CTP

$PP_i$

CMP-NeuAc

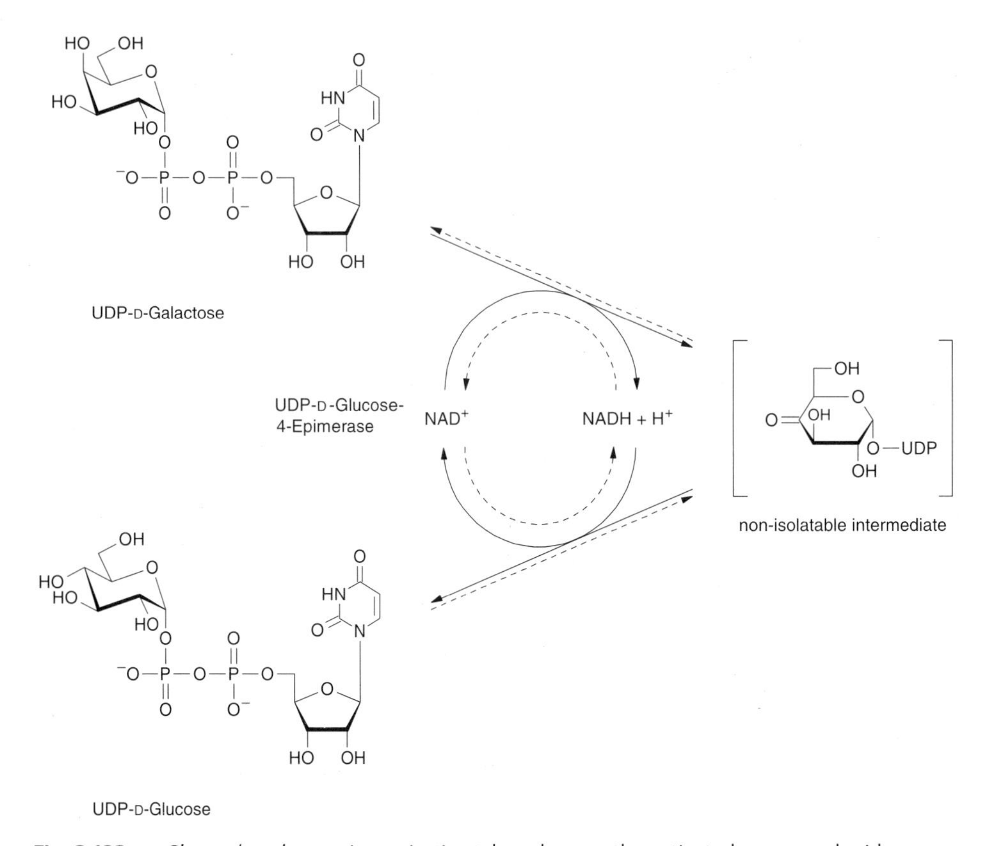


**Fig. 2.132** *D-Glucose/D-galactose isomerisation* takes place on the activated monosaccharides.

Uronic acids are constituents of many polysaccharides and glycoconjugates. Specifically, *D-glucuronic acid* serves additionally for the detoxification of chemical compounds poisonous to the body. All glycuronides originate with the corresponding nucleoside diphosphate monosaccharides, which with $NAD^+$ and a dehydrogenase are transformed into activated uronic acids, which for their part serve as donor substrates for the biosynthesis of uronic acid derivatives (Fig. 2.133).

Structural polysaccharides in plants often contain the pentoses D-xylose and L-arabinose. Both monosaccaharides are produced in the form of their nucleoside diphosphate, UDP-D-xylose and UDP-L-arabinose, formed from UDP-D-glucuronate and UDP-D-galacturonate, respectively. The enzymes responsible are UDP-D-glucuronate decarboxylase and UDP-D-galacturonate decarboxylase (Fig. 2.134).

Similar to the 4-epimerisation, the decarboxylation requires $NAD^+$ as a cofactor. Activation is obtained through the 4-keto intermediate which, as a $\beta$-keto carboxylic acid, readily evolves carbon dioxide. Stereospecific reduction of the decarboxylated 4-keto intermediate concludes the reaction sequence. For the reversible 4-epimerisation of UDP-D-xylose into UDP-L-arabinose there is an enzyme in plants comparable to the UDP-D-glucose 4-epimerase which, as this one, uses $NAD^+$ as a cofactor.

The physiological *de novo* biosynthesis of oligo- and polysaccharides proceeds, fundamentally, from activated forms of the monosaccharide. In general these are the nucleoside diphosphate monosaccharides as glycosyl donors. This does not exclude transglycosylation as a synthetic principle for the linking of monosaccharides nor reversal of the oligosaccharide or polysaccharide cleavage through hydrolysis, phosphorolysis or alcoholysis. Such reactions are either insignificant or biological exceptions.

There are very few naturally occurring oligosaccharides. When one thinks how large the possibilities are for combination of monosaccharides, this is remarkable. Human milk is a good source of a series of complex oligosaccharides, especially such which contain *N*-acetyl-D-glucosamine in addition to D-ga-

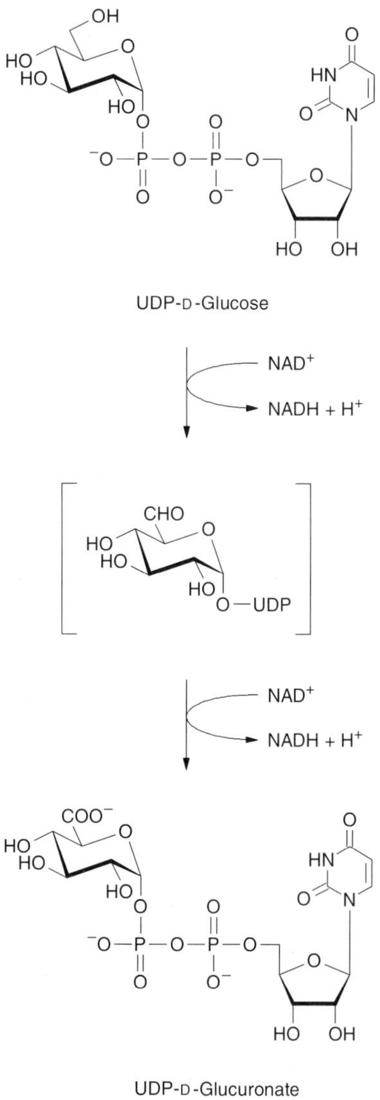

**Fig. 2.133** Dehydrogenation of D-glucose to *D-glucuronate* takes place on the UDP-derivative.

lactose and D-glucose. These nitrogen containing compounds amount to at least around 5% of the main oligosaccharide of the milk, at 70 g/L, lactose. Cow's milk contains only lactose as oligosaccharide.

The biosynthesis of lactose in the mammary glands of mammals takes place through an enzyme distributed in all vertebrates, galactosyltransferase, which in this form normally is responsible for the galactosylation of *N*-acetyl-D-glucosaminyl end groups in glycoconjugates. The galactosyltransferase changes its acceptor specifity during lactation in the mammary glands, since together with α-lactalbumin it forms lactose synthase. D-Glucose now becomes the favoured acceptor for the D-galactosyl residue from UDP-D-galactose (Fig. 2.135).

While lactose is limited in its distribution, as the exclusive product of mammals, this is not valid for the two other natural disaccharides, *sucrose* and *trehalose*, whose function as transport and storage carbohydrates, in plants and in insects, yeast and fungi, respectively, was considered in section 2.1.4.1 (p. 106) and section 2.1.4.2 (p. 108).

UDP-D-Glucose is the D-glucosyl donor for the biosynthesis of the main oligosaccharide of the plants, *sucrose*, and D-fructose 6-phosphate serves as acceptor. The primary product, sucrose phosphate, loses

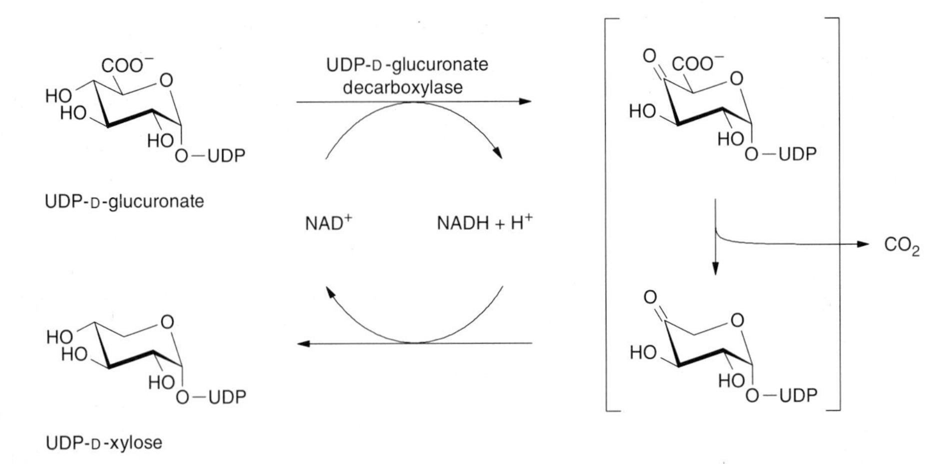


**Fig. 2.134** UDP-D-Xylose is formed from UDP-D-glucuronate after activation to facilitate the decarboxylation.

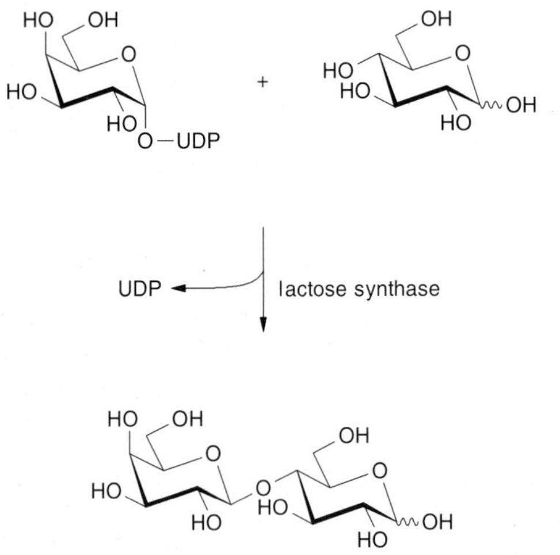

**Fig. 2.135** *Lactose synthesis* is typical for the formation of an oligosaccharide.

the phosphate residue immediately after its formation. Lactose synthesis (Fig. 2.135) which proceeds in an analogous manner, lacks this step. The equilibrium of the total process of sucrose synthesis lies fully on the side of free sucrose, because of the irreversibility of the last component step, so that from relatively low concentrations of precursors, high concentrations of sucrose can be obtained. In many plants the glycosylation of free fructose is possible with a relatively evenly balanced position of equilibrium. With high concentrations of sucrose, the re-synthesis of UDP-Glc is also possible with this enzyme. However, this pathway is not of particular importance.

### 2.3.4 Hydrolysis of the glycosidic bond

Glycosides, oligosaccharides, polysaccharides and glycoconjugates arising by transglycosylation, as components of living organisms, are liable to regular exchange just as are the biomolecules of other classes of natural substances. The life time of a certain type of compound varies according to function between several years for many proteoglycans in connective tissue and some hours for serum glycoproteins or even fractions of a second in the case of "trimming" processes in the Golgi apparatus of the mammalian cell. These modifications of molecules within the organism itself includes the degradation of food carbohydrates in the digestive apparatus and the huge quantities of carbohydrates in dead tissue, especially of plants, which are decomposed through microorganisms and so are returned to the cycle of living material.

All of these compounds are formal condensation products of monosaccharides, either with one another, or with compounds of other classes of natural products, which arise through elimination of water. The condensation products are glycosides, that is mixed acetals, and through hydrolysis they are split up into their components. Nature has so arranged things that the glycosidic bond does not normally undergo spontaneous hydrolysis under physiological conditions. The relative stability is a result of the inductively stabilising neighbouring group to the glycosidic carbon atom which generally is always present, either an hydroxy group or an *N*-acylamino group. The hydrolytic cleavage of the glycosidic compound is catalysed by more or less specific glycoside hydrolases, which are also called glycosylases and, in the case of polysaccharides, glycanases. According to their capability to cleave a polysaccharide chain internally or only from glyconic ends, they are distinguished as endo- and exoglycanases. Exoglycoside hydrolases cleave, in general, only one monosaccharide residue, sometimes also two, from the glyconic end. Recently, a bacterial enzyme has been discovered which is said to attack a maltodextrin from the aglyconic reducing end. Many exoglycoside hydrolases cannot transform polymers and many require a minimum molecular size in order to become active. The hydrolytic cleavage of a glycosidic compound is a reaction which may be reckoned amongst the most common in the whole biosphere.

The mechanism of enzyme-catalysed cleavage of glycosides has been investigated for decades and although today there are still questions in dispute, the course of enzymatic hydrolysis of glycosides can in the meantime be very well described. Glycoside hydrolases are, in general, very specific concerning the glyconic part or the glyconic side of the bond undergoing cleavage and some are also even specific with respect to the aglyconic part. This is true especially for endoglycanases, but above all for the highly specific exoglycoside hydrolases of the trimming process in the Golgi apparatus. The specificity of all glycoside hydrolases is almost absolute, as far as can be said today, with regard to the anomeric configuration of the bond being cleaved. However, this applies only to natural substrates and to the proven group of pure enzymes.

For a chemist it is self-evident that an enzyme-catalysed reaction must follow plausible chemistry, that is, in principle, also by chemistry that could proceed by non-enzymatic pathways. There are numerous investigations into glycoside hydrolysis[158] which involve the proton-catalysed reaction. Few, but weighty reasons support the argument that of the two possibilities for hydrolysing a glycosidic bond, that one which involves a cyclic oxocarbonium ion (glycosyl cation) as an intermediate (extreme cases, e. g. *tert*-butyl glycosides excepted) is the correct one. In spite of contradicting arguments which arise from time to time, it can be assumed that the enzyme-catalysed reaction also occurs without ring opening during the hydrolytic step[159].

Although the glyconic part of the cleaved glycoside as a *free* sugar is subject to spontaneous mutarotation, and it is known indeed that some glycosides are hydrolysed with inversion of configuration, most glycosides are hydrolysed with retention. Since the reaction partner water is normally present in a great excess, enzymatic hydrolyses of glycosides under natural conditions are quantitative. It is true that in some cases transglycosylations can be performed in a synthetic manner if the enzyme is able to bind preferentially potential acceptors, such as for example other sugars or alcohols, as opposed to water. Such syntheses are then kinetically favoured (see section 2.4.1.1, p. 236). However, under physiological conditions they are insignificant.

All enzyme-catalysed hydrolyses of glycosides may proceed very similarly apart from the mentioned differences in specificity. To this day, countless investigations on very different glycoside bonds have given no reason not to regard as generally valid the hitherto single, structure-based mechanism which has been explained in detail. Fine modifications do not change the fundamental process. The high resolution, three-dimensional structure of a complex between lysozyme, a bactericidal enzyme widespread in the animal kingdom, and a non-cleavable inhibitor, led to the description of the aforementioned cleavage mechanism[160], which is applied here in a general form to the cleavage of a disaccharide. A prerequisite for binding is recognition of the glyconic portion A and in this case, possibly also of the aglyconic part B of the substrate which is to be split (Fig. 2.136, p. 188).

The liberated bond energy is partwise transformed into a ring deformation of the glyconic monosaccharide moiety, a requirement which is still debated today. The ring deformation fulfils two important purposes for the catalysis:

1. It raises the free energy of the substrate in the ground state and lowers thereby the energy difference to the transition state, that is the activation energy.
2. In the case of equatorial aglyconic substituents, it allows a quasi anti-periplanar arrangement of an orbital with a free electron pair on the ring oxygen atom to the leaving group.

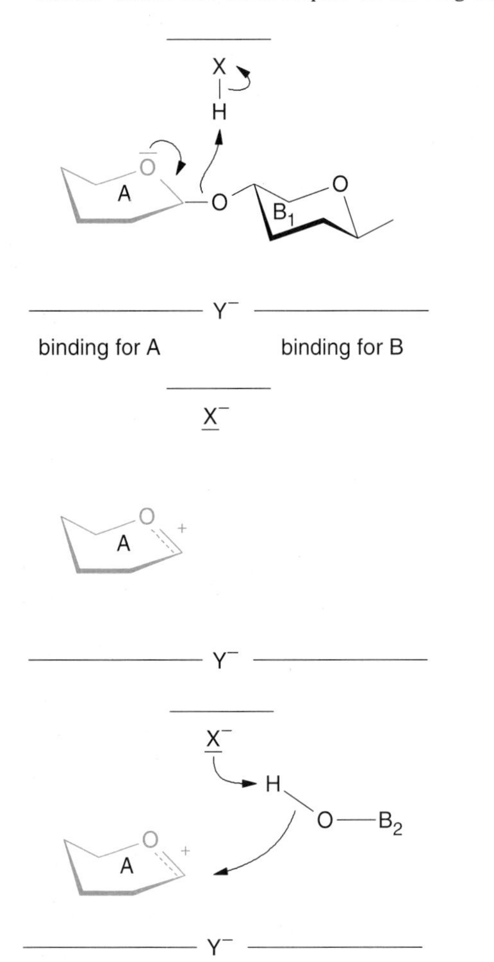


1. Deformation of the ring of A, protonation of the glycosidic oxygen.

2. The nucleofuge group $B_1$-O-H is separated. The intermediate $A^+$ remains bound as glycosyl-enzyme complex.

3. The reaction is carried out with the acceptor $B_2$-O-H. If $B_2$-O-H is water, a hydrolysis occurs.

**Fig. 2.136** The hydrolytic cleavage of a $\beta$-linked disaccharide represented here is related to the *lysozyme model*. Fundamentally, this model is applicable to each enzyme catalysed transglycosylation. An electrophilic donor A can be transferred to a nucleophilic acceptor B. The binding region of the enzyme decides on the acceptance of A and B and on the anomeric configuration – both of the substrate and also of the product. The region depicted as $Y^-$ is an additional stabilising or binding group for the glycosyl residue.

A proton donor group, usually a carboxyl group – Glu or Asp – activates the glycosidic oxygen atom through proton transfer. The glycosidic bond is broken and the so-formed glycosyl cation is stabilised by the anionic charge. A molecule of water binds in the position of the dissociated aglyconic residue and reacts with the glycosyl cation. The return of the proton onto the conjugate base of the acid and dissociation of the enzyme-product complex completes the reaction cycle.

A large portion of the catalytic power results from the fact that the reaction occurs in the active site of the enzyme with exclusion of the water medium. The reacting groups are therefore not solvated and thereby deactivated. It is still debated whether, instead of a cyclic glycosyl cation, an acyclic one occurs as an intermediate in the fission process. The question cannot be clarified absolutely unambiguously. However, numerous indirect indications suggest that in the case of the proton-catalysed reaction a cyclic intermediate is more likely.

Enzyme-catalysed glycoside hydrolysis is a special case of all transglycosylations and can surely serve, as regards the mechanism, as a model for all reactions in which the bond to the anomeric carbon atom is broken and again reformed, and also for the reactions which lead to the synthesis of oligosaccharides, polysaccharides and glycoconjugates. Differences lie only in the ability of the enzyme as *receptor* to bind and to orientate specifically different reaction partners in order to catalyse a reaction, which is always similar in its chemistry, by the same means.

## 2.3.5 Mobilisation and storage

Nearly all living things have the possibility to lay by reserves of energy and of raw materials, which in times of deficiency or excessive needs can become mobilised. These deposits consist of triglycerides, polyhydroxybutyric esters, polyamino acids, or polysaccharides. It is important that the reserve substances are osmotically inactive which means that they must be either giant molecules or water insoluble compounds such as the triglycerides. The reserve polysaccharide of many animals, fungi and many prokaryotes is glycogen, whose structure and properties were considered in section 2.1.3.2 (p. 101).

Glycogen metabolism takes place in liver and muscle cells. Its extraordinary importance for mammals, and thereby humans also, caused it to become in general one of the most fundamentally researched metabolic processes. The necessary maintenance of blood glucose at a constant level requires a very fine regulation of the equilibrium between blood glucose and its storage form, glycogen (Fig. 2.137). As shown in the example of glycolysis and of gluconeogenesis in section 2.3.2 (p. 174), the synthesis and degradation of biomolecules always occurs in different ways. Only in this way can such contra-running processes be controlled and the particular requirements of the cells and of the whole organism be adjusted. Glycogen is no exception.

In *glycogen*, similarly as with starch, a part of the potential energy of D-glucose is conserved as glycosidic binding energy and is transmitted on degradation by phosphorylase in the form of the phosphate ester bond of *glucose 1-phosphate*. Phosphorylase exists in liver, associated with its substrate glycogen, as active *phosphorylase* a, and *phosphorylase* b with less activity. Both forms can be interconverted. Through a hormone regulated, very complex enzyme cascade, the details of which will not be discussed here, phosphorylase b is converted into the active phosphorylase a, as the end point of this cascade, by phosphorylation with ATP and phosphorylase b kinase. By transfer of a terminal $\alpha$-(1→4)-D-glucopyranosyl residue to inorganic phosphate according to an extensively investigated transglycosylation mechanism, phosphorylase a releases the product *$\alpha$-D-glucopyranosyl 1-phosphate* (Fig. 2.138).

The phosphorylase degrades a maltodextrin chain up to about 4 glucosyl residues before a branch point. Two auxillary enzymes, a transglycosylase and an $\alpha$-glucosidase, ensure that the phosphorylase can bring about further degradation by chain transfer of the $\alpha$-(1→4) linked chain stub, which has become too short, onto a free end of an $\alpha$-(1→4) linked chain and by removal of the remaining $\alpha$-(1→6) linked glycosyl residue. Accordingly, besides glucose 1-phosphate, some glucose is always set free during glycogen breakdown.

The conversion of glucose 1-phosphate into *glucose 6-phosphate* occurs through *phosphoglucomutase*, an enzyme which requires glucose 1,6-diphosphate as cofactor. The reversible reaction includes the transfer of a phosphate residue from the 1-position of the cofactor onto the 6-position of glucose 1-phosphate. In this way the cofactor is continually regenerated with formation either of glucose 6-phosphate or, in the reverse direction, of glucose 1-phosphate. Glucose 6-phosphate can then be supplied for its various functions according to demand (Fig. 2.139).

If there exists no requirement for glucose 6-phosphate or free glucose, respectively, or if food uptake provides for a surplus in both these metabolites, then glycogen breakdown is brought to rest through a suspension of the hormone action, which in the end means the conversion of phosphorylase a into phosphorylase b through a phosphatase, and simultaneously glycogen synthesis is switched on. The switching on results from activation of glycogen synthase which, as the phosphorylase, also exists in an active (a) and a less active (b) form. However, the active glycogen synthase a is the non-phosphorylated form and glycogen synthase b the phosphorylated form. Dephosphorylation of glycogen synthase by phosphatase has the effect, therefore, of activation of glycogen synthesis with simultaneous deactiva-

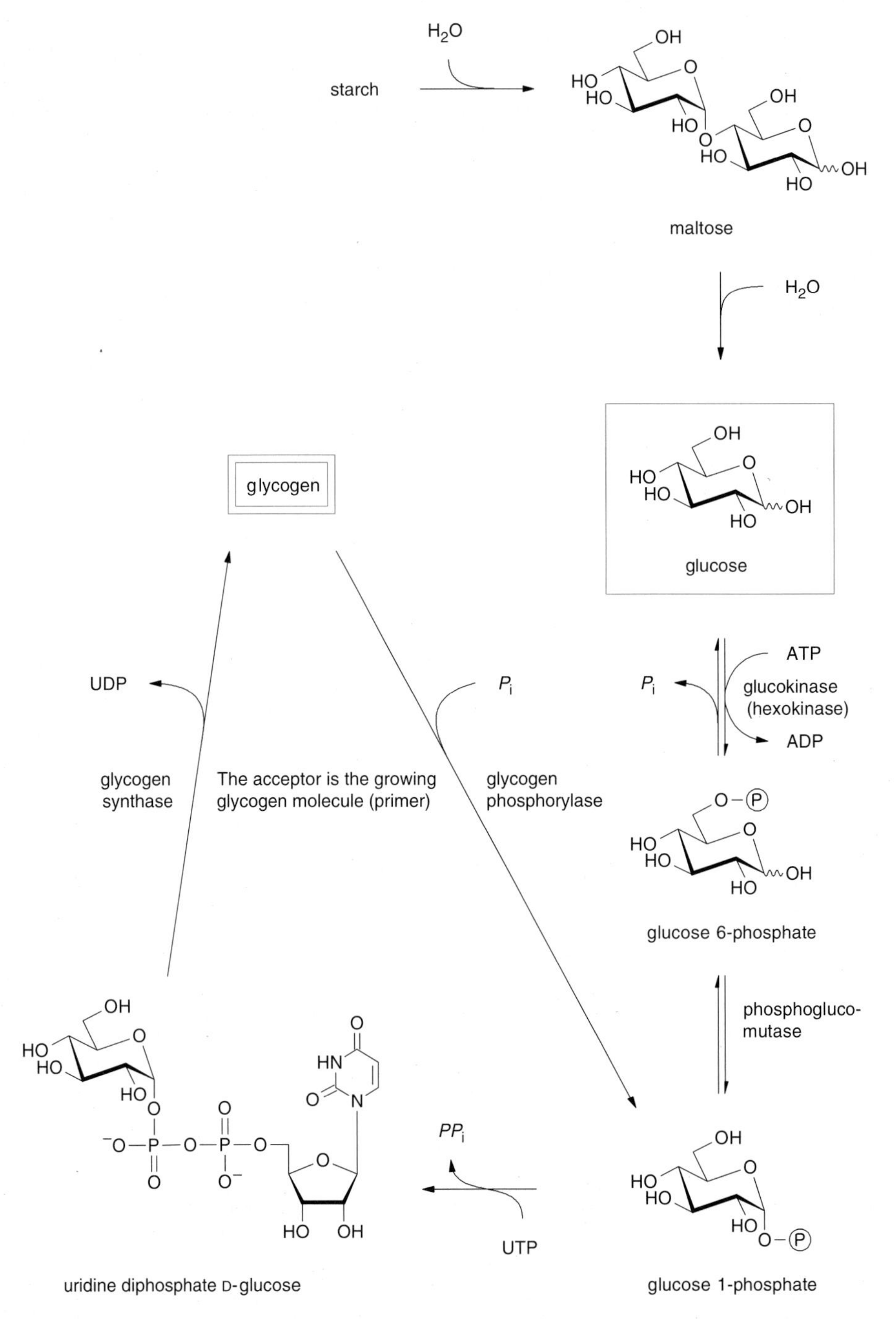


**Fig. 2.137** *Synthesis and degradation of glycogen*. A relationship exists between the two depending on the utilisation of starch from food.

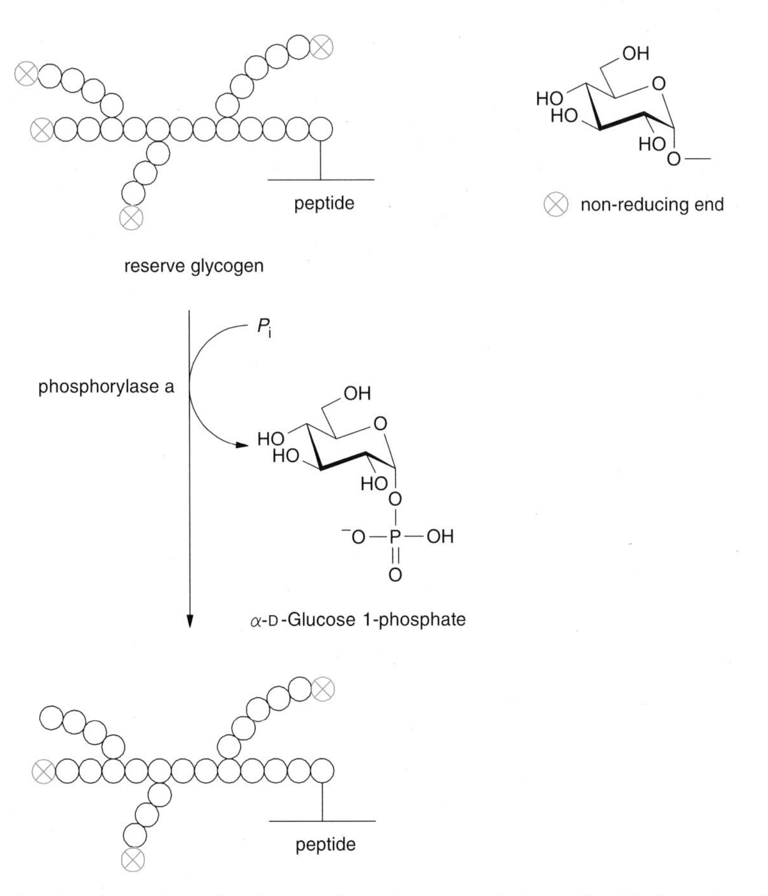


**Fig. 2.138** *Phosphorylase a* degrades glycogen from the non-reducing ends with formation of *glucose 1-phosphate.*

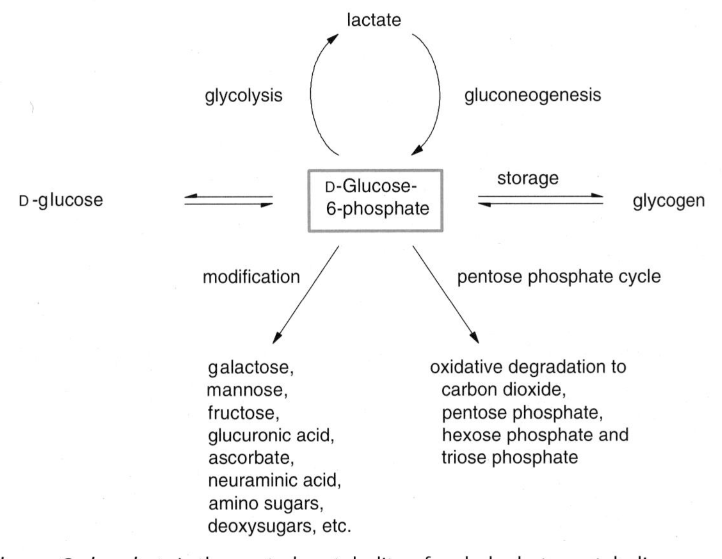


**Fig. 2.139** *Glucose 6-phosphate* is the central metabolite of carbohydrate metabolism.

tion of the phosphorylase a. This reciprocal principal prevents simultaneous synthesis and degradation of glycogen occurring in the same cell with equal intensity.

Besides the activation and deactivation through hormonal regulated enzyme cascades, specific effectors also regulate directly the synthesis and degradation of glycogen through binding to the corresponding enzyme. Thus, AMP (deficiency in ATP) activates significantly the inactive phosphorylase b, and glucose 6-phosphate stimulates the inactive *glycogen synthase b*. The substrate of glycogen synthase is UDP-D-glucose which acts as glucosyl donor for the α-(1→4)-chain extension on a protein bound "primer". The typical, tree-like structure of glycogen is brought about by a *glucano-α-(1-6)-transferase*, which from time to time attaches long maltodextrin chains by transglycosylation and thereby liberates new end groups for chain lengthening by glycogen synthase (Fig. 2.140). In this manner, the tree-like structure of glycogen arises (see section 2.1.3.2, p. 101). The hormones *adrenalin* and *glucagon* stimulate phosphorylase activity and reduce synthase activity. Insulin, on the other hand, promotes glycogen synthesis through activation of glycogen synthase.

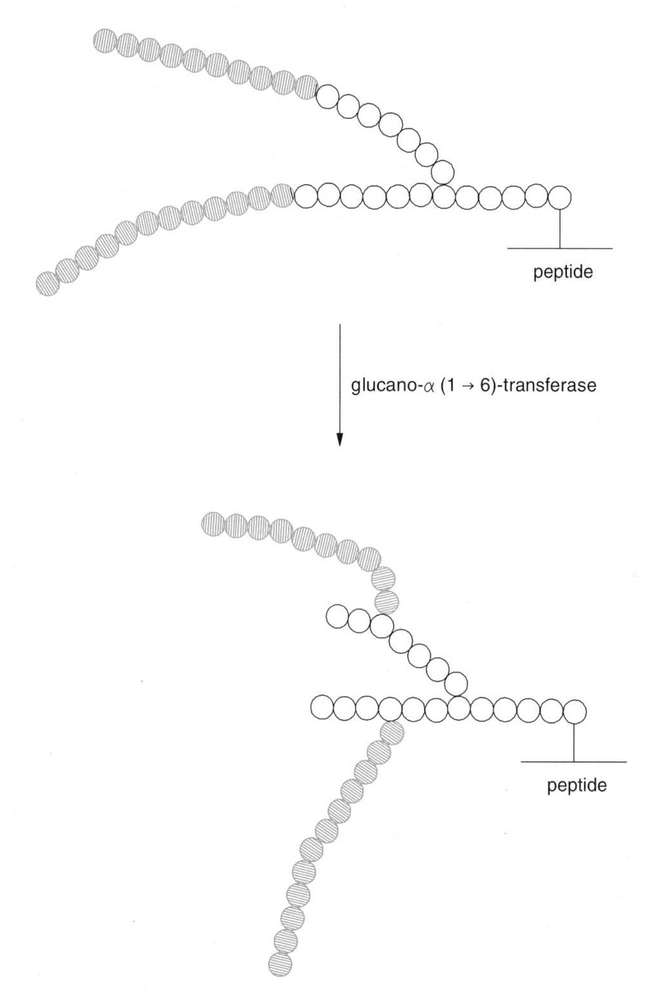


**Fig. 2.140** By the attachment of maltodextrin chains through transglycosylation, new chain ends are produced and, thereby, the degree of branching is increased.

*Starch*, as the reserve polysaccharide of plants, is also degraded by a phosphorylase and built up by a synthase. As glucosyl donor though, plants employ ADP-D-glucose; GDP-D-glucose, however, is used for the synthesis of cellulose and UDP-D-glucose for sucrose synthesis. In this way it is possibile for the regulation of polysaccharide metabolism in plants to act at the stage of the activated sugar.

## 2.3.6 Hexose-pentose interconversions and the Calvin cycle

Besides glycolysis and gluconeogenesis, a further universal combination of reactions exist in which pentoses and hexoses are particularly concerned and whose central part is the so-called pentose phosphate cycle. Closely connected with photosynthetic $CO_2$-fixation, the pentose phosphate cycle has added significance as a producer of NADPH and pentoses. NADPH is necessary for synthetic metabolism (anabolism) and the pentose D-ribose, which only can arise in this way, is used for nucleotide and nucleic acid synthesis.

The pentose-hexose conversions are thus less, as was originally thought, a possibility for the complete combustion of glucose with recovery of energy, but are rather a coordinating point for synthetic metabolism and therewith not competition, but rather a functional supplement to glycolysis. The relatively complex relationships are shown in the following scheme (Fig. 2.141).

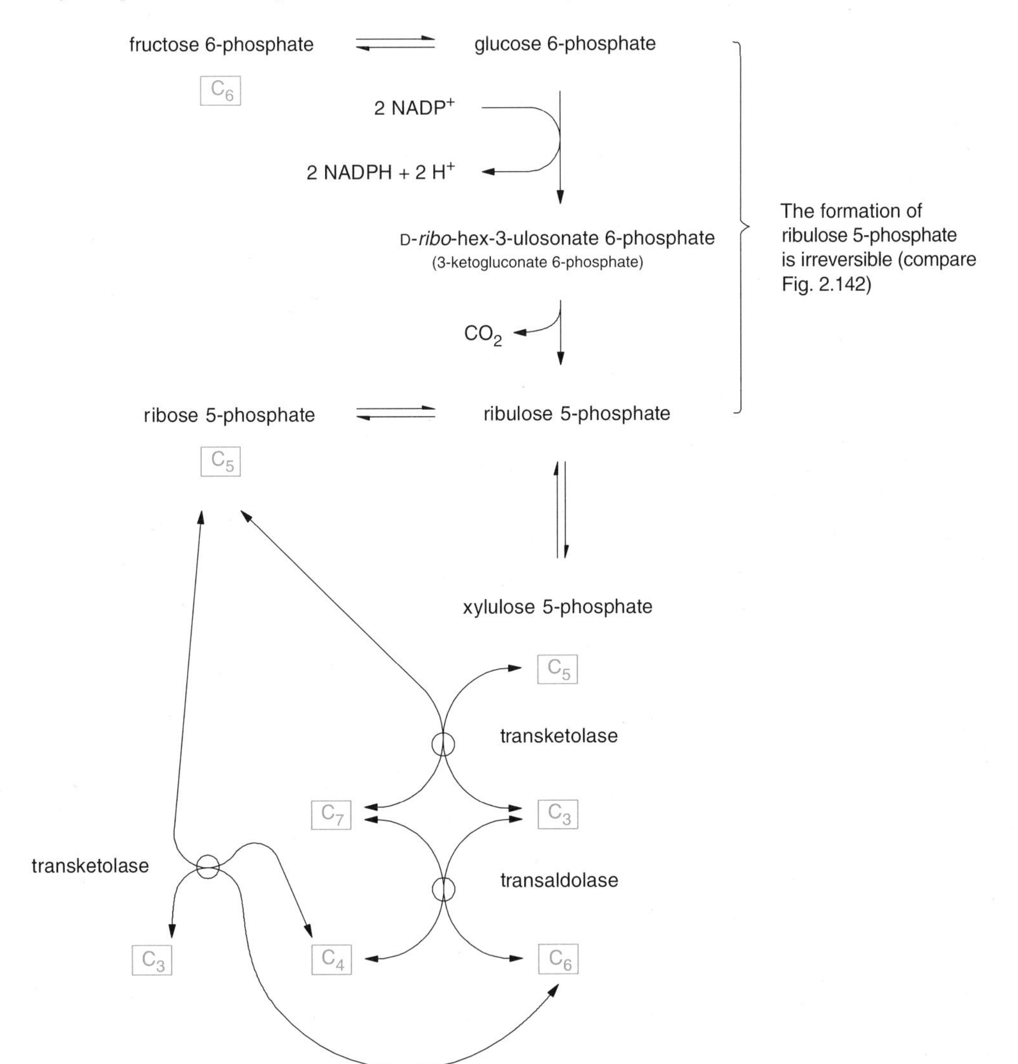


**Fig. 2.141** Mutual transformations in the *pentose phosphate pathway*. The transformations are simplified by the use of number symbols without naming the compounds produced.

D-Glucose 6-phosphate is dehydrogenated with $NADP^+$ to *D-gluconolactone 6-phosphate.* After action of lactonase, *D-ribo-hex-3-ulosonate 6-phosphate* is formed in a second dehydrogenation step with $NADP^+$, and this spontaneously loses carbon dioxide to form D-ribulose 5-phosphate (Fig. 2.142).

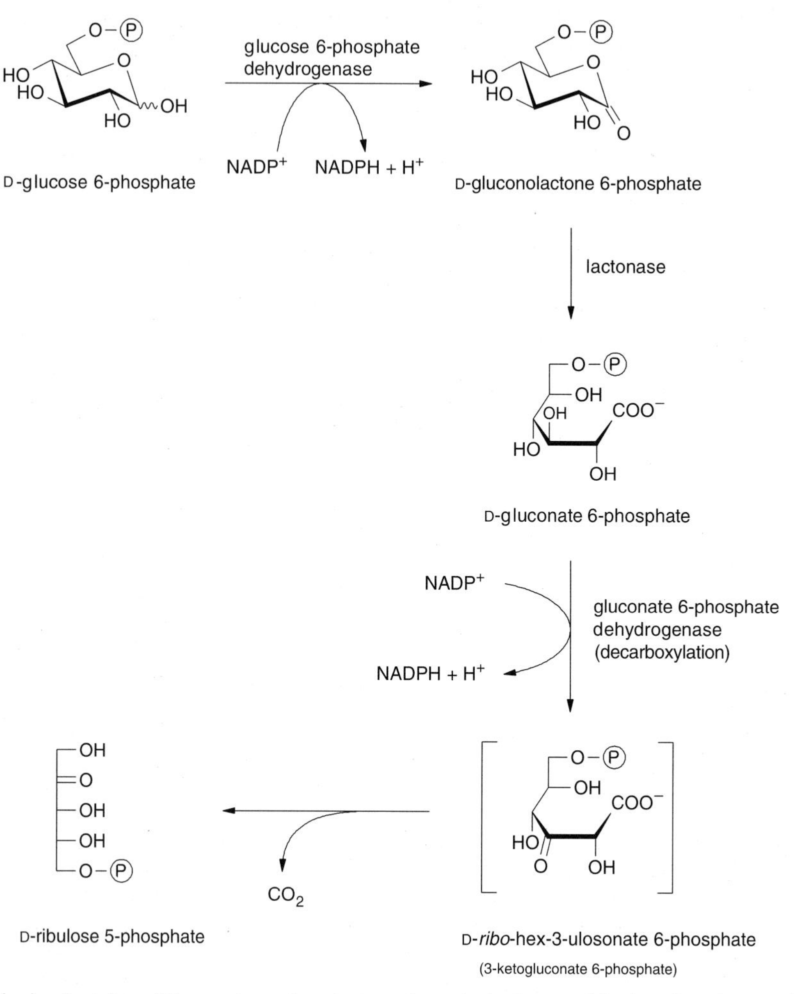


**Fig. 2.142** The beginnining of the *pentose phosphate pathway* is the irreversible decarboxylation of D-*ribo*-hex-3-ulosonate 6-phosphate (3-ketogluconate 6-phosphate).

This first phase of the process, irreversible because of the decarboxylation, is followed by some very interesting and, for the pentose phosphate pathway, typical disproportionation steps, which are catalysed by two enzymes, transketolase and transaldolase. Auxillary enzymes are isomerases which provide for the preparation of necessary substrates. Thus D-ribulose 5-phosphate, after epimerisation to *D-xylulose 5-phosphate* (C5), is subject to a type of retro-aldol reaction through the enzyme transketolase, which contains thiamine pyrophosphate (TTP) as co-factor. This reaction is designated as a transketolase reaction (Fig. 2.143, p. 195).

The C2-fragment remains as the so-called *active* glycolaldehyde, for the time being attached to the cofactor, and never becomes free, while the other product, *D-glyceraldehyde 3-phosphate* (C3), is liberated. The activated glycolaldehyde can now be combined with a second C5 unit after isomerisation of D-ribulose 5-phosphate into *D-ribose 5-phosphate* with the help of an aldol-type reaction to give *sedo-*

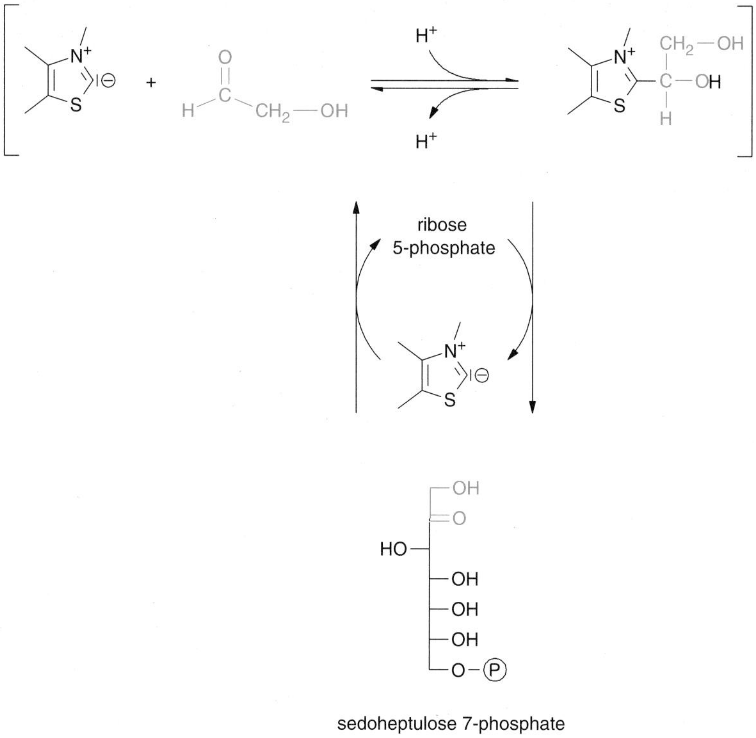

**Fig. 2.143** Reversible formation of *sedoheptulose 7-phosphate* by reaction of the glycolaldehyde, activated with thiamine pyrophosphate (TTP), with ribose 5-phosphate.

*heptulose 7-phosphate* (C7). The same reaction brings about the disproportionation C5 + C4 = C3 + C6, if for example the active glycolaldehyde reacts with *D-erythrose 4-phosphate* (Fig. 2.144, p. 196).

D-Erythrose 4-phosphate arises from sedoheptulose 7-phosphate (C7) with the help of the transaldolase through a *retro-aldol reaction*. The remaining nucleophile (C3) is not set free as such, but is used at once for bond formation through an aldol reaction, for example with D-glyceraldehyde 3-phosphate (C3). In this way, *D-fructose 6-phosphate* (C6) arises according to the scheme C7 + C3 = C4 + C6 (Fig. 2.145, p. 196).

All reaction pathways of these isomerisations and disproportionations are reversible, so that all compounds of the general metabolism remain constantly available, according to demand. Requirements exist, as already mentioned, above all for ribose and NADPH. The balance of the pentose phosphate pathway,

$$\text{D-Glucose 6-phosphate} + H_2O + 2\ NADP^+ \rightarrow \text{D-Ribose 5-phosphate} + CO_2 + 2\ NADPH + 2\ H^+$$

emphasises the significance of these important metabolic reactions. Except for D-ribose, the pentoses in mammals are almost without importance; however, some very interesting mutual transformations of monosaccharides of this type are mentioned here, especially as pentoses can be constituents of food.

The transformations all occur through reduction-oxidation. The two pentoses D-xylulose and L-xylulose come from two different sources. The former, as the 5-phosphate, is a component of the pentose phosphate pathway, into which the sugar may be again incorporated through action of a kinase; the latter arises from D-glucuronate metabolism. Through reduction-oxidation, both of them, because of their special stereochemistry, are connected not only with one another, but also with D-xylose (Fig. 2.146, p. 197).

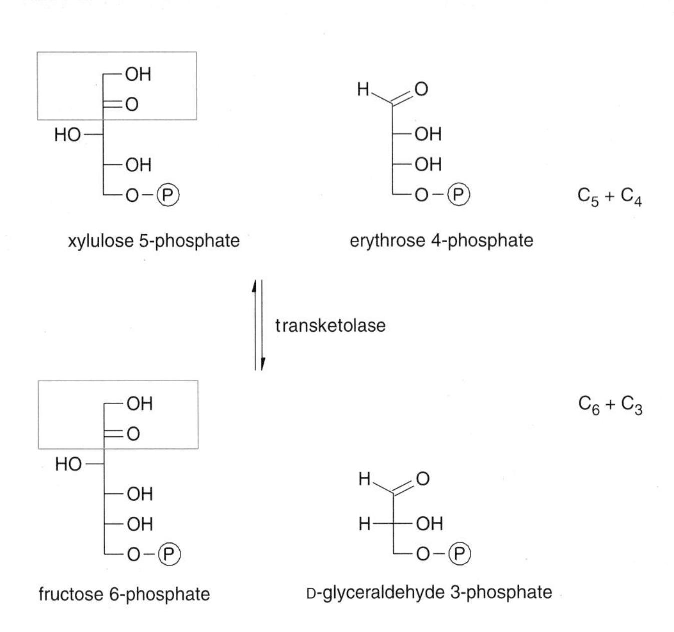


**Fig. 2.144** Disproportionation of the monosaccharide framework catalysed by transketolase.

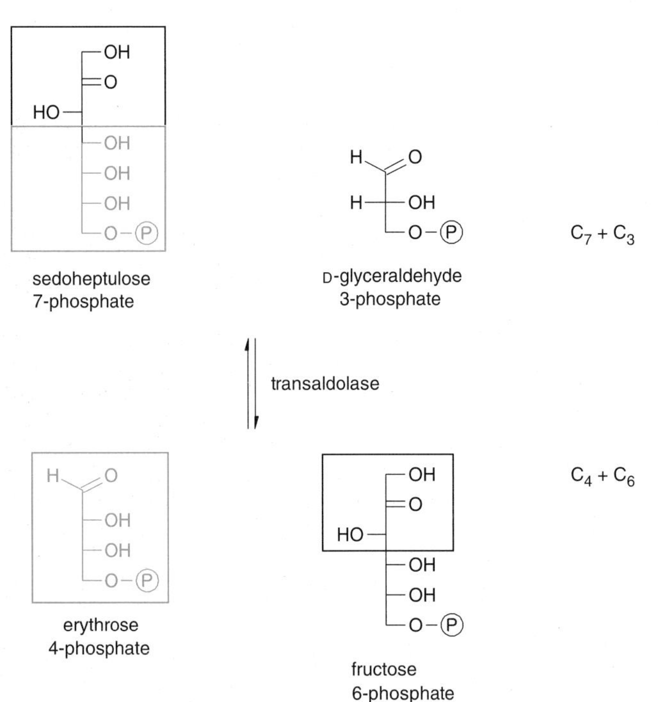


**Fig. 2.145** Disproportionation of the monosaccharide framework catalysed by transaldolase.

Since L-arabinose may be converted into *L-xylulose* via their reduction product L-arabinitol, the last of the pentoses occurring in nature (lyxose does not occur in nature) is consequently also made accessible in general metabolism.

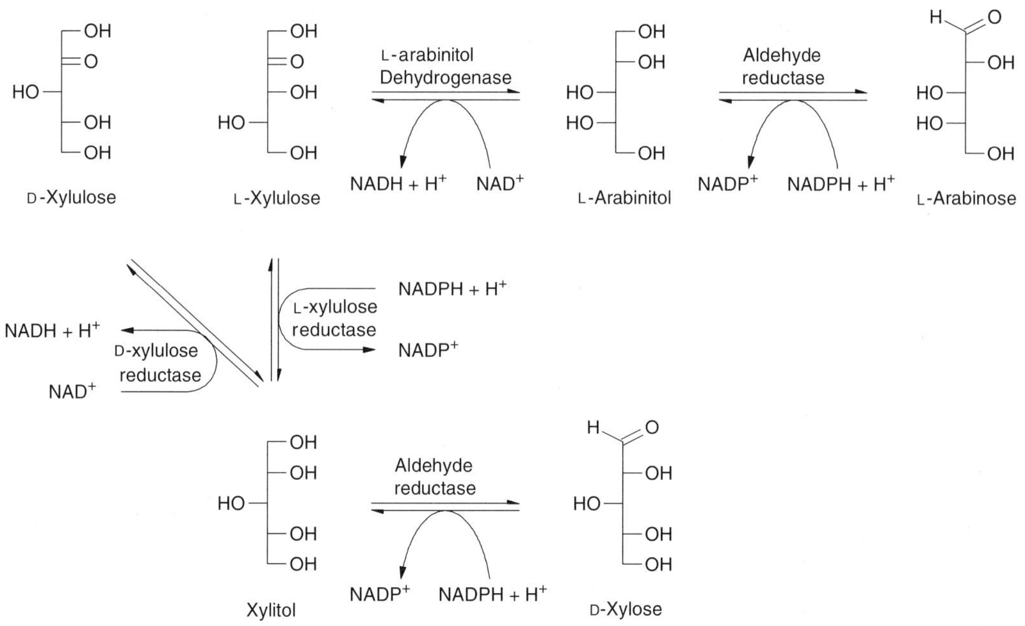

**Fig. 2.146** Isomerisation of the pentoses and pentuloses takes place by reduction-oxidation sequences.

The *de novo* biosynthesis of organic materials, as occurs in the so-called carbon dioxide-assimilation or photosynthesis in green plants, can be seen perhaps as the most important biochemical single process, because without it there would be no living things as we know them on the earth.

Photosynthesis consists of an extraordinarily complicated biophysical part, which permits the transformation of light energy into chemical energy, that is ATP (photophosphorylation) as well as the conversion of the protons of water into hydride ions of NADPH, and a biochemical part, which comprises the cyclic process of the conversion of carbon dioxide into monosaccharides. The two parts are called accordingly *light reaction* and *dark reaction*. Only the dark reaction is discussed here.

As in many biologically synthetic processes, nature makes use of a reaction cycle for the reduction of carbon dioxide to carbohydrate which, for biochemical reactions, begins with a very unusual transformation. *D-Ribulose 1,5-diphosphate* reacts as a nucleophilic enediol with carbon dioxide and forms a hypothetical 3-ketocarboxylic acid. This undergoes a hydrolytic cleavage into two moles of *3-phosphoglycerate*, a reaction which corresponds to the *acid cleavage* of 3-ketoesters. The 3-phosphoglycerate is reduced with NADPH with consumption of ATP to D-glyceraldehyde 3-phosphate (Fig. 2.147, p. 198).

The latter is imported into the already discussed pentose phosphate pathway from which arise hexoses, that is to say starch or sucrose, in balance as photosynthesis products, and simultaneously the acceptor for carbon dioxide, D-ribulose 1,5-diphosphate, is again formed (Fig. 2.148, p. 198).

This cyclic metabolic sequence of events was clarified by an American chemist Melvin Calvin and his research group, and for this reason is generally called[161] the *Calvin Cycle*.

## 2.3.7 Biochemical modification of monosaccharides

A great part of all condensed carbohydrates is built up out of the few, neutral monosaccharides, which arise directly out of the main metabolism. However, many also contain monosaccharides or monosaccharide derivatives with modified structures. Above all, the 2-amino-2-deoxyhexoses, the 6-deoxyhexoses and the uronic acids occur in considerable amounts as components of glycosides, polysaccharides, and glyconjugates. On the other hand, many monosaccharides can be described as relatively rare. They are not generally widespread and are often typical for only one type of organism. Microorganisms and fungi especially and also plants have proved to be an almost inexhaustible source of such sub-

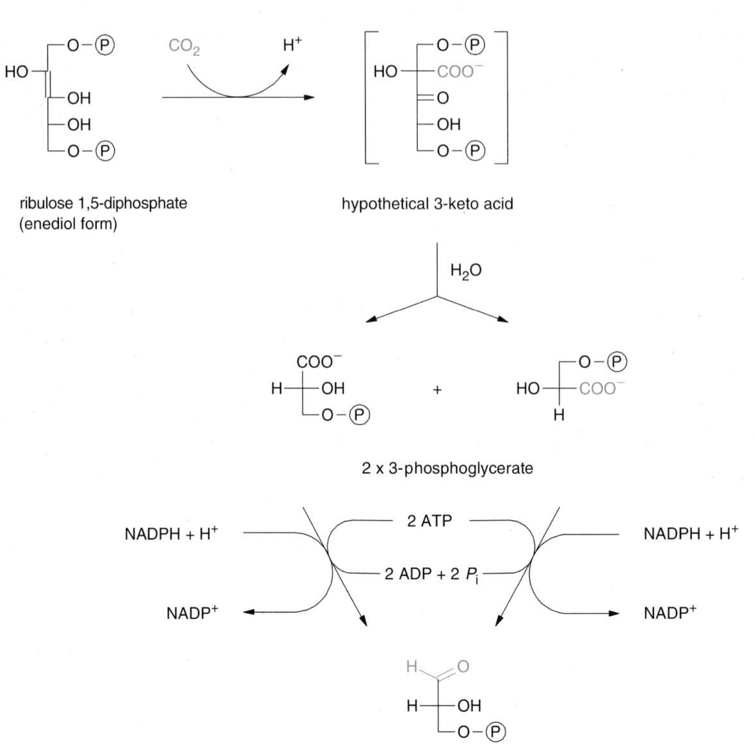


**Fig. 2.147** The *dark reaction* of the photosynthetic process.

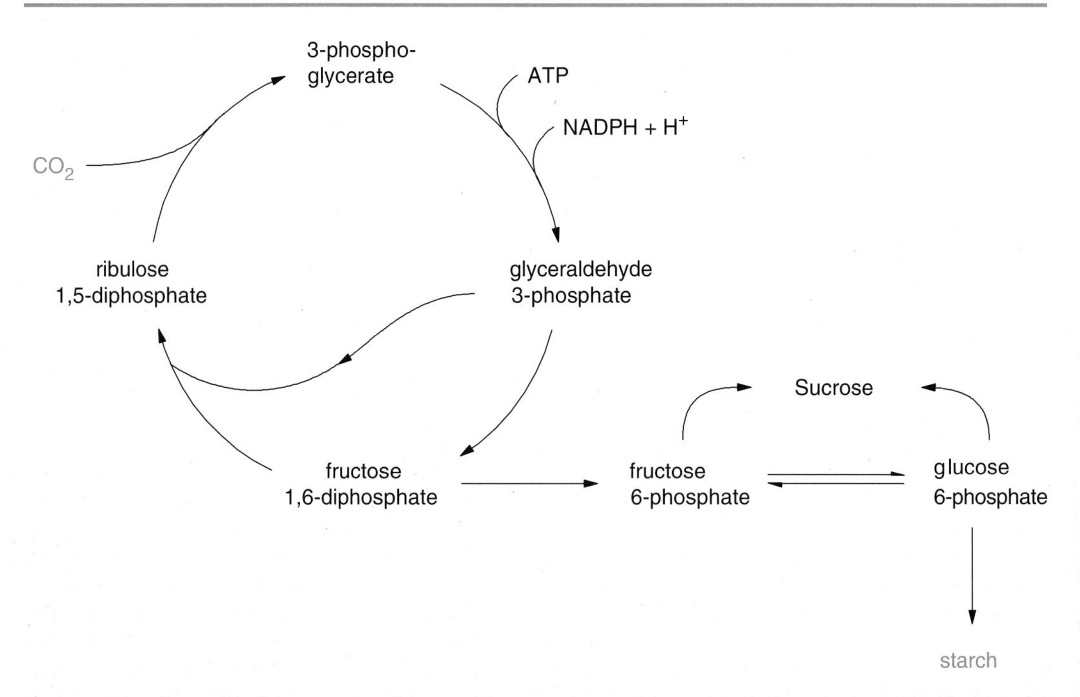


**Fig. 2.148** Relationship between $CO_2$ fixation (photosynthetic dark reaction), the pentose phosphate cycle, and the *synthesis* of *starch* and *sucrose.*

stances, with the improvement in separation and analytical methods. In the following section many of the biochemical transformations of monosaccharide frameworks are shown, which can lead to a wealth of compounds, which no longer conform to the general overall formula for carbohydrates of $C_nH_{2n}O_n$. The conversions can be classified under just a few reaction types, oxidation, reduction, substitution, alkylation, and acylation.

### 2.3.7.1 Oxidation

**Glyconic and glycuronic acids.** We have seen in earlier sections that many degradative metabolic pathways begin with the oxidation of monosaccharides either to *glyconic acids* or to *glycuronic acids*. Such acidic compounds are also components of numerous polysaccharides and glycoconjugates. Both D-glucuronic acid and D-galacturonic acid are biosynthesised as a UDP derivative from UDP-D-glucose – first of all by dehydrogenation, then in the case of D-galactonuronic acid through 4-epimerisation (Fig. 2.149) – and then are built into the corresponding polysaccharide by transglycosylation.

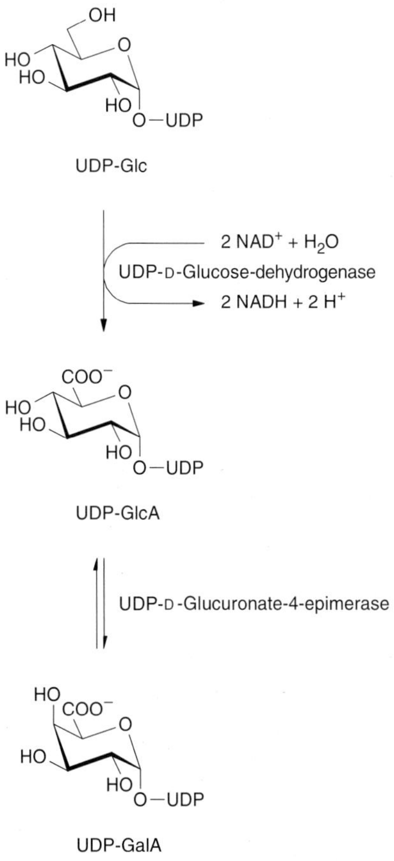

UDP-Glc

2 $NAD^+$ + $H_2O$
UDP-D-Glucose-dehydrogenase
2 NADH + 2 $H^+$

UDP-GlcA

UDP-D-Glucuronate-4-epimerase

UDP-GalA

**Fig. 2.149** The conversion of UDP-D-glucose into *UDP-D-uronic acids*.

D-Glucuronic acid is found especially in the glycosaminoglycans hyaluronic acid and chondroitin sulphate of connective tissue, and D-galacturonic acid or its methyl ester in plant polysaccharides such as alginic acids and pectins. Catalysed by UDP-D-glucuronyl transferase, UDP-D-glucuronic acid reacts with numerous alcoholic or phenolic acceptors and also with aromatic or branched chain aliphatic carboxylic acids, with formation of *β-D-glucuronides*. Conjugation with glucuronic acid is an important process occurring in the liver for detoxication of the organism. The highly water-soluble glucuronides, always having a *β*-configuration, are excreted in the urine. Of the glyconic acids of the naturally occurring aldohexoses, only gluconic acid (which as the 6-phosphate is an important product of carbohydrate metabolism) is found in nature. It should be noted that D-gluconic acid lacks the hemiacetal group necessary for the formation of conjugates.

5-Acetamido-3,5-dideoxy-D-*glycero*-D-*galacto*-non-2-ulosonic acid (*N*-acetylneuraminic acid – NeuAc) and 3-deoxy-D-*manno*-oct-2-ulosonic acid (*3-deoxy-2-keto-D-manno-octonic acid* – Kdo) possess both the carboxy group of a glyconic acid, and also the carbonyl group of a monosaccharide. However, neither compound is obtained through biochemical oxidation but both are formed through an aldol reaction between pyruvic acid (2-oxopropanoic acid) and an aldose.

Neuraminic acid, always as *N*-acylneuraminc acid, occurs in almost all glycoconjugates of mammals and is widely distributed in the animal kingdom and also in microorganisms . Its biological function is still not today unequivocally clear. It can be concluded that both the position as peripheral end group in glyconjugates as well as the free carboxyl group are significant for its function. Neuraminic acid, besides 3-deoxy-D-*manno*-oct-2-ulosonic acid (Kdo) which is found only in bacteria, is the only glyconic acid not limited only to its role in metabolism. Its carbon skeleton arises through an aldol reaction between *N*-acetyl-D-mannosamine 6-phosphate and phosphoenolpyruvate, catalysed by *N*-acetylneuraminate 9-phosphate synthase (Fig. 2.150).

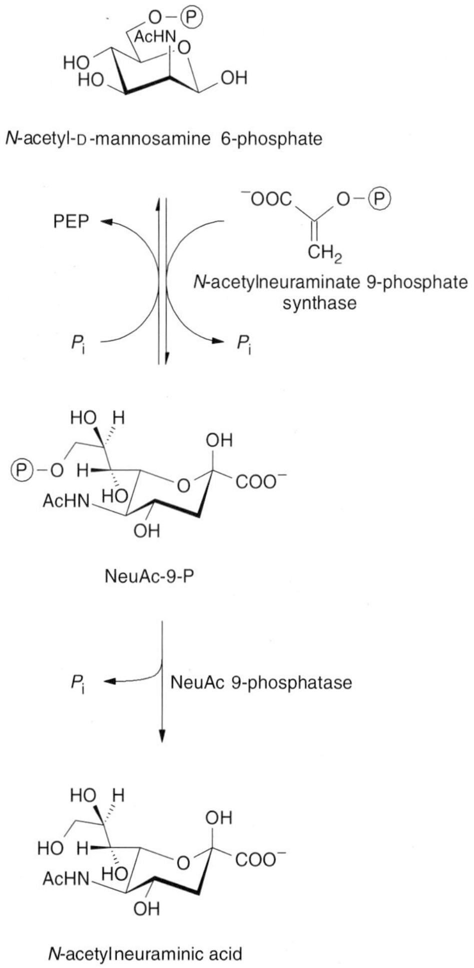

*N*-acetyl-D-mannosamine 6-phosphate

PEP

$^{-}$OOC, O–Ⓟ, $CH_2$

*N*-acetylneuraminate 9-phosphate synthase

$P_i$

NeuAc-9-P

NeuAc 9-phosphatase

*N*-acetylneuraminic acid

**Fig. 2.150** *N-Acetylneuraminic acid* arises through an aldol-type reaction from ManNAc-6-P and PEP. The dephosphorylation of NeuAc-9-P is responsible for displacement of the equilibrium.

After removal of the phosphate group by a specific phosphatase, activation takes place with CTP and CMP-sialate synthase. It should be noted that the notation *sialic acids* for various *O*- or *N*-acylated neuraminic acids is customary. It has been suggested that the different acylation of neuraminic acids also has a biological purpose. In nature, neuraminic acid occurs neither as the free sugar nor with a free amino group.

Transfer of the sialyl residue to a specific glycosyl residue in a glycoconjugate takes place through very specific sialyl transferases. Numerous oligosaccharides in the colostrum also contain *N*-acetylneuraminic acid. The biological significance for this is unclear, as is the occurrence of a homopolysaccharide from NeuAc, the *colominic acid* found in the cell wall of some bacteria.

Kdo is a normal component of the core region of bacterial lipopolysaccharides. As with NeuAc, Kdo derives its carboxyl group from pyruvic acid, which reacts as phosphoenolpyruvate with D-arabinose 5-phosphate. In bacterial cell walls, still additional 3-deoxy-aldulosonic acids have been found[162].

### 2.3.7.2 Reduction

**Alditols and anhydroalditols (C-glycosides)** do not possess the carbonyl, hemiacetal or acetal group which are, in general, typical for carbohydrates. However, as reduction products of monosaccharides, they are both structurally and also biogenetically closely related to them. The cyclitols, a widespread class of substances found mainly in plants, are polyfunctional cyclohexanes and their derivatives. Their biogenesis is discussed in section 2.2.1.3 (p. 132).

It is certain that most sugar alcohols found in organisms are chemically resistant transport and storage forms of reducing sugars. This applies especially for the hexitols *sorbitol* (*D-glucitol*), *D-mannitol*, and *galactitol* (*dulcitol*). Higher sugar alcohols from plants are also known.

The $NADP^+$ dependent reduction of hexoses to hexitols through dehydrogenases in the lens and in the vitreous humor of the eye plays a special role. High glucose levels in the blood, for example in the case of diabetes, lead to increased formation of sorbitol and thereby to precipitation and to corresponding turbidity of the lens. The hexitols, through $NAD^+$-dependent dehydrogenation, can be converted into ketoses and consequently brought back again into general metabolism.

The so-called *C-glycosides*, better named anhydroalditols, are a widespread and, with respect to the aglycone, very heterogeneous class of substances. The description C-glycoside is misleading, since in general glycosides should be cleaved through hydrolysis, which C-glycosides are not. Their biogenesis, however, suggests a comparison with the normal glycosides Today, it appears certain that compounds with a D-glucopyranosyl residue attached to a C-atom arise, as normal *O*-glycosides, through a transfer reaction, for example from UDP-D-glucose, to a corresponding acceptor (Fig. 2.151). Nearly all naturally occurring C-glycosides carry aglyconic residues with nucleophilic functional groups, so that transfer onto a C-atom is not explicable by the absence of better alternatives.

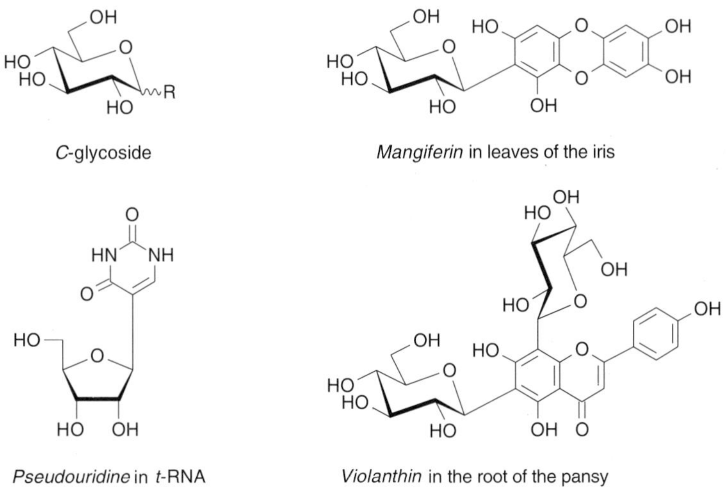

*C*-glycoside

*Mangiferin* in leaves of the iris

*Pseudouridine* in *t*-RNA

*Violanthin* in the root of the pansy

**Fig. 2.151** *C-Glycosides* occur relatively frequently in nature. They arise by glycosylation of electron-rich aromatic compounds.

### 2.3.7.3 Substitution

Very often in nature monosaccharides are found in which one or more hydroxy groups are missing, replaced by amino groups, hydrogen atoms, or alkyl residues. In a formal chemical sense, their formation could be regarded as a substitution, although biochemically such a direct exchange does not take place. In the case of branched chain sugars, a hydrogen atom can also be replaced by an alkyl residue, in which case the hydroxy group remains.

■ **Aminosugars** like the uronic acids, are important components of polysaccharides and glycoconjugates. Reference to aminosugars generally implies the 2-amino-2-deoxy-hexoses *D-glucosamine*, *D-galactosamine*, and *D-mannosamine*, which are present mostly in the *N*-acylated form and most frequently in the *N*-acetylated form. The *N*-acetyl groups are ascribed a special function, the building of stable hydrogen bridges between glycoconjugates and proteins.

All 2-amino-2-deoxy-hexoses arise from D-fructose 6-phosphate in a transamination reaction with glutamine. It is assumed that the initially formed imine undergoes an *Amadori rearrangement* into D-glucosamine 6-phosphate (Fig. 2.152). *N*-Acetylation then occurs with **acetyl-*S*-CoA**. *N*-Acetyl-D-

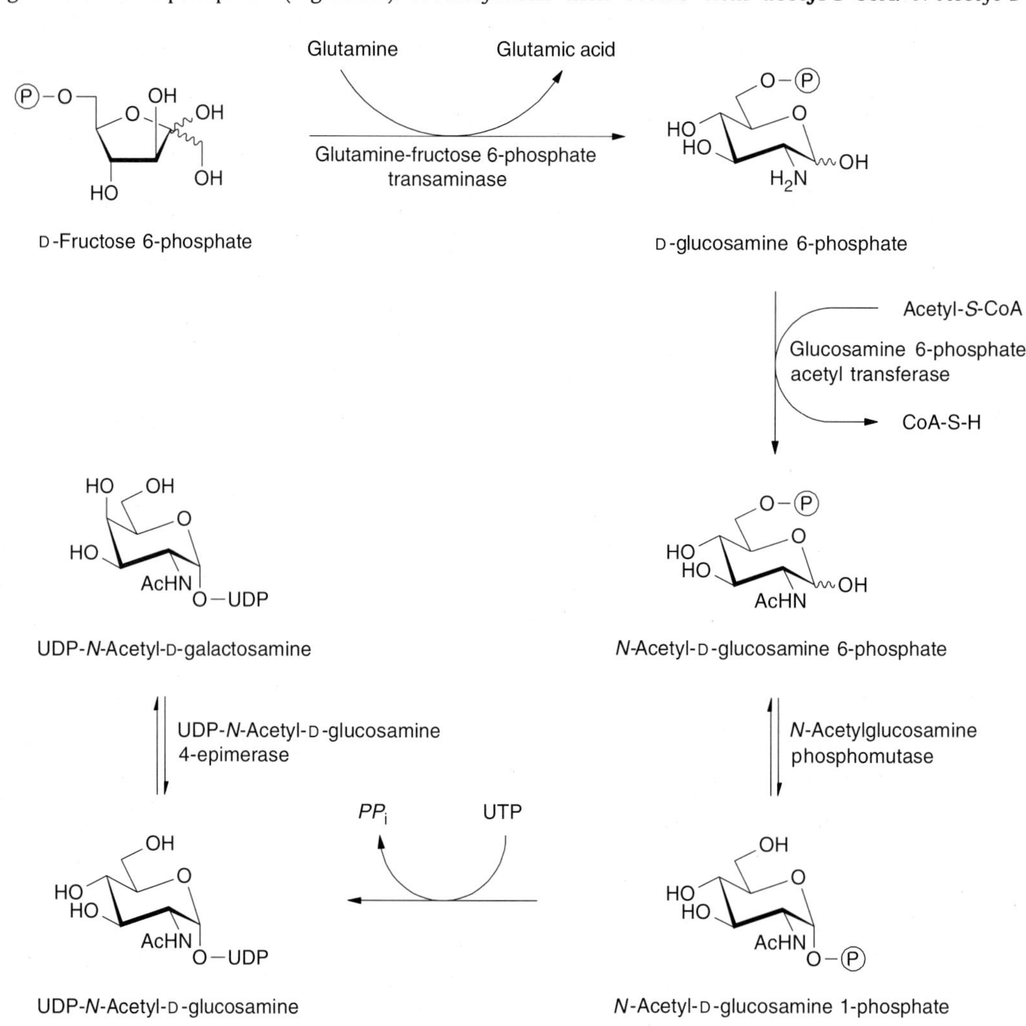


**Fig. 2.152** *Biosynthesis* of *N-acetyl-D-glucosamine* and *N-acetyl-D-galactosamine* from D-fructose 6-phosphate occurs through an Amadori rearrangement.

glucosamine 6-phosphate is then transformed in the known phosphomutase reaction (see section 2.3.2, p. 174) into the 1-phosphate and this is activated with UTP.

UDP-*N*-Acetyl-D-glucosamine is converted with the aid of the corresponding 4-epimerase into *UDP-N-acetyl-D-galactosamine* (Fig. 2.152). Both activated sugars are then ready for transglycosylation. 2-Epimerisation of the *N*-acetyl-D-glucosamine occurs from UDP-GlcNAc with loss of UDP. The *N*-acetyl-D-mannosamine so-produced is needed by the organism exclusively for the biosynthesis of *N*-acetylneuraminic acid. ManNAc itself is not found in nature, either free or conjugated. An abundance of amino-deoxy monosaccharides have been isolated in recent years in the search for antibiotically effective substances from different microorganisms. In contrast to the generally widespread glycosamines, they carry mostly a non-acetylated amino or alkylamino group in random positions of the monosaccharide skeleton and in general occur only as low molecular weight glycosides. The biosynthesis is known only in a few cases. Many of these strong bases are extraordinarily effective competitive inhibitors of glycosidases (see section 2.3.8.8, p. 230).

■ **Deoxysugars.** Besides the special case of 2-deoxy-D-ribose of DNA, which will not be considered here, it is primarily *L-fucose* (6-deoxy-L-galactose) as a component of glycoconjugates and polysaccharides in the animal and plant kingdom as well as *L-rhamnose* as a component of many plant glycosides and polysaccharides whose biogenesis has been examined and elucidated. It may be assumed that the numerous other naturally occurring deoxy sugars – the lipopolysaccharides of bacteria are a rich source of all types of deoxy sugars – arise in a similar manner by similar mechanisms.

L-Fucose is formed as the GDP derivative from the corresponding derivative of D-mannose. Similar to the cycloaldolase reaction, the activation of the substrate begins with a dehydrogenation brought about in the 4-position by enzyme bound NAD⁺. The carbonyl group facilitates the dehydration of the substrate to the non-isolable 5,6-unsaturated intermediate (Fig. 2.153).

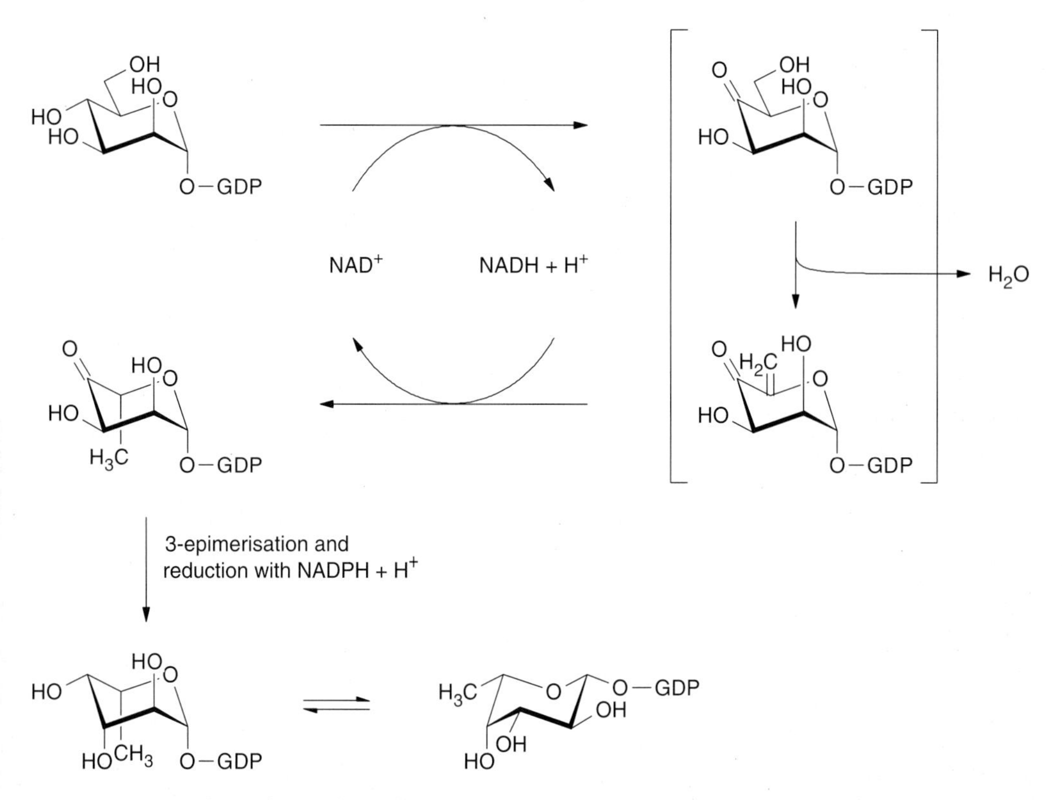


**Fig. 2.153** *Biosynthesis* of *GDP-L-fucose* from GDP-D-mannose.

The C,C-double bond is hydrogenated stereospecifically with the originally formed NADH and finally the process, after epimerisation at the 3-position, is fixed and concluded through the in turn stereospecific reduction of the carbonyl group with NADPH (Fig. 2.153). By this principle it is also possible to replace an hydroxy group by an H-atom in other positions of the monosaccharide skeleton.

**Branched chain sugars.** Usually, monosaccharides have an unbranched skeleton. Exceptions to this rule are *apiose* and *hamamelose*. Both discovered as glycosides in certain plants, they were still in the 1940's the sole members of their class. The number of branched chain sugars increased rapidly with the investigation of antibiotically active substances from microorganisms. Apiose also occurs, as is now known, in numerous plants as a constituent of polysaccharides.

The biogenesis of chain branching occurs in general at the stage of the nucleoside diphosphomonosaccharide, either through rearrangement or through alkylation, corresponding to activated early stages. The biosynthesis of D-apiose from UDP-D-glucuronic acid as precursor is now well investigated and understood.

Dehydrogenation at the 4-position by $NAD^+$ is followed by decarboxylation and ring contraction. The rearrangement is concluded by reduction of the branched chain intermediate with the NADH formed at the start (Fig. 2.154). Methyl groups are found very frequently in branched chain monosaccharides. The

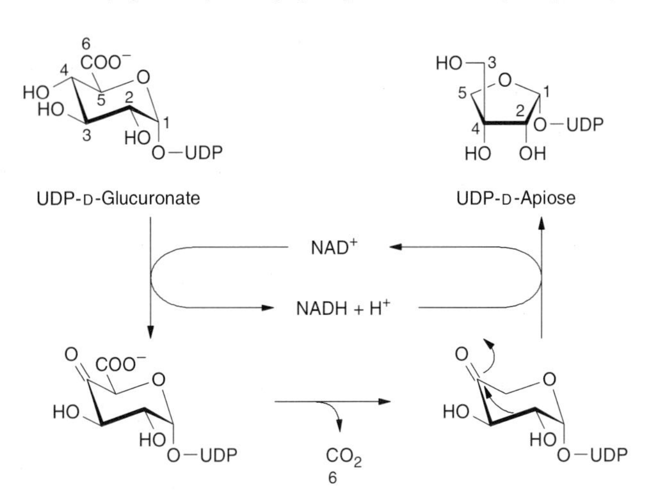


**Fig. 2.154** *Biosynthesis* of *UDP-D-apiose* from UDP-D-glucuronic acid.

methyl group, as far as yet investigated, is transferred from *S*-adenosylmethionine onto a nucleoside diphosphosugar, activated through $NAD^+$ dehydrogenation. Alkylation then occurs at the α-position to the carbonyl group, very likely through an enolate. Often epimerisation accompanies this reaction and, as in the biosynthesis of *nucleoside diphospho-L-mycarose* from nucleoside diphospho-D-glucose (Fig. 2.155, p. 205), dehydration and reduction of the C,C-double bond by NADPH (see Fig. 2.153). Also here the reduction is completed with NADH formed at the outset of the reaction sequence.

**Sugar sulphonic acids.** In the chloroplast membranes (less in other tissues) of all photosynthesising plants and correspondingly widely distributed, is found a glycolipid whose sugar component carries a C-sulphonyl group. This functional group, extremely rare in nature, makes the compound *6-deoxy-6-sulpho-D-glucose* or its di-*O*-acylglyceryl-α-D-glycoside one of the strongest acids occurring in nature. Although there is a range of hypotheses about the function of the lipid in photosynthesising tissues, reliable information is missing. Further, to this day, the biosynthesis is not proven beyond doubt. Certainly, independent evidence makes the biosynthetic pathway shown (Fig. 2.156) seem likely. The synthesis of the natural sulpholipid was achieved with a nucleoside diphospho derivative of the 6-deoxy-6-sulpho-D-glucose and an enzyme preparation and thereby an indication was obtained that the suggested synthetic scheme was correct.

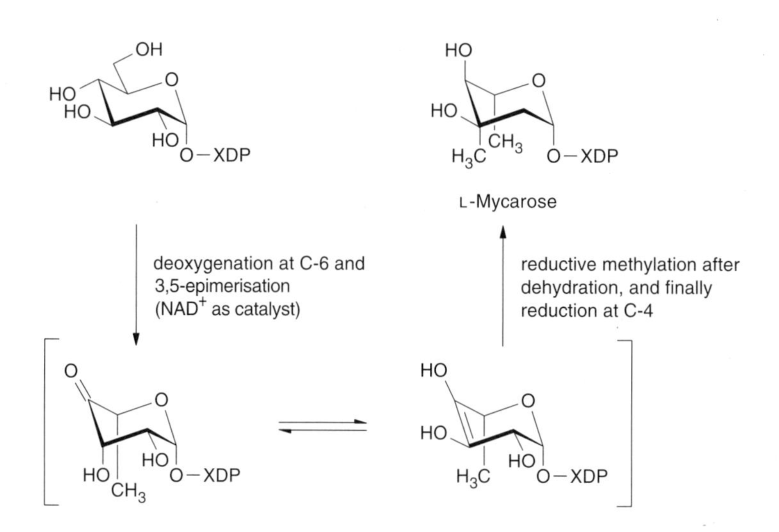


**Fig. 2.155** *Biosynthesis* of *L-mycarose* from D-glucose. Methylation of an enediol intermediate at the 3-position is brought about by *S*-adenosylmethionine (compare Fig. 2.153).

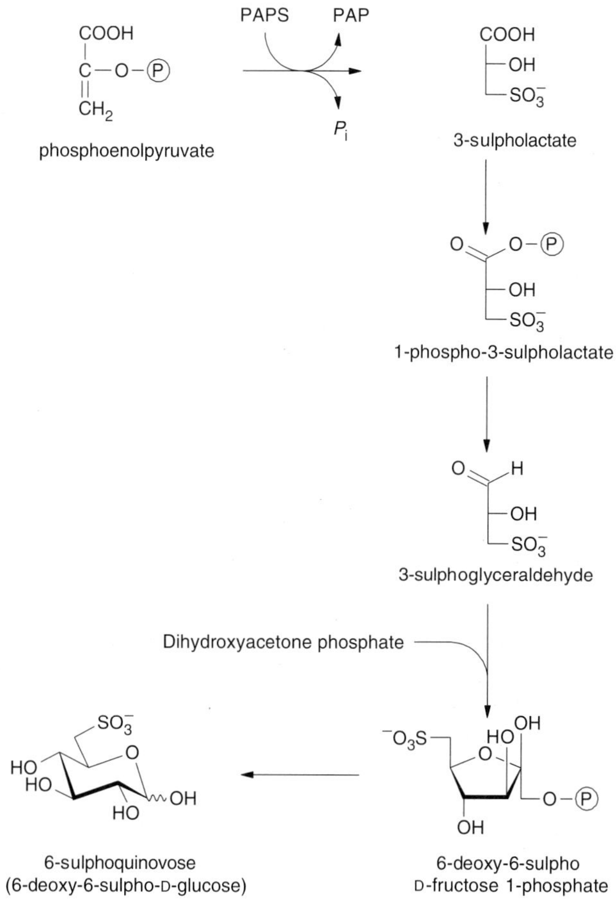

PAPS PAP

COOH
C–O–(P)
$CH_2$

phosphoenolpyruvate

$P_i$

COOH
OH
$SO_3^-$

3-sulpholactate

O O–(P)
OH
$SO_3^-$

1-phospho-3-sulpholactate

O H
OH
$SO_3^-$

3-sulphoglyceraldehyde

Dihydroxyacetone phosphate

$^-O_3S$ HO OH O O–(P) OH

6-deoxy-6-sulpho
D-fructose 1-phosphate

$SO_3^-$ O HO HO OH HO

6-sulphoquinovose
(6-deoxy-6-sulpho-D-glucose)

**Fig. 2.156** Biosynthesis pathway for the formation of *6-sulphoquinovose*, one of the strongest acids occurring in nature.

### 2.3.7.4 Alkylation, acetalation and acylation

In all living things, the post-biosynthetic modification of carbohydrates is a means, after successful biosynthesis, to bring about structural changes, mostly in completed polysaccharides or glycoconjugates. For what purpose these changes are intended is in many cases unknown at the present time and is subject only to speculation. In bacteria, in whose cell surface polysaccharides the most modified monosaccharides probably occur[3], the modification is best explained as an easily obtained camouflage effect against a hostile outside world. In the case of higher organisms and with mammals, the chemical modifications of the sugars are mostly confined to the sulphation of the polysaccharides and it is very difficult to judge how far these post-biosynthetic modifications occur in a strictly regiospecific manner.

Principally, the modification is performed at three stages of the biogenesis of a polysaccharide or glycoconjugate: before the activation of the monosaccharide, at the stage of the activated sugar and also of activated oligosaccharide, and after the polysaccharide or glycoconjugate is complete. It is not always possible to find the right order. *O-Methylations*, so far as have been demonstrated, are all carried out through the universal co-factor *S*-adenosylmethionine (SAM). Few examples of the related *N-methylation* of aminosugars are known. For *O*-methylation there is apparently no preferred position, since methylations have been observed on all available hydroxy groups and also multiple *O*-methylation of a monosaccharide unit[163].

The alkylation of hydroxy groups of a monosaccharide unit with a lactic acid residue has functional importance in bacteria. The introduction of a lactyl residue *[O-(1-carboxyethyl)]* in the 3-position of a GlcNAc residue giving *N-acetylmuramic acid* is the essential prerequisite for the cross-linking of murein through short peptide sequences (see section 2.1.2.1, p. 73). However, the lactyl residue plays a role not only in this connection. Some capsular polysaccharides of bacteria contain glucose or mannose residues which carry the lactyl residue in the 4- or in the 3-position, with both the (*R*)- and (*S*)-configuration.

Alkylation with the lactyl residue always occurs with phosphenolpyruvate, whose phosphate residue is substituted by a hydroxy group of the sugar. Reduction of the C,C-double bond with NADPH affords specifically the (*R*)- or (*S*)-*O*-(1-carboxyethyl) group. Through a change in the reaction sequence, acetalisation can take place if a sterically favourably-oriented hydroxy group adds to the intermediate enol ether. The cyclic acetals of pyruvic acid (2-oxopropanoic acid) are found in the most varied, terminal monosaccharides of numerous bacterial polysaccharides.

*Acylation*, above all *acetylation* but also glycoylation and *formylation*, is found very often as a modification at the stage of the completed polysaccharide or glycoconjugate. As far as acetyl groups are concerned, all reactions must have taken place with *acetyl-S-coenzyme A* and with more or less specific transacetylases. The *O*-acetylation of different hydroxy groups found relatively frequently on the peripheral neuraminic acid in glycoconjugates of higher animals is sometimes associated with the specific recognition of such structural elements in cell-cell communication.

The glycosaminoglycans of mammals are in part highly sulphated, and in them the sulphate groups are linked to an O- and also, less often, to an N-atom. Sulphation occurs always with phosphoadenosine phosphosulphate (PAPS) in the already biosynthesised polysaccharide and, so far as can be proved, after an acylation pattern typical for each polysaccharide. The sulphate residues are especially important for the function of the glycosaminoglycan concerned. On the one hand these are aimed at specific recognition and on the other, for the general function of binding water in the connective tissue. With a deficiency of PAPS, insufficient sulphation leads to a drastic diminution in the volume of the connective tissue and with it to the stunting of the limbs in the case of mice. It is very possible that sulphate groups, which are also found in plant polysaccharides, have as their main function the general binding of a hydration sheath. With the very heterodisperse structure of such sulphated polysaccharides, it is just difficult to accept that sulphation occurs here according to a programmed plan. Unfortunately, there are hardly any unequivocal research results on this subject.

## 2.3.8 Biosynthesis of condensed, high molecular weight carbohydrates

In no other area of carbohydrate metabolism, indeed even in the whole of glycobiology, was so much pioneering research carried out in the past 20 years as in the field of polysaccharides and glycoconjugates of all types[164,165]. It can be stated with some confidence, that today the most important metabolic pathways which lead to the formation of polymeric and oligomeric carbohydrate structures in pure polysaccharides, peptidoglycans, and proteogycans as well as in glycolipids and glycoproteins, are known in principal. It has been learnt that the polysaccharides and peptidoglycans of prokaryotes have a fundamentally different biogenesis to that of the polysaccahrides and glycocojugates of the eukaryotes and that also, in the case of eurokaryotes, a distinct evolution is observed with increasing differentiation of the organism, especially of the glycoconjugate biosynthesis.

Although the biological role above all of the glycoconjugates of mammals and of humans is mostly still obscure, this does not hold for the biosynthesis of these complex structures. In the following section, the biosynthesis of the polysaccharides will be considered first, such as those of bacteria, and then those of the so-called glycoconjugates. It must be understood that the term *polysaccharide* embraces the most different types of structures, and that probably there is hardly a pure polysaccharide, that is one that consists of only monosaccharides. The term glycoconjugate is easier to define. Glycoconjugates contain, by way of comparison, small carbohydrate structures which are covalently bonded to either lipids or to proteins. The quantitative relation between the carbohydrate and lipid part in glycolipids or the protein part in glycoproteins can vary greatly.

In a polysaccharide and also in a lipopolysaccharide or proteoglycan, the carbohydrate part is dominant but in a glycoconjugate that is not so.

### 2.3.8.1 Bacterial polysaccharides[13,166]

Investigations into the biosynthesis of the peptidoglycan murein were initiated essentially by the discovery of penicillin[167]. Two observations helped to put researchers on the right track in their investigations. When treated with penicllin, *Staphylococcus aureus* accumulates in the cytoplasm large quantities of what were at that time new types of nucleotides, whose occurrence seemed inexplicable. One of the main nucleotides, a *UDP-MurAc-pentapeptide* (Fig. 2.157), was then found very much later as a compound, whose amino acid residues were to be found also in the peptidoglycan of cell walls[168].

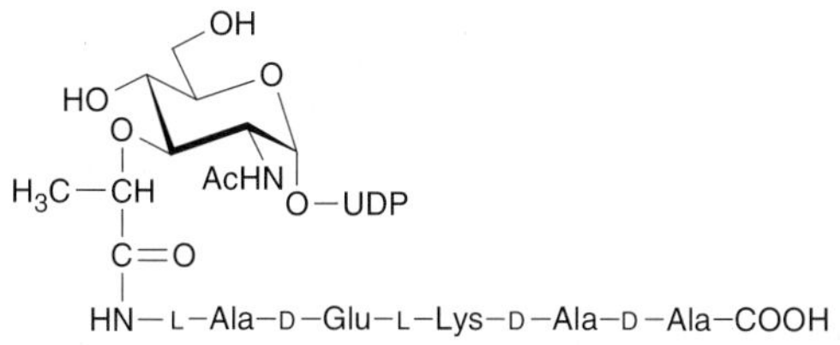

**Fig. 2.157** *UDP-MurAc-pentapeptide* accumulates in the cytoplasm of *Staphylococcus aureus* on treatment with penicillin.

Elucidation of the biosynthetic connection regarding the peptidoglycan of prokaryotes is due in a very large part, along with a series of other laboratories, to the research group of J. L. Strominger, whose main research subject was *Staphylococcus aureus*. Accordingly, most of the results considered here are drawn from this gram-positive bacterium, although the biosynthetic principle is comparable for all mureins. The entire process can be subdivided into three independent synthetic sequences.

■ **Murein.** The first series of reactions occurs exclusively in the cytoplasm and leads finally to the activated and peptide-derivatised muramic acid intermediate shown in Fig. 2.157. As the sequence shown in Fig. 2.158 makes clear, the formation of the UDP-MurAc-pentapeptide is the production of a larger building element, a ready made building block, which then can be used further in another place. This economic principle is found generally in the synthesis of polysaccharides which are assembled from repeating units.

All reactions forming the peptide bond require ATP and $Mn^{2+}$. The correct sequence is guaranteed through the high specificity of the enzymes concerned and does not take place on a template as in *normal* protein biosynthesis. A peculiarity is the attachment of L-Lys onto the $\gamma$-carboxy group of D-glutamic acid. The latter, and also the terminal D-Ala residues are peculiarities of the system. The two D-Ala-residues are transferred as a dipeptide, which is also unusual.

In the second stage of murein biosynthesis, also called the *undecaprenyl cycle*, the site of action shifts to the cytoplasmic membrane. In this connection it should be mentioned that the elucidation of this second part is bound up with the discovery of the first of the so-called lipid "carriers" for active carbohydrates, *undecaprenyl phosphate*, and also *bactoprenyl phosphate*, which belong to the group of polyprenyl phosphates which are commonly recognised today[169]. The *lipid carrier* only makes the hydro-

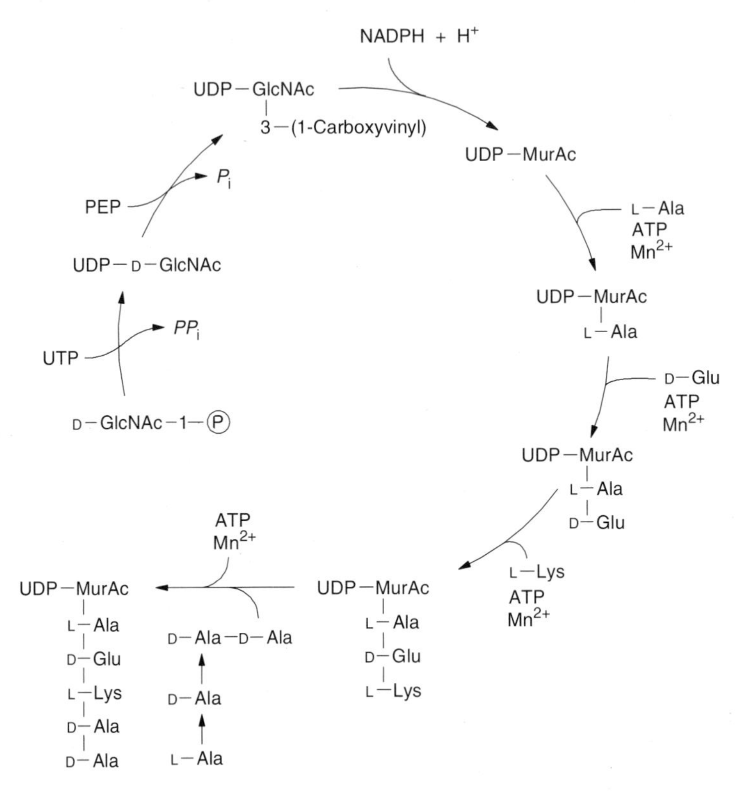


**Fig. 2.158** Biosynthesis of the first building block, *UDP-MurAc-pentapeptide*, occurs in the cytoplasm.

philic carbohydrate compatible with the lipid membrane. The discovery of the *undecaprenyl-PP-MurAc pentapeptide* is therefore almost as meaningful as the discovery of the first nucleoside diphosphate monosaccharide.

From the UDP-MurAc-pentapeptide arising in the cytosol, the 1-phosphate is transferred with liberation of UMP to the undecaprenyl phosphate. This *translocation*, catalysed by the enzyme translocase, is one of the most important steps in the biosynthesis of most polysaccharides and also of other condensed saccharides. The step is inhibited by the antibiotic *tunicamycin*, which imparts a great importance to this inhibitor in the investigation of the biosynthesis of complex carbohydrates because the transition to lipid bound activated sugar is of general occurrence and is not confined to the biosynthesis of the peptidoglycans.

The undecaprenyl cycle is continued by transfer of GlcNAc from UDP-GlcNAc onto the 4-position of the lipid bound MurAc residue, the conversion of D-glutamic acid into α-D-glutamic acid amide and finally the stepwise acylation of the free ε-amino group in the L-lysine with in total five glycyl residues (Fig. 2.159). In this case a specific glycyl-tRNA is the donor. The next step leads to chain lengthening of the polysaccharide according to the insertion mechanism shown in Fig. 2.160. The chain grows from inside to outside, a principle which deviates from that generally common in transglycosylation, on the non-reducing end (compare the biosynthesis of starch).

The acceptor is always only a single repeat unit. Each lengthening step liberates Un-PP, which through a membrane bound phosphatase is converted into Un-P and thereby the undecaprenyl cycle is again supplied with one of its key components. The antibiotic bacitracin inhibits this step (see section 2.3.8.8, p. 230).

When a certain chain length is reached, the lipid bound polysaccharide is displaced to the outside of the membrane. Here the murein chain can be extended and cross-linking to the murein-sacculus occurs. The cross-linking involves exclusively linking of peptides by peptide transfer (transpeptidation)

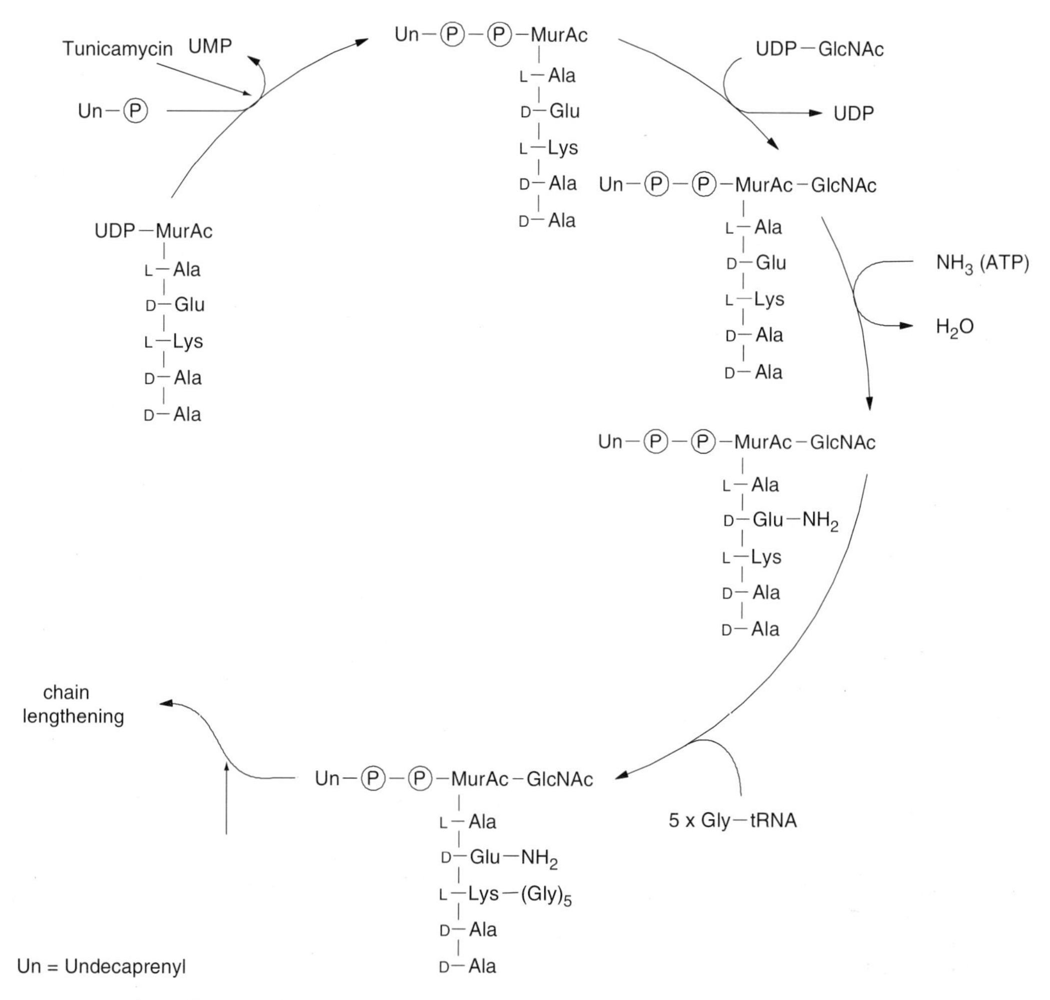


**Fig. 2.159** The *undecaprenyl cycle* serves for the completion of the lipid-bound repeat unit -$\beta$-D-GlcNAc-(1→4)-MurAc- as well as for the elaboration of the peptide framework.

(see section 2.1.2.1, p. 73). This essential biosynthetic step is irreversibly inhibited by penicillin (Fig. 2.161, p. 210 and Fig. 2.162, p. 211).

■ **Teichoic acids.** In contrast to the peptidoglycans, which only display structural variations with respect to the peptides but hardly any in relation to the carbohydrate part of murein, such variations are frequently encountered in the *teichoic acids*. The common structural element is the phosphodiester linkage. The alcoholic components are polyhydroxy compounds of different types, most frequently however glycerol or ribitol. Donors for the chain extension are in these cases *CDP-L-ribitol* or *CDP-D-glycerol*. Almost certainly the first acceptor for formation of the growing chain is a *GlcNAc-PP-lipid*. After sufficient chain lengthening, the head end-group GlcNAc-P is transferred with liberation of the lipid phosphate (also here a undecaprenyl phosphate – Un-P) to the 6-position of a MurAc residue of the murein chain (Fig. 2.163, p. 211).

■ **Lipopolysaccharide.** The biosynthesis of the core region of the lipopolysaccharides, especially that of *lipid A*, which guarantees the anchoring of the growing polysaccharide into the membrane, has been

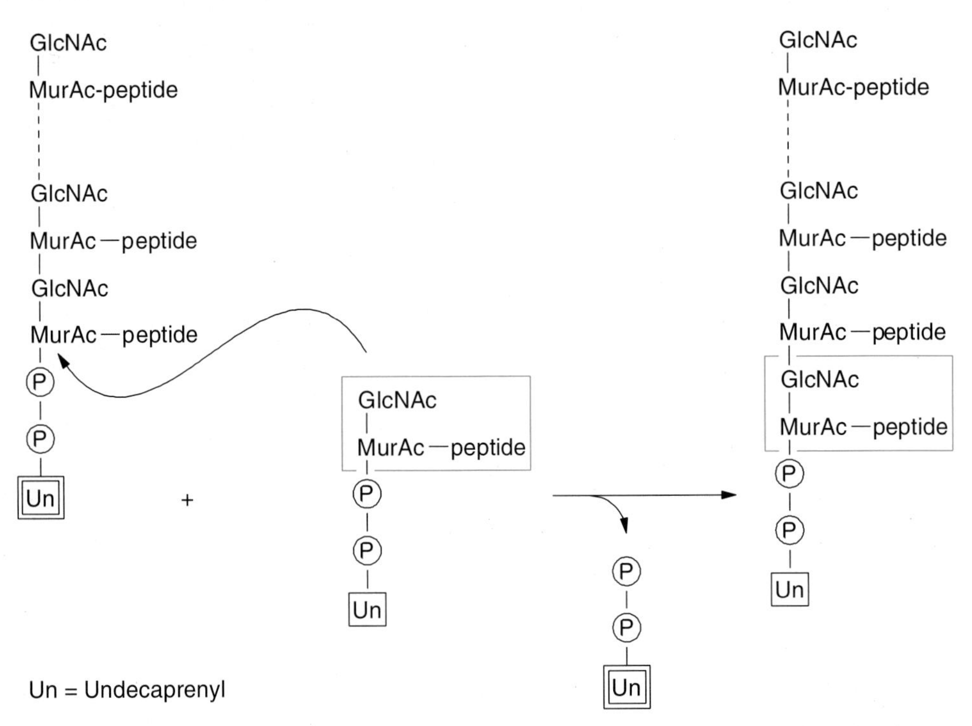


**Fig. 2.160** Chain lengthening of murein with the repeat unit -MurAc(peptide)-GlcNAc takes place by an insertion mechanism, whereby the single disaccharide unit is the acceptor and the already available chain is the donor.

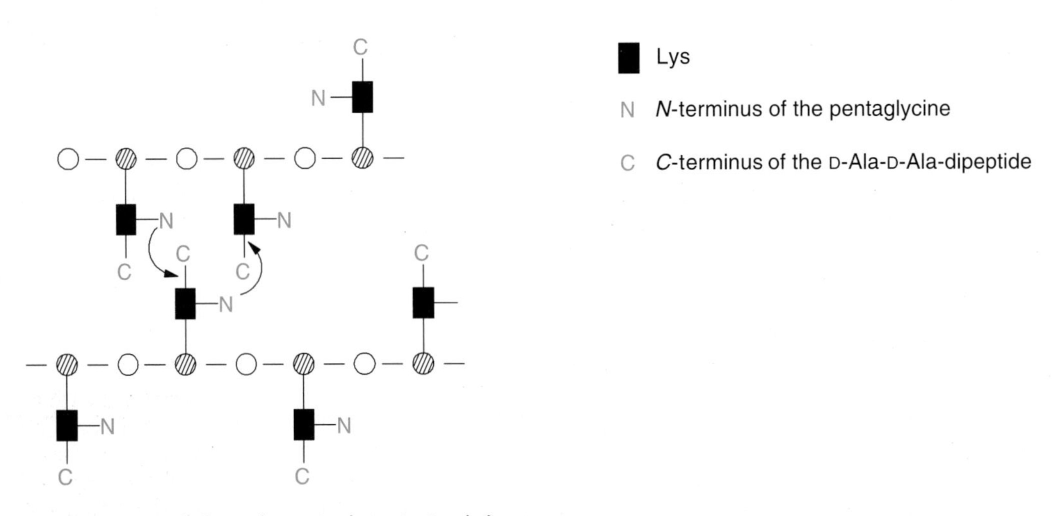


**Fig. 2.161** *Crosslinking* of *murein chains* in *Staphylococcus aureus*.

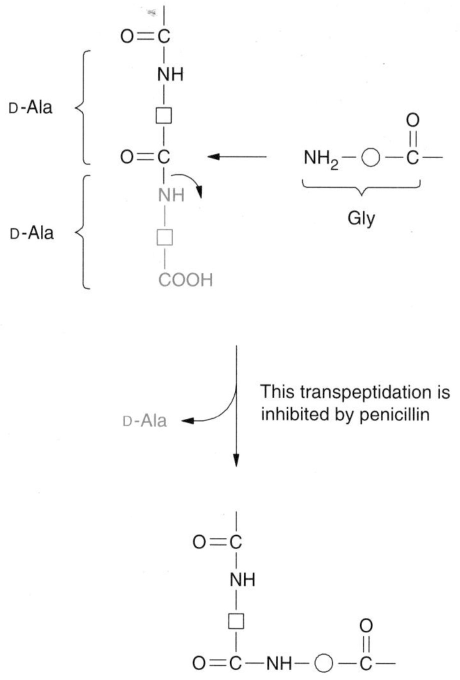

**Fig. 2.162** The amino end of the fifth glycine residue displaces, through transpeptidation, the terminal D-Ala from its own peptide bond.

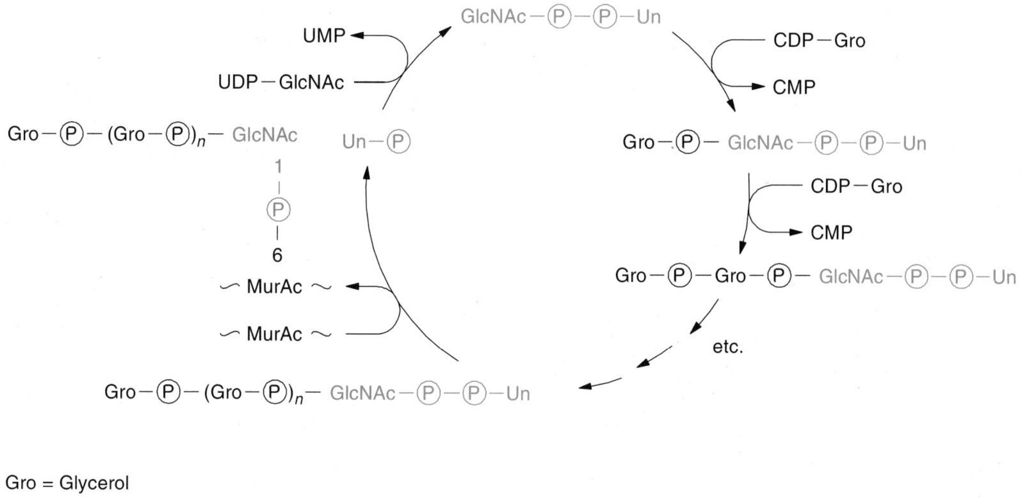

**Fig. 2.163** Biosynthesis of *teichoic acids* takes place through the transfer of a polyol phosphate onto the lipid *primer* (GlcNAc-PP-Un) followed by further lengthening. On reaching the required degree of polycondensation, the teichoic acid is attached with its head group GlcNAc-P- to the 6-position of the MurAc in the peptidoglycan.

described[170]. The synthesis of the inner and outer core region occurs in the usual manner through transglycosylation from nucleoside diphosphate sugars. Noteworthy is the use of CMP-Kdo for the formation of the typical $(Kdo)_3$ unit in lipid A. The use of nucleoside monophosphates as glycosyl donors for the ulosonic acids Kdo and NeuAc is typical.

The biosynthesis of the so-called *O*-specific polysaccharide chains is especially well researched in the *Salmonella* species[171]. Here a large range of variation in structures exists with the same principle of synthesis. Repeat units of different sizes (see section 2.1.2.1, p. 73), often with monoglycosyl branches, are preprepared on the lipid carrier. The polycondensation is the same as already described for the peptidoglycan, and transfer to the core structure on lipid A concludes the biosynthesis (Fig. 2.164). Exceptions to the typical insertion reaction in the chain lengthening of the *O*-chain have been demonstrated[172] in the case of *E. coli*. The structure of the end product shown here is that of the lipopolysaccharide from *S. typhimurium*[173].

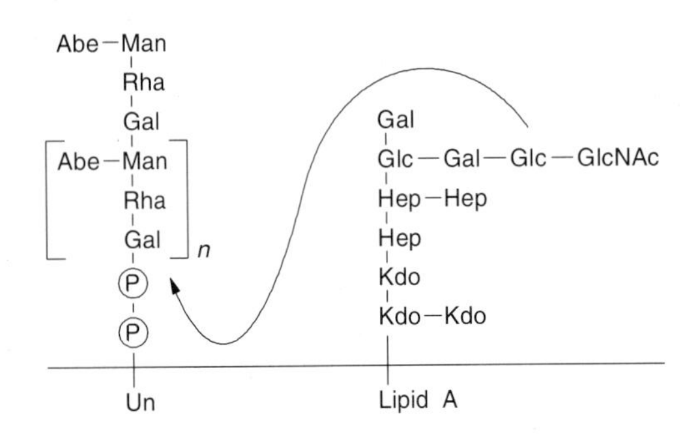


**Fig. 2.164** The *O-antigen polysaccharide chains* arise from repeat units on the lipid carrier Un-PP. From there they are transferred to the core region, anchored in the membrane with lipid A. The transfer reactions are very similar to those of the peptidoglycan synthesis.

A good many bacteria produce so-called capsular polysaccharides. They differ from the lipopolysaccharides through their composition (mostly they contain acidic monosaccharide components) and through their location as a voluminous zone outside the cell wall. In many, diacylglycerol phosphate could be identified as an aglyconic end group. An extraordinary capsular polysaccharide produced by many *E. coli* strains is *colominic acid*, a polysaccharide from NeuAc units[174,175]. The biosynthetic pathways of the capsular polysaccharides are comparable to those of the lipopolysaccharides, with the exception that they lack the transfer step to the anchored core structure. Some bacteria, *Leuconostoc mesenteroides* or *Streptococcus mutans*, are in the position, with an extracellular supply of sucrose, to transform their monosaccharides into either polyglucans[176] (*dextran*) or polyfructans (*levan*) through transglycosylation without the participation of activated sugars. The biosynthesising enzyme, dextran sucrase or levan sucrase, is found outside the cell wall.

### 2.3.8.2 Cellulose and glycogen

Of the numerous polysaccharides, which were discussed partly as structural polysaccharides and partly as reserve polysaccharides in sections 2.1.2 (p. 72) and 2.1.3 (p. 98), respectively, the biosynthetic pathways of the homoglucans cellulose and glycogen should be discussed as representative examples, so far as they have been elucidated. Both polymers are products of eukaryotic cells; only a very few bacteria can synthesise cellulose. A part of the *in vitro* investigations were undertaken in bacteria (*Acetobacter xylinium*), which indicate UDP-Glc as donor substrate. The biosynthesis of *cellulose* in plants, the main source of this most abundant organic compound, is to this day disputed. Above all the problem lies in the almost invincible difficulties in constructing an unequivocal, general, reliable test for the enzymatic chain extension and its initiation reaction.

One of the main issues of contention is the question over the use of *UDP-Glc* or *GDP-Glc* as glucosyl donors. From both radio-labelled donors, obvious incorporation of labelled glucose into cellulose is observed with enzyme systems from many different sources. Meantime it is thought that both sources are used for glycosylation at different times of the polymer growth. Further complications are introduced by the observation that glycosylated lipids are also concerned in the complex process of cellulose biosynthesis. In green algae, dolichyl diphosphoglucose is found (Fig. 2.165) and also lipid bound longer oligosaccharides. *Dolichol* is typical for eukaryotes. It differs from undecaprenol of the bacteria not only through the chain length, but also through a saturated isoprene unit[177]. The lipid bound oligosaccharides draw their glucose from UDP-glucose, as was proved with radio-labelled donor substrate.

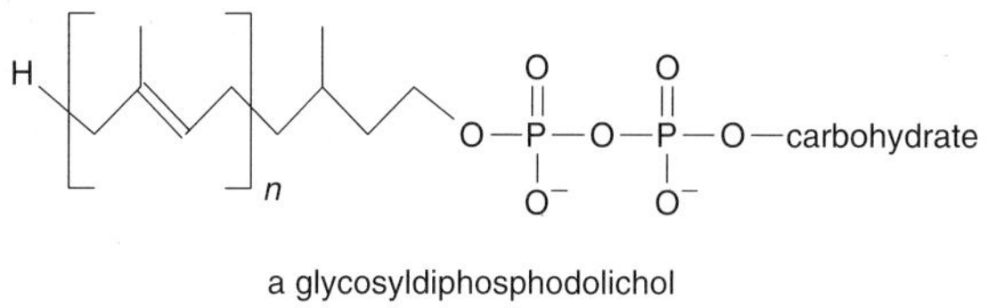

a glycosyldiphosphodolichol

**Fig. 2.165** *Dolichols* (n = 14 – 24), related to undecaprenol, are longer than the latter and possess a saturated $C_5$ unit at the hydroxyl end. In functional terms, there is no difference between the lipid-carrier of bacteria and the dolichols.

Dolichyl phosphate plays a key role in the synthesis of condensed high molecular weight carbohydrates in eukaryotes. In the case of cellulose, however, it was made clear that the insoluble, native polysaccharide is not condensed on a lipid carrier but probably exists as a proteoglycan. Although conclusive proof is lacking, the hypothetical course of cellulose biosynthesis is shown here which takes all known experimental results into account (Fig. 2.166). It should be noted that GDP-Glc, discussed at the beginning as a glycosyl donor, is only involved in further chain extension at the stage of the proteoglycan.

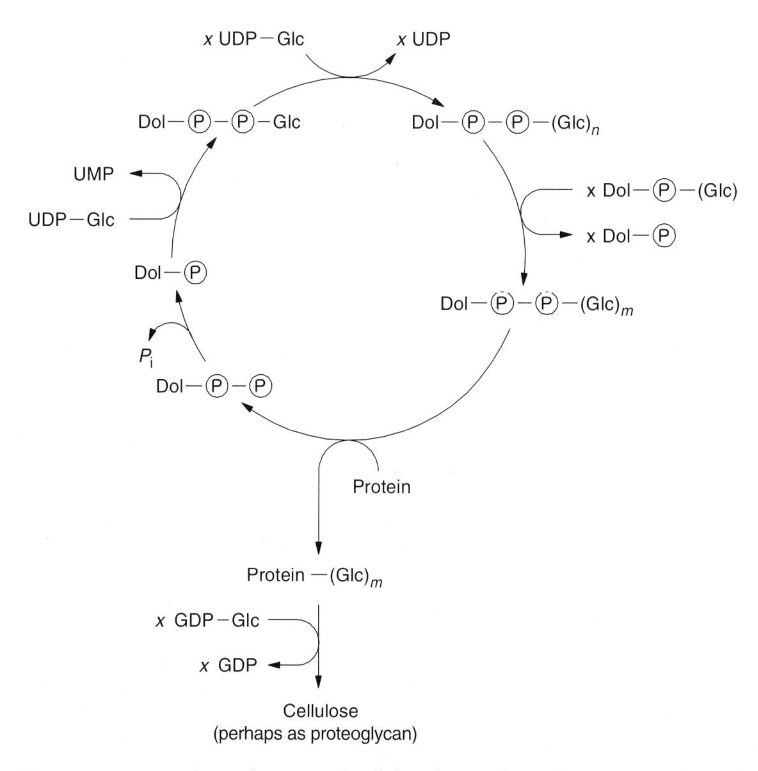


**Fig. 2.166** Hypothetical course of *cellulose biosynthesis.*

Like cellulose, *glycogen* is a typical eukaryotic polysaccharide, although bacteria also employ glycogen as a *reserve substance* under limiting growth conditions with sufficient provision of a C-source[178]. The special significance of glycogen for mammals is its synthesis and degradation and above all its regulation in one of the best-researched metabolic processes. It need not be emphasised again that a strict separation of the pathways exists.

Two enzymes are involved in glycogen biosynthesis – UDP-glucose-pyrophosphorylase, which makes UDP-Glc from UTP and GlC-1-P, is not included here:

**1.** *Glycogen synthase* (UDP-glucose: 1,4-α-D-glucan-4-α-D-glucosyl transferase) and

**2.** *Glycogen branching enzyme* (1,4-α-D-glucan: 1,4-α-D-glucan-6-α-D-glucosyl transferase).

The chain extension catalysed by glycogen synthase is again a classical transfer reaction (see section 2.3.4, p. 186). UDP-Glc is cleaved in the first reaction step with liberation of the aglycone UDP (Fig. 2.167). The glucosyl-enzyme intermediate then binds to the end of a maltodextrin chain and transfers the glucopyranosyl residue to the terminal 4-position. The acceptor in this reaction must always be an already existing maltodextrin chain.

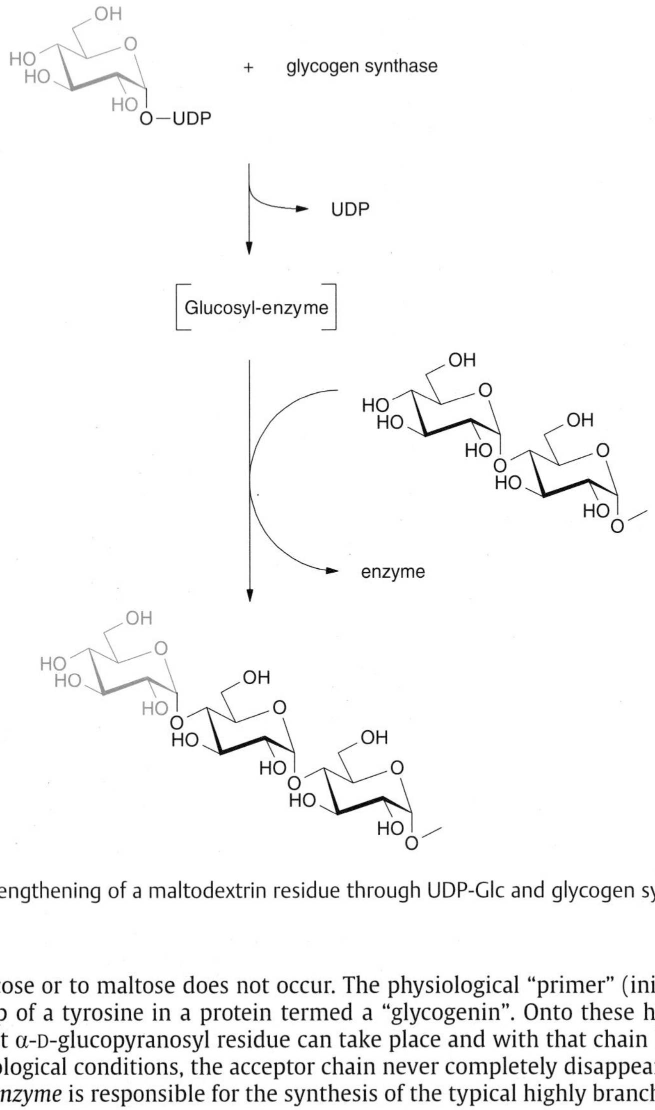

**Fig. 2.167** Chain lengthening of a maltodextrin residue through UDP-Glc and glycogen synthase.

Transfer to glucose or to maltose does not occur. The physiological "primer" (initiator molecule) is the hydroxy group of a tyrosine in a protein termed a "glycogenin". Onto these hydroxy groups the transfer of the first α-D-glucopyranosyl residue can take place and with that chain lengthening. However, under physiological conditions, the acceptor chain never completely disappears.

The *branching enzyme* is responsible for the synthesis of the typical highly branched tree-like structure of glycogen. Its failure or under production leads to a dangerous storage illness. The reaction catalysed by the branching enzyme is unusual in as much that the transfer reaction starts off as a (1→4)-linkage and ends up as a (1→6)-linkage, which is brought about by a special mechanism regarding donor- and acceptor-binding. Criteria regarding the length of the donor chain and of the length of the transferred glucanoyl residue are also relatively rigorous. The donor chain should be a maltodextrin consisting of at least 11 glucopyranosyl residues [always (1→4)-α-D-linked] and the transferred chain has in general a length of 7 glucopyranosyl residues. Thence, a high and thick degree of branching results (Fig. 2.168).

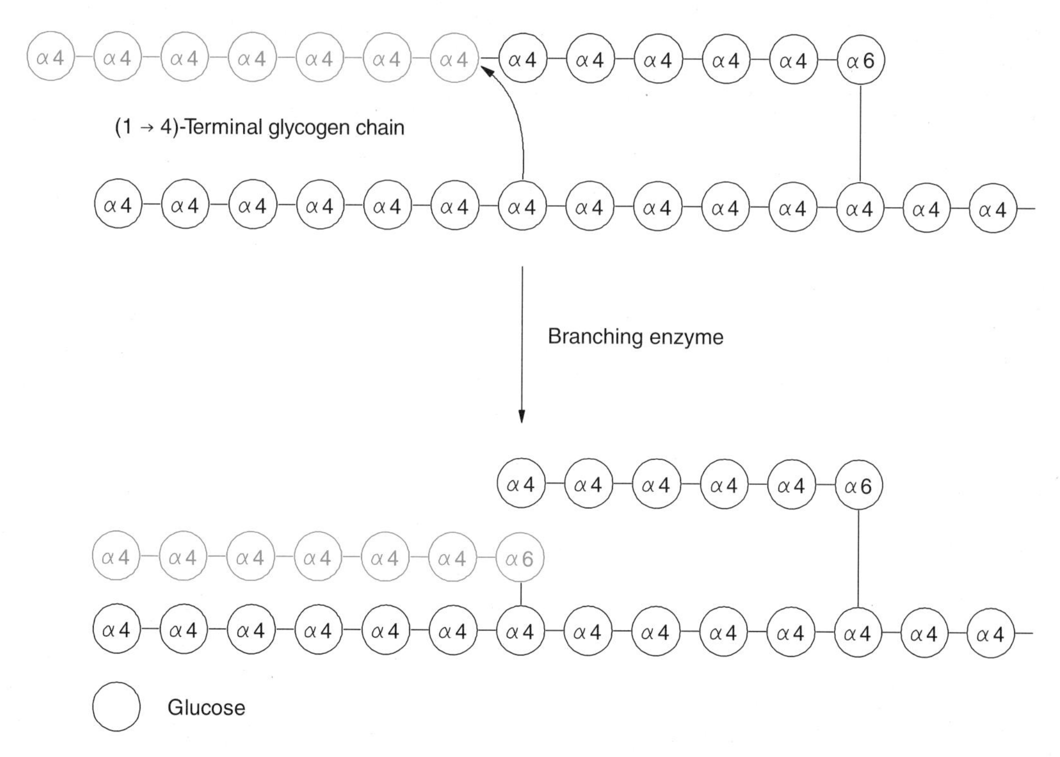


**Fig. 2.168** Branching enzyme provides a high and dense ramification of glycogen. The (1→4) to (1→6) transfer is irreversible.

It should be noted that the unique specificity of the transfer reaction prohibits reversibility. The degradation of glycogen takes place, accordingly, in a completely different manner (see section 2.3.5, p. 189).

### 2.3.8.3 Mannans

Mannans, better called proteomannans, are essential components of the cell walls of yeasts[179]. *Baker's yeast* (*Saccharomyces cerevisiae*), belonging to the primitive eukaryotes, is one of the best known and extensively researched cell types of this stage of development. Regarding their biosynthesis, the yeast mannans are, in some ways, in a hybrid stage, on the one hand between the carbohydrate structures of bacteria and the higher organisms, and on the other they link up with the known biosynthetic pathways of the glycoproteins of vertebrates. With the latter, they have the differentiation between *N*- and *O*-glycosidic conjugates in common as well as similar lipid-bound core structures, like perhaps $Glc_3$-$Man_9$-$GlcNAc_2$-, bound (and thereby activated) on dolichyl pyrophosphate. Transfer of this oligosaccharide to the aspargine side chain of the acceptor protein and further processing, comparable to that occurring in the Golgi system of mammal cells, are also to be found (see section 2.3.8.5, p. 218).

The mannose part of the so-called tetradeca-oligosaccharide is produced either by transfer from the donor substrate GDP-Man or from Dol-P-Man. In the case of the sole $\beta$-linked mannose residue, which is attached to the chitobiosyl unit, GDP-Man is established as donor. After attachment of the three glucosyl residues, which probably acts as a signal, the whole oligosaccharide is transferred from dolichyl pyrophosphate to the protein (Fig. 2.169). The process has great similarity with the biosynthesis of the structurally similar, activated core oligosaccharide occurring in the endoplasmic reticulum of mammals and is likewise inhibited by tunicamycin. After removal of the triglucosyl signal sequence the difference to the "processing" and "trimming" occurring in the Golgi apparatus of higher organisms becomes apparent. The yeast cells not only leave the $Man_9$-core portion intact, but increase it still more by up to about 8 mannosyl residues. Thereon, a massive mannosylation takes place with GDP-Man as the donor. In this way the highly branched mannans are formed with very different degrees of polycondensations (Fig. 2.169). Linkages are firstly (1→6), but then also (1→2) and (1→3), always with α-configuration (see section 2.1.2.2, p. 80). It is thought that there are about 10 different mannosyl transferases for the synthesis of the cell wall mannans in baker's yeast.

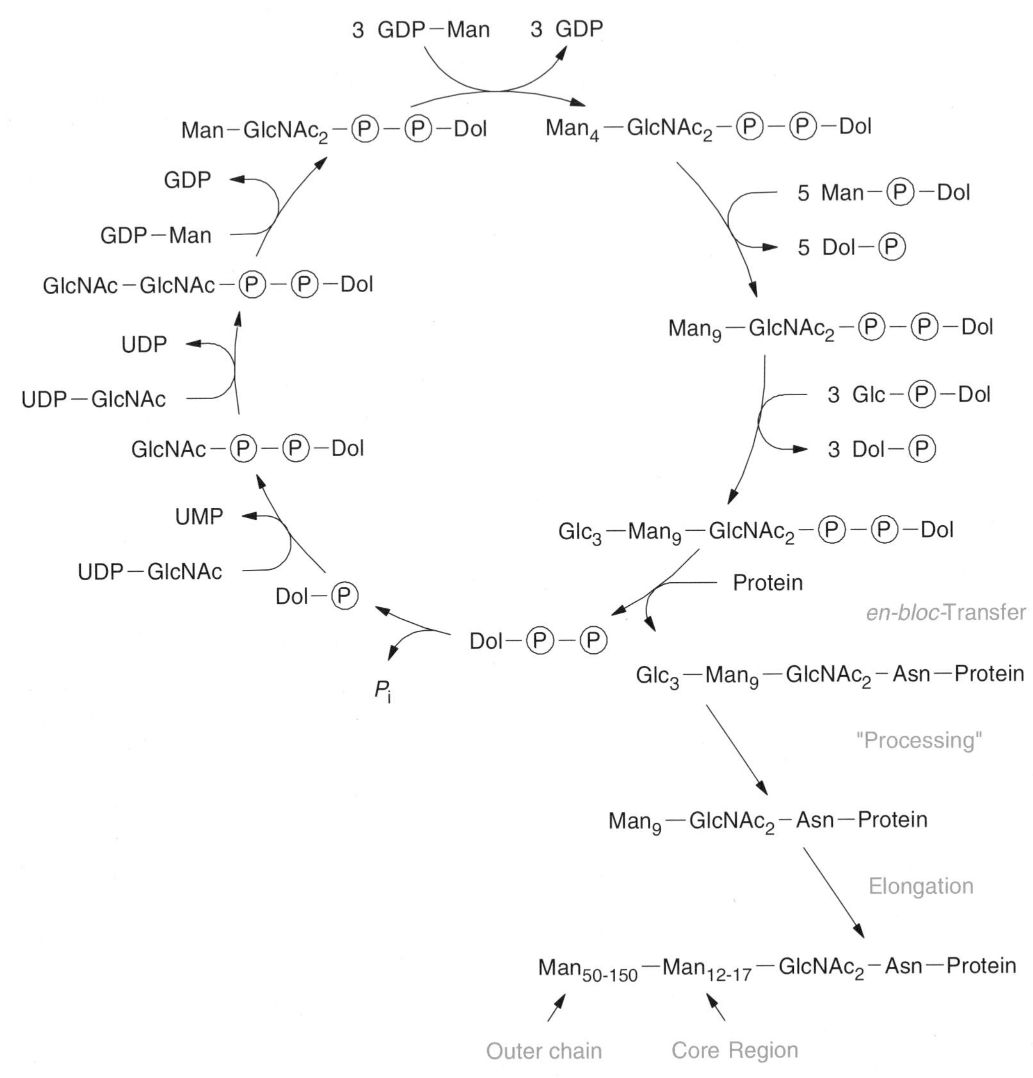


**Fig. 2.169** Biosynthesis of the *cell wall mannan* of yeast has in the initial stage large similarity with the synthesis of the core structures in the endoplasmic reticulum of the mammalian cell (see section 2.3.8.5, p. 218).

A large part of the yeast mannans are *O*-glycosidically bound to proteins. They have no core region and are smaller than the *N*-glycosidic mannans. Probably, Man-P-Dol is the donor for the mannosylation of serine or threonine. Further mannosyl residues are then transferred onto the bridgehead mannosyl residue, with GDP-Man as donor. Possibly the same transferases are responsible for peripheral mannosylation in the case of *N*- and *O*-glycosidic structures.

### 2.3.8.4 Glycosaminoglycans

With the exception of hyaluronic acid (conclusive clarification is still missing here) all glycoaminoglycans[37,180] and also proteoglycans are of very different size. Their biosynthesis up to the *mature* polysaccharide is complex. It requires, besides numerous synthetic and modifying enzymes, also an ingenious transport system which effects both the displacement of the growing polymer inside the cell, from ER to the Golgi apparatus and from there to the cell surface, and also the intercellular transport in the tissue. Many details of these events are still unclear, yet the relatively few sure results allow a rough description, which is given in outline here for the identified proteoglycans.

Of the three different types of glycosyl residues found in proteoglycans, the *N*-glycosidic (section 2.3.8.5, p. 218), the *O*-glycosidic or the mucin type (section 2.3.8.6, p. 226), and the carbohydrates bound through β-D-xylosyl serine, the latter is typical for glycosaminoglycans (keratan sulphate is excluded here; it possesses an *N*-glycosidic core region and, unique for a glycosylaminoglycan, no glucuronic acid, but galactose in the repeat unit). It is conjectured that *N*-glycosidic oligosaccharides are addresses for the translocation of the proteoglycans.

The *biosynthesis* begins, apparently, with the very favoured xylosylation of a serine in the protein sequence, which is especially rich in glycine. Probably the place of this initial step is the *cis*-compartment of the Golgi apparatus, which requires a membrane bound, very specific xylosyl transferase. Some researchers point out that the control of which serine is xylosylated arises not so much from the supposed signal sequence Gly-Ser-Gly, but from definite regions of the tertiary structure of the peptide fragment to be xylosylated [181]. Likewise in the *cis*-Golgi, further synthesis of the core region takes place with two β-D-galactosylations by two different galactosyl transferases, and certainly here the β-D-xylose residue is to be considered as just a lead signal for the first galactosylation step (Fig. 2.170).

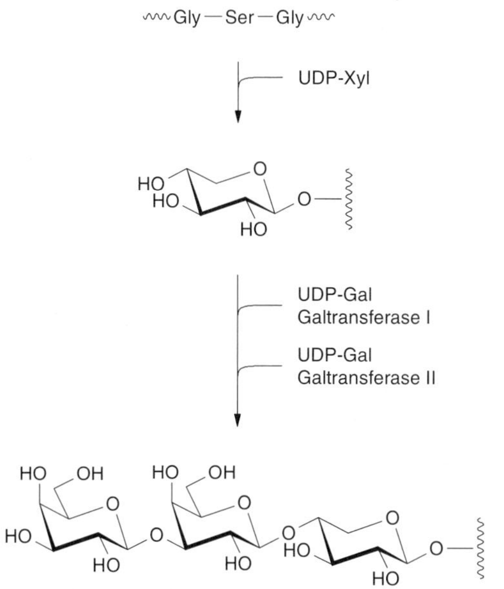

**Fig. 2.170** Biosynthesis of all *proteoglycans* begins with the shown trisaccharide-core region on the protein portion.

Following the biosynthesis of the core region, the synthesis of the actual polysaccharide chain takes place from disaccharide repeat units. The first monosaccharide residue of the chain – in each case a β-D-glucuronyl unit, bound to the 3-position of the galactose – is often also still counted as the core region, because the transferase responsible is not identical with that of the further chain extension. The repeat unit is built up through a pair of transferases (in part the activities are combined in one protein) working hand in hand. Their acceptor specificity is confined to the last monosaccharide in the chain.

Since the different glycosaminoglycans as proteoglycans have the same core region at their disposal, yet in spite all similarities possess differently structured repeating units, there still remains the unanswered question concerning the signal for the synthesis of a definite glycosaminoglycan. Likewise, the regulation of chain lengthening is an open question. When and how is chain growth finished? It seems certain that the *native* polymer chain grows very quickly since no unfinished parts are found and the post-biosynthetic modification – *N*- and *O*-sulphation through PAPS (phosphoadenosine phosphosulphate) as well as 5-epimerisation of the uronic acid residues – follows just as fast (Fig. 2.171).

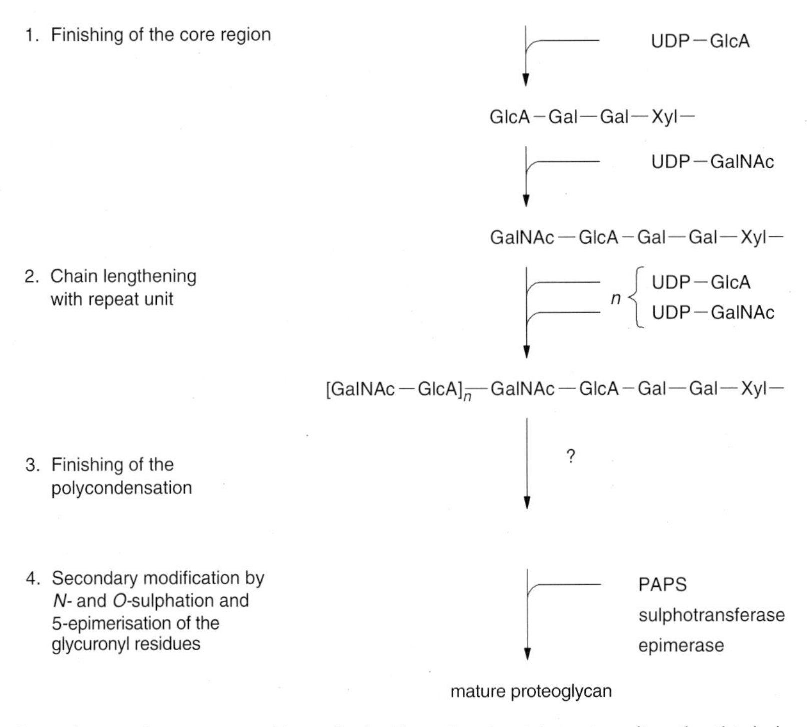


**Fig. 2.171** Four phases of *proteogycan biosynthesis*. There is uncertainty regarding the third phase.

Least is known about the biosynthesis of *hyaluronic acid*, above all the initial step. Only the molecular size is extraordinary and makes investigation difficult. To be certain, hyaluronic acid is in almost every respect a case on its own. It lacks the protein, it lacks the sulphate residues, and a core region has never been found. A very unusual mechanism of biosynthesis which is based on numerous, meticulous investigations was published by Prehm[182]. Pecularities are an alternating insertion of UDP-GlcNAc and UDP-GlcA from the reducing end of the growing chain, a synthesis mechanism reminiscent of the bacterial polysaccharides (see 2.3.8.1, p. 207), whereby the last introduced UDP-residue always forms the end of the chain (Fig. 2.172).

The chain should grow out of the membrane which, without an anchor, is difficult to imagine. For the insertion of polar UDP-hexoses outside the cell membrane, an additional transport system would be necessary.

### 2.3.8.5 N-Glycosidic glycoproteins

The application of specific inhibitors of the different biosynthetic stages in the formation of the so-called *N*-glycosidic linked oligosaccharides of the glycoproteins is mainly responsible for the extensive elucidation of their biogenesis. Since the molecules in question have relatively low molecular weights, the technical difficulties, which have to be overcome in corresponding investigations of the glycosaminoglycans and other polysaccharides, are avoided. Instead, however, the uniquely elaborate manner with which the *mature N*-glycosidic proteins[9,183] of the cell are produced can be regarded as special hindrance on the way to elucidating the biochemical relationships. Certainly, in the past 20 years or so,

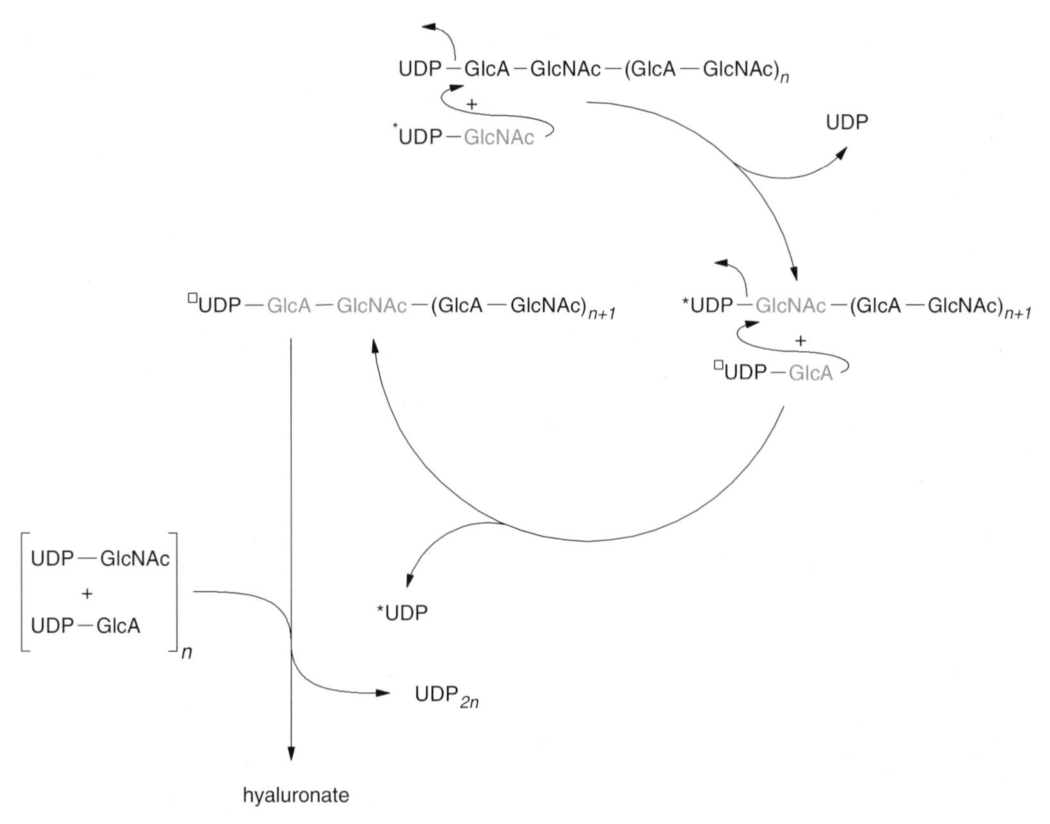


**Fig. 2.172** An unusual suggestion from P. Prehm for *hyaluronate biosynthesis*, with some uncertainties. The *insertion mechanism* occurs, otherwise, only in the biosynthesis of bacterial polysaccharides.

in which special scientific attention was devoted to just this group of carbohydrates, their potential biological significance was the driving force for intensive research activity. The probably more than one thousand structurally different *N*-linked oligosaccharides found up to the present day appear to actually confirm their specific task to serve for cell-cell recognition, cell matrix recognition, or generally as addresses for organelles and even also whole cells[184].

The biosynthesis of *N*-glycosidic oligosaccharides in glycoproteins can be subdivided into two aspects, one concerning the timing of events and the other concerning the place of action:

1. The synthesis of a lipid-bound oligosaccharide precursor as well as its complete transfer to the growing polypeptide chain which occurs in the rough endoplasmic reticulum (RER).
2. The processing of the oligosaccharide structure through the degradation (glycosidases) and synthesis (glycosyl transferases) taking place in the different compartments of the Golgi apparatus, until the apparently mature glycoconjugate is formed.

The first step in the biosynthetic sequence is probably the same for all eukaryotes or very similar (see 2.3.8.3., p. 215). For the second step, there are indeed commonalities, though at this step the structural individualites are elaborated, which can be typical for a particular cell, for a certain tissue type, or for a particular organism.

The *dolichol phosphate cycle* begins with the first step in the RER-membrane. UDP-GlcNAc : dolichol phosphate GlcNAC-1-phosphotransferase catalyses the transfer of a GlcNAC-1-P onto Dol-P, giving rise to dolichyl diphosphate *N*-acetyl-D-glucosamine (Dol-PP-GlcNAc) (Fig. 2.173, p. 220). This step is almost completely inhibited by the antibiotic tunicamycin, so that after longer treatment no more lipid bound

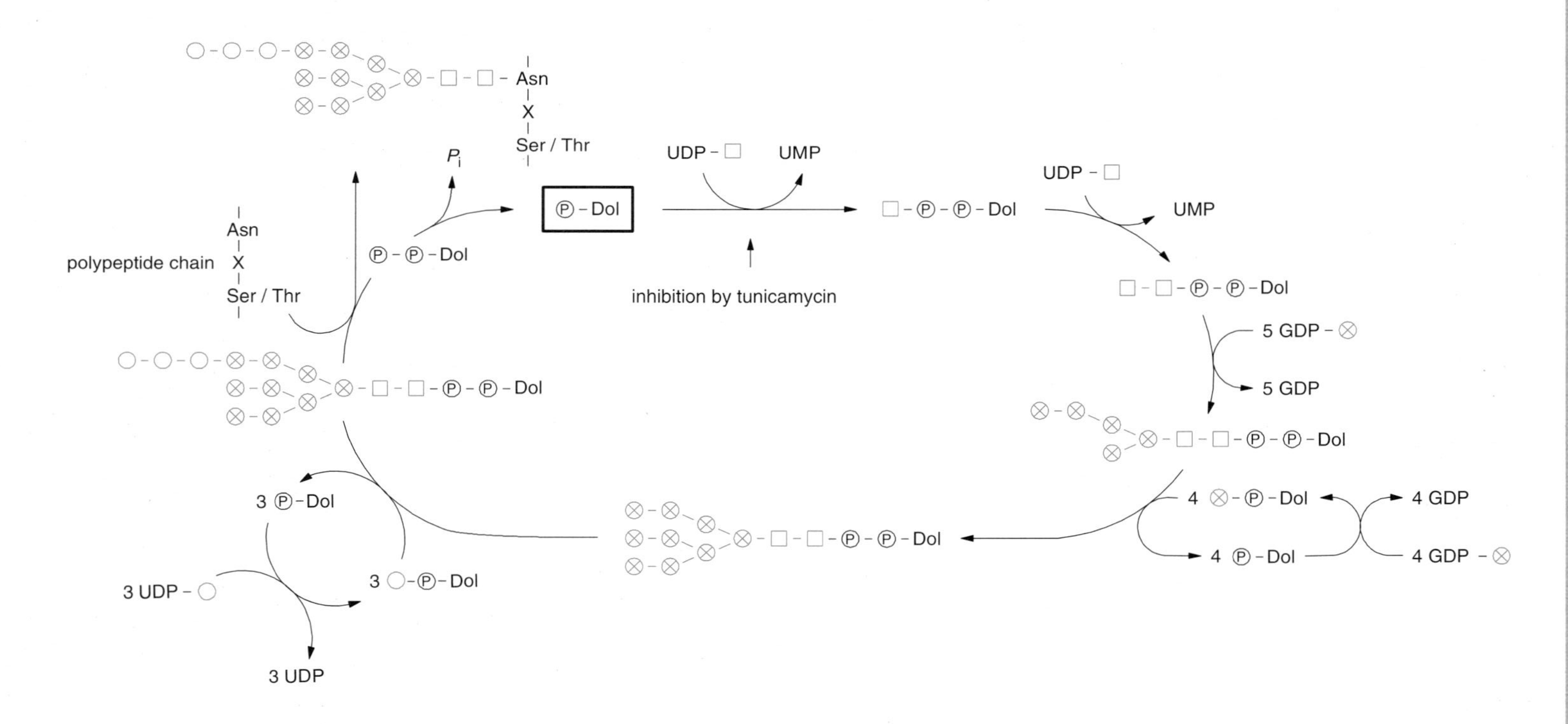


○ Glucose ⊗ Mannose □ *N*-Acetylglucosamine Dol Dolichol

**Fig. 2.173** In the *dolichol cycle* the precursor glycolipid for N-glycosylation originates in the membrane system of the RER. The cycle begins with the reaction of the dolichol phosphate □. This step is completely inhibited by *tunicamycin*.

oligosaccharides can be detected. With that, *N*-glycosylation of protein is also no longer possible. Treatment with tunicamycin is the most important proof for the biogenesis of an *N*-glycosidic glycoprotein.

The lipid portion of Dol-PP-GlcNAc is anchored in the membrane of the RER, above all in the cytosol side. At a certain point in time of the oligosaccharide synthesis, very probably at the stage of the Dol-PP-heptasaccharide, comprising the chitobiosyl unit and five mannose residues, the shift to the lumen side occurs[185]. It has indeed been demonstrated that transport systems exist for nucleotide sugars which guarantee their availability also in the RER lumen, yet these donors are apparently not used for the first steps of the Dol-PP-oligosaccharide synthesis. Step by step, synthesis of the lipid bound oligosaccharide takes place through transfer reactions from the specific nucleotide sugars, UDP-GlcNAc, GDP-Man, and UDP-Glc, in which in part (the four last Man residues and the three Glc residues) a preceding transfer takes place to Dol-P.

Not all eukaryotes make the complete lipid precursor with the tetradecaosyl residue. Even the three glucosyl residues, originally regarded as a universal signal, are not necessary in the case of many lower organisms for the final transfer of the oligosaccharide to the protein[186]. Also, other structural variations have been noted as exceptional cases. Obviously, such incomplete precursor structures ought also to be transferable to the protein. The normal case in mammalian cells is, though, the course of the *dolichol cycle* shown in Fig. 2.173.

After the first phase is concluded, the transfer of the oligosaccharide residue to the protein is initiated as the intermediate phase. Apparently this process is signalled by the tri-glucosyl sequence in the glycolipid. Transfer is performed by a specific oligosaccharyl transferase[187], whose identification has still not been accomplished up to now. The enzyme, perhaps also an enzyme complex, recognises not only the donor but also, as acceptor sequence in the continuously biosynthesised protein, the tripeptide -Asn-X-Ser/Thr, in which the amino acid X may be any one except proline. These facts are in agreement with the supposition that only a special conformation, stabilised by a hydrogen bond between the α-NH of Asn in the peptide chain and the oxygen of a Ser/Thr, is recognised (Fig. 2.174).

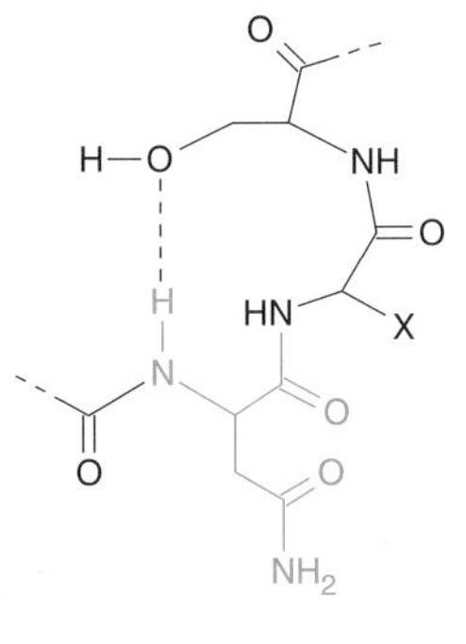

**Fig. 2.174** The conformation of -Asn-X-Ser/Thr-, supposedly necessary for *N-glycosylation*, stabilised through a hydrogen bond.

This could also explain why many Asn-residues, despite a correct signal sequence, are not glycosylated. The self-forming tertiary structure of the protein perhaps does not permit the special conformation to be adopted. For the time being the glycosylated protein remains in the lumen of the RER and there it is reduced in size by three again very specific glycosidases, α-glucosidase I and α-glucosidase II as well as the ER-α-mannosidase, by the three peripheral glucosyl residues and one of the in total nine mannosyl residues (Fig. 2.175 and 2.176). The specificity, above all of the ER-α-mannosidase, is quite astonishing, if one remembers that there is a choice of three terminal (1→2)-linked α-mannosyl residues.

ER-α-mannosidase is inhibited by *1-deoxymannonojirimycin* (see section 2.3.8.8, p. 230). The Golgi α-mannosidase I found in the *cis-Golgi vesicle* is somewhat less specific. As could be shown with test substrates, it is also able to remove hydrolytically, beside the mannose physiologically already separated in the ER, the other (1→2)-α-mannosyl residues. For this reason the biological function of the ER-α-mannosidase is not directly recognisable. Hypotheses have been discussed in the literature[188].

The cleavage of the mannosyl residues occurs through α-mannosidase I, as already stated, in the *cis*-Golgi, through α-mannosidase II in the medial Golgi and ends in general with a core oligosaccharide with three mannosyl residues and a GlcNAc end-group (**A**), which originates previously by transglycosylation (Fig. 2.177) at an intermediate stage. The intermediate stage is the triantennary oligosaccharide

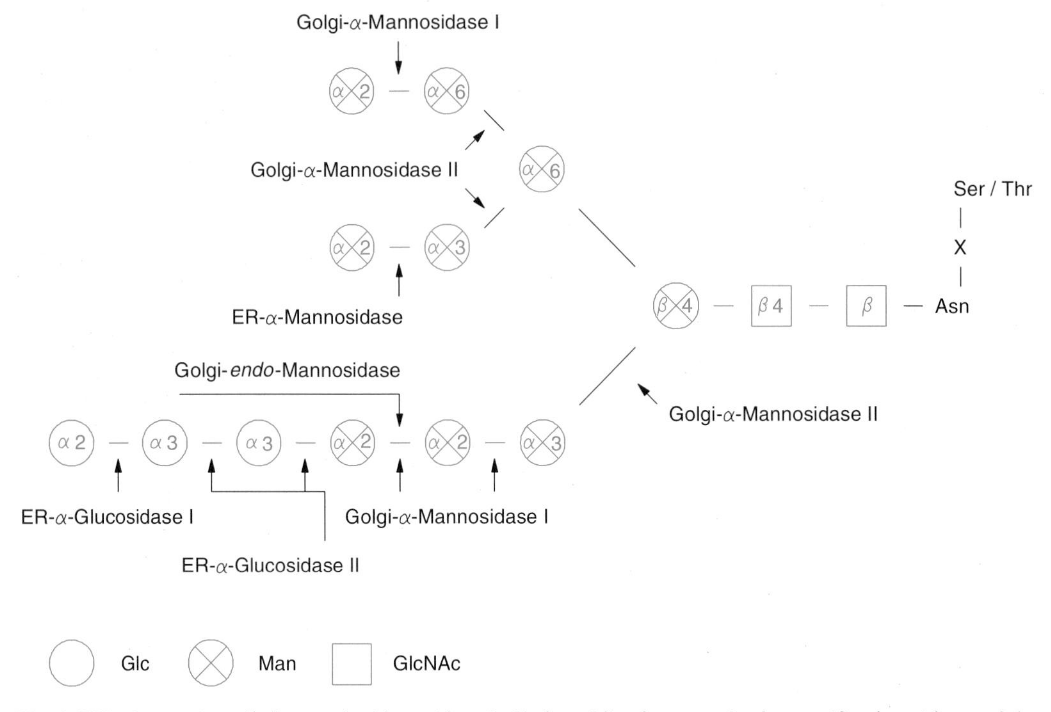


**Fig. 2.175** Processing of oligosaccharide residues in *N*-glycosidic glycoproteins by specific glycosidases of the endoplasmic reticulum and Golgi apparatus.

(**B**) still containing five mannosyl residues. The original octamannosyl derivative (**C**) discharged from the ER is the starting compound for the so-called "high mannose" glycoprotein. Further degradation can be prevented through phosphorylation of the (1→3)-linked α-mannosyl residue. Mannose phosphorylation, however only of (1→6)-linked α-mannosyl residues, serves as an "address" for lysosomal enzymes (see section 2.2.2.3, p. 140), but can also block access for mannosidases and thereby the further cleavage of mannosyl residues. In addition to the exoglycosidases already mentioned, an endomannosidase has also been found in the *medial Golgi system*, whose significance lies perhaps in the removal of peripheral chains with glucosyl ends that have *slipped through*.

In the medial Golgi system the synthesis of the *N*-glycosidic glycoproteins begins by re-glycosylation with the aid of the in part very specific glycosyl transferases. Their substrates, besides the acceptors, are the usual nucleoside diphospho-monosaccharides for whose transport in the Golgi vesicles special transport systems are responsible. After the action of the mannosidases, the three *N*-glycosidic bound acceptor oligosaccharides **A**, **B**, and **C**, already named above, remain in the medial Golgi system (**A** and **B**) and the *cis*-Golgi system (**C**), respectively (Fig. 2.176 and 2.177.).

The component **C** belongs to the "high mannose" glycoconjugates. From them, arise, for example, the mannans through further mannosylation (see section 2.3.8.3, p. 215). From the intermediate stage **B**, formed with the help of the GlcNAc-transferase 1 in transferring only one GlcNAc residue to the shorter of the (1→3)-α-mannosyl antennae, arises the starting material for the biosynthesis of the "hybrid type" glyconjugates and the initial acceptor (**A**) for the biosynthesis of all "complex type" glycoproteins.

In the medial Golgi system, besides the GlcNAc-transferase I already described, four further highly specific GlcNAc-transferases and a Fuc-transferase are at work (Fig. 2.178). In this manner an abundance of different acceptors can result for subsequent glycosylation in the *trans-Golgi system.* It can be assumed that the key to the guidance of the biosynthesis of specific oligosaccharide structures is to be found in the action of the different GlcNAc-transferases.

In the *trans*-Golgi system, the biosynthesis is completed. Galactosyl-, sialyl- , fucosyl-, and again a GlcNAc-transferase (this for the synthesis of polylactosylamine chains) produce through further specific

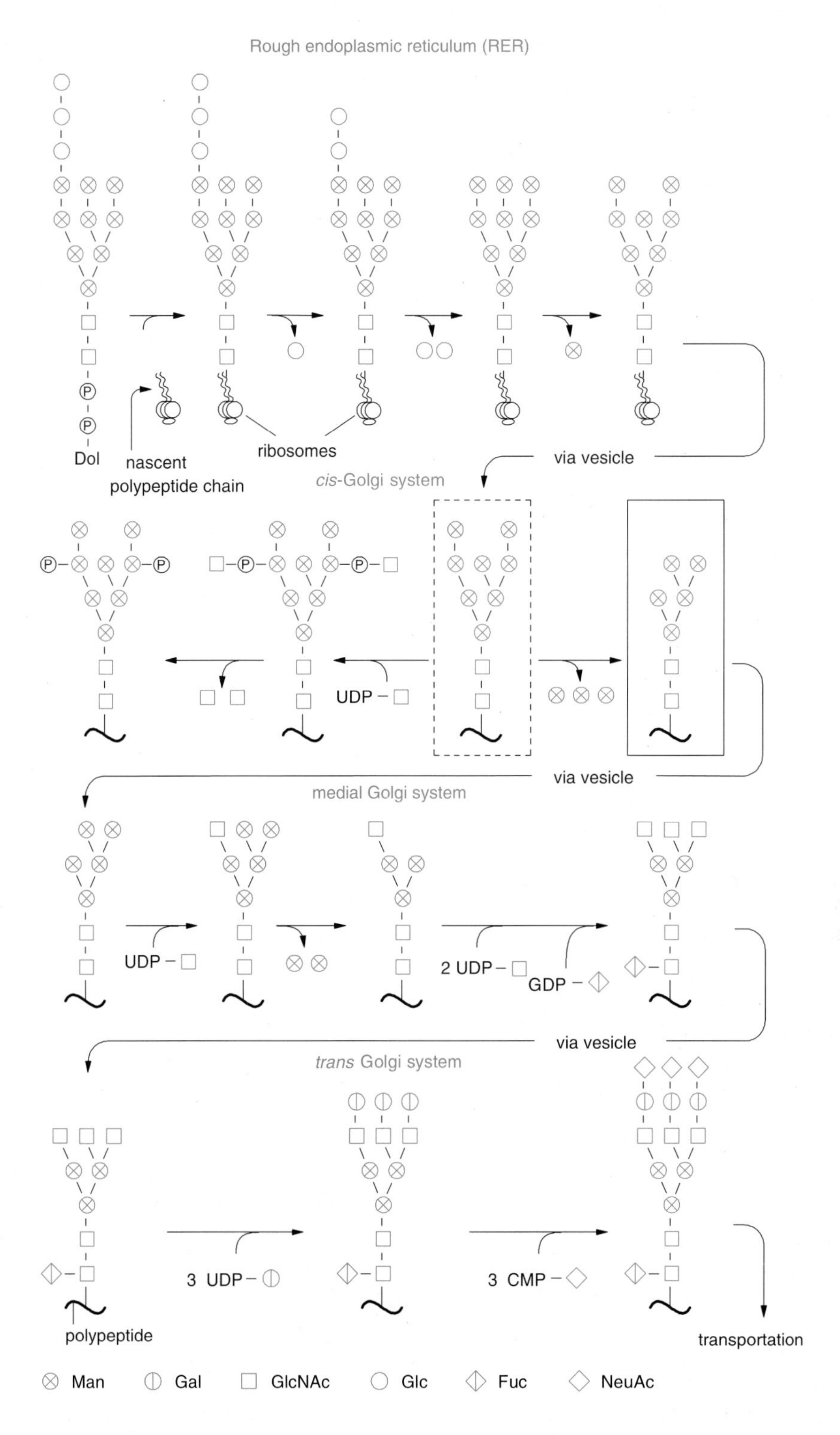


**Fig. 2.176** The *octamannosyl oligosaccharide* released from the ER and the remaining core oligosaccharide after action of α-mannosidase 1 are acceptor substrates. The former can be used for phosphorylation or further mannosylation; the latter are transformed and built up in the Golgi system by glycosyl transferases and glycosidases.

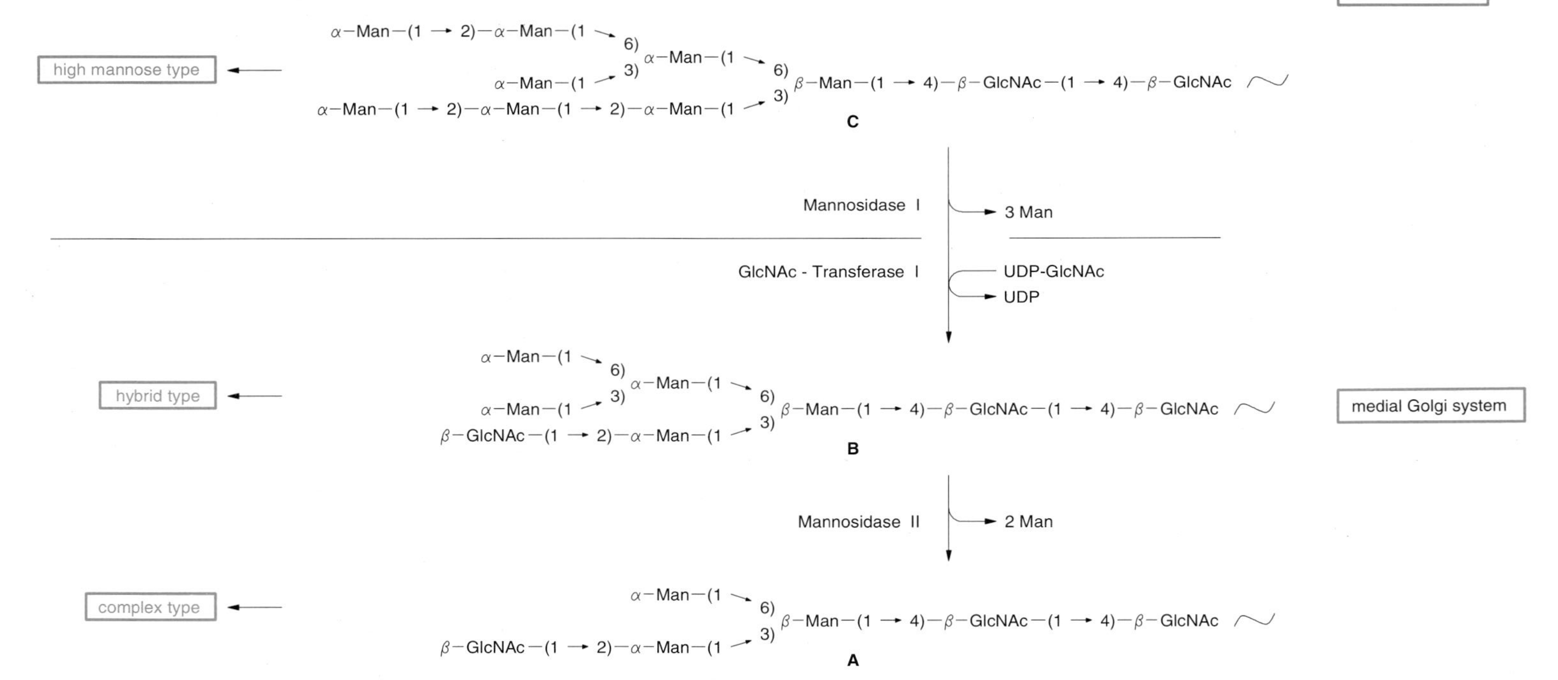


**Fig. 2.177** In the *cis*- and medial Golgi system the pathways divide for the formation of the three different types of oligosaccharide.

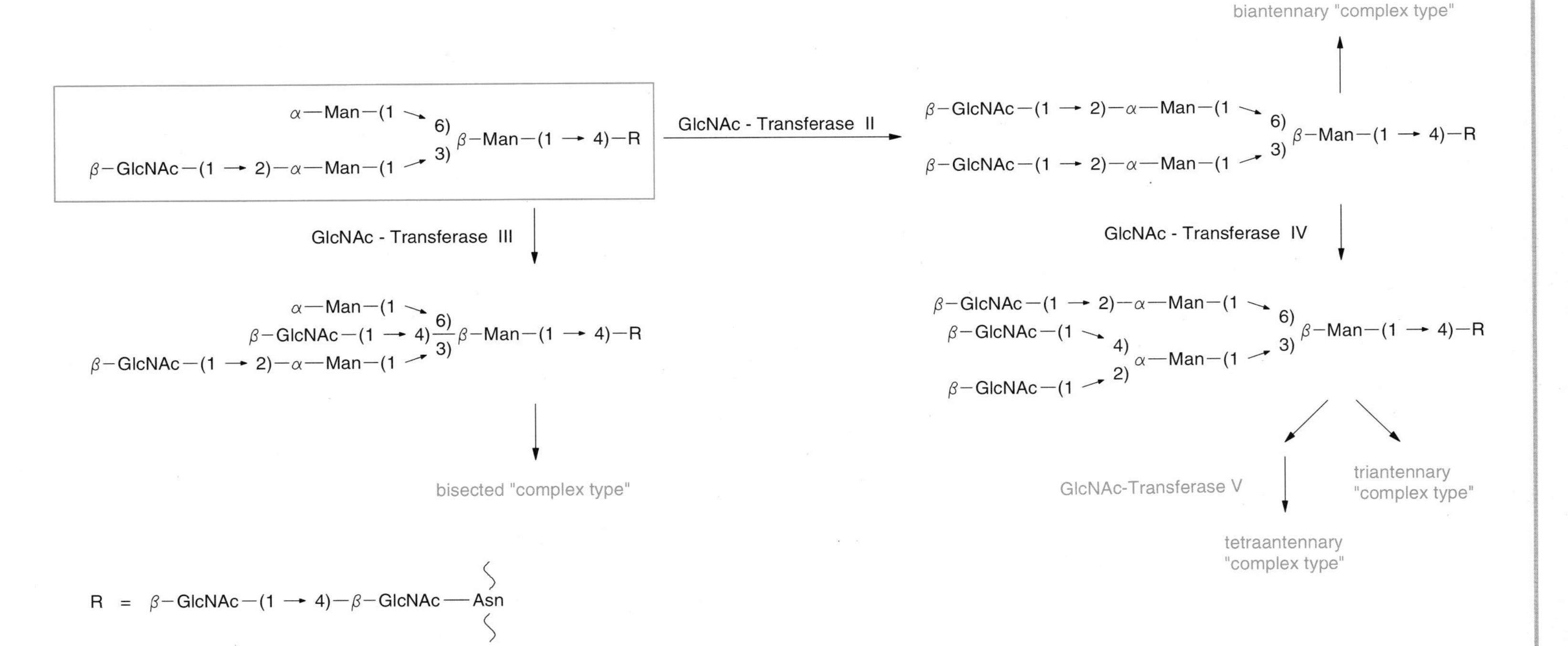


**Fig. 2.178** The "complex type" oligosaccharide of the *N*-glycosidic proteins originate from a hexasaccharide core structure

reactions the innumerable structures of the *N*-glycosidic bound oligosaccharides. As more or less mature final stages, the glycoproteins finally leave the Golgi system and are brought by vesicle transport to their respective destinations.

The biosynthesis of specific *N*-glycosidic oligosaccharides is controlled first of all through the provision of certain glycosyl transferases. Additionally, the possibility arises to control the synthesis of the glycoconjugate by way of the availability of acceptors, which for their part are formed through the deployment of specific glycoside hydrolases. Certainly, the at first sight seemingly complicated and elaborate treatment of the original precursor in the Golgi apparatus has its biological purpose. Research in the coming years will have to clarifiy the still many unsolved problems, above all in the area of overriding control and in regulation of place of the biosynthetic processes.

### 2.3.8.6 O-Glycosidic glycoproteins

Although indications exist that in many cases *O*-glycosidic glycoproteins[39,189] can work biologically in a similar manner to their *N*-glycosidic counterparts (see section 2.2.2.5, p. 146), nevertheless *O*-glycosidic bound carbohydrate structures (mucins) must rather befit general modulating functions (see section 2.1.5, p. 110). The lack of inhibitors of the biosynthesis of the core structure, similar perhaps to ***tunicamycin*** which prevents the formation of *N*-glycosidic glycoproteins, hinders the elucidation of specific biological functions of *O*-glycosidic glycoproteins.

*O*-Glycosidic glycoproteins have a comparatively direct biogenesis , which is dictated exclusively by the specificity of glycosyl transferases. The availability and activity of these enzymes decides the synthesis of a certain structure and it can be said with some confidence that a certain glycosyl transferase can make only a specific glycosyl bond. This includes anomeric configuration, type of acceptor monosaccharide (also oligosaccharide) and position on the acceptor monosaccharide. In contrast to the *N*-glycosidic glycoproteins, several core structures occur in the *O*-glycosidic glycoproteins, which with very few exceptions all begin with α-GalNAc linked to Ser/Thr (Fig. 2.179). The general chain elongation takes place from this core structure. At least the formation of the core structure takes place in the lumen of the Golgi apparatus. Numerous investigations have considered the initiating step and its specificity in relation to the acceptor peptide. The enzyme responsible, UDP-GalNAc: polypeptide α-*N*-acetylgalactosaminyl transferase, transfers to serine or threonine, preferably in a segment of chain containing proline residues. Apparently, a transport system exists for UDP-GalNAc in the Golgi membrane. It is to be assumed that the other nucleotides used for the core region arrive in the Golgi vesicles in a similar way.

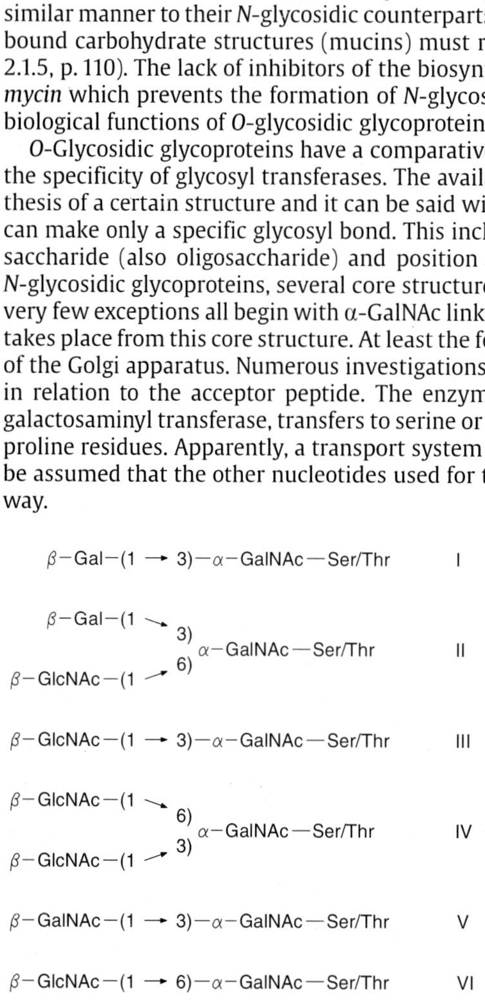

β−Gal−(1 → 3)−α−GalNAc—Ser/Thr    I

β−Gal−(1 → 3)<br>α−GalNAc—Ser/Thr    II<br>β−GlcNAc−(1 → 6)

β−GlcNAc−(1 → 3)−α−GalNAc—Ser/Thr    III

β−GlcNAc−(1 → 6)<br>α−GalNAc—Ser/Thr    IV<br>β−GlcNAc−(1 → 3)

β−GalNAc−(1 → 3)−α−GalNAc—Ser/Thr    V

β−GlcNAc−(1 → 6)−α−GalNAc—Ser/Thr    VI

**Fig. 2.179** There are at least 6 core structures (I – VI) in *O*-glycosidic glycoproteins. With very few exceptions they begin at the aglyconic end with an α-GalNAc residue.

The step following the initial step in the formation of the core structure I has been very well investigated. The galactosyl transferase which specifically transfers β-configured Gal onto the 3-position of GalNAc requires UDP-Gal as the donor substrate (Fig. 2.180). The enzyme, which is associated with

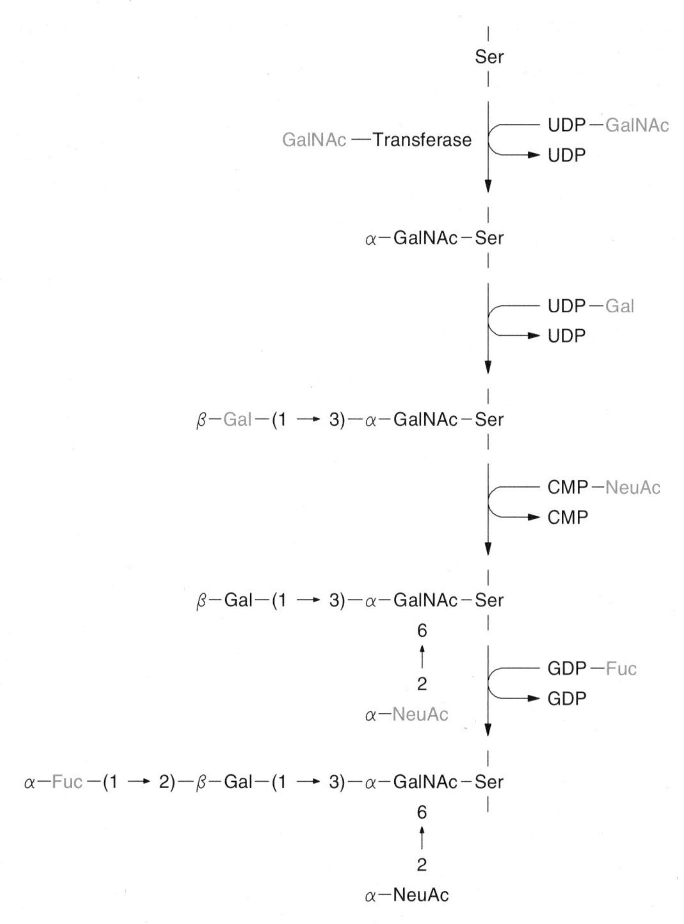


**Fig. 2.180** *Biosynthesis of the oligosaccharide from the submaxillar-mucin of the dog.* The formation of *O*-glycosidic oligosaccharide structures is often concluded without further chain lengthening with the attachment of typical peripheral monosaccharide residues such as fucose or *N*-acetylneuraminic acid.

the membrane, has been discovered in many different sources. It can be obtained in highly enriched form and therefore acceptor specificity can be established.

The extension of the core region leads to more or less expanded oligosaccharide structures, in which β-linked Gal and GlcNAc residues occur preferentialy. Linkage takes place through the 3-, 4-, and 6-positions. Many of the enzymes involved can be isolated cell-free and their specificity investigated.

Mostly the biosynthesis is interrupted with the formation of certain peripheral structures, which are also to be found in *N*-glycoproteins and in sphingolipids. Little is known about the control of these reactions and this knowledge is in part contradictory. Besides the already mentioned L-fucose and sialic acid, D-galactose is also considered a peripheral sugar. The glycosyl residues are mostly α-linked. The biosynthesis of the *O*-glycosidic oligosaccharides in the submaxillar mucin of the dog ends without further chain extension with the fucosylation and sialylation of the core structure 1 (Fig. 2.180).

It is an amazing fact, illustrated by several examples, that *N*-glycosidic and *O*-glycosidic oligosaccharide structures are to be found in one and the same protein and, accordingly, the biosyntheses take place under the same conditions. Hence, differentiation must be programmed already with the help of the signal sequence of the protein, or more probably with the signal tertiary structure. The fundamentally different acceptor structures, here a single GalNAc unit, there multiantennary oligomannosyl residues are also defined by it. In this manner, through the high specificity of the glycosyl transferases, the heterogeneity of both oligosaccharide types is guaranteed, although the donor substrates used for both pathways are, for the most part, the same. It will be seen that the very many great similarities of the core structure between O-glycosidic glycoproteins and the glycosphingolipids can lead to large similarities of the peripheral region. In this context the blood group determinants should be remembered, which are to be found both in secreted and membrane bound O-glycosidic glycoproteins and also in glycosphingolipids of the cell membrane.

Despite all of the progress of past years, a large number of questions remain open, above all in respect of the control of O-glycoside biosynthesis. It is unknown which factors influence the biosynthesis of the specific glycosyl transferases, which in turn control the structures of the oligosaccharides formed. With regard to such structures, significant variations are found in relation to the pathological conditions such as cancer or inflammation illnesses. Nowhere are there so large structural differences in the oligosaccharides, even in the same protein. A good many structures can be described already as polymers, many are branched, many linear. In the case of long chains, the question arises about the signal for chain termination. Branching may, or may not occur on identical chain segments. Factors that determine which core structure should be synthesised need to be addressed. Much is still unclear in the field of O-glycoprotein biosynthesis and further research is required.

### 2.3.8.7 Glycosphingolipids

All glycosphingolipids[78] contain *ceramide* as the aglyconic, hydrophobic component, that is N-acylated sphingosine, whose hydrophobic N-acyl chain can vary, as can also the hydrophobic chain of sphingosine itself. However, the countless sphingolipids can be distinguished by their carbohydrate structures. Compounds whose head group is a $\beta$-galactosyl residue (galactosphingolipids are small molecules, often sulphated in the carbohydrate part), can be rejected as relatively unimportant. All others may be subdivided somewhat arbitrarily into three groups, which differ with respect to their biogenesis at a relatively early stage (Fig. 2.181, p. 229). The first two steps involving glycosylation of the ceramides are identical in each case, so that all complex glycosphingolipids show the same core structure, a $\beta$-lactosyl residue. The core structure probably arises in the Golgi apparatus, similarly to the formation of the O-glycosidic glycoprotein. The glycosyl transferases are specific, above all in relation to the ceramide residue, whereby up to a minimum size the length of all of the N-acyl residues appear to be of no significance. Just as specifically, the transfer of galactose to the $\beta$-D-glucosylceramide residue then proceeds (Fig. 2.181). The enzymes are all membrane-associated and correspondingly difficult to handle, and up to the present have not been isolated in pure form. Donor substrates are the corresponding nucleotide derivatives. A lipid mediated synthesis of the glycosphingolipids could not be proved as yet and is also certainly unlikely.

Probably they separate according to cell type at the third biosynthetic step of the pathway. When a further galactosyl residue is transferred (path 1), relatively large neutral structures (*globo series* and *isoglobo series*) are formed, which above all find themselves again on the erythrocyte membranes. It should be noted that the second galactosyl residue in the chain is transferred with the $\alpha$-configuration which underlines the peculiarity of the catalysing galactosyl transferase. Chain elongation occurs through the action of GalNAc transferases. Above all, the linking is generally (1→3)-$\beta$. The *Forssman antigen* (Fig. 2.182) is an example of how the biosynthesis of a glycosphingolipid is concluded with a (1→3)-$\alpha$ linked GalNAc end group (see section 2.3.8.6, p. 226).

To the second group of glycosphingolipids belong those which, after conclusion of their biosynthesis carry, for example, blood group active structures and apparently also can offer long polysaccharide chains. The structures stand out also because of the peripheral fucosyl residues.

Path II is initiated by the transfer of a $\beta$-GlcNAc residue to the 3-position of the terminal galactose. A (1→4)-$\beta$-galactosylation follows so that most representatives of the glycosphingolipids built by path II exhibit the (1→3)-lactosaminyl-lactose core region. This can either lead to polymers by repetition of the series of transfers with the lactosamine-repeat unit or, through galactosylation and fucosylation, to the numerous blood group antigens. Here it can be very clearly recognized that comparable acceptor structures with the help of the available glycosyl transferases can lead to the same peripheral determinants with different glycoconjugates, so that glycosphingolipids and O-glycosidic glycoproteins, with regard to their oligosaccharides, often differ only in the core region.

A group of glycosphingolipids which appear especially important are the gangliosides, for whose structure one or more sialic acid residues are typical and which therefore can also be named acidic

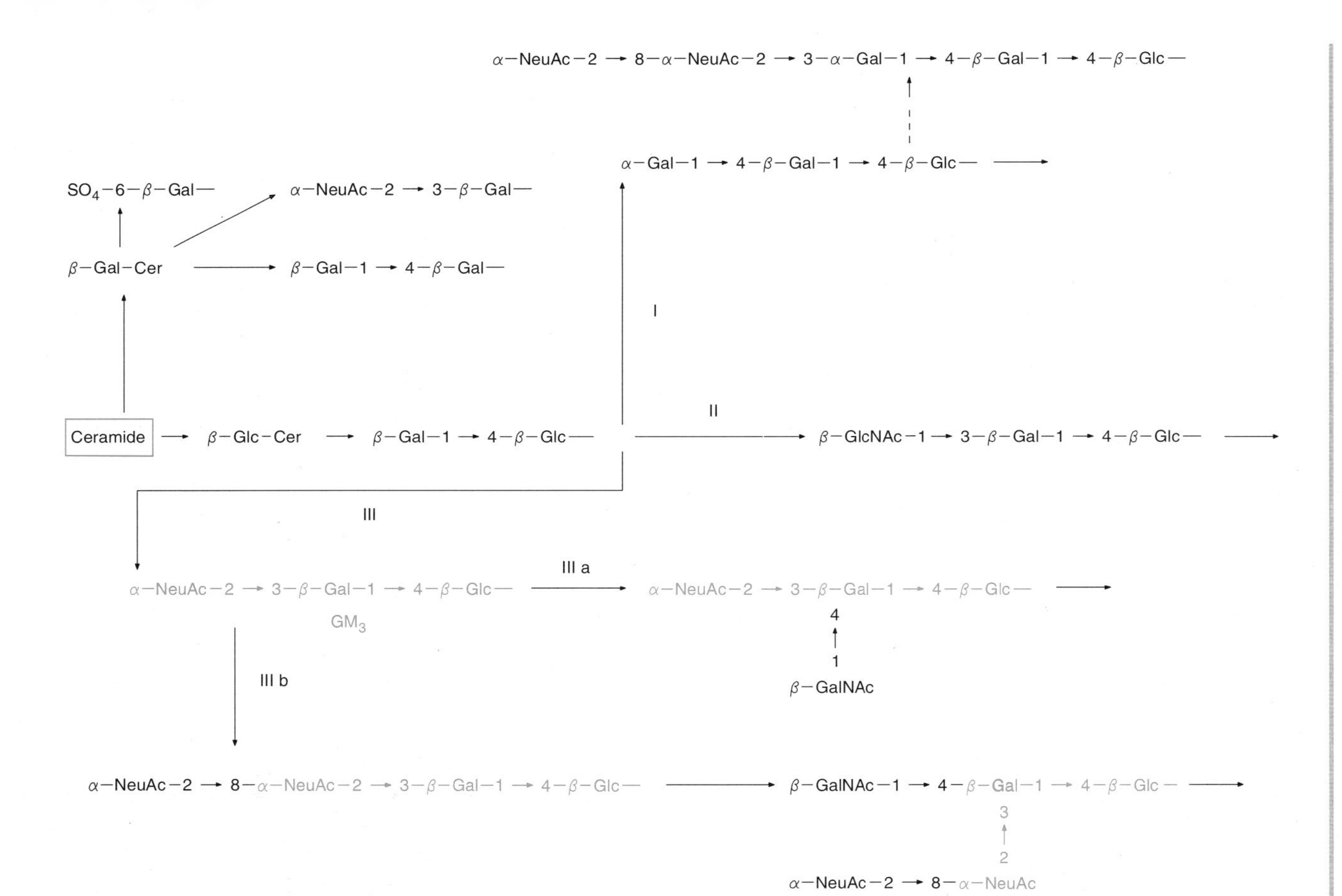


**Fig. 2.181** The three pathways for the *biosynthesis* (I-III) of *complex glycosphingolipids* with glucose as the bridgehead monosaccharide separate with the third transfer step. Each produces a special group of carbohydrate structures. The simple cerebrosides with Gal as the bridgehead monosaccharide branch off previously. Probably the synthesis of the gangliosides of the central nervous system proceeds on two physiologically separated pathways IIIa and IIIb starting from a common preliminary stage $GM_3$.

α−GalNAc−(1 → 3)−β−GalNAc−(1 → 3)−α−Gal−(1 → 4)−β−Gal−(1 → 4)−β−Glc−Cer

**Fig. 2.182** The *Forßman antigen.*

glycosphingolipids. As can be seen from Fig. 2.181, Path III leads to this group of compounds. The biosynthetically active, specific glycosyl transferases were isolated from brain tissue and used biosynthetically *in vitro* with the corresponding nucleotide derivatives as glycosyl donors for the synthesis of the various gangliosides. The biosynthetic pathway of the whole class of substances was initiated by the transfer of α-NeuAc from CMP-NeuAc to the 3-position of galactose in lactosyl ceramide. The enzyme is a very specific sialyl transferase.

The product $G_{M3}$ can now either be supplied (path IIIa) with a GalNAc residue or with a second NeuAc residue (path IIIb). Although both paths can lead after passage through more transfer steps to a common product, with of course different intermediate steps, it is thought that *in vivo* a separation of both paths exists (Fig. 2.181).

The gangliosides often have several NeuAc residues in various positions on the base-chain comprising neutral sugars. Frequently, several NeuAc residues are linked together mostly with (2→8)-α-links. Fucosyl residues, linked to the neutral sugar chain, open up a structural variation which it is difficult to overlook. It is understandable that with the ever increasing number of gangliosides isolated, the biogenetic pathways are frequently still unclear.

### 2.3.8.8 Inhibitors of the biosynthesis of glycoconjugates

The frequently reported findings, whereupon certain glycosylation patterns in glycoconjugates[190] go together with definite biological functions[191] and pathological conditions[192], makes it seem worthwhile to be able to artificially intervene in the biosynthesis of such structures. Above all, the intervention should at the same time be targetted, that is to say be as specific as possible. So far as is known today, the synthesis, or rather the expression of distinct glycoforms, is linked directly through glycosyl transferases and glycosidases and eventually indirectly through the transport system for glycosyl donors, acceptors, and mature products. According to the hypothesis for the transferases put forward by H. Schachter of *one bond = one enzyme*, which partly is valid also for the glycosidases of the "trimming" system (see section 2.3.8.5, p. 218), the number of the available enzymes must be very large. All the more difficult is it therefore to influence definite activities by inhibition. Competitive and also irreversible enzyme inhibitors are an important means for the investigation of biogenesis mechanisms, but they are possibly applicable also for a modulation – not to say blocking – of the biosynthesis of glycoforms *in vivo*.

It is in the nature of glycosyl transferases, that an influence on their action can only occur specifically by way of an inhibition of the binding of acceptor substrates, since only this restricts the special specificity of many of the transferring enzymes. If the donor binding was to be influenced, which hitherto has almost exclusively been the case, it would affect, relatively unspecifically, possibly several enzymes of the same donor specificity. All sialyl transferases – it is suggested that alone in humans ten different ones exist – use CMP-NeuAc as donor substrate. The use of inhibitors which are mostly analogues of the substrate, can definitely show action *in vitro* but would indeed be unsuited to a selective therapy. Similar considerations apply also to the other glycosyl transferases. In general the inhibition constants are not particularly good (mM) and the chances of achieving an effective inhibition *in vivo*, thanks to the normally high affinity of the physiological donor substrate, are certainly estimated to be small.

The synthesis of a so-called *bisubstrate inhibitor* for the very effective inhibition of the β-D-galactoside-α-(1→2)-L-fucosyl transferase, lying in the $\mu$M region, shows a new approach in this area[193] (Fig. 2.183). The compound binds not only to the donor binding site through the nucleotide portion but also via β-D-galactosyl residue to the acceptor binding site.

Kinetic investigations were performed with isolated enzyme preparations. Limitations are to be expected with *in vivo* systems. Noteworthy is the strong increase in the affinity compared with *normal* donor analogues. This synergy is designated as a "clustering" effect[194], that is an increase in affinity resulting from binding in more than one binding site through structural elements covalently bound to one another. A similar effect can be established[195] in the competitive inhibition of (1→4)-β-D-galactosyl transferase from bovine milk with acceptor analogues. Further experiments to obtain the selective

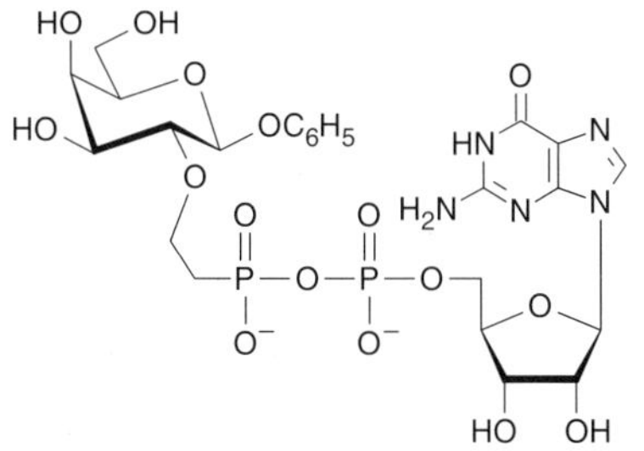

**Fig. 2.183** The compound is a combination of a donor and an acceptor analogue. It effectively inhibits $\beta$-D-galactoside-$\alpha$-(1→2)-L-fucosyl transferase by occupation of both the donor and acceptor binding sites.

blocking of a specific glycosyl transferase – in this case the GlcNAc-transferase V – by acceptor analogues were successful[196]. Inhibition constants lie in the region of about 0.1 mM. The syntheses of acceptor analogues, in which the position to be glycosylated in a trisaccharide is blocked, (Fig. 2.184), are very time consuming.

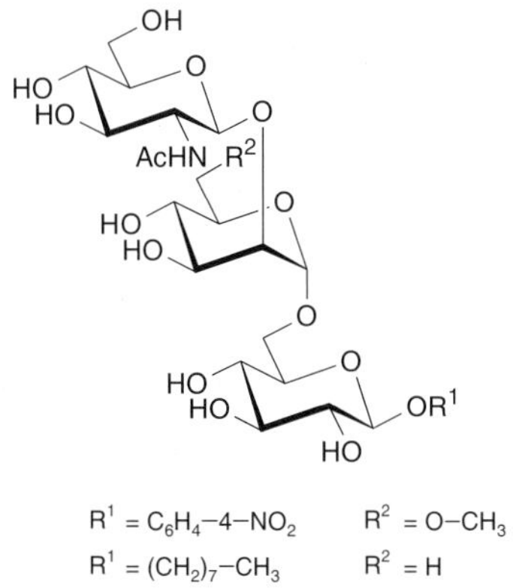

**Fig. 2.184** Structures of two acceptor analogues for the specific inhibition of the GlcNAc-transferase V. The 6-position in the $\alpha$-mannosyl residue is blocked towards glycosylation.

$R^1 = C_6H_4{-}4{-}NO_2$ $R^2 = O{-}CH_3$

$R^1 = (CH_2)_7{-}CH_3$ $R^2 = H$

The difficulties demonstrated here of the selective inhibition of certain glycosyl transferases have, to an extent, greatly hindered the elucidation of the mechanism of biogenesis for *O*-glycosidic glycoproteins and glycosphingolipids, whose specific synthesis depends indeed only on the transferases, and they are also the reason for the comparatively poor record of success up till now of research in this area.

The oligosaccharides of *N*-glycosidic glycoproteins arise from synthetic pathways, which allow intervention with inhibitors which are not only those of the usual glycosyl transferase reactions. Especially sensitive targets are the *dolichol cycle* in the RER, whose function is a requirement for the formation of all *N*-glycoproteins, and the very specific degradation reactions through glycosidases in the Golgi system (see 2.3.8.5, p. 218). The inhibition of the dolichol cycle is naturally universal, that of the glycosidases can be relatively specific. The use of inhibitors in this area has contributed very much to the elucidation of the biogenesis of *N*-glycosidic glycoproteins. The great scientific progress of the past years has its roots here.

The most highly active inhibitors are antibiotics, whose importance was first recognised in the blocking of cell wall growth in bacteria. The *tunicamycins* are important and, in less measure, *bacitracin* (Fig. 2.185).

The tunicamycins (isolated from streptomyces species) differ from one another only in the length of the hydrophobic *N*-acyl residue. They block the first step of the dolichol cycle through inhibition of the transfer of GlcNAc-1-P from UDP-GlcNAc to the dolichyl monophosphate with the remarkable inhibition constant of $7 \times 10^{-9}$ M. With the example of the tunicamycins, one of the principles of strong competitive inhibition can be very nicely illustrated. Donor UDP-GlcNAc and acceptor Dol-P are imitated by two structural parts, the acyl chain and hydrophilic carbohydrate-nucleoside residue, and bound on the basis of the similarity. The fact, that in the inhibitor both parts are covalently bound to one another, has a "clustering" effect and thereby results in high affinity (Fig. 2.185).

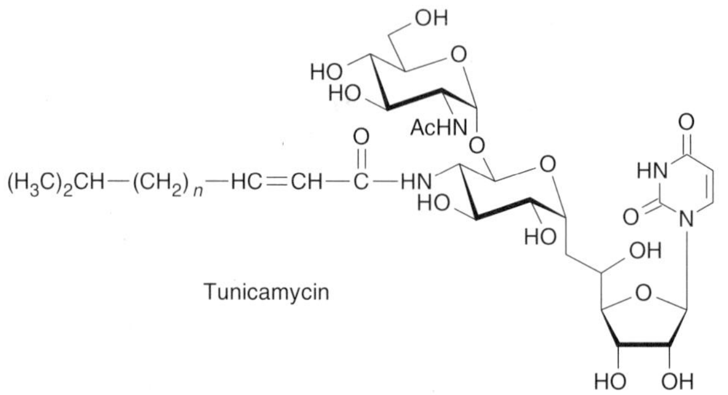

Tunicamycin

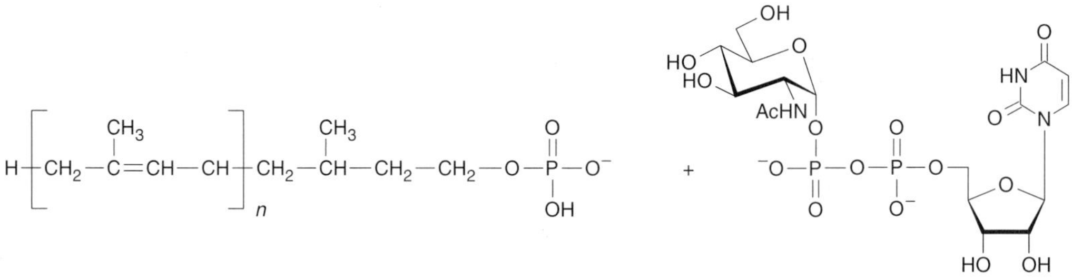

Dolichol phosphate

UDP-GlcNAc

**Fig. 2.185** High *inhibitory activity* of the *tunicamycins* may probably be attributed to the cooperative interaction of their hydrophilic and hydrophobic structural elements, which resemble UDP-GlcNAc and dolichol, respectively.

*Bacitracin*, a dodecapeptide partly consisting of *unnatural* amino acids, works by blocking dolichyl pyrophosphate phosphatase and prevents the preparation of dolichyl monophosphate, necessary for the cycle. The inhibitor is not transported through the cell membrane and therefore is harmless for eukaryotes but is highly active against bacteria (see section 2.3.8.1, p. 207203). Some further compounds which inhibit the Dol-P-cycle are known, part peptide with hydrophobic residues, part nucleoside analogue, but have never attained the significance of tunicamycin.

Most recently, the influence of the "trimming" processes by glycosidases was especially given great attention. There exist a series of natural inhibitors of relatively simple structure which are easily chemically imitated and modified. Most of these compounds have structures similar to monosaccharides and are distinguished by basic (cationic) properties as a result of a non-acylated amino N-atom[159]. Of the numerous known compounds – most are chemically synthesised analogues of natural products – only the most important can be introduced here. As the formulae show, these basic compounds are hydroxylated, partly bicyclic, tetrahydropyridines or pyrrolidines, which by reason of their structure are similar to monosaccharides or glycosides and, accordingly, to the substrates or products (Fig. 2.186). The high inhibitory potential which many of the compounds show can be ascribed not only to their conformations being *related* to those of monosaccharides but, in addition, to the presence of secondary or tertiary amino groups which can build stable salt bridges with acid residues in the active site of the particular enzyme.

The high specificity of many inhibitors for very special glycosidases still makes it necessary to consider differentiating factors, which probably are to be found in fine differences in the binding region of the glycoside hydrolases. In Fig. 2.187 are shown some of the transformations which underlie the processing of the *N*-glyconjugates in the Golgi system and the inhibitors which block specific hydrolytic reactions.

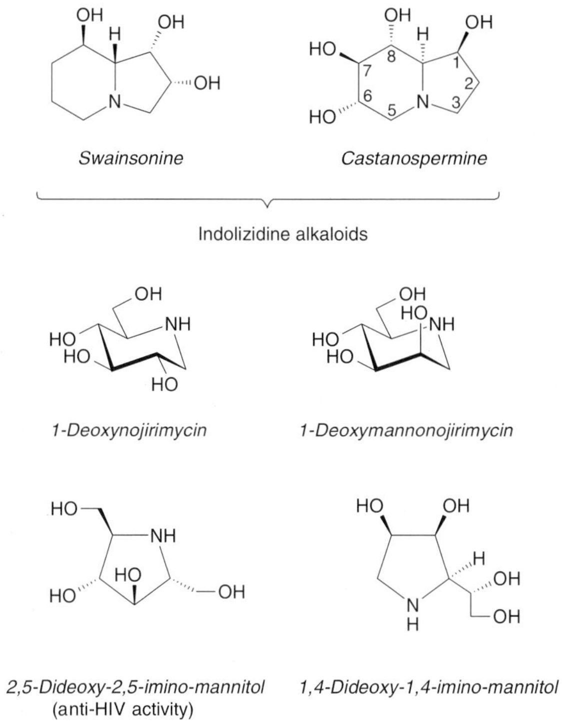

Swainsonine

Castanospermine

Indolizidine alkaloids

*1-Deoxynojirimycin*

*1-Deoxymannonojirimycin*

*2,5-Dideoxy-2,5-imino-mannitol*
(anti-HIV activity)

*1,4-Dideoxy-1,4-imino-mannitol*

**Fig. 2.186** Some basic glycosidase inhibitors.

The synthesis of new inhibitors forms the research of many laboratories. Those systems which reflect the hypothetical transition state of the enzymatic glycoside cleavage seem to be very effective. The compounds in general can adopt a flattened conformation in the region of the anomeric centre, which mimics the situation which leads to the glycosyl cation (Fig. 2.188).

In this connection, it should not be overlooked that many of the countless compounds tested only work on isolated enzyme preparations. It is possible that for some, because of their high polarity, passage through cell membranes as well as other membranes, and thereby an action *in vivo*, is prevented. Many inhibitors may not withstand also the hydrolytic enzymes in the interior of the cell and thus quickly lose their inhibitory properties through biological degradation. However, it has been shown with classical experiments on intact cells with the inhibitors *castanospermine* and *swainsonine*[197], that it is indeed possible to influence glycoprotein biosynthesis with competitive inhibitors even in organisms.

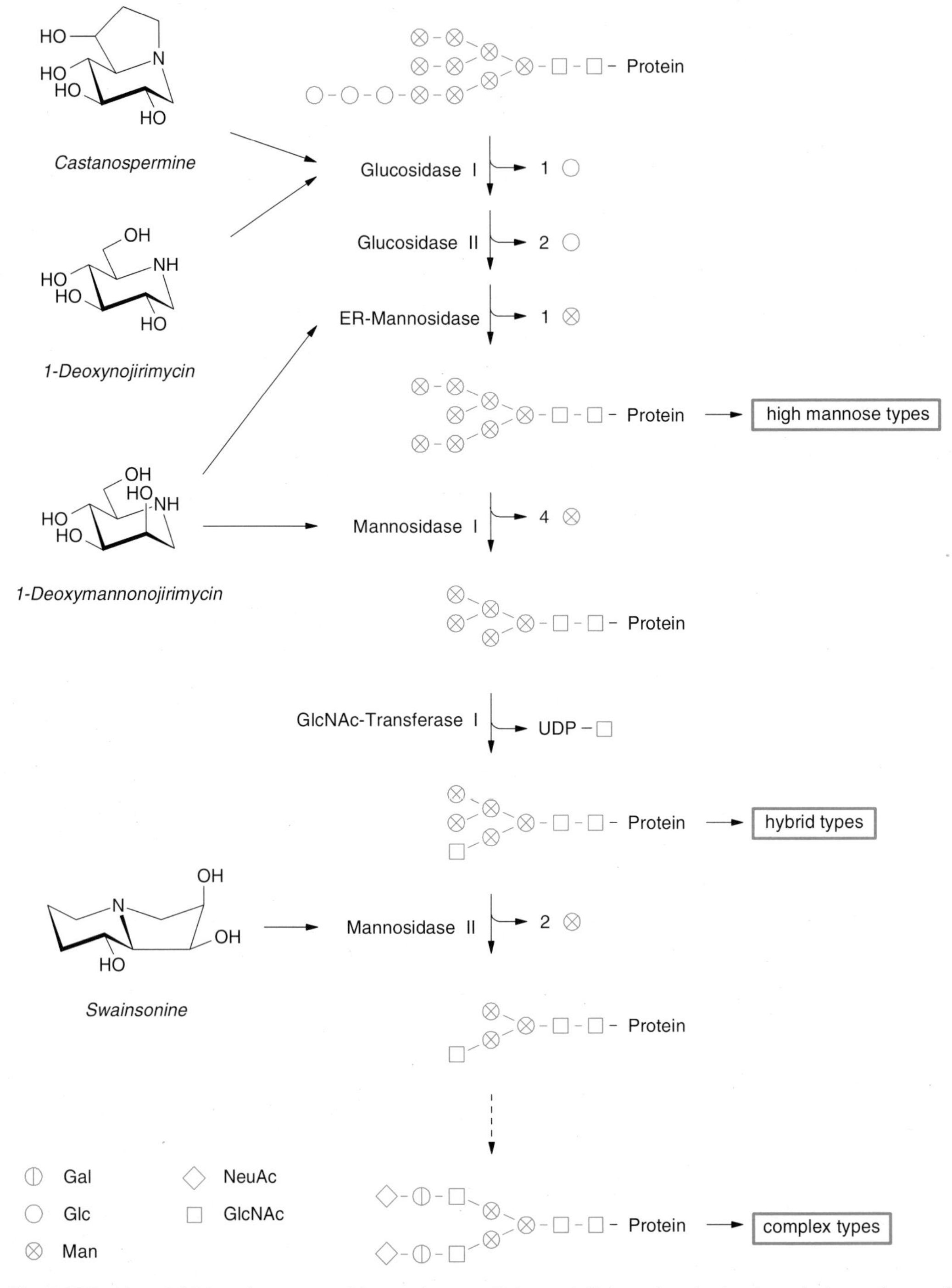


**Fig. 2.187** Many inhibitors intervene with amazing specificity at definite points in the degradative pathway of the *N*-glycosidic glycoproteins.

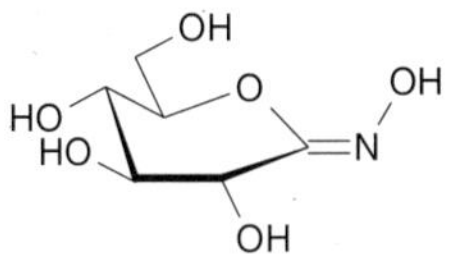

**I** Lactone oxime

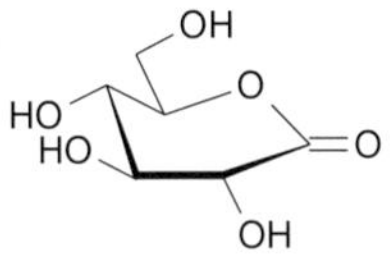

**II** Lactone

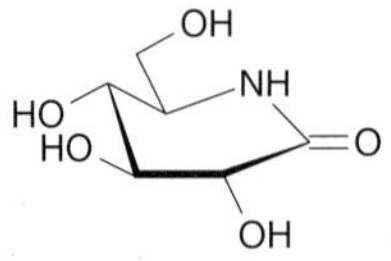

**III** Lactam

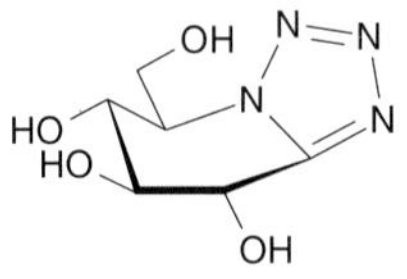

**IV** Tetrazole

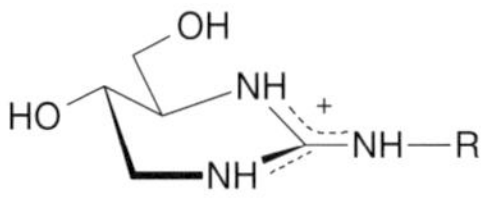

**V** Guanidinium ion

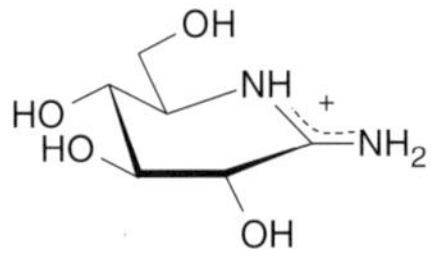

**VI** Amidinium ion

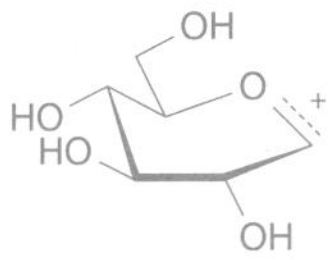

A

**Fig. 2.188** Some of the synthetic inhibitors of glycoside cleaving enzymes, which are designated as transition state analogues. They are characterised, as is the hypothetical glycosyl cation **A**, by a flattened conformation in the region of the anomeric centre.

**I** Beer, D., Vasella, A.: *Helv. Chim. Acta* **1986**, *69*, 267–270,
**II** Leaback, D. H.: *Biochem. Biophys. Res. Comm.* **1968**, *32*, 1025–1030,
**III** Legler, G., Sinnott, M. L., Withers, S. G.: *J. Chem. Soc. Perkin Trans. 2* **1980**, 1376–1383,
**IV** Ermert, Ph., Vasella, A.: *Helv. Chim. Acta* **1991**, *74*, 2043–2053,
**V** Lehmann, J., Rob., B.: *Justus Liebigs Ann. Chem.* **1994**, *8*, 805–809,
**VI** Ganem, B., Papandreou, G.: *J. Am. Chem. Soc.* **1991**, *113*, 8984–8985.

# 2.4 Biochemical methods for the synthesis and interconversion of carbohydrates[198]

A knowledge of the interrelationships in general carbohydrate metabolism – aldol reactions, condensations, hydrolyses – should form the basis for a meaningful application of enzymes as organic chemical reagents. Above all it should show the limitations to which the application of enzymes must be subject. In this regard it is astonishing to what extent many enzymes are able to tolerate the unphysiological conditions of organic chemistry. Perhaps it will be possible in the future to make use of enzymes which have been designed and produced by human hands for a specific purpose. *Protein engineering*, the genetic modulation and modelling of proteins with catalytic activity or the production of catalytic antibodies, points in this direction.

The application of biocatalysts for the production of natural substances has a very long history especially in the area of carbohydrates. Probably one of the oldest indications of the possibility of using degradative enzymes such as *maltase* from yeast for the synthesis of *maltose* from D-glucose dates from the last century[199]. The synthesis of *gentiobiose* from *D-glucose* with help from *almond emulsin*, as it was described some years later[200], remains a good method to obtain this disaccharide. It is obvious that a detailed knowledge of carbohydrate metabolism, the availability of the corresponding purified enzymes or required cultivated microorganism, allied to highly developed technical apparatus, have secured biochemical methods for obtaining carbohydrates a high status under the general preparative methods of organic chemistry.

The advantages are obvious: reactions run under mild conditions which are in general regioselective and almost always stereospecific. Accordingly, the use of protecting groups can be avoided. Immobilised enzymes make possible the use of these biocatalysts in continuous synthetic processes. Biochemical methods are frequently a valuable supplement to the procedures of organic chemistry. Thus, combined chemical/enzymatic syntheses are developed in increasing measure for carbohydrate and non-carbohydrate substances.

It is obvious that biochemical methods must remain confined to relatively few basic transformations and that they may not be regarded as competitive with the abundance of reactions possible today in organic chemistry. Biocatalysed reactions for the production of carbohydrates and their derivatives comprise above all condensation reactions, for example glycosylation and acylation, as well as the corresponding hydrolysis reactions. C,C-Bond forming reactions depend exclusively on the principle of the aldol reaction and find widespread application for the synthesis of unusual monosaccharides. Reductions and oxidations serve for the modification of monosaccharides.

Since the fermentation procedures in execution and objectives do not differ all that much from those with isolated enzymes, although they are limited in their possibilities, they should be incorporated, or at least mentioned, in the corresponding sections.

## 2.4.1 Oligosaccharides and glycosides

Principally there are four different possibilities for producing oligosaccharides or glycosides through by enzymatic glycosylation:

1. Reversal of the glycoside hydrolase reaction.
2. Transglycosylation, starting from glycosidic precursors or glycosyl fluorides.
3. Transglycosylation, starting from nucleoside diphosphate sugars.
4. Glycosylation, starting from unsaturated glycosyl donors.

The abundance of literature in this connection is restricted almost exclusively to the synthesis of analytically detectable quantities of substances. It is therefore only necessary to refer to the few procedures which are really useful on a preparative scale.

### 2.4.1.1 Reversal of the glycoside hydrolase reaction

Glycoside hydrolases (glycosidases) have the biological function to cleave hydrolytically glycosides or oligosaccharides. They are in general absolutely specific with respect to the glyconic part of the substrate and its anomeric configuration, yet often unspecific with respect to the aglyconic component. Fundamentally, the enzymatic hydrolysis of a disaccharide is reversible and, by proper choice of concentration of the monosaccharides, the reverse reaction can be used for synthesis. It is to be noted, of course, that the glycosyl residue transferred from the enzyme can react with all hydroxy groups of the acceptor, so that in the case of a hexopyranose, which is simultaneously donor and acceptor, the formation of five different linked disaccharides is possible. In general, however, the (1→6)-linked product, formed by transfer to the primary hydroxy group, predominates and its isolation is therefore worthwhile. The method stands out through its simplcity, above all if the enzymes are commercially available or else are easily accessible.

The classical method for producing *gentiobiose* was modified with modern aids[201]. It can serve as an example with the comment that the other disaccharides are capable of being produced in a comparable manner and that the somewhat antiquated method of chromatography on active carbon-Kieselguhr can be replaced by preparative gel chromatography or high pressure liquid chromatography (HPLC). The equilibrium position is reached after several days, after which work up yields about 6 grams of pure gentiobiose from an original 60 grams of glucose.

The *reversal* of *oligosaccharide hydrolysis*, as represented here as a synthetic method, has some weaknesses, as can easily be recognised. The yields are small on account of the unfavourable thermodynamic position of equilibrium. In addition there are very long reaction times and sometimes separation problems which are difficult to solve. On the other hand, however, the simplicity of the method is attractive.

A distinct improvement can be achieved with immobilised enzymes. The oligosaccharide formed is separated by circulation of the solution away from the region of the immobilised catalyst and is continuously isolated from the equilibrium mixture with a suitable adsorbent. In this way and manner, a series of disaccharides can be produced in very good yields through the action of β-galactosidases on monosaccharide solutions, from either D-glucose, 2-acetamido-2-deoxy-D-glucose, or D-fructose, each as acceptor and D-galactose as donor[202]. The immobilised catalyst and the adsorbent (active charcoal) for the oligosaccharides are accommodated in a columns arranged one after another. The yields in the favoured disaccharide amount to well above 10% after one day (Fig. 2.189)[202].

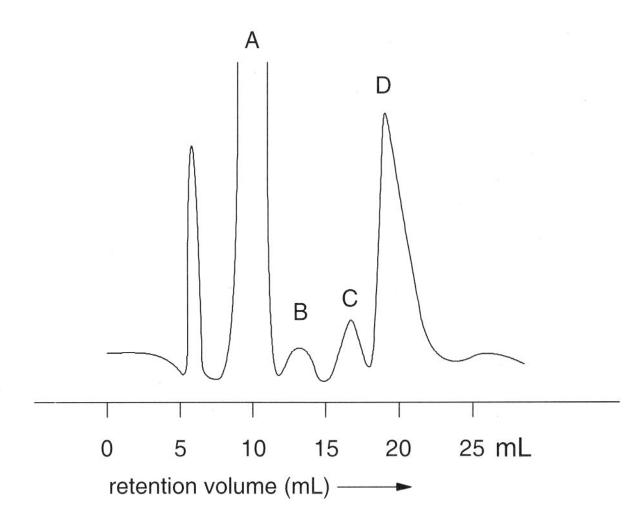


**Fig. 2.189** HPLC of the 50% ethanolic eluate from the activated charcoal column. The products resulted after circulation for 24 hours of an aqueous solution of 10% galactose (donor) and 30% *N*-acetyl-D-glucosamine (acceptor) through a system of two columns, one with immobilised β-galactosidase (*E. coli*) and one with active carbon. **C** corresponds to the (*N*-acetyllactosamine), **D** to the (1→6)-linked disaccharide.

Non-reducing trisaccharides with a trehalose unit were produced both by the procedure described above with immobilised glycosidases and active charcoal columns connected one after another and also in solution with the help of graded substrate specificity[203]. For example, isomeric β-D-galactosyl trehaloses, synthesised in an initial step with β-galactosidase (*A. orycae*), were hydrolysed again with the help of another β-galactosidase (*E. coli*) to leave, after hydolysis of all the rest, the trisaccharide *O*-β-D-galactopyranosyl-(1→3)-α-D-glucopyranosyl-α-D-glucopyranoside.

### 2.4.1.2 Transglycosylations

**Glycosidic precursors or glycosyl fluorides.** A synthesis from free monosaccharides is not suitable for the preparation of mixed oligosaccharides in reasonable yields. A certain selectivity and higher yields can be obtained if one starts with oligosaccharides as glycosyl donors. A fine example of an enzyme catalysed positional isomerisation is the conversion of the easily obtained lactose into *allo-lactose* ($\beta$-D-galactopyranosyl-[1→6]-D-glucose). The principle of transglycosylation also applies here. Attainment of thermodynamic equilibrium is not expected since the transfer reaction is accomplished under kinetic control. In a systematic investigation it was found that the translocation of the D-galactosyl residue from the 4- to 6-position of the D-glucose can take place without the latter leaving the acceptor binding site. The typical reaction profile (Fig. 2.190), obtained by HPLC analysis of the incubation mixture over short time intervals, shows the product distribution in an incubation mixture of lactose with $\beta$-galactosidase from *E. coli* [204].

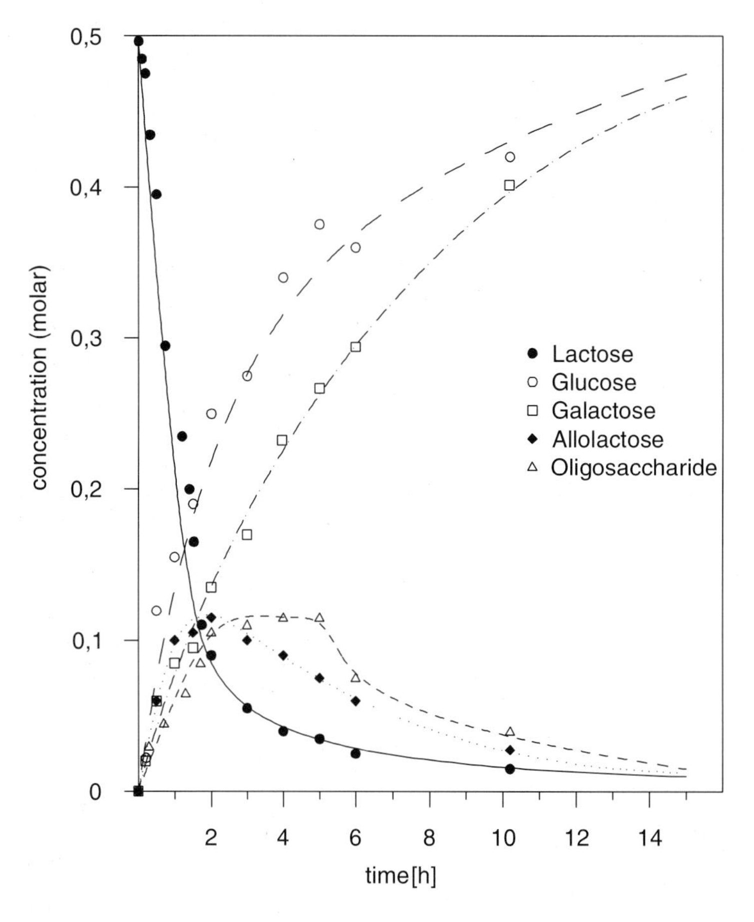


**Fig. 2.190** Rate of lactose consumption and of product formation in an experiment involving the incubation of a 0.5 M lactose solution with $\beta$-galactosidase from *E. coli*. The term oligosaccharide includes all tri- and tetra-saccharides.

Although the relative quantity of *allolactose* can be relatively high and the reaction time relatively short under optimum reaction conditions, the preparative separation of the complex mixture for the production of larger amounts of the disaccharide makes for considerable difficulty.

While in general the glyconic residue of a reducing disaccharide is recognised and bound very specifically by a glycoside hydrolase, there is great tolerance regarding the aglyconic moiety. This applies not only for the hydrolysis, the transfer of glycosyl residue to water, but also for the transfer to a monosaccharide with formation of a disaccharide. The relatively high affinity of a monosaccharide for the acceptor binding site is certainly also responsible for the fact that under kinetic control in dilute solution, in general, a synthesis of oligosaccharides is possible with the aid of glycoside hydrolases. In contrast to polyhydroxy compounds, simple aliphatic alcohols or phenols have only moderate affinity for the acceptor binding site, so that a synthesis of glycosides with such acceptors, under physiological conditions in any case, is very inefficient[205]. In contrast, aromatic glycosides are good donor substrates for two reasons (Fig. 2.191). On account of the nucleofugic properties of phenols (HO-Aglyc$_1$) the intermediate glycosyl enzyme is formed quickly ($k_1$ is relatively large). The practically non-existent affinity of the phenol for the acceptor binding site compared with a monosaccharide prevents the resynthesis ($k_2 << k_1$) of the aromatic glycoside and promotes the synthesis of an oligosaccharide ($k_3 >> k_2$). If the monosaccharide as acceptor is present in higher concentration, the equilibrium is shifted further.

$$\textbf{a}\quad \text{Glyk—O—Aglyk}_1 + \text{Enz—H} \underset{k_2}{\overset{k_1}{\rightleftharpoons}} \text{Enz—Glyk} + \text{HO—Aglyk}_1$$

$$\textbf{b}\quad \text{Enz—Glyk} + \text{HO—Aglyk}_2 \underset{k_4}{\overset{k_3}{\rightleftharpoons}} \text{Enz—H} + \text{Glyk—O—Aglyk}_2$$

**Fig. 2.191 a,b** Simplified reaction scheme for the reaction of a donor substrate (Glyc-O-Aglyc$_1$) with a corresponding glycoside hydrolase (Enz-H) in the presence of two different acceptors (HO-Aglyc$_1$ and HO-Aglyc$_2$).

Allolactose, with phenyl β-D-galactoside as donor substrate, D-glucose as acceptor, and β-galactosidase from *E. coli*, is formed in 25% yield[206].

The enzymatic synthesis of oligosaccharides with glycoside hydrolases, starting from phenolic glycosides, is generally applicable. The methodology gains in importance, as also that of the hydrolytic reversal, with the development of efficient separation methods[207]. Mostly, a transglycosylation with glycoside hydrolases occurs preferentially on a primary hydroxy group if one is present in the acceptor molecule, which naturally limits the general applicability of the method.

Some years ago, the interesting observation was made that structural alteration of the acceptor at its anomeric centre can influence the regioselectivity of the transfer reaction. A series of investigations probed this hypothesis[208]. The regioselectivity of the transfer reaction changes according to aglycone of the acceptor glycoside and its anomeric configuration. This can go as far as to drive the usual production of (1→6)-linked compounds in favour of glycoside bond formation at other positions. Yields of 20% and

**Table 2.4** Combination possibilities for the enzymatic synthesis of oligosaccharide glycosides

| Enzyme | Glycosyl donor | Glycosyl acceptor | Main product | yield (%) |
|---|---|---|---|---|
| α-D-Galactosidase (coffee) | α-Gal–OPhNO$_2$-*p* | α-Gal-OCH$_3$ | α-Gal-(1→3)-α-Gal-OCH$_3$ | 27 |
| | | | α-Gal-(1→6)-α-Gal-OCH$_3$ | < 2 |
| | | β-Gal-OCH$_3$ | α-Gal-(1→3)-β-Gal-OCH$_3$ | 9 |
| | | | α-Gal-(1→6)-β-Gal-OCH$_3$ | 18 |
| | | α-Gal-OPhNO$_2$-*p* | α-Gal-(1→2)-α-OPhNO$_2$-*p* | 2 |
| | | | α-Gal-(1→3)-α-OPhNO$_2$-*p* | 16 |
| | α-Gal-OPhNO$_2$-*o* | α-Gal-OPhNO$_2$-*o* | α-Gal-(1→2)-α-OPhNO$_2$-*o* | 6 |
| | | | α-Gal-(1→3)-α-OPhNO$_2$-*o* | < 1 |
| β-D-Galactosidase (*E. coli*) | β-Gal-OPhNO$_2$-*o* | α-Gal-OCH$_3$ | β-Gal-(1→6)-α-Gal-OCH$_3$ | 14 |
| | | | and other isomers | < 1 |
| | | β-Gal-OCH$_3$ | β-Gal-(1→6)-β-Gal-OCH$_3$ | 3 |
| | | | β-Gal-(1→3)-β-Gal-OCH$_3$ | 22 |
| α-D-Mannosidase (jack beans) | α-Man–OCH$_3$ | α-Man–OCH$_3$ | α-Man–(1→2)-α-Man–OCH$_3$ | 18 |
| | | | α-Man–(1→6)-α-Man–OCH$_3$ | 4 |
| | α-Man–OPhNO$_2$-*p* | α-Man–OPhNO$_2$-*p* | α-Man–(1→2)-α-Man–OPhNO$_2$-*p* | 8 |
| | | | α-Man–(1→6)-α-Man–OPhNO$_2$-*p* | < 0,1 |
| | | | and other isomers | 0,4 |

more of a single oligosaccharide with (1→2)- or (1→3)-linkages can be obtained in this manner. The methodology is suitable both for dissolved and also for immobilised enzymes. The hidden potential of this technique is illustrated by the results[208] presented in Table 2.4. There are also possibilities for the enzymatic production of glycosides with monosaccharide and disaccharide structures[209] (Table 2.5).

**Table 2.5** Combination possibilities for the enzymatic synthesis of glycosides and oligosaccharide-glycosides

| Enzyme | Glycosyl donor | Glycosyl acceptor | Main products |
|---|---|---|---|
| β-D-Galactosidase (*E. coli*) | Lactose | Allyl alcohol | β-D-Gal–O–$CH_2$–CH=$CH_2$ |
| | | | β-D-Gal–(1→3)–β-D-Gal–O–$CH_2$–CH=$CH_2$ |
| | | | β-D-Gal–(1→6)–β-D-Gal–O–$CH_2$–CH=$CH_2$ |
| | | Benzyl alcohol | β-D-Gal–O$CH_2$Ph |
| | | | β-D-Gal–(1→3)–β-D-Gal–O$CH_2$Ph |
| | | | β-D-Gal–(1→6)–β-D-Gal–O$CH_2$Ph |
| | | 2-Trimethylsilyl-ethanol | β-D-Gal–O$(CH_2)_2$Si$(CH_3)_3$ |
| | | | β-D-Gal–(1→3)–β-D-Gal–O$(CH_2)_2$Si$(CH_3)_3$ |
| β-D-Galactosidase (bovine testes) | β-Gal–OPh$NO_2$-*p* | α-GalNAc–OEt | β-Gal–(1→3)–β-GalNAc–OEt |
| | β-Gal–OPh$NO_2$-*p* | β-GalNAc–OEt-2-Br | β-Gal–(1→3)–β-GalNAc–OEt-2-Br |
| | | β-GalNAc–OEt-2–Si$(CH_3)_3$ | β-Gal–(1→3)–β-GalNAc–OEt-2-Si$(CH_3)_3$ |
| | β-Gal–OPh$NO_2$-*p* | β-GlcNAc–O$CH_3$ | β-Gal–(1→3)–β-GlcNAc–O$CH_3$ |
| | | | β-Gal–(1→4)–β-GlcNAc–O$CH_3$ |
| α-D-Galactosidase (coffee) | Raffinose | Allyl alcohol | α-D-Gal–O$CH_2$CH=$CH_2$ |
| | α-Gal-OPh$NO_2$-*p* | α-Gal-O–$CH_2$CH=$CH_2$ | α-D-Gal–(1→3)–α-D-Gal–O–$CH_2$–CH=$CH_2$ |
| | | α-GalNAc–OEt | α-D-Gal–(1→3)–α-GalNAc–OEt |
| β-*N*-Acetylglucos-aminidase (jack beans) | β-GlcNAc–OPh$NO_2$-*p* | β-Gal–O$CH_3$ | β-GlcNAc–(1→6)–β-Gal–O$CH_3$ |
| | | β-Man–O$CH_3$ | β-GlcNAc–(1→6)–β-Man–O$CH_3$ |

The production of undesired positional isomers of disaccharides can be supressed, as already indicated in section 2.4.1.1 (p. 236), and thereby a part of the separation difficulties removed[210]. The method again depends on the differing aglycone specificity of two β-galactosidases, that from bovine testes and the other from *E. coli*. While the former brings about hydrolysis of (1→3)-linked disaccharides relatively easily, but also brings about their synthesis by transfer to a good degree, (1→3)-linked disaccharides are particularly stable in the presence of the latter enzyme and all other isomers are quickly hydrolysed. The correct choice of conditions is important. The results of synthesis using β-galactosidase from bovine testes is represented graphically in Fig. 2.192. The synthesis is especially productive in concentrated solution. Hydrolyses of the undesired disaccharides are then performed in dilute solution.

It has been known for a long time that cyclodextrinase [1,4-α-D-glucan: 4-α-D-(1,4-α-D-glucano)-transferase (cyclising)], for example from *Bacillus macerans*, besides other transglucosylations, catalyses the synthesis of the so-called homologous *maltodextrins* from cyclodextrins and a D-glucopyranosyl derivative[211]. The limitation is that, except for the 1-position and partially also the 6-position, the acceptor must display free hydroxy groups. The use of cyclodextrins as glucosyl donors and an α- or β-D-glucopyranoside or also another D-glucopyranosyl derivative as acceptor allows the synthesis of *malto-oligosaccharide glycosides* in an elegant manner and almost without loss, which is not possible in other ways. The highly developed separation methods of the present day make the preparation of pure maltodextrin glycosides possible with up to 10 glucosyl residues[212].

Sucrose has a greater potential for group transfer than other disaccharides and oligosaccharides since on its hydrolysis there is a greater decrease in free energy than with the hydrolysis of other disaccharides. This is true for both the transfer of the β-D-fructofuranosyl and also for the α-D-glucopyranosyl residue. This fact permits the synthesis of even *glucose 1-phosphate* with the help of certain phosphorylases at the expense of sucrose, which as a glycosylated acid is more energy rich than sucrose.

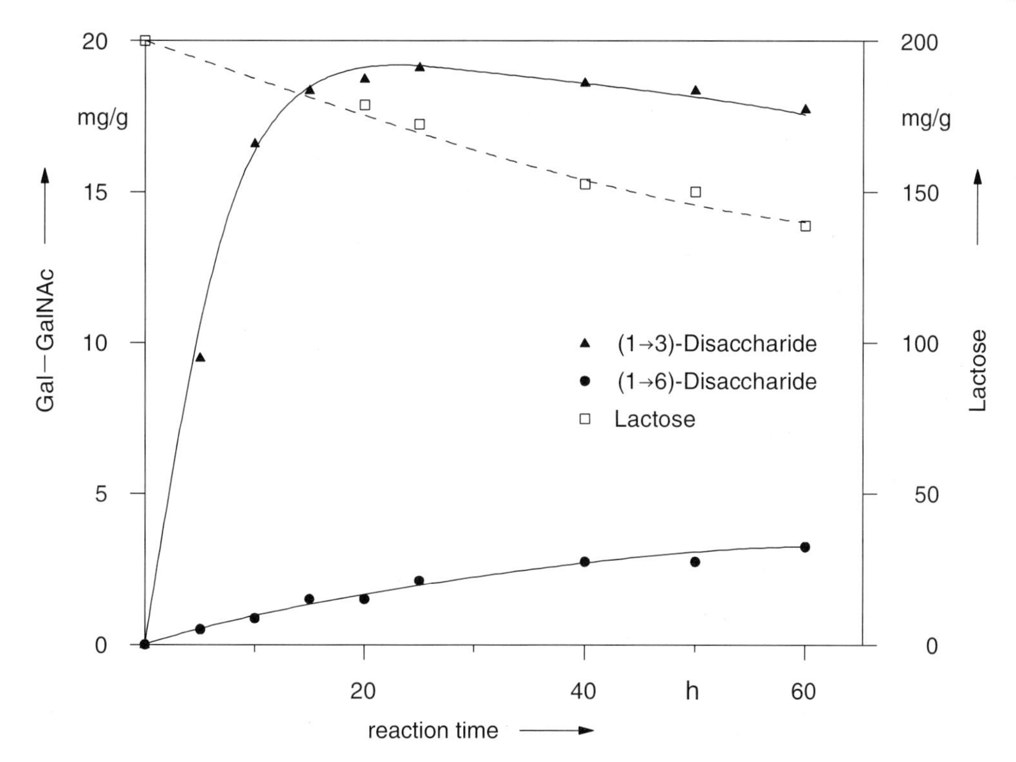


**Fig. 2.192** Formation of *O-β-D-galactopyranosyl-(1→3)-N-acetyl-D-galactosamine* and *O-β-D-galactopyranosyl-(1→6)-N-acetyl-D-galactosamine* during transglycosylation catalysed by β-galactosidase from bovine testes. The solution contains 5% N-acetyl-D-galactosamine as acceptor and 20% lactose as donor.

Sucrases, as other glycosidases, can not only hydrolyse their substrates but also isomerise them. *Isomaltulose (6-O-α-D-glucopyranosyl-D-fructose)*, generally also known as *palatinose*, is a non-cariogenic sucrose substitute, just as is also palatinitol obtained therefrom by catalytic hydrogenation. There are numerous procedures, also with immobilised cells[213], for the production of palatinose.

Glycosyl fluorides are prominent glycosyl donors in glycoside hydrolase catalysed transfer reactions[214] for the production of oligosaccharides. They are suited also for enzyme catalysed synthesis of *cyclodextrins*, which can be thought of as non-reducing oligosaccharides. The enzyme employed is the previously mentioned cyclodextrinase (CGTase) which, in contrast to the glycoside hydrolases, possesses no hydrolytic activity. Both α-maltosyl fluoride and also α-D-glucopyranosyl fluoride can be used as donors, whereby an immobilised enzyme raises the yield[215].

■ **Nucleoside diphosphate sugars.** In general, the formation of glycosidic bonds occurs in nature with glycosyl transferases (see sections 2.3.3, p. 178 and 2.3.8, p. 206) whose donor substrates are nucleoside diphosphate sugars (NDPX), polyprenyl phosphate sugars, or polyprenyl diphosphate sugars. There are important reasons why living organisms make use of these relative complicated substrates. Particularly, these are the high specificity of the transfer reactions as well as the possibility of a quantitative transformation.

Clearly, specific and quantitative transformation would also be very important for the enzyme catalysed synthesis of oligosaccharides on a preparative scale. Unfortunately, there are obstacles to the broad application of the *physiological* synthesis of oligosaccharides. Nucleoside diphosphate sugars – the polyprenyl derivatives are out of the question on account of the fact they are difficult to handle – are accessible only through costly chemical synthesis or by isolation from microorganisms, e. g. yeast, and correspondingly are very expensive. UDP-Glc, by far the cheapest compound, costs (in 1997) about £ 17.00 per 100 mg while, on the other hand, UDP-Gal and GDP-Man cost £ 331.00 and £ 162.40, respectively, for the same amount. Further, there are very few, relatively easily obtainable glycosyl transferases. However, a series of technically and economically realisable syntheses has been developed and these syntheses are a valuable addition to the simpler procedures described hitherto and above all faciltate access to the more unusual oligosaccharides.

Sucrose synthase is one of those relatively easily accessible glycosyl transferases which can not only be isolated by those laboratories requiring it (it is also commercially available) but which also utilises as donor substrate UDP-Glc, one of the cheaper compounds of this type. Above all, transformations are suited for the synthesis of chemically modified sucroses[216], because the sucrose synthetase as a D-glucosyl transferase indeed converts only the donor substrate UDP-Glc, but it tolerates modified fructose derivatives as acceptors (Fig. 2.193).

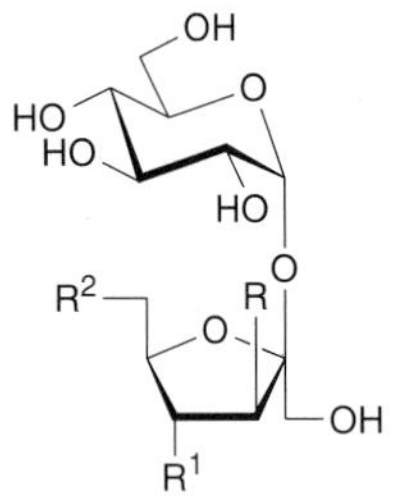

**Fig. 2.193** Sucrose derivatives chemically modified in the fructose moiety.

1 R = $N_3$; $R^1$ = $R^2$ = OH
2 $R^2$ = H; R = $R^1$ = OH
3 $R^2$ = H; R = $R^1$ = OH
4 $R^1$ = F; $R^2$ = R = OH

The coupled enzymatic synthesis of 6'-deoxy-6'-fluorosucrose proceeds in very good yield (Fig. 2.194). Glucose isomerase establishes an equilibrium between 6-deoxy-6-fluoro-D-glucose and the corresponding D-fructose derivative. The latter, although present only in small concentration, is trapped through irreversible glucosylation.

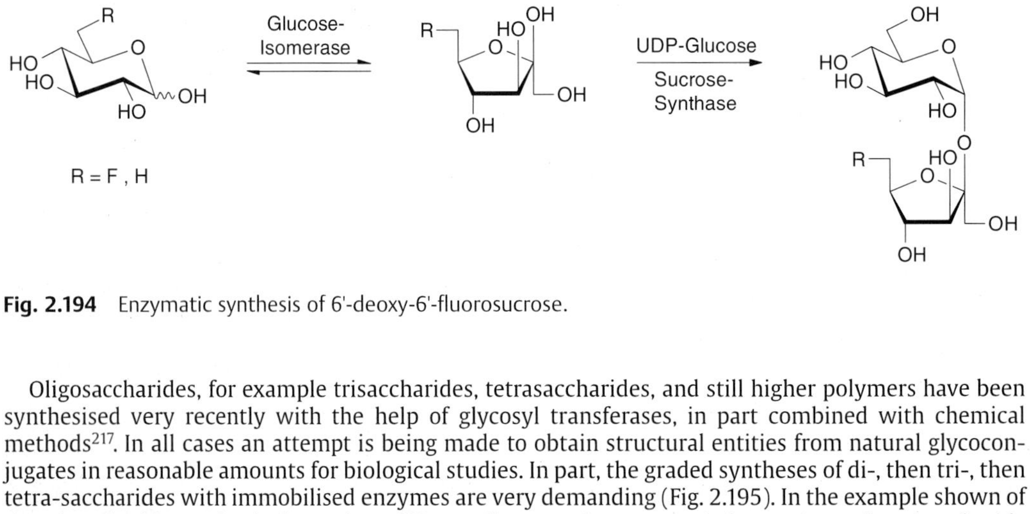

Glucose-Isomerase
UDP-Glucose
Sucrose-Synthase
R = F , H

**Fig. 2.194** Enzymatic synthesis of 6'-deoxy-6'-fluorosucrose.

Oligosaccharides, for example trisaccharides, tetrasaccharides, and still higher polymers have been synthesised very recently with the help of glycosyl transferases, in part combined with chemical methods[217]. In all cases an attempt is being made to obtain structural entities from natural glycoconjugates in reasonable amounts for biological studies. In part, the graded syntheses of di-, then tri-, then tetra-saccharides with immobilised enzymes are very demanding (Fig. 2.195). In the example shown of the trisaccharide synthesis, formation of *N*-acetyllactosamine was carried out prior to the trisaccharide synthesis using 6 immobilised enzymes[218].

The immobilisation of the enzyme can take place – there are other possibilities – through covalent coupling[219] (Fig. 2.196).

More recently, in a preparation which indicates potential future developments in many oligosaccharide syntheses, *N-acetyllactosamine* has been produced by a regenerative process with a mixture of immobilised enzymes. As already just described for the production of a trisaccharide, the use of a stoichiometric quantity of an activated monosaccharide is unnecessary for the essential step of transglycosylation. In a cyclic process the very expensive UDP-D-galactose is permanently replenished. Two inexpensive starting materials, phosphoenolpyruvate (PEP) and D-glucose 6-phosphate, permanently

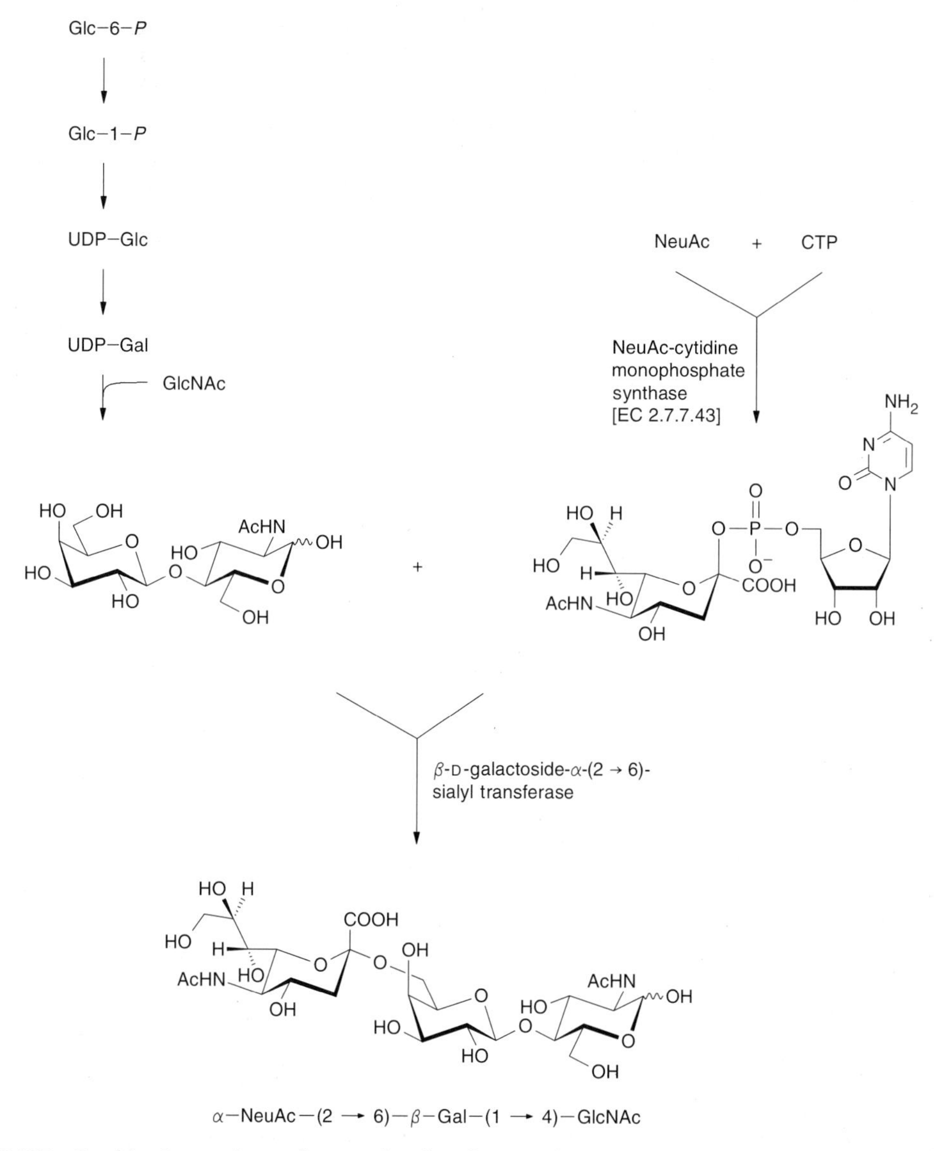


**Fig. 2.195** Combination synthesis of a *trisaccharide* with *N-acetylneuraminic acid* as the non-reducing (glyconic) end group.

provide for the supply of the reaction cycle (Fig. 2.197). Since the transgalactosylation is not reversible, it also does not matter if the epimerase equilibrium between UDP-Glc and UDP-Gal lies on the side of the former compound. The yields of 70% calculated on GlcNAc are surprisingly good and the reactions can actually be carried out on a preparative scale.

■ **Unsaturated glycosyl donors**. Glycoside hydrolases are able to catalyse the sterospecific addition of alcohols onto enolic C,C-double bonds of glycals with the corresponding configuration[220] or of 2,6-anhydro-1-deoxyhept-1-enitols[221]. In the former case *2-deoxy-glycopyranosides* are formed, in the latter *1-deoxy-2-ketopyranosides*. The formation of *1-deoxyglycerin-1-yl-2-deoxy-β-D-lyxo-hexopyranoside* takes place in about 20% yield. The starting material is D-galactal and the catalyst β-galactosidase

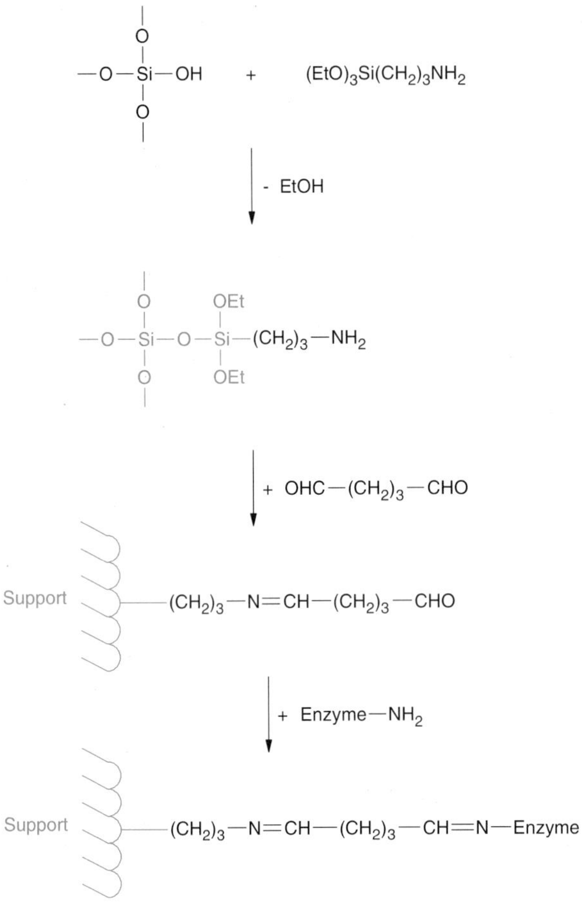

$-O-Si(O-)(O-)-OH$ + $(EtO)_3Si(CH_2)_3NH_2$

↓ - EtOH

$-O-Si(O-)(O-)-O-Si(OEt)_2-(CH_2)_3-NH_2$

↓ + $OHC-(CH_2)_3-CHO$

Support $-(CH_2)_3-N{=}CH-(CH_2)_3-CHO$

↓ + $Enzyme-NH_2$

Support $-(CH_2)_3-N{=}CH-(CH_2)_3-CH{=}N-Enzyme$

**Fig. 2.196** Covalent coupling of an enzyme on a polymeric support.

from *E. coli.* The dimerisation of D-glucal to *1,5-anhydro-2-deoxy-3-O-(2-deoxy-β-D-arabino-hexopyranosyl)-D-arabino-hex-1-enitol* with a yield of nearly 50% is catalysed by emulsin β-glucosidase (Fig. 2.198).

## 2.4.2 Redox reactions

As alcohols or carbonyl compounds, carbohydrates are well suited for structural modification through dehydrogenation (oxidation) or reduction. It is especially epimerisation which affords in this way access to rarer occurring compounds. In addition, there is the possibility of transforming non-carbohydrates, for instance hydrocarbons, into carbohydrates or compounds similar to carbohydrates. Oxidases, dehydrogenases, oxygenases, and reductases are widely distributed and are in part very unspecific enzymes, which can only be beneficial for their general application. However, it must be noted that comparatively few cases have been published of the preparative, biological conversion of any substrates into carbohydrates through oxidation or reduction.

### 2.4.2.1 Dehydrogenation and oxidation

There is no difference in terms of chemical formalism between an oxidation and a dehydrogenation. The mechanism is, however, different. The biochemical systems which are employed are also different . The typical enzymatic dehydrogenation requires a cofactor, as for example $NAD^+$ or $NADP^+$, which accepts

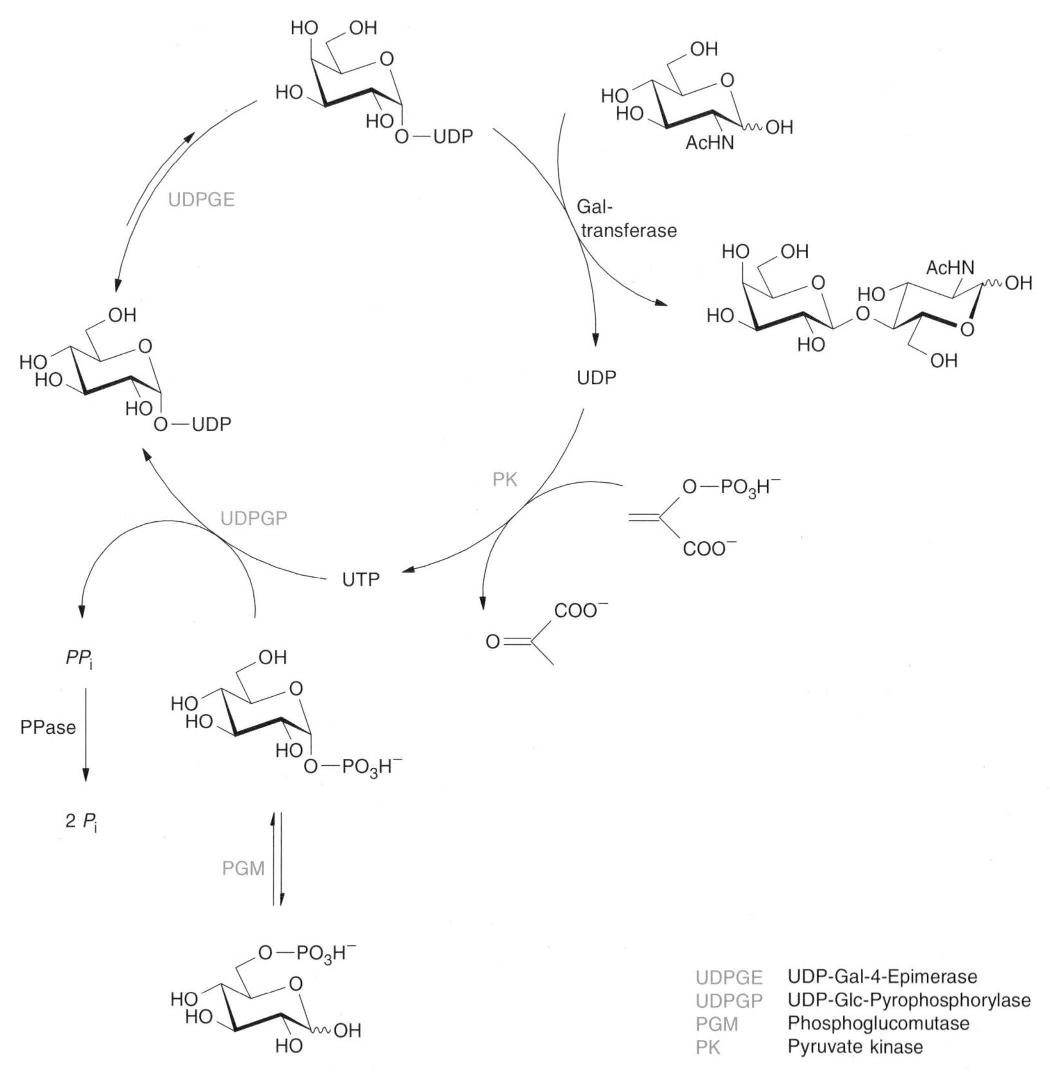


**Fig. 2.197** Preparative production of *N-acetyllactosamine* is carried out with immobilised enzymes by a regenerative process.

the hydrogen from the substrate. In general, the hydrogen is converted finally into water in the so-called respiratory chain (see sections 2.3.2, p. 174 and 2.3.6, p. 193). The substrate is oxidised with the dehydrogenation step. A typical dehydrogenation, for instance, is the oxidation of ethanol to acetaldehyde with the help of alcohol dehydrogenase (ADH) and nicotinamide adenine dinucleotide ($NAD^+$). In the case of *oxidations*, oxygen is reduced directly, mostly to $H_2O_2$ or with incorporation into the substrate. Differentiation is made between the catalysing enzymes, the dioxygenases and the monooxygenases. Also in the case of oxidation, cofactors or co-substrates are necessary.

Without differentiating between the different mechanistic processes and going into them in detail, some reactions employed on a preparative scale are presented here, which are to be regarded formally as oxidations. The regioselective oxidation of a polyol can, for example, rarely be used for the preparation of a monosaccharide. Such reactions exist in the metabolism of a great number of microorganisms, and not only there but also in higher organisms. In human liver there even exists an $NAD^+$ dependent

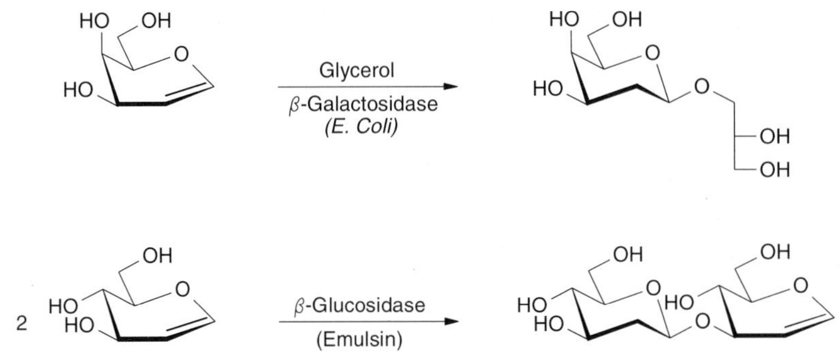

Glycerol
$\beta$-Galactosidase
*(E. Coli)*

$\beta$-Glucosidase
(Emulsin)

**Fig. 2.198** Many unsaturated monosaccharide derivatives, such as glycals or 2,6-anhydro-1-deoxyhept-1-enitols are suitable as glycosyl donors if they can be activated by glycosyl hydrolases.

sorbitol dehydrogenase, which has the ability to oxidise also (2*R*,3*R*)-2,3-butanediol in addition to other polyols[222]. A classical reaction is the production of L-sorbose from D-glucitol by the action of *Acetobacter suboxydans*[223]. The reaction is an important step in the large scale production of *vitamin C*. A comparable reaction produces in high yield L-fructose from *L-mannitol* and *D-sorbose* from L-glucitol, though with the isolated enzyme from a microorganism, which up to the present time has not yet been identified[224]. A very similar reaction with the known *Gluconobacter oxydans* (ATCC 15178)[225] and related strains on the chemically synthesised precursor 2-deoxy-D-*arabino*-hexitol affords the outstanding sweetening agent *5-deoxy-D-threo-hexulose*. The yield is about 65%.

The partly low specificity of the biological oxidation is underlined by the diversity in the substrates utilised. *Gluconobacter oxydans* is not only able to oxidise open chain polyols to monosaccharides but also monosaccharides or monosaccharide derivatives to the correspondong glyconic acid derivatives. Even compounds with basic amino functions, as for instance the 2-epimer of the known glycosidase inhibitor *nojirimycin*, mannonojirimycin, are oxidised to the lactam (Fig. 2.199). Both compounds, *mannonojirimycin* and the lactam are, incidentally, very strong competitive inhibitors, both of an $\alpha$-mannosidase, which is to be expected, and also of a $\beta$-glucosidase[226].

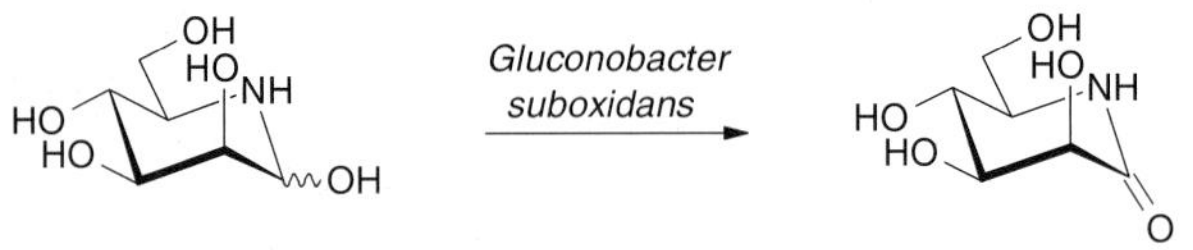

*Gluconobacter suboxidans*

*Mannonojirimycin*

**Fig. 2.199** Basic monosaccharide derivatives may also be oxidised with the help of microbial oxidation. The example here is the conversion of mannonojirimycin into the corresponding lactam.

In the same way, only on a more than tenfold scale, the production of crystalline D-gluconic acid $\delta$-lactam from crude nojirimycin was described in the same publication. D-Galactose oxidase is one of those enzymes which utilise oxygen as co-substrate. The Cu(II)-metalloprotein catalyses the conversion of D-galactose into *D-galactodialdose*, whereby oxygen is reduced to hydrogen peroxide. The reaction has proved useful above all for the radiolabelling of D-galactosyl end groups in glycoconjugates with tritium. The glycoconjugate is firstly treated with D-galactose oxidase and then reduced with tritium labelled sodium borohydride.

The enzyme is by no means specific for D-galactose alone, but apparently oxidises primary hydroxy groups, as long as there is a neighbouring (*R*,*R*)-*threo*-grouping. The structural unit evidently required for enzyme activity[227] is represented in Fig. 2.200.

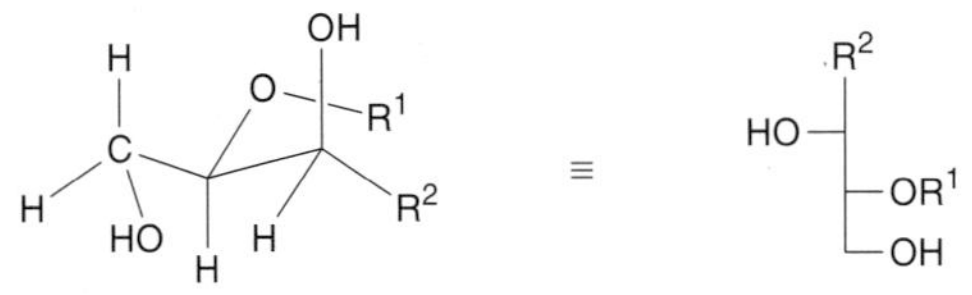

**Fig. 2.200** Structural requirements for the *oxidation* of a primary hydroxy group with D-galactose oxidase.

Since with polyols the aldehyde formed is partly removed in an equilibrium involving cyclisation, probably less product inhibition occurs here than with the oxidation of a galactoside[228]. The oxidation offers above all access to the L-series of the aldoses (Fig. 2.201).

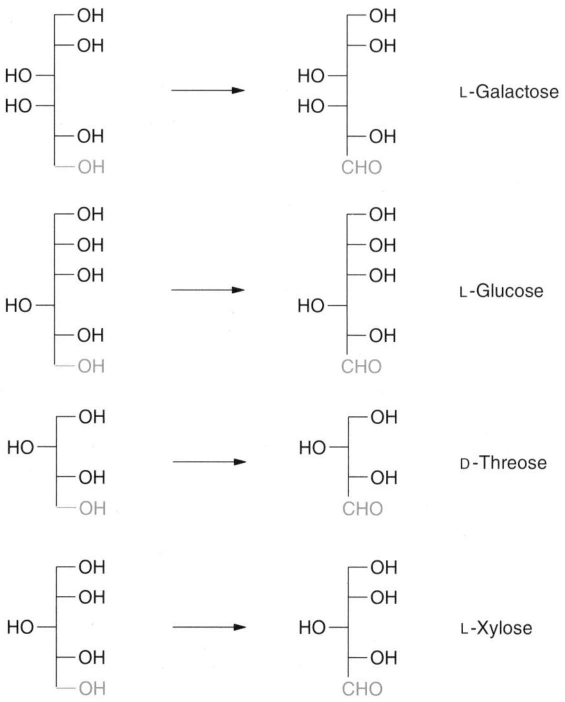

L-Galactose

L-Glucose

D-Threose

L-Xylose

**Fig. 2.201** The synthesis of some monosaccharides by oxidation of polyols with galactose oxidase.

The microbial oxidation of aromatic compounds, especially of benzene derivatives, gives access to enantiomerically pure polyhydroxy carbocycles and therewith to many different classes of natural products[229]. Although carbohydrates cannot be obtained directly in this manner, since additional chemical reactions are still necessary to a greater or lesser extent, nevertheless attention should be drawn here to this very elegant possibility for carbohydrate synthesis through a combination of biochemical and chemical methods. Benzene or substituted benzenes as prochiral compounds are oxidised through genetically manipulated strains of *Pseudomonas putida*[230] to the corresponding 5,6-dihydroxycyclohexa-1,3-dienes. Both L- and D-erythrose may be produced from chlorobenzene[231] (Fig. 2.202).

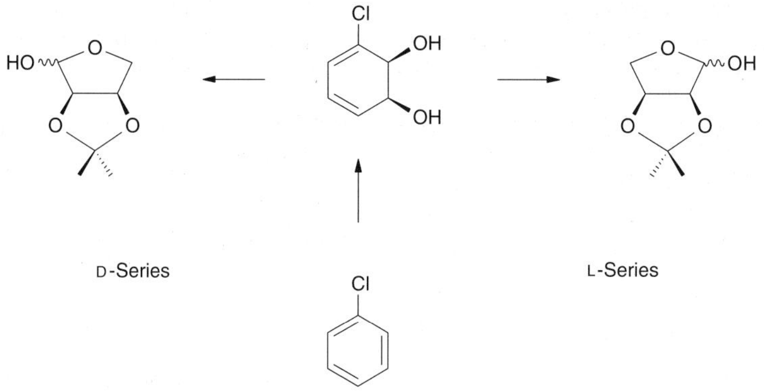

D-Series

L-Series

**Fig. 2.202** Possible chemical transformation of *(5S,6S)-1-chloro-5,6-dihydroxycyclohexa-1,3-diene*, obtained by the microbial oxidation of chlorobenzene, into derivatives of the D- or L-erythrose.

The product of microbial oxidation, after reaction with acetone, was oxidatively degraded by ozonolysis to 2,3-*O*-isopropylidene-D-erythruronic acid lactone (Fig. 2.203). The latter was then converted into 2,3-*O*-isopropylidene-L-erythrose by selective reductions. An additional reaction step is necessary for the production of the enantiomer.

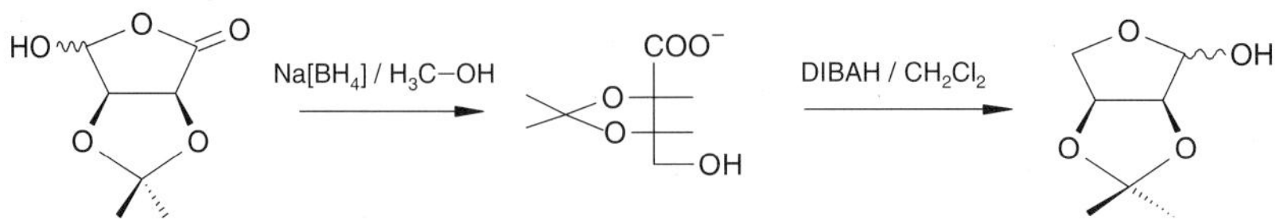

**Fig. 2.203** Conversion of 2,3-*O*-Isopropylidene-D-erythruronolactone into *2,3-O-isopropylidene-L-erythrose.*

With six carbon atoms, the parent compound of (5*S*,6*S*)-1-chloro-5,6-dihydroxycyclohexa-1,3-diene, *cis*-5,6-dihydroxycyclohexa-1,3-diene, is also a suitable synthon for the synthesis of unusual hexose[232] and cyclitol derivatives[233].

### 2.4.2.2 Reduction

Biological reduction serves in general for the production of chiral alcohols from ketones[234]. In the case of selected, suitably protected ketones, as for instance the three compounds shown in Fig. 2.204, unusual monosaccharide derivatives may be prepared stereospecifically. In all three cases, Baker's yeast reduces the carbonyl group in the vicinity of an acetoxy group, diastereoselectivily and enantioselectively, to the corresponding *erythro* compound[235]. Comparable reactions as shown here were also undertaken on other ketones. Chemical deprotection led finally to the desired *deoxyhexoses* or *deoxypentoses*. In a very similar manner, branched chain monosaccharides may be prepared.

### 2.4.2.3 Isomerisation

The mutual isomerisation of ketoses and aldoses and the epimerisation of secondary chiral alcohol groups in carbohydrates are methods often used to interconvert monosaccharides biochemically. Since such isomerisations are formally redox reactions they are mentioned here at the end of this section. The metabolism of all organisms exhibit numerous such reaction sequences. Here especially the triose phosphate isomerisation, the glucose phosphate and mannose phosphate isomerisation, and the UDP-D-glucose and UDP-D-galactose isomerisation should be named (see sections 2.3.2, p. 174 and 2.3.3, p. 178).

Whereas these conversions in general metabolism all involve the phosphate esters of the carbohydrates, with microorganisms reactions with free monosaccharides are also known. Of great industrial importance, for instance, is the enzymatic isomerisation of D-glucose into D-fructose. The so-called *high-fructose corn syrup* (HFCS) is produced on a ton scale from starch hydrolysate with immobilised bacterial glucoisomerase[236].

The isomerisation seldom serves for the direct production of unusual carbohydrates, but at all events it is suited for the transformation of precursors produced previously by aldolase reactions[237] or such synthesised by chemical reactions. For example, 6-substituted D-glucose derivatives, as for instance 6-fluoro-D-glucose, can be first transformed to the fructose derivative with glucose isomerase and this then, with the help of UDP-Glc and sucrose synthetase, to the corresponding sucrose derivative.

The enzymatic 4-epimerisation of D-galactose and D-glucose is a further important isomerisation reaction of general metabolism. It takes place only on the corresponding uridine diphosphohexoses. For preparative purposes, it is only of interest if the UDP-Gal for transgalactosylation, produced from relatively cheap UDP-Glc, may be obtained on a preparative scale, as for instance with the production of *N-acetyllactosamine*. The high price of the galactosyl donor UDP-Gal demands combined reactions with the participation of several enzymes.

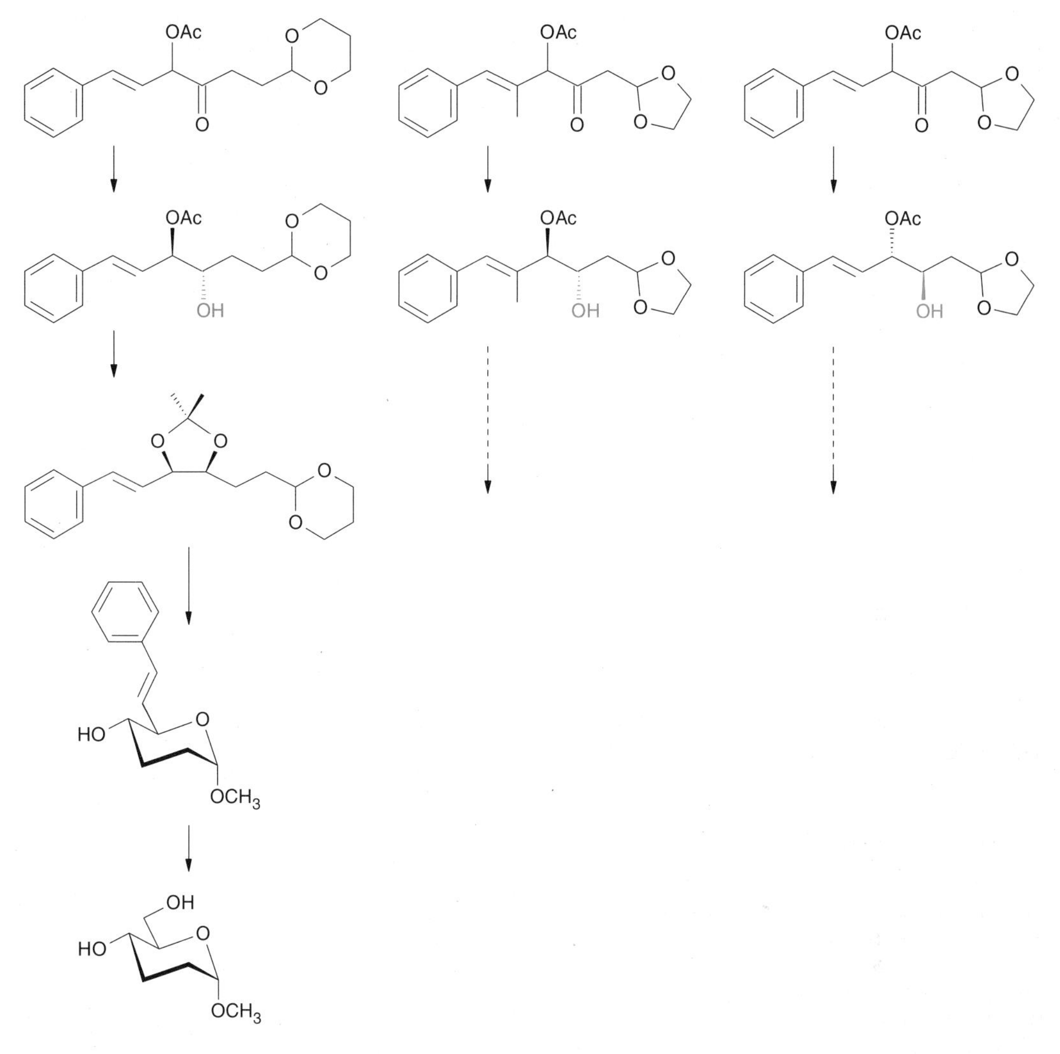


**Fig. 2.204** Formation of chiral centres through biological reduction of ketones opens up synthetic pathways for the preparation of unusual monosaccharides.

## 2.4.3 Formation of C,C-bonds

Synthesis in organic chemistry cannot be contemplated without the linkage of structural elements through the formation of C,C-bonds. The same applies to biochemical syntheses. The carboxylation of ribulose diphosphate as the first step in the Calvin cycle is surely the most common single reaction in the living world. No less significant are the following steps of disproportion of monosaccharide phosphates with three to seven carbon atoms in the hexose phosphate-pentose phosphate pathway (see section 2.3.6, p. 193). The aldol reaction is the most important and most universal metabolic process for the production of naturally occurring monosaccharides through C,C-coupling (see section 2.3.2, p. 174).

#### 2.4.3.1 The aldolase reaction

Aldolases are the most important of the enzymes which catalyse C,C-bond formation. Their practical application for *in vitro* synthesis of monosaccharides has been well developed by numerous research groups today and is based on earlier, frequently overlooked work of Meyerhof et al.[238], in which the preparative production of non-physiological products through aldolase catalysed reactions of 1,3-dihydroxyacetone phosphate with numerous aldehydes was described. The enolate of a ketone always reacts stereospecifically with an aldehyde. Two classes of aldolases are distinguished based on the nature of their active centre. Aldolases of the higher plants and animals activate the keto component through formation of an imine with the ω-amino group of a lysine, but the aldolases of bacteria and fungi employ $Zn^{2+}$ for that purpose.

The various possibilities to link an aldehyde as an electrophile with a ketone are described in comprehensive review articles[239]. The most frequently utilised enzyme is the aldolase from rabbit muscle, whose natural reaction is the linkage of 1,3-dihydroxyacetone phosphate with D-glyceraldehyde phosphate to give fructose-1,6-diphosphate, or alternatively its cleavage. The stereospecificity of the reaction is absolute. The newly formed glycol group always has the D-*threo*-configuration (Fig. 2.205).

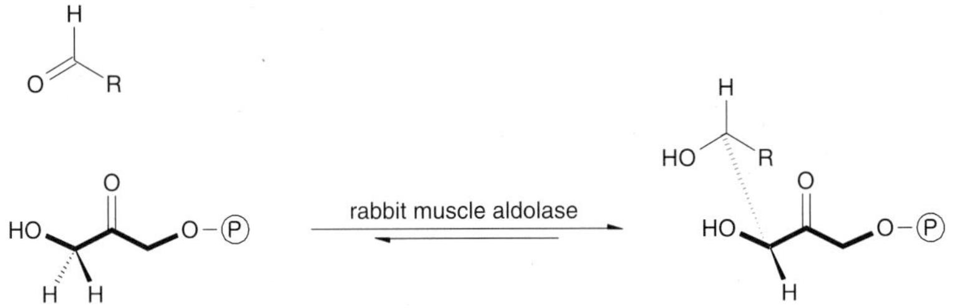

**Fig. 2.205** Stereochemistry of the aldol reaction which results from using the aldolase from rabbit muscle.

While the aldehyde component can be structurally very variable[240] – really only very strongly hindered aldehydes such as pivalylaldehyde (2,2-dimethylpropanal) or α,β-unsaturated aldehydes do not react at all – the reaction remains confined regarding the ketone component to ***1,3-dihydroxyacetone phosphate***, except for two exceptions known today. In general it has proved worthwhile to prepare the 1,3-dihydroxyacetone phosphate *in situ* from fructose 1,6-diphosphate. Besides the aldolase, triose phosphate isomerase is needed, which places the 3-phosphoglyceraldehyde in equilibrium with the 1,3-dihydroxyacetone phosphate required for the reaction (Fig. 2.206).

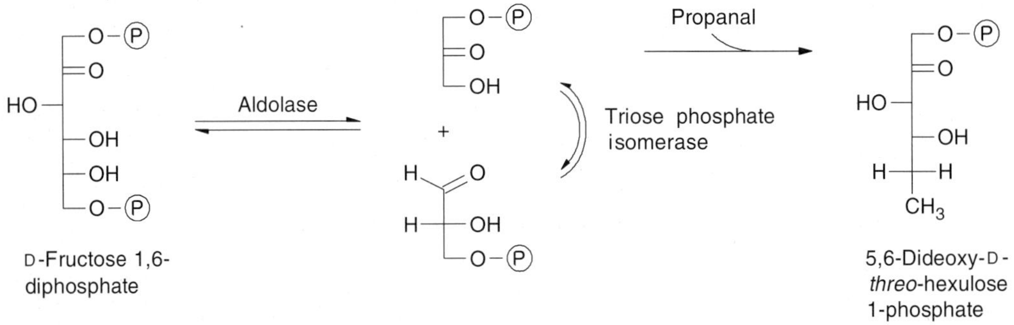

**Fig. 2.206** The aldolase reaction may be used in the presence of the triose phosphate isomerase. The preparation of ketones from D-fructose 1,6-diphosphate has proved useful.

The linking of 1,3-dihydroxyacetone phosphate with propanal under the action of immobilised enzyme affords, for example, in over 70% yield *5,6-dideoxy-D-threo-hexulose 1-phosphate*, from which the free sugar can be obtained with acid phosphatase. The methodology is similar for almost all corresponding reactions with other aldehydes as substrate. Numerous applications of the aldolase reaction on diverse, varied aldehydes[241] and also aldoses[242] have been published.

A variant of the reaction described above is well suited for the preparation of nitrogen containing monosaccharides[243]. Use of 3-azido-3-deoxyglyceraldehyde as the aldehyde component yields corresponding hexuloses by the aldolase reaction, which can be chemically transformed in elegant manner into *1-deoxynojirimycin* and *1-deoxymannonojirimycin*.

Fructose 1,6-diphosphate aldolase is indeed the most frequently applied enzyme for biochemical C,C-bond formation, but by no means the only one. An aldolase which has a very specific importance in metabolism, especially in animals, is that which brings about reaction of phosphoenolpyruvate with *N*-acetyl-D-mannosamine to give *N*-acetylneuraminic acid. Different *O*- and *N*-substituted neuraminic acids are named sialic acids. The pattern of sialic acids of glycoconjugates is often organ specific, which suggests a corresponding biological function. Molecular biological research into glycoconjugates is above all directed to the investigation of structure-activity relationships of peripheral sialic acids in structural variants of neuraminic acid. Since the chemical syntheses are very costly, *in vitro* biochemical syntheses are increasingly employed.

The sialic acid-aldolase (acylneuraminate-pyruvate-lyase, EC 4.1.3.3. – Sigma) catalyses the aldol reaction between the pyruvic acid (2-oxopropanoic acid) as the keto component and a D-mannosamine derivative as the aldehyde component. Not only is the biochemical production of natural NeuAc of commercial interest, but above all the synthesis of modified derivatives[244]. It is noteworthy that the sialyl aldolase accepts not only modified amino sugars as substrates but also neutral monosaccharides. Thus, pyruvic acid may be coupled with a series of monosaccharides, for example D-mannose, D-arabinose, or even 2-deoxy-2-C-phenyl-D-mannose, in good yields[245].

### 2.4.3.2 The transketolase reaction

Disproportionations of monosaccharides are important reactions for the biological production of pentoses (see section 2.3.6, p. 193). Besides the transaldolase reactions, which differ from the classical aldolase reactions only regarding the aldehyde component, the transketolase reactions play a crucial role in disproportionations. Thiamine pyrophosphate (TPP) as cofactor activates with its 1,3-thiazolium system the carbonyl component, perhaps a ketose or an α-ketocarboxylic acid, so that with splitting of a C,C-bond, an aldehyde component is formed, nucleophilc at the α-carbon atom (Fig. 2.207).

The activated aldehyde can then be added to the carbonyl group of another aldehyde with C,C-bond formation to give an acyloin, which in the biochemical case is a ketose (Fig. 2.208).

Transketolases are widely occurring enzymes, easily isolated from bacteria. Although not so abundant as the aldolases, they are used for the *in vitro* synthesis of ketosugars. The relatively non-specific nature of transketolases with regard to the aldehyde component – an enzyme from *Bacillus pumilus* is able to transfer the activated acetaldehyde from pyruvate to both D- and L-glyceraldehyde[246] – seems to be general. So, for example, both D-allose 6-phosphate and also D-glucose 6-phosphate could be reacted with β-hydroxypyruvate, catalysed by an enzyme from yeast, to give the isomeric octuloses[247]. Enzymatic C,C-bond formation is especially useful for the synthesis of *carbohydrates regiospecifically labelled with isotopes* and may be carried out directly and smoothly with the transketolase[248]. In the case shown here, *DL-(2,3-$^{13}$C)-serine* acts as carrier of the isotopic label (Fig. 2.209).

## 2.4.4 Deprotection and protection with acylases and lipases

Preparative carbohydrate chemistry cannot be conceived without protection and deprotection reactions. The accumulation of similarly reactive hydroxy groups makes regioselective reactions a fundamental problem, which is often difficult to overcome. It can well be imagined that the introduction of protecting groups enzymatically was considered very early on and also their removal. Lipases appear to be relatively unspecific enzymes predestined to cleave hydrolytically carbohydrate esters[249]. Their ready accessibility was a further reason for a relatively frequent use. That lipases or general acylases still were not utilised so often preparatively probably lies in the lack of competitiveness in comparison

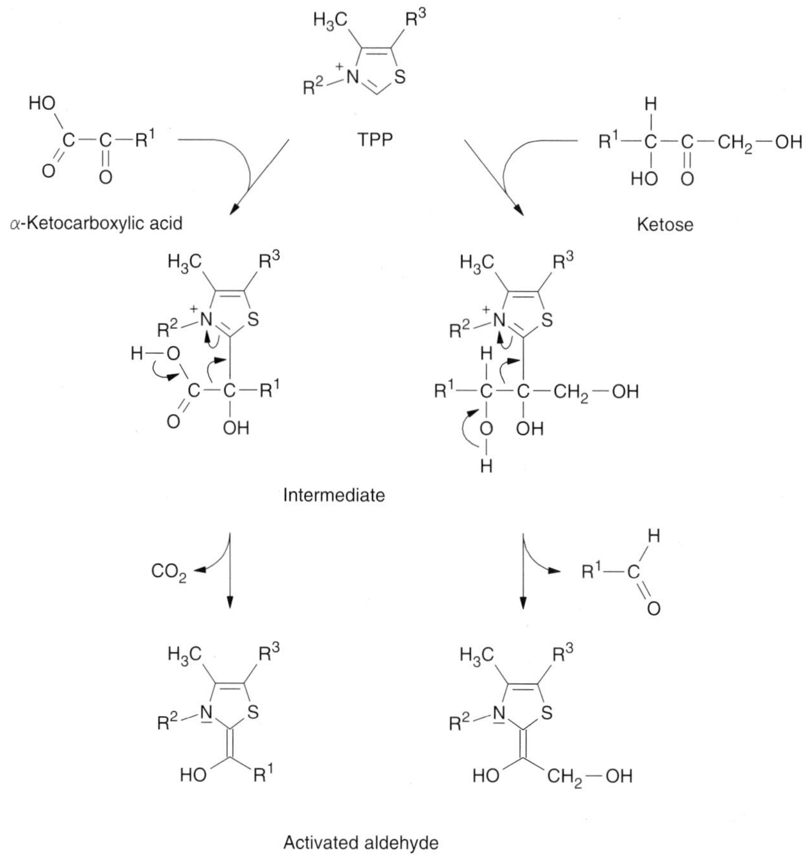

Fig. 2.207 Activation of a keto component by the 1,3-thiazolium system in thiamine pyrophosphate.

with the chemical methods, refined over decades, and the wide range of different, tested protecting groups.

Up to the present, the use of enzymatic methods has been confined almost exclusively to acyl groups (regioselective and stereoselective opening of oxiranes by enzymatic means is one of the few exceptions[250]), whereby typically the regioselective introduction of acyl residues is given more importance than regioselective deprotection.

■ **Deprotection.** Already in the early 1960's, impure enzyme extracts had been used for the deacetylation of esterified monosaccharides and disaccharides, without the method acquiring preparative significance[251]. The regioselectivity of the deacylation of glucose, maltose, and cellobiose esters were thoroughly investigated later with purified wheat germ lipase. Distinct gradations in the rates of deesterification of different acyl residues were found. For example, *acetyl* residues were removed about 25 times more quickly than *benzoyl* residues. Also, regioselectivity depends on the relative rates of deesterification of similar acyl residues. Removal of acetyl groups at the anomeric centre is favoured and somewhat less favoured at the primary 6-position. With the distinctly slower reacting secondary groups, the following order of decreasing reaction velocity is observed: 4 > 3 > 2. With peracetylated disaccharides, reaction at comparable positions always occurs faster at the one in the non-reducing monosaccharide unit. Recently, these earlier observations have been confirmed. The regioselective removal on a preparative scale of 1-*O*-acyl residues in per-*O*-acylated sugars[252] was successfully achieved as well as the deprotection of the primary position in glycosides[253].

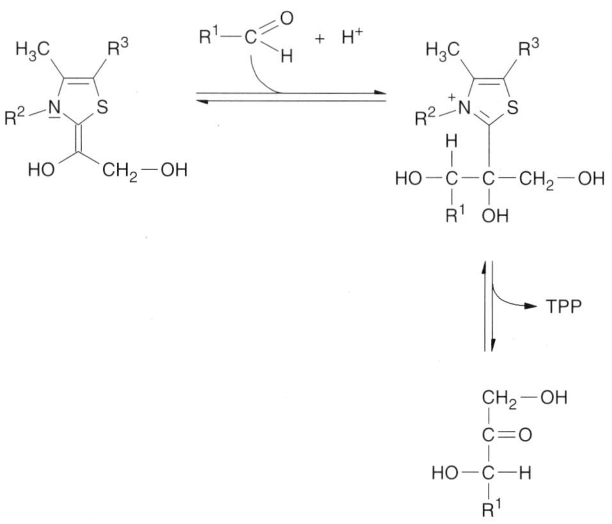

**Fig. 2.208** Addition of the TPP adduct (activated aldehyde) to a carbonyl component with formation of an acyloin (ketose).

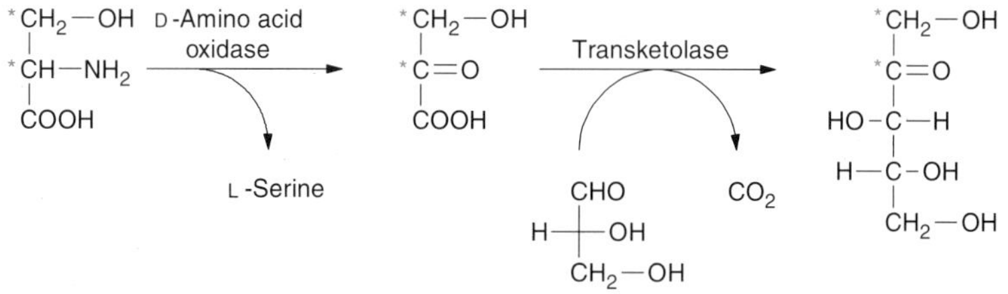

**Fig. 2.209** Ketoses, regioselectively labelled with isotopes can be obtained with the help of the transketolase reaction. In the example shown here, the ketose is obtained directly from *DL-(2,3-$^{13}$C)-serine* after biochemical oxidative deamination of the latter to provide the substrate for the formation of the activated aldehyde.

It really needs no special mention that acyl residues in differently protected carbohydrates can also be completely and very gently cleaved by enzymatic hydrolysis[254]. Something similar applies for α-aminoacyl residues, which have been examined as enzymatically removable protecting groups[255].

Among the secondary *O*-acyl residues in 1,6-anhydro-2,3,4-tri-*O*-butanoyl-*β*-D-galactopyranose, stepwise removal occurs first at the axial ester group in the 2-position, then at the equatorial 4-position. The reaction can be performed with soluble and immobilised lipases. The preparation of *1,6-anhydro-3,4-di-O-butanoyl-β-D-galactopyranose* proceeds in 90% yield. The 3-*O*-butanoyl compound is formed in 65% yield if the reaction is allowed to run for 52 h rather than 24 h[256].

■ **Protection.** Numerous publications in recent years have been concerned with the selective, enzyme catalysed acylation of carbohydrates. Especially interesting is the observation that lipases can be employed for acylation with higher efficiency and selectivity in organic solvents, for example pyridine[257]. As has been observed many times, the preferential acylation of primary hydroxy groups over secondary groups with activated carboxylic acids can rarely be accomplished in a chemical manner. Table 2.6 shows the notable results obtained by acylation of different monosaccharides with trichloroethyl esters of carboxylic acids, in water free pyridine as solvent, catalysed with lipase from porcine pancreas[258].

**Table 2.6** Selective protection by enzyme catalysed acylation with 2,2,2,-trichloroethyl esters of carboxylic acids

| D-Monosaccharide (mmol) | | Trichloroethyl ester | Regio-selectivity | Product | yield (g) |
|---|---|---|---|---|---|
| Glucose | (33) | Butyrate | 82 | 6-O-Butyryl-Glc | 2,1 |
| Glucose | (11) | Acetate | 85 | 6-O-Acetyl-Glc | 0,8 |
| Glucose | (11) | Caprylate (Octanoate) | 84 | 6-O-Capryloyl-Glc | 0,7 |
| Glucose | (11) | Laurate (Dodecanoate) | 95 | 6-O-Lauryl-Glc | 1,4 |
| Galactose | (3,9) | Acetate | 95 | 6-O-Acetyl-Gal | 0,5 |
| Mannose | (33) | Acetate | 100 | 6-O-Acetyl-Man | 2,4 |
| Fructose | (11) | Acetate | 100* | 1-O-Acetyl-Fru | 0,5 |
| | | | | 6-O-Acetyl-Fru | 0,2 |

* with reference to the sum of both acetylated hydroxyl groups

*N-Acetyl-9-O-acetylneuraminic acid* can be produced in a similar manner just as can also 5'-*O*-acylated derivatives of pyrimidine nucleosides. The regioselective acylation of oligosaccharides is astonishing, as for instance of maltotriose to more than 95% in the 6"-position. The reaction, which is also suited for the monoacylation of glycosides and nucleosides, is catalysed by the protease *subtilisin* in water-free *N,N*-dimethylformamide[259].

Yields depend on the structure of the carbohydrate and on the chain length of the reacted fatty acid[260]. The acylation of of the 6-position in ethyl D-glucopyranoside (no indication of the anomeric configuration is given in the original literature) appears to take place especially well and with high regioselectivity with long chain fatty acids. It is interesting that with free glucose, hardly any transformation occurs and methyl α-D-glucopyranoside is transformed only to 50%. Unfortunately, in this work the experimental details are only fragmentary but it is worth mentioning that the glucoside can be condensed directly with the corresponding fatty acid and a thermostable lipase. The regioselective reactions with palmitic acid, stearic acid and oleic acid all lie clearly over 90%.

It might well be expected that enzymatic acylation as well as chemical acylation under widely different conditions and with very different substrates is always selective for primary hydroxy groups, if this possibility exists. However, in many cases selectivity is also obtained for acylation of secondary alcohol functions. A series of pentosides were compared for their preferred position of enzymatic acylation[261]. Differences are noted, above all, on comparing anomeric glycosides. It is very welcome to find that the chemical and enzymatic methods are complementary. Positions which are difficult to acylate chemically, as for instance an axial 4-hydroxy group, are preferentially acylated by lipases. Fig. 2.210 shows the selective acylation obtained on the anomeric methyl glycosides of L-arabinopyranoside and on methyl α-D-xylopyranoside with the lipase from *Candida cylindracea*, with ethyl acetate as acetyl donor. Comparable reactions were carried out with 6-deoxyhexoses[262].

Differential blocking was successful with carboxylic acid esters of the vinyl alcohols (ethenols)[263]. Vinyl esters of phenoxyacetic acid, methoxyacetic acid, and 4-oxo-4-phenylbutanoic acid can be used as acyl donors for transacylation with lipases. Glycals, blocked at the 6-position, can be protected regioselectively through acylation with lipase from *Pseudomonas*. Acylation proceeds at the allylic position in over 90% yield.

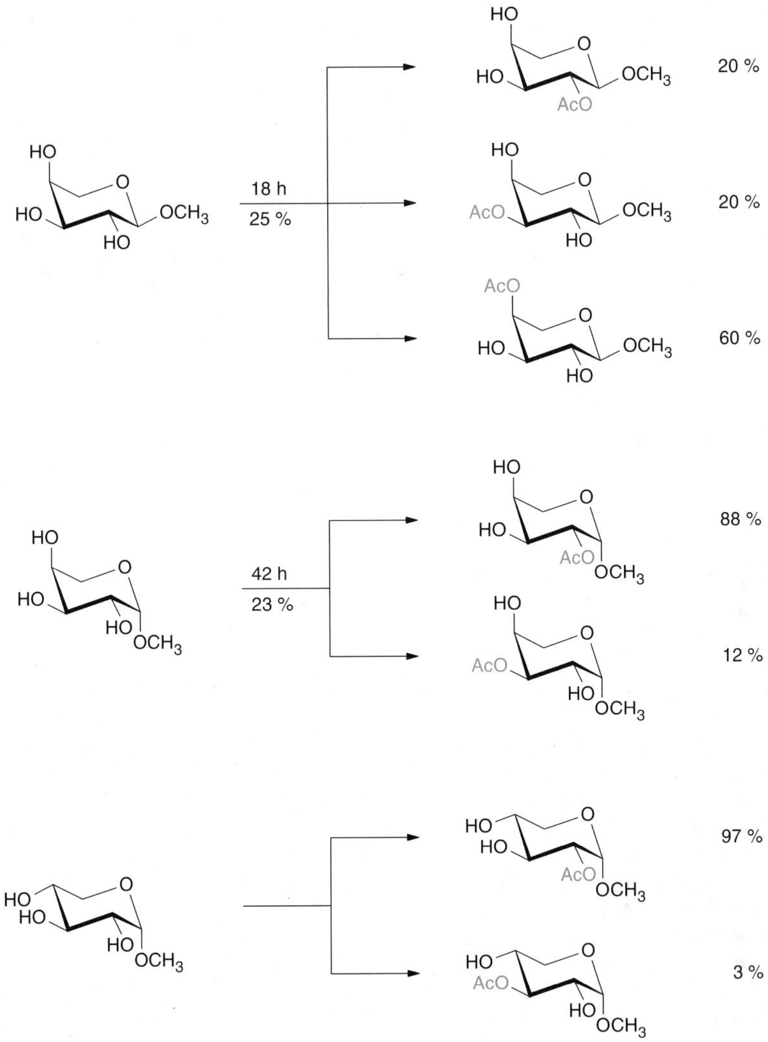

18 h
25 %

20 %
20 %
60 %

42 h
23 %

88 %
12 %

97 %
3 %

**Fig. 2.210** Regioselectivity in the lipase-catalysed acetylation of pentosides.

# Literature

## Sources and supplementary literature

### Organic chemistry

Carey, F. A.; Sundberg, R. J. *Advanced Organic Chemistry; 3rd edition;* Plenum: New York, 1990.

### Biochemistry

Voet, D.; G. Voet. J. *Biochemistry, 2nd edition;* Wiley: New York, 1995.

### Molecular Biology

Alberts, B.; Bray, D.; Lewis, J.; Raff, M.; Roberts, K.; Watson, J. D. *Molecular Biology of the Cell, 3rd edition;* Freeman: New York, 1995.
Darnell, J.; Lodish, H.; Baltimore, D. *Molecular Cell Biology, 2nd edition;* Freemann: New York, 1990.

### Carbohydrates

Specialist Periodical Reports; The Royal Society of Chemistry
Collins, P.; Ferrier, R. *Monosaccharides – Their Chemistry and Their Roles in Natural Products;* Wiley: Chichester, 1995.

## References


1 Gates, D. M. *Sci. Am.* **1971**, *225 (3)*, 88.
2 Sarmiento, J. L. *Chem. Eng. News* **1993**, *71*, 30.
3 Kenne, L.; Lindberg, B. In *The Polysaccharides*, Vol. 2, Aspinall, G. O., Academic Press: London, 1983, Chapter 5.
4 IUPAC-IUB Joint Commission on Biochemical Nomenclature (JCBN), Nomenclature of Carbohydrates (Recommendations 1996), *Pure Appl. Chem.* **1996**, *68*, 1919; *Carbohydr. Res.* **1997**, *297*, 1.
5 Abbreviated Terminology of Oligosaccharide Chains, *J. Biol. Chem.* **1982**, *257*, 3347.
6 Nomenclature of Glycoproteins, Glycopeptides and Peptidoglycans, *J. Biol. Chem.* **1987**, *262*, 13.
7 The Nomenclature of Lipids, *Eur. J. Biochem.* **1977**, *79*, 11.
8 Kornfeld, S. In *The Glycoconjugates*, Vol. 3, Horowitz, M. I.; Academic Press: London, 1982, Chapter 1.
9 Beyer, T. A.; Hill, R. L. In *The Glycoconjugates*, Vol. 3, Horowitz, M. I.; Academic Press: London, 1982, Chapter 2.
10 C. P. Stowell, Y. C. Lee, *Adv. Carbohydr. Chem. Biochem.* **1980**, *37*, 225.
11 Gram, C. *Fortschr. Med.* **1884**, *2*, 185.
12 Cheng, K.-J.; Costerton, J. W. *J. Bacteriol.* **1977**, *129*, 1506.
13 Salton, M. R.; *The Bacterial Cell Wall*, Elsevier: Amsterdam, 1964.
Rogers, H. J.; Perkins, H. R.; Ward, J. B. *Microbial Cell Walls and Membranes*, Chapman & Hall: London, 1980.
14 Munson, R. S.; Glaser, L. In *Biology of Carbohydrates*, Vol. 1, Ginsburg, V.; Robbins, P.; John Wiley & Sons: New York, 1981, Chapter 3.
15 Duckworth, M. In *Surface Carbohydrates of the Procaryotic Cell*, Sutherland, I. W.; Academic Press: New York, 1977.
16 Archibald, A. R.; Baddiley, J. *Adv. Carbohydr. Chem.* **1966**, *21*, 323.
17 Jann, K.; Westphal, O. In *The Antigens*, Vol. 3, Sela, M.; Academic Press: New York, 1975, Chapter 1.
18 Rietschel, E. Th. *Molecular Aspects of Inflammation* Springer: Berlin, 1991.
19 Galanos, C.; Lüderitz, O.; Rietschel, E. T.; Westphal, O. *Int. Rev. Biochem.* **1977**, *14*, 239.
20 Aman, P.; Franzén, L.-E.; Darvill, J. E.; McNeill, M.; Darvill, A. G.; Albersheim, P. *Carbohydr. Res.* **1982**, *103*, 77.
21 Bartnicki-Garcia, S. *Annu. Rev. Microbiol.* **1968**, *22*, 87,
22 Gorin, P. A. J.; Barreto-Bergter, E. In *The Polysaccharides*, Vol. 2, Aspinall, G. O.; Academic Press: New York, 1983, Chapter 6.
23 Rosinski, M. A.; Campana, R. J. *Mycologia* **1964**, *56*, 738.
24 Muzzarelli, R. A.A. In *The Polysaccharides*, Vol. 3, Aspinall, G. O.; Academic Press: New York, 1985, Chapter 6.
25 Schopf, J. W. *Sci. Am.* **1978**, *239* (3), 84.
26 Lüttge, U.; Kluge, M.; Bauer, G. *Botanik*, VCH: Weinheim, 1989.
27 Painter, T. J. In *The Polysaccharides*, Vol. 2, Aspinall, G. O.; Academic Press: New York, 1983, Chapter 4.
28 Rees, D. A.; Morris, E. R.; Thom, D.; Madden, J. K. In *The Polysaccharides*, Vol. 1, Academic Press, Aspinall, G. O.; New York, 1982, Chapter 5.
29 Andresen, I.-L.; Skipnes, O.; Smidsrod, O.; Ostgaard, K.; Hemmer, P. C. *ACS Symp.Ser.* **1977**, *48*, 361.
30 Haug, A. *Methods Carbohydr. Chem.* **1965**, *5*, 69.
31 Smidsrod, O.; Haug, A. *Acta Chem. Scand.* **1968**, *22*, 1989.
32 Grant, G. T.; Morris, E. R.; Rees, D. A.; Smith, P. J. C.; Thom, D. *FEBS Lett.* **1973**, *32*, 195.
33 Marchessault, R. H.; Sundararajan, P. R. In *The Polysaccharides*, Vol. 2, Aspinall, G. O.; Academic Press: New York, 1983, Chapter 2.
34 McNeil, M.; Darvill, A. G.; Fry, St. C.; Albersheim, P. *Annu. Rev. Biochem.* **1984**, *53*, 625.
35 Muzzarelli, R. A. A. *Chitin*, Pergamon: Oxford, 1977.
36 Alberts, B.; Bray, D.; Lewis, J.; Raff, M.; Roberts, K.; Watson, J. D. *Molecular Biology of the Cell, 3rd edition;* Freeman: New York, **1995**.
37 Hascall, V. C. In *Biology of Carbohydrates*, Vol.1, Ginsburg, V.; Robbins, P.; John Wiley & Sons: New York, 1981, Chapter 1.
38 Fransson, L.-A. In *The Polysaccharides*, Vol.3, Aspinall, G. O.; Academic Press: New York, 1985, Chapter 5.

39 Sadler, J. E. In *Biology of Carbohydrates*, Vol.2, Ginsburg, V.; Robbins, P.; John Wiley & Sons: New York, 1984, Chapter 4.
40 Hill, H. D.; Reynolds, J. A.; Hill, R. L. *J. Biol. Chem.* **1977**, *252*, 3791.
41 Preiss, J.; Walsh, D. A. In *Biology of Carbohydrates*, Vol.1, Ginsburg, V.; Robbins; P.; John Wiley & Sons: New York, 1981, Chapter 5.
42 Guilbot, A.; Mercier, C. In *The Polysaccharides*, Vol. 3, Aspinal, G. O.; Academic Press: Orlando, 1985, Chapter 3.
43 Yamaguchi, M.; Kainuma, K.; French, D. *J. Ultrastruct. Res.* **1979**, *69*, 249.
44 Gagnaire, D.; Perez, S.; Tran, V. *Carbohydr. Polym.* **1982**, *2*, 171.
45 Banks, W.; Greenwood, C. T. *Starch and its Components*, Edinburgh Univ. Press, Edinburgh 1975.
46 Carlson, T. L. G.; Larsson, K. *Stärke* **1979**, *31*, 222.
47 Peat, S.; Whelan, W. J.; Thomas, G. J. *J. Chem. Soc.* **1956**, 3025.
48 French, D. *MTP Int. Rev. Sci. Biochem.* **1975**, *5*, 267.
49 Robin, J. P.; Mercier, C.; Duprat, F.; Charbonniere, R.; Guilbot, A. *Stärke* **1975**, *27*, 36.
50 Meyer, K. H.; Bernfeld, P. *Helv.Chim. Acta* **1940**, *23*, 875.
51 Gunja-Smith, Z.; Marshall, J. J.; Mercier, C.; Smith, E. E.; Whelan, W. J. *FEBS Lett.* **1970**, *12*, 101.
52 Geddes, R.; Harvey, J. D.; Wills, P. R. *Eur. J. Biochem.* **1977**, *81*, 465.
53 Kobata, A. In *The Glycoconjugates*, Vol.1, Horowitz, M. I.; Pigman, W.; Academic Press: New York, 1977, Chapter 5, Section 3.
54 Egge, H. *In Oligosaccharide, Proteine und Glykoproteine in Frauenmilch und im Urin von Frühgeborenen*, Kunz, Cl.; Thieme , Stuttgart 1994.
55 Hayes, M. L.; Castellino, F. J. *J. Biol. Chem.* **1979**, *254*, 8777.
56 Kao, K.-J.; Pizzo, S. V.; McKee, P. A. *J. Biol. Chem.* **1980**, *255*, 10134.
57 Osuga, D. T.; Feeney, R. E. *J. Biol. Chem.* **1978**, *253*, 5338.
58 Geoghegan, K. F.; Osuga, D. T.; Ahmed, A. I.; Yeh, Y.; Feeney, R. E. *J. Biol. Chem.* **1980**, *255*, 663.
59 Feeney, R. E.; Osuga, D. T.; Ward, F. C.; Rearick, J. I.; Glasgow, L. R.; Sadler, J. E.; Hill, R. L. *Am. Chem. Soc. Abstracts of Papers* **1978**, Biol *013*, 176.
60 Glabe, C. G.; Hanover, J. A.; Lennarz, W. J. *J. Biol. Chem.* **1980**, *255*, 9236.
61 Marchesi, V. T.; Furthmayr. H.; Tomyta, M. *Annu. Rev. Biochem.* **1976**, *45*, 667.
62 Furthmayr, H. In *Biology of Carbohydrates*, Vol. 1, Ginsburg, V.; Robbins, P.; John Wiley & Sons: New York, 1981, Chapter 4.
63 Thomsen, O.; Friedenreich, V.; Worsae, E. *Acta Pathol. Microbiol. Scand.* **1930**, *7*, 157.
64 Schachter, H.; Michaels, M. A.; Tilley, C. A.; Crookston, M. C.; Crookston, J. H. *Proc. Natl. Acad. Sci. USA* **1973**, *70*, 220.
65 Simons, K.; Warren, G. *Adv. Protein Chem.* **1984**, *36*, 79.
66 Simons, K.; Garoff, H.; Helenius, A. *Spektrum der Wiss.* **1982**, *4*, 52.
67 Montefiori, D. C.; Robinson, W. E., Jr.; Mitchell, W. M. *Proc. Natl. Acad. Sci. USA* **1988**, *85*, 9248.
68 Geyer, H.; Holschbach, C.; Hunsmann, G.; Schneider, J. *J. Biol. Chem.* **1988**, *263*, 11760.
69 Mizouchi, T.; Spellman, M. W.; Larkin, M.; Solomon, J.; Basa, L. J.; Feizi, T. *Biochem. J.* **1988**, *254*, 599.
70 Hilkens, J.; Ligtenberg, M. J. L.; Vos, H. L.; Litvinov, S. V. *TIBS* **1992**, *17*, 359.
71 Jentoft, N. *TIBS* **1990**, *15*, 291.
72 Hartmann, E.; Messner, P.; Allmeier, G.; König, H. *J. Bacteriol.* **1993**, *175*, 4515.
73 Hynes, R. O. *Spektrum der Wiss.* **1986**, *8*, 80.
74 Martin, G. R.; Timpl, R. *Annu. Rev. Cell Biol.* **1987**, *3*, 57.
75 Hakomori, S. *Sci. Am.* **1986**, *254 (5)*, 32.
76 Weigandt, H. *Adv. Neurochem.* **1982**, *4*, 149.
77 Sandhoff, K.; Quintern, L. *Naturwissenschaften* **1988**, *75*, 123.
78 Kundu, S. K. In *Glycoconjugates*, Allen, H. J.; Kisailus, E. C.; Dekker: New York, 1992, Chapter 8.
79 Hakomori, S. *Annu. Rev. Biochem.* **1981**, *50*, 733.
80 Ferguson, M. A. J.; Williams, A. F. *Annu. Rev. Biochem.* **1988**, *57*, 285.
81 Low, M. G.; Saltiel, A. R. *Science* **1988**, *239*, 268.
82 Iozzo, R. V. *TIGG* **1991**, *3*, 327.
83 Kindl, H., Wöber, G. *Biochemie der Pflanzen*, Springer: Berlin; 1975, p. 312 ff.
84 Blithe, D. L. *TIGG* **1993**, *5*, 81.
85 Wold, F. *Annu. Rev. Biochem.* **1981**, *50* , 783.
86 Goldstein, I. J.; Hayes, C. E. *Adv. Carbohydr. Chem.* **1978**, *35*, 127.
87 Gabius, H.-J.; Rüdiger, H.; Uhlenbruck, G. *Spektrum der Wissenschaften* **1988**, Nov. 50.
88 Ashwell, G.; Harford, J. *Annu. Rev. Biochem.* **1982**, *51*, 531.
89 Ashwell, G.; Morell, A. G. *Adv. Enzymol.* **1974**, *41*, 99.
90 Hudgin, R. L.; Pricer, W. E., Jr.; Ashwell, G.; Stockert, R. J.; Morell, A. G. *J. Biol. Chem.* **1974**, *249*, 5536.
91 Kawasaki, T.; Ashwell, G. *J. Biol. Chem.* **1977**, *252*, 6536.
92 Drickamer, K. In *Biology of Carbohydrates*, Vol.3, Ginsburg, V.; Robbins, P. W.; JAI Press: London, 1991, Chapter 3.
93 Kuhlenschmidt, T. B.; Kuhlenschmidt, M. S.; Roseman, S.; Lee, Y. C. *Biochemistry* **1984**, *23*, 6437.
Lee, Y. C. In *Carbohydrate Recognition in Cellular Function*, Bock, D.; Harnett, S.; John Wiley & Sons: Chichester; 1989, Chapter 5.
94 Drickamer, K. *J. Biol. Chem.* **1988**, *263*, 9557.
95 Bevilaqua, M. P.; Stengelin, S.; Gimbrone, M. A. J.; Seed, B. *Science* **1989**, *243*, 1160.
Siegelman, M. H.; Van de Rijn, M.; Weissman, I. L. *Science* **1989**, *243*, 1165.
96 Weis, W. I.; Drickamer, K.; Hendrickson, W. A. *Nature (London)* **1992**, *360*, 127.
97 Drickamer, K. *Nature (London)* **1992**, *360*, 183.

98 Distler, J. J.; Guo, J.; Jourdian, G.; Shrivastava, O. P.; Hindsgaul, O. *J. Biol. Chem.* **1991**, *266*, 21687.
99 Tomoda, H.; Ohsumi, Y.; Ichikawa, Y.; Shrivastava, O. P;. Kishimoto, Y,; Lee, Y. C. *Carbohydr. Res.* **1991**, *213*, 37.
100 Björk, I.; Lindahl, U. *Mol. Cell Biochem.* **1982**, *48*, 161.
101 Marcam, J. A.; Rosenberg, R. D. In *Biology of Carbohydrates*, Vol. 3, Ginsberg, V.; Robbins P. W.; JAI Press: London, 1991, Chapter 2.
102 Lindahl, U., Thunberg, L., Bäckström, G., Riesenfeld, J., Nordling, K., Björk, I. *J. Biol. Chem.* **1984**, *259*, 12368.
103 Mourey, L., Samama, J. P., Delarue, M., Choay, J., Lormeau, J. C., Petitou, M., Moras, D. *Biochimie* **1990**, *72*, 599.
104 Wassarman P. M.; In *Carbohydrate Recognition in Cellular Function*, Bock, D.; Harnett John S.; Wiley & Sons: Chichester, 1989, Chapter 8.
105 Litscher, E. S.; Wassarman, P. M. *TIGG* **1993**, *5*, 369.
106 Florman, H. M.; Wassarrman, P. M. *Cell* **1985**, *41*, 313.
107 Yurewicz, E. C.; Pack, B. A.; Sacco, A. G. *Mol. Reprod. Dev.* **1991**, *30*, 126.
108 Noguchi, S.; Hatanaka, Y.; Tobita, T.; Nakano, M.; *Eur. J. Biochem.* **1992**, *204*, 1089.
109 Miller, D. J.; Macek, M. B.; Shur, B. D. *Nature (London)* **1992**, *357*, 589.
Litscher, E. S.; Wassarman, P. M. *TIGG* **1993**, *5*, 369.
110 Alberts, B.; Bray, D.; Lewis, J.; Raff, M.; Roberts, K.; Watson, J. D. *Molecular Biology of the Cell, 3rd edition;* Freeman: New York, **1995**, p. 831.
111 Risse, H. J.; Rössler H. H. In *The Glycoconjugates*, Vol. 3, Horowitz, M. I.; Academic Press: New York, 1982, Chapter 2, Section 1.
112 Springer, W. R. *TIGG* **1991**, *3*, 91.
113 Ziska, S. E.; Henderson, E. J. *Proc. Natl. Acad. Sci. USA* **1988**, *85*, 817.
114 Yoshida, M.; Matsui, T.; Fuse, G.; Ouchi, S. *FEBS Lett.* **1993**, *318*, 305.
115 Hynes, R. O. *Sci. Am.* **1986**, *254 (6)*, 32.
Yoshida, M.; Matsui, T.; Fuse, G.; Ouchi, S. *FEBS Lett.* **1993**, *318*, 305.
116 Yamada, K. M. In *The Glycoconjugates*, Vol. 3, Horowitz, M. I.; Academic Press: New York, 1982, Chapter 3, Section 2.
117 Tanzer, M. L.; Dean, J. W.; Chandrasekaran, S. *TIGG* **1991**, *3*, 302.
118 Liotta, L. A.; Rao, C. N.; Wewer, U. M. *Annu. Rev. Biochem.* **1986**, *55* ,1037.
119 Liotta, L. A. *Sci. Am.* **1992**, *266 (2)*, 34.
120 Akiyama, St. K. *TIGG* **1992**, *4*, 346.
121 Chammas, R.; Veiga, S. S.; Line, S.; Potocnjak, P.; Bretani, B. *J. Biol. Chem.* **1991**, *266*, 3349.
122 Phillips, M. L.; Nudelman, E.; Gaeta, F. C. A.; Perez, M.; Singhal, A. K.; Hakomori, S.-I.; Paulson, J. C. *Science* **1990**, *250*, 1130.
123 Goelz, S. E. *TIGG* **1992**, *4*, 14.
124 Lasky, *Science* **1992**, *258*, 964.
125 Mulligan, M. S.; Paulson, J. C.; DeFrees, S.; Zheng, Z.-L.; Lowe, J. B., Ward, P. A. *Nature (London)* **1993**, *364*, 149.
126 Karlsson, K.-A. *Annu. Rev. Biochem.* **1989**, *58*, 309.
127 Schulze, I. T.; Manger, I. D. *Glycoconjugate J.* **1992**, *9*, 63.
128 Karlson, G. B.; Butters, T. D.; Dwek, R. A.; Platt, F. M. *J. Biol. Chem.* **1993**, *268*, 570.
129 Matthews, T. J.; Weinhold, K. J.; Lyerly, H. K.; Lanlois, A. J.; Wigzell, H.; Bolognesi, D. P. *Proc. Natl. Acad. Sci. USA* **1987**, *84*, 5424
130 Montefiori, D. C.; Robinson, W. E., J.; Mitchell, W. M. *Proc. Natl. Acad. Sci USA* **1987**, *85*, 9248.
131 Weis, W.; Brown, J. H.; Cusack, S.; Paulson, J. C.; Skehel, J. J.; Wiley, D. C. *Nature (London)* **1988**, *333*, 426.
132 v. Itzstein, M.; Wu, W.-Y.; Kok, G. B.; Pegg, M. S.; Dyason, J. C.; Jin, B.; VanPhan, T.; Smythe, M. L.; White, H. F.; Oliver, S. W.; Colman, P. M.; Varghese, J. N.; Ryan, D. M.; Woods, J. M.; Bethell, R. C.; Hotham, V. J.; Cameron, J. M.; Penn, C. R. *Nature (London)* **1993**, *363*, 418.
133 Karlsson, K.-H.; Strömberg, N. *Methods Enzymol.* **1987**, *138*, 220.
134 Koshland, D. E., Jr.; *Annu. Rev. Biochem.* **1981**, *50*, 765.
135 Bourret, R. B.; Borkovich, K. A.; Simon, M. I. *Annu. Rev. Biochem.* **1991**, *60*, 401.
136 Gardina, P.; Conway, C.; Kossman, M.; Manson, M. *J. Bacteriol.* **1992**, *174*, 1528.
137 Spurlino, J. C.; Lu, G.-Y.; Quiocho, F. A. *J. Biol. Chem.* **1991**, *266*, 5202.
138 Dazzo, F. B. In *Cell Interaction and Development*, Herausg. Yamada, K. M.; John Wiley & Sons: New York; 1982, Chapter 10.
139 Barondes, S. H.; *Annu. Rev. Biochem.* **1981**, *50*, 207.
140 Wang, J. L.; Schindler, M.; Ho, S.-C. *TIGG* **1993**, *5*, 331.
141 Brooks, C. J. W.; Watson, D. G. *Natural Product Reports* **1985**, 427.
142 Keen, N. T. *Science* **1975**, *187*, 74.
143 Cline, K.; Albersheim, P. *Plant Physiol.* **1981**, *68*, 221.
144 Albersheim, P.; Darvill, A. *Sci. Am.* **1985**, *253 (3)*, 44.
145 Albersheim, P.; Darvill, A.; Augur, C.; Cheong, J.-J. Eberhard, St.; Hahn, M. G.; Mafra, V.; Mohnen, D. O'Neill, M. A.; Spiro, M. D.; York, W. S. *Acc. Chem. Res.* **1992**, *25*, 77.
146 Tran Thanh Van, K.; Toubart, P.; Cousson, A.; Darvill, A. G.; Gollin, D. J.; Chelf, P.; Albersheim, P. *Nature (London)* **1985**, *314*, 615.
147 Reitz, A. B. *ACS Symp. Ser.* **1991**, *463*.
148 Berridge, M. J.; Irvine, R. F. *Nature (London)* **1989**, *341*,197.
149 Majerus, P. W.; Connolly, T. M.; Bansal, V. S.; Inhorn, R. C.; Ross, T. S.; Lips, D. L. *J. Biol. Chem.* **1988**, *263*, 3051.
150 IUB Nomenclature Committee, *Biochem. J.* **1988**, *258*, 1.

151 Davies, M. B.; Austin, J.; Partridge, D. A. *Vitamin C, Its Chemistry and Biochemistry*, The Royal Society of Chemistry Paperbacks: Cambridge, 1991.
152 Cerami, A.; Vlassara, H.; Brownlee, M. *Sc. Am.* **1987**, *256 (5)*, 82.
153 Ellis, G. P. *Adv. Carbohydr. Chem.* **1959**, *14*, 63,
154 Horiuchi, S.; Shiga, M.; Araki, N.; Takata, K.; Saitoh, M.; Morino, Y. *J. Biol. Chem.* **1988**, *263*.
155 See for example: Voet, D.; Voet, J. G.: *Biochemistry, 2nd edition;* Wiley: New York, 1995.
Zubay, G. *Biochemistry, 3rd edition;* Wm. C. Brown Publishers: Dubuque, Iowa, 1993.
156 Neufeld, E. F.; Hassid, W. Z. *Adv. Carbohydr. Chem.* **1963**, *18*, 309.
Nikaido, H.; Hassid, W. Z. *Adv. Carbohydr. Chem. Biochem.* **1971**, *26*, 352.
157 Caputto, R.; Leloir, L. F.; Cardini, C. E.; Paladini, A. C. *J. Biol. Chem.* **1950,** *184*, 333.
158 BeMiller, J. N. *Adv. Carbohydr. Chem.* **1967**, *22*, 25.
Capon, B. *Chem. Rev.* **1969**, *69*, 407.
159 Legler, G. *Adv. Carbohydr. Chem. Biochem.* **1990**, *48*, 319.
160 Blake, C. C. F.; Johnson, L. N.; Mair, G. A.; North, A. T. C.; Phillips, D. C.; Sarma, V. R. *Proc. R. Soc. London, Ser. B* **1967**, *167*, 378.
161 Bassham, J. A.; Calvin, M.: *The Path of Carbon in Photosynthesis*, Prentice – Hall, Englewood Cliffs, N. J. 1957.
162 Shibaev, V. N. *Adv. Carbohydr. Chem. Biochem.* **1986**, *44*, 277.
163 Lindberg, B. *Adv. Carbohydr. Chem. Biochem.* **1990**, *48*, 279.
164 James, D. W. Jr.; Preiss, J.; Elbein, A. D. In *The Polysaccharides*, Vol. 3, Aspinall, G. O.; Academic Press: New York, 1985, Chapter 2.
165 Stoddart, R. W.: *The Biosynthesis of Polysaccharides*, Croom Helm Ltd., Beckenham 1984.
166 Lüderitz, O.; Staub, A. M.; Westphal, O. *Bacteriol. Rev.* **1966,** *30*, 192.
167 Fleming, A. *Br. J. Exp. Pathol.* **1929**, *10*, 226.
168 Park, J. T. *J. Biol. Chem.* **1952,** *194*, 897.
Park, J. T.; Strominger, J. L. *Science* **1957**, *125*, 99.
169 Higashi, Y.; Strominger, J. L.; Sweely, C. C. *J. Biol. Chem.* **1970**, *245*, 3697.
170 Raetz, C. R. H. *J. Bacteriol.* **1993**, *175*, 5745.
171 Wright, A.; Dankert, M.; Robbins, P. W. *Proc. Natl. Acad. Sci. USA* **1965**, *54*, 235.
172 Kanegasaki, S.; Jann, K. *Eur. J. Biochem.* **1979**, *95*, 287.
173 Lüderitz, O.; Westphal, O.; Staub, A. M.; Nikaido, H.: *Microbial Toxins*, Vol. 4, Academic Press, New York 1971.
174 Schauer, R. *Adv. Carbohydr. Chem. Biochem.* **1982**, *40*, 132.
175 Orskov, F.; Orskov, J. *J. Exp. Med.* **1979,** *149*, 669.
176 Hehre, E. J. *Science* **1941**, *93*, 237.
177 Pont Lezica, R.; Romero, P. A.; Dankert, M. A. *Plant Physiol.* **1976**, *58*, 675.
178 Sigal, N.; Cattaneo, J.; Segel, I. H. *Arch. Biochem. Biophys.* **1964**, *108*, 440.
179 Ballou, C. E. *Adv. Microbiol. Physiol.* **1976**,*14*, 93.
180 Garg, H. G.; Lyon, N. B. *Adv. Carbohydr. Chem. Biochem.* **1991**, *49*, 239.
181 Mann, D. M.; Yamaguchi, Y.; Bourdon, M. A.; Ruoslahti, E. *J. Biol. Chem.* **1990**, *265*, 5317.
182 Prehm, P. *Biochemical J.* **1983**, *211*, 191.
183 Snider, M. D. In *Biology of Carbohydrates*, Vol. 2, Ginsburg, V.; Robbins, P. W.; Wiley: New York 1984, Chapter 3.
Schachter, H. In *The Glycoconjugates*, Vol. 2, Horowitz, M. I.; Academic Press: New York and London 1978, Chapter 2.
184 Cummings, R. D. In *Glycoconjugates*, Allen, H. J.; Kisailus, E. C.; Dekker: New York, 1992, Chapter 10.
185 Abejon, C.; Hirschberg, C. B. *TIBS* **1992**, *17*, 32.
186 Previato, J. O.; Mendelzon, D. H.; Parodi, A. J. *Mol. Biochem. Parasitol.* **1986**, *18*, 343.
187 Kornfeld, R.; Kornfeld, S. *Annu. Rev. Biochem.* **1985**, *54*, 631.
188 Bischoff, J.; Moremen, K.; Lodish, H. F. *J. Biol. Chem.* **1990**, *265*, 17110.
189 Schachter, H.; Brockhausen, I. In *Glycoconjugates*, Allen, H. J.; Kisailus, E. C.; Dekker: New York, 1992, Chapter 9.
190 Khan, S. H.; Matta, K. L. In *Glycoconjugates*, Allen, H. J.; Kisailus, E. C.; Dekker: New York, 1992, Chapter 11.
Elbein, A. D. In *Biology of Carbohydrates*, Vol. 3, Ginsburg, V.; Robbins, P. W.; JAI Press: London 1991, Chapter 4.
191 Rademacher, T. W.; Parekh, R. B.; Dwek, R. A. *Annu. Rev. Biochem.* **1988**, *57*, 785.
Dwek, R. A. *Chem. Rev.* **1996**, *96*, 683.
Varki, A. *Glycobiology*, **1993**, *3*, 97.
192 Anderson, B.; Davis, L. E.; Venegas, M. *Adv. Exp. Med. Biol.* **1988**, *228*, 601.
Radoux, V.; Menard, H. A.; Begin, R.; Decary, F.; Koopmann, W. J. *Arthritis Rheum.* **1987**, *30*, 249.
Morgan, K. L. *Lancet* **1987**, *1*, 1017.
Morse, M. L. *Lancet* **1986**, *1*, 625.
193 Palcic, M. M.; Heerze, L. D.; Shrivastava, O. P.; Hindsgaul, O. *J. Biol. Chem.* **1989**, *264*, 17174.
194 Connolly, D. T.; Townsend, R. R.; Kawaguchi, K.; Bell, W. R.; Lee, Y. C. *J. Biol. Chem.* **1982**, *257*, 939.
195 Ats, S.-Cs.; Lehmann, J.; Petry, St. *Carbohydr. Res.* **1992**, *233*, 141.
196 Palcic, M. M.; Ripka, J.; Kaur, K. J.; Shoreibah, M.; Hindsgaul, O.; Pierce, M. *J. Biol. Chem.* **1990**, *265*, 6759.
Khan, S. H.; Abbas, S. A.; Matta, K. L. *Carbohydr. Res.* **1990**, *205*, 385.
197 Pan, Y. T.; Hori, H.; Saul, R.; Sanford, B. A.; Molineux, R. J.; Elbein, A. D. *Biochemistry* **1983**, *22*, 3975.
Tulsiani, D. P.R.; Harris, T.; Touster, O. *J. Biol. Chem.* **1982**, *257*, 7936.
198 David, S.; Augé, C.; Gautheron, C. *Adv. Carbohydr. Chem. Biochem.* **1991**, *49*, 176.
Bednarski, M. D.; Simon, F. S., *ACS Symp. Ser.* **1991**, *466*.
Wong, C.-H; Halcomb, R. L.; Ischikawa, Y; Kapimoto, T. *Angew. Chem.* **1995**, *107*, 453, 569.

199 Croft Hill, A. *J. Chem. Soc.* **1898**, 634.
200 Bourquelot, E.; Hérissey, H.; Coirre, J. *C. R. Acad. Sci.* **1913**, *157*, 732.
201 Peat, S.; Whelan, W. J.; Hinson, K. A. *Nature (London)* **1952**, *170*, 1056.
B. Helferich, J. F. Leete, *Org. Synth.* **1942**, *22*, 53.
Goldstein, I. L.; Whelan, W. J. *Methods Carbohydr. Chem.* **1962**, *1*, 313.
202 Ajisaka, K.; Fujimoto, H.; Nishida, H. *Carbohydr. Res.* **1988**, *180*, 35.
203 Ajisaka, K.; Fujimoto, H. *Carbohydr. Res.* **1990**, *199*, 227.
204 Huber, R. E.; Kurz, G.; Wallenfels, K. *Biochemistry* **1976**, *15*, 1994.
205 Wallenfels, K.; Weil, R. In *The Enzymes*, 3rd ed., Vol. 7, Boyer, T. P.; Academic Press: New York, 1972, p. 617.
206 Huber, R. E.; Wallenfels, K.; Kurz, G. *Can. J. Biochem.* **1975**, *53*, 1035.
207 Kyosaka, S.; Murata, S.; Tsuda, Y.; Tanaka, M. *Chem. Pharm. Bull.* **1986**, *34*, 5140.
Nilsson, K. G. *Carbohydr. Res.* **1989**, *188*, 9.
208 Nilsson, K. G. *Carbohydr. Res.* **1987**, *167*, 95; **1988**, *180*, 53.
209 Nilsson, K. G. *Carbohydr. Res.* **1989**, *188*, 9.
210 Hedbys, L.; Johansson, E.; Mosbach, K.; Larsson, P.-O.; Gunnarsson, A.; Svensson, S. *Carbohydr. Res.* **1989**, *186*, 217.
211 French, D.; Levine, M. L.; Norberg, E.; Nordin, P.; Pazur, J. H.; Wild, G. M. *J. Am. Chem. Soc.* **1954**, *76*, 2387.
212 Wallenfels, K.; Földi, P.; Niermann, H.; Bender, H.; Linder, D. *Carbohydr. Res.* **1978**, *61*, 359.
213 Cheetham, P. S. J.; Imber, C. E.; Isherwood, J. *Nature (London)* **1982**, *299*, 628.
214 Okada, G.; Hehre, E. J. *Carbohydr. Res.* **1973**, *26*, 240.
Okada, G. Genghof, D. S.; Hehre, E. J. *Carbohydr. Res.* **1979**, *71*, 287.
Kitahata, S.; Brewer, C. F.; Genghof, D. S.; Hehre, E. J. *J. Biol. Chem.* **1981**, *256*, 6017.
215 Hehre, E. J.; Mizokami, K.; Kitahata, S. *J. Jpn. Soc. Starch Sci.* **1983**, *30*, 76.
Treder, W.; Thiem, J.; Schlingmann, M. *Tetrahedron Lett.* **1986**, *27*, 5605.
216 Card, P. J.; Hitz, W. D.; Ripp, K. G. *J. Am. Chem. Soc.* **1986**, *108*, 158.
217 Sabesan, S.; Paulson, J. C. *J. Am. Chem. Soc.* **1986**, *108*, 2068.
Augé, C.; David, S.; Matieu, C.; Gautheron, C. *Tetrahedron Lett.* **1984**, *25*, 1467.
Thiem, J.; Treder, W. *Angew. Chem.* **1986**, *98*, 110.
218 Wong, C. H.; Haynie, S. L.; Whitesides, G. M. *J. Org. Chem.* **1982**, *47*, 5418.
219 Wheetall, H. H. *Methods Enzymol.* **1976**, *44*, 134.
220 Lehmann, J.; Schröter, E. *Carbohydr. Res.* **1972**, *23*, 359; ibid. **1977**, *58*, 65.
221 Brockhaus, M.; Lehmann, J. *Carbohydr. Res.* **1977**, *53*, 21; *58*, 65.
222 Maretu, W.; Auld, D. S. *Biochemistry* **1988**, *27*, 1622.
223 Lockwood, L. B. *Methods Carbohydr. Chem.* **1962**, *1*, 151.
224 Dhawale, M. R.; Scarek, W. A.; Hay, G. W.; Kropinski, A. M. B. *Carbohydr. Res.* **1986**, *155*, 262.
225 Tiwari, K. N.; Dhawale, M. R.; Scarek, W. A.; Hay, G. W.; Kropinski, A. M. B. *Carbohydr. Res.* **1986**, *156*, 19.
226 Niwa, T.; Tsuruoka, T. Goi, H.; Kodama, Y.; Itoh, J. Inouye, S.; Yamada, Y.; Niida, T.; Nobe, M.; Ogawa, Y. *J. Antibiotics (Tokyo)* **1984**, *37*, 1579.
227 Root, R. L.; Durrwachter, J. R.; Wong, C. H. *J. Am. Chem. Soc.* **1985**, *107*, 2997.
228 Schlegel, R. A.; Gerbeck, C. R.; Montgomery, R. *Carbohydr. Res.* **1968**, *7*, 193.
229 Hudlicky, T.; Luna, H.; Barbieri, G.; Kwart, L. D. *J. Am. Chem. Soc.* **1988**, *110*, 4735.
230 Gibson, D. T.; Hensley, H.; Yoshioka, M. T.; Mabry, J. *Biochemistry* **1970**, *9*, 1626.
231 Hudlicky, T.; Luna, H.; Price, J. D.; Rulin, F. *Tetrahedron Lett.* **1989**, *30*, 4053.
232 Lehmann, J.; Moritz, A. *Justus Liebigs Ann. Chem.* **1991**, 937.
233 Ley, S. V.; Parra, M.; Redgrave, A. J.; Sternfeld, F.; Vidal, A. *Tetrahedron Lett.* **1989**, *30*, 3557.
Ley, S. V.; Sternfeld, F. *Tetrahedron* **1989**, *45*, 3463.
234 Casati, P.; Fuganti, C.; Grasselli, P.; Servi, S. Spreafico, S.; Zitotti, C. *J. Org. Chem.* **1984**, *49*, 4087.
235 Fronza, G.; Fuganti, C.; Grasselli, P.; Servi, S. *Tetrahedron Lett.* **1985**, *26*, 4961.
236 Lehmann, J.; Rapp, K. *Ullmann's Encyclopedia of Industrial Chemistry: Carbohydrates*, Vol. A 5, VCH: Weinheim, 1986, p. 79.
237 Jones, J. K. N.; Sephton, H. H. *Can. J. Chem.* **1960**, *38*, 753.
238 Meyerhof, O.; Lohmann, K.; Schuster, Ph. *Biochem. Z.* **1936**, *286*, 301.
239 Toone, E. J.; Simon, E. S.; Bednarski, M. D.; Whitesides, G. M. *Tetrahedron* **1989**, *45*, 5365.
Whitesides, G. M.; Wong, C.-H. *Angew. Chem.* **1985**, *97*, 617.
240 Bednarski , M. D.; Simon, E. S.; Bischofsberger, N.; Fessner, W.-D.; Kim, M.-J.; Lees, W.; Saito, T.; Waldmann, H.; Whitesides, G. M. *J. Am. Chem. Soc.* **1989**, *111*, 627.
241 Borysenko, C. W.; Spaltenstein, A.; Straub, J. A.; Whitesides, G. M.*J. Am. Chem. Soc.* **1989**, *111*, 9275.
242 Bednarski, M. D.; Waldmann, H. J.; Whitesides, G. M. *Tetrahedron Lett.* **1986**, *27*, 5807.
Durrwachter, J. R.; Sweers, H. M.; Nozaki, K.; Wong, C.-H. *Tetrahedron Lett.* **1986**, *27*, 1261.
243 Straub, A.; Effenberger. F.; Fischer, P. *J. Org. Chem.* **1990**, *55*, 3926.
244 Augé, C.; David, S.; Gautheron, C.; Malleron, A.; Cavayé, B. *New J. Chem.* **1988**, *12*, 733.
245 Augé, C.; Gautheron, C. David, S.; Malleron, A.; Cavayé, B.; Bouxom, B. *Tetrahedron* **1990**, *46*, 201.
246 Yokota, A.; Sasajima, K. *Agric. Biol. Chem.* **1984**, *48*, 149.
247 Kapuscinsky, M.; Franke, F. P.; Flanigan, I.; MacLeod., J. K.; Williams, J. F. *Carbohydr. Res.* **1985**, *140*, 69.

248 Demuynck, C.; Bolte, J.; Hecquet, L.; Samaki, H. *Carbohydr. Res.* **1990**, *206*, 79.

249 Fink, A. L.; Hay, G. W. *Can. J. Biochem.* **1969**, *47*, 353.

250 Barili, P.-L.; Berti, G.; Catelani, G.; Colonna, F.; Mastrorilli, E. *J. Chem. Soc., Chem. Commun.* **1986**, 7.

251 Frohwein, Y. Z.; Leibowitz, J. *Enzymologia* **1961**, *23*, 202 u. 208; **1962**, *25*, 297; **1963**, *26*, 193.

252 Shaw, J. F.; Klibanov, A. M. *Biotech. Bioeng.* **1987**, *29*, 648.

253 Sweers, H. M.; Wong, C. H. *J. Am. Chem. Soc.* **1986**, *108*, 6421.
Hennen, W. J.; Sweers, H. M.; Wang, Y. F.; Wong, C. H. *J. Org. Chem.* **1988**, *53*, 4939.

254 Klosterman, M.; Mosmuller, E. W. J.; Schoemaker, H. E.; Meijer, E. M. *Tetrahedron Lett.* **1987**, *28*, 2989.

255 Tamura, M.; Kinomura, K.; Tada, M.; Nakatsuka, T.; Okai, H.; Fukui, S. *Agric. Biol. Chem.* **1985**, *49*, 2011.

256 Ballesteros, A.; Bernabé, M.; Cruzado, C.; Martin-Lomas, M.; Otero, C. *Tetrahedron* **1989**, *45*, 7077.

257 Zaks, A.; Klibanov, A. M. *Proc. Natl. Acad. Sci. USA* **1985**, *82*, 3192.

258 Therisod, M.; Klibanov, A. M. *J. Am. Chem. Soc.* **1986**, *108*, 5638.

259 Riva, S.; Chopineau, J.; G. Kieboom, A. P.; Klibanov, A. M. *J. Am. Chem. Soc.* **1988**, *110*, 584.

260 Björkling, F.; Godtfredsen, S. E.; Kirk, O. *J. Chem. Soc., Chem. Commun.* **1989**, 934.

261 Carpani, G.; Orsini, F.; Sisti, M.; Verotta, L. *Gazz. Chim. Ital.* **1989**, *119*, 463.

262 Ciuffreda, P.; Colombo, D.; Ronchetti, F.; Toma, L. *J. Org. Chem.* **1990**, *55*, 4178.

263 Holla, E. W. *J. Carbohydr. Chem.* **1990**, *9*, 113.

# Index

## D

## E



## F

## N



## O

T



U